Introduction to Matrix-Analytic Methods in Queues 2

This book is dedicated to my parents:
Mrs. P.S. Rajalakshmi and Mr. P.S.S. Raghavan;
to my professors:
Dr. Marcel F. Neuts and Dr. K.N. Venkataraman;
to my wife, son, and daughter-in-law:
Jayanthi, Arvind, and Vina;
and to His Holiness
Sri Maha Periyava (Sri Chandrasekharendra Saraswati Mahaswamigal)
of Kanchi Kamakoti Peetham

Series Editor
Nikolaos Limnios

Introduction to Matrix-Analytic Methods in Queues 2

Analytical and Simulation Approach – Queues and Simulation

Srinivas R. Chakravarthy

iSTE

WILEY

First published 2022 in Great Britain and the United States by ISTE Ltd and John Wiley & Sons, Inc.

ISTE Ltd
27-37 St George's Road
London SW19 4EU
UK

www.iste.co.uk

John Wiley & Sons, Inc.
111 River Street
Hoboken, NJ 07030
USA

www.wiley.com

Library of Congress Control Number: 2022935181

British Library Cataloguing-in-Publication Data
A CIP record for this book is available from the British Library
ISBN 978-1-78630-823-8

Contents

List of Notations

Symbols

e	column vector, with all elements equal to 1, of appropriate dimension
e^T	row vector, with all elements equal to 1, of appropriate dimension
e_i	column vector with 1 in the ith position and 0 elsewhere
I	identity matrix
$\otimes$	Kronecker product
$\oplus$	Kronecker sum
$\Delta(.)$	diagonal matrix with entries given in the parentheses

Abbreviations

BMAP	batch Markovian arrival process
BP	busy period
CDF	(cumulative) distribution function
CPH	continuous phase type
CTMC	continuous-time Markov chain
D-BMAP	discrete-time BMAP
DPH	discrete phase type
DRI	directly Riemann integrable
DTMC	discrete-time Markov chain

EMC	embedded Markov chain
LST	Laplace-Stieltjes transform
LT	Laplace transform
MAM	matrix-analytic method(s)
MAP	Markovian arrival process
MC	Markov chain
MRP	Markov renewal process
PDF	probability density function
PGF	probability generating function
PH	phase type
PMF	probability mass function
TPM	transition probability matrix
VMPP	versatile Markovian point process

Preface

The introduction of the phase type (PH) distributions in the early 1970s by Marcel Neuts opened up a wide range of possibilities in Applied Probability modeling and ushered in the idea that finding computable, numerical solutions was an acceptable and desirable goal in analyzing stochastic models. Furthermore, he popularized incorporating the computational aspects in the study of stochastic models. It gave researchers a powerful new tool that enabled them to move beyond the traditional models limited to exponential processes for analytical convenience to studying more realistic stochastic models with algorithmic solutions and simple, elegant probabilistic interpretations. The goal of building models with computationally tractable solutions rather than the abstract transform-based solutions took root. This rapidly led to an entirely new area of research on the study of stochastic models in queues, inventory, reliability, and communication networks using matrix-analytic methods (MAM). The versatile Markovian point process (VMPP) was introduced by Neuts in the late 1970s. This process was used in the study of a single-server queueing system with general services by one of Neuts's students, V. Ramaswami, for his PhD dissertation. In 1990, this VMPP was studied differently as a batch Markovian arrival process (BMAP) by Neuts and his students David Lucantoni and Kathy Meier-Hellstern. At that time it was thought that VMPP was a special case of BMAP, but it was proved that BMAP and VMPP are the same. However, the compact and transparent notations with which BMAP is described allowed the readers to understand this versatile point process with relative ease, and since then VMPP is referred to as BMAP in the literature. In the case of single arrivals, the process is referred to as a Markovian arrival process (MAP).

The study of stochastic models possessing matrix-geometric solutions (thus extending the geometric solution result for the scalar case for Poisson arrivals and exponential services in a single server queue) by Neuts in the late 1970s and the introduction of PH distributions, BMAP, and an emphasis on the algorithmic approach, paved the way for Neuts to the introduction of MAM. Ever since, these methods have been extensively studied both theoretically and computationally in the context of a variety of stochastic models useful in many applied areas. A handful of books starting with Neuts's two classical books, *Matrix-Geometric Solutions in Stochastic Models: An Algorithmic Approach*, originally published in 1981, and *Structured Stochastic Matrices of M/G/1 Type and Their Applications* in 1989, to the latest one, *The Theory of Queuing Systems with Correlated Flows* by Dudin, Klimenok and Vishnevsky in 2020, have appeared in the literature that deal with MAM. The other books published from 1989 to 2020 include *Introduction to Matrix Analytic Methods in Stochastic Modeling* by Latouche and Ramaswami (1999); *Numerical Methods for Structured Markov Chains* by Bini et al. (2005); *Queueing Theory for Telecommunications* by Alfa (2010); *Fundamentals of Matrix-Analytic Methods* by He (2014); and *Matrix-Exponential Distributions in Applied Probability* by Bladt and Nielsen (2017).

All the texts mentioned above provide an excellent foundation of a variety of stochastic models in general and of theoretical properties and applications of MAM to those models. The present work takes a different approach by covering the basics of MAM but focusing on clearly illustrating its use in analyzing many stochastic models. It is also my strong belief that the art of model building and analysis is better learned by studying carefully constructed examples and by practicing those skills on other models. A text that incorporates the mathematical ideas of MAM along with clearly illustrated examples on the use of these methods in analyzing interesting stochastic models makes MAM more accessible to current researchers. I believe this juxtaposition reinforces the power of MAM, enables one to appreciate and get better in the art of model building, and helps in improving "probabilistic thinking" of models and solutions. It is also for these reasons that I have included a large collection of exercises, most of which are computational in nature, for the reader to practice, experiment and get the experience in the algorithmic and computational procedures. It should be pointed out that the illustrative examples and the exercises are deliberately made more generic so as to let the readers modify them for their areas of applications. This approach was motivated by the feedback that I have had from numerous graduate students and fellow researchers in the past.

Furthermore, the text also contains exploration of simulation as an integral tool in studying stochastic models, which don't admit analytical or numerical solutions. Finally, the text provides detailed explanations on the mathematics and the applications of MAM to enable graduate students with a strong background in probability and mathematics.

Thus, the text is a useful source of reference for researchers established in this field and, more importantly, a valuable, inviting guide for those venturing into this rich research area.

This two-volume book is organized as follows: Volume 1 has nine chapters. Chapter 1 reviews basic concepts in probability and matrix theory. Chapter 2 has a brief review of discrete-state-discrete-time and discrete-state-continuous-time Markov chains. These two chapters are not meant to be an exhaustive and thorough review of those topics but hopefully sufficient enough to understand the rest of the materials in the two-volume book. In Chapter 3, discrete-time phase type distribution is presented. Chapter 4 focuses on the continuous-time phase type distributions. Chapters 5 and 6, respectively, cover discrete and continuous time BMAP. The basics of MAM from both discrete time (through embedded epochs) and continuous time points of view along with many examples and exercises are, respectively, presented in Chapters 7 and 8. A brief summary of the applications of queueing (and in turn MAM) is given in Chapter 9. The presence of numerous detailed solutions and exercises benefit students by piquing their interest in MAM, helping them learn and understand basic concepts, and succeed in constructing and solving models of their own in their research. The solutions to these exercises can be found at the following link: www.iste.co.uk/chakravarthy/queues2.zip.

Volume 2 contains seven chapters. In Chapters 1, 2, and 3, respectively, single-server queues are studied by looking at departure, arrivals, and at arbitrary epochs. Chapter 4 focuses on the busy periods in queues. Selected multi-server queues are studied in Chapter 5, and finite capacity queues are the focus in Chapter 6. Finally in Chapter 7, we present analysis of queues via simulation using ARENA, a powerful simulation software. In this chapter we also provide a very brief introduction to ARENA. All these chapters have exercises.

This two-volume book can be used in a number of settings. Senior undergraduate students (with sufficient background in probability) and Master's level graduate students could use Volume 1 to get an understanding of the fundamentals of MAM. Research scholars pursuing MPhil and PhD degrees can start with Volume 1 and, after going through the basics covered there, move on to finishing Volume 2. For research scholars pursuing MPhil and PhD degrees, the two volumes would constitute a two- to three-semester course.

Writing this book has been lot of fun but also a challenge. However, my family, friends, and mentors helped me to meet that challenge. I take a great pleasure in acknowledging them. This book project would not have been possible without the educational foundation, moral support, encouragement, and critical analysis of teachers, friends, and families.

Specifically, I want to acknowledge the following people who made a positive difference.

– My (late) father, P.S.S. Raghavan, for being a role model. My mother, P.S.S. Rajalakshmi, for her encouragement. Both my parents made many sacrifices that enabled me to first go to college and, later on, to leave for the United States to pursue higher studies.

– My sister, Vasumathi Parthasarathy, for exposing me to mathematics at a very young age.

– My (late) Professors Marcel F. Neuts and K.N. Venkataraman. While K.N.V. gave me an opportunity to learn probability theory under him while in India, M.F.N. showed me the path to MAM. I owe a debt of gratitude to him for what I am now and for his important role in shaping my career as a teacher and a researcher.

– My college teachers, Prof. D. Ratnasabapathi (Presidency College in Madras) and Prof. K. Suresh Chandra (University of Madras), who not only taught me statistics but also were a source of encouragement to pursue higher studies.

– I benefited a lot through interacting with my friends and colleagues V. Ramaswami, D.M. Lucantoni, Kathy Meier-Hellstern and S. Kumar, during my days at Delaware.

– R. Parthasarathy (Kent State University), whom I knew from my college days in India and who has always been there to give moral support since those days.

– My research colleagues who played key roles in my career, notably A.S. Alfa, A.N. Dudin, A. Krishnamoorthy, and A. Rumyantsev.

– My students: Serife Ozkar, who visited from Turkey to finish up her doctoral thesis with me at Kettering, and Shruti Goel, who attended the workshops I conducted in India. The questions these students, along with a countless others, including Alka Choudhry, a doctoral student at the Central University of Rajasthan, India, raised provided the impetus for this book. Furthermore, I am thankful to both Serife and Shruti for helping me put the bibliography in the format required by the publishers.

– Several of my colleagues at Kettering, notably (the late) Duane McKeachie, Petros Gheresus, and T.R. Chandrupatla (who retired from Rowan University recently after serving nearly two decades at Kettering) for their friendship and encouragement throughout my career at Kettering.

– Kettering University for its support of my sabbatical which was instrumental in completing this book project.

– The ISTE team for their continued and timely help during the production process of this two-volume book.

– Finally, the most important people in my life, my wife, Jayanthi Chakravarthy, son Arvind Chakravarthy and his beloved wife, Vina Harji Chakravarthy. Since

Jayanthi and Arvind came into my life, their understanding, love and support have helped me to focus on my research and career without any distraction. They along with Vina have been a source of constant inspiration to finish the book. No words are adequate to express my sincere appreciation to them.

Srinivas CHAKRAVARTHY
April 2022

1

Single-Server Queues Embedded at Departure Epochs

Wherever we go and whatever we do, we invariably have to wait for our turn to get service or attention or execute our job. When we look at the total time spent in executing the job, it includes the actual time for executing the job plus the waiting time (the time spent before our job gets into service). The study of such waiting line problems both from the customers' and the service facilities' points of view, so as to optimize the time in the system and to best utilize the resources, falls under the general topic of queueing models.

Queueing theory plays an important role in many areas, notably telecommunications, production and manufacturing engineering, computer science and other service offering facilities. Queueing theory was developed originally in the context of telephone traffic congestion by Danish Mathematician A.K. Erlang in 1909. He noticed that the telephone systems were generally characterized by either Poisson input and exponential holding times and multiple channels or by Poisson input and constant holding times and a single channel. Erlang was responsible for the concept of stationary equilibrium. E.C. Molina, in 1927, published *Applications of the Theory of Probability to Telephone Trunking Problems*. In the early 1930s, F. Pollaczek did some fundamental work on Poisson input, arbitrary output and single/multiple channel problems. At the same time, A. Kolmogorov and A. Khintchine (from Russia), C.D. Crommelin (France) and C. Palm (Sweden) did some additional work. Other additional contributions include the works by D.R. Lindely, L. Takacs, D.G. Kendall, S. Karlin, J.L. McGregor, J.D.C. Little, M.F. Neuts, N.U. Prabhu, U.N. Bhat, T.L. Saaty and R. Syski. Modern queueing theory incorporating the computational aspects is attributed to Neuts through his pioneering work on matrix-analytic methods (MAM) that started in the 1970s.

Thus, it is safe and fair to say that queueing theory was developed in three stages since the identification and the solution of simple practical queueing problems in the early 20th century. In the first stage, queueing system analysis developed along with the growth of the area of stochastic processes into topics such as birth and death processes and their analysis using generating functions, Laplace transforms and integral equation techniques. The second stage started with the introduction of the imbedded Markov chain technique by David Kendall in order to handle non-Markovian arrival and service processes. However, all of the analysis used a complex variable approach (using Laplace transforms and/or generating functions) and explicit results, if there were any, gave formulas for perhaps the steady-state mean queue length or waiting time. Others involved rather complicated formulas for Laplace-Stieltjes or other transforms. The third stage of development of queueing theory research spearheaded by Neuts in the form of MAM is central to the theme of this two-volume book.

It is customary, as can be seen in many books on queueing theory (see Alfa (2016); Bhat (2015); Dudin et al. (2020); Gross and Harris (1998); Kleinrock (1975)), to present queueing models starting with the most simple ones, such as $M/M/1$, $M/M/c$ and $M/M/\infty$, and then move on to more complex ones. It should be pointed out that the book by Alfa (2016) focuses on discrete-time and hence the discrete analog of $M/M/1$ is $Geo/Geo/1$ ("Geo" stands for geometric distribution). However, here we take a different approach. This is mainly to show how one can unify many queueing models and also establish different techniques to solve the same queueing model.

In this second volume, we focus on the study of queues using MAM techniques, which were presented in Volume 1. As mentioned earlier, we freely refer to the materials, where appropriate, from Volume 1. Thus, it is highly encouraged that the readers go through Volume 1 (or at least refer to the appropriate sections when needed) to fully understand the materials presented here. Most of the chapters in this volume depend heavily on the materials seen in Volume 1. However, readers familiar with the basic concepts in MAM may find this volume to be self-contained. Otherwise, readers should ensure that they fully understand the materials and notations from Volume 1 before continuing.

In this chapter, after briefly describing the basics of queueing theory, we focus on the study of queueing models by looking at the departure epochs. It is worth mentioning that Chapter 9 in Volume 1 discusses briefly queueing theory applications. Queueing models play an important role in many industries, notably the service sectors. Some of these services are offered online. Thus, the arrivals to the service sectors (as well as to other industries) are from different sources and hence the assumption that the inter-arrival times are independent is a very novice idea. How do we build correlation into the inter-arrival times but at the same time make the queueing models amenable to practical implementation? This was one of the main concerns that motivated Neuts to lay the foundation for MAM through introducing phase type (PH) distributions (see Chapters 3 and 4 of Volume 1) and batch Markovian arrival processes (see Chapters 5 and 6 of Volume 1).

This chapter relies heavily on the materials presented in Chapters 7 and 8 of Volume 1. Astute readers might want to know the difference between those materials and the ones in this chapter. They are as follows:

1) In those chapters in Volume 1, the (matrix) elements appearing in the transition probability matrix (TPM) or the generator (depending on the context) are assumed to be known. In this chapter, as well as the rest of the book, these (matrix) elements are obtained based on the model under study.

2) Also in Volume 1, there is no prior assumption about the connection, if any, between the matrices $\{A_n\}$ and $\{B_n\}$ (see sections 7.2 and 7.4). However, in this book they are connected and hence result in additional insights as well as simplifications.

A queueing model is characterized by the following key quantities (see Bhat (2015); Gross and Harris (1998)):

1) *Input process (arrival process)*: Here the arrival pattern of customers (or jobs) is described. One specifies the following: single or multiple (group) arrivals and the distribution of the inter-arrival times between successive arrivals.

2) *Service process*: This describes the number of servers (or machines) in the system as well as the distribution of the service times. In the case of more than one server, the servers may be identical or may have different service time distributions. Servers may offer services singly or in groups. In the case of multiple servers, they may be in parallel or in series or each customer has to be attended by more than one server at a time.

3) *Service discipline*: This plays a major role. The mechanism of serving the customers is described here. There are many such schemes and some of these are first come, first served; last come, first served; service in random order; and round robin services. In some applications, customers are grouped into various priorities and within each priority the customers will be served according to some order. There are many priority schemes. Some of which are dynamic or static; preemptive or non-preemptive; or a combination of some of these. In this book, we focus only on a first come, first served (FCFS) basis.

4) *Buffer size (waiting room)*: Arriving customers are queued up before being offered services. The waiting room size could be 0, any positive finite number or infinite.

5) *Performance measures*: Any system, especially a queueing system, should be evaluated by one or more performance measures. These include (1) mean, (2) median, (3) mode, (4) standard deviation and (5) percentiles of various stationary

queue length densities as well as stationary waiting time distributions. Furthermore, the measures like the system idle probability, the server utilization, the mean number of busy servers, the mean busy period and the mean number of customers served during a busy period play an important role in managerial decisions.

6) *Busy Period (BP)*: The BP can be defined in a number of ways. In this book, we define a *partial busy period* to be the duration of the time interval that begins with an arrival of a customer to an empty system and ends with the system becoming empty again at their departure. In queueing literature (see Artalejo and Lopez-Herrero (2001); Daley and Servi (1998)), many authors use full and partial BPs when dealing with multiple-server systems. A full BP is one in which a BP starts from all servers becoming busy and ends when at least one server becomes free. Note that in a single-server queueing system the partial and full BPs are one and the same. Note that a partial BP comprises the full period and the duration where at least one server is free and at least one server is idle. The type of BP used in this volume should be clear from the context in which it is used.

There are many variations to the type of queueing systems just described. It is impossible to describe all of them here. Every year hundreds of papers appear in the literature on queueing and related fields. Some key journals that focus on the practical use of queueing models are (not in any specific order) *Operations Research*, *Management Science*, *Stochastic Models*, *Queueing Systems*, *Performance Evaluation*, *IEEE Transactions on Communications*, *European Journal of Operational Research*, *Computers and Industrial Engineering*, *Computers and Operations Research*, *Naval Research Logistics*, *Queueing Models and Service Management*, *Applied Mathematical Modeling*, *Applied Mathematics and Computation*, *Opsearch* and *IIE Transactions*.

The following is the standard notation, due to David Kendall, in queueing literature. $GI/G/s/K$ refers to an s-server queueing model in which the inter-arrival times are i.i.d random variables; the service times are i.i.d random variables; and there is a finite waiting room of size K. In the case when $K = \infty$, the notation will be $GI/G/s$. In the case when the inter-arrival times are exponential, GI will be replaced by M. When the service times are exponential, G will be replaced by M. For example, $M/M/1$ is a single-server queue with Poisson arrivals, exponential service times and infinite buffer size. When we look at an $M/PH/1$ queueing model, we are looking at a single-server infinite capacity queue with Poisson arrivals and phase type services. The queueing models can be studied either in discrete-time or in continuous-time. Here, our focus is on continuous-time and we refer the reader to Alfa (2016) for queues in discrete-time.

Before we discuss specific queueing models, we register some general observations. Toward this end, let $\{a_n : n \geq 1\}$ denote the inter-arrival times of the customers, assumed to be i.i.d; $X(t)$ denote the number of customers in the system at time t; $N(t)$ denote the number of customers arriving by time t (i.e. number arriving during $[0,t)$); and T denote the total time a customer spends in the system before leaving. We assume that the limit, $\lim_{n\to\infty} P(X(t) = k)$, exists and is given by $\pi_k, k \geq 0$. Suppose, for $k \geq 0$, we denote by u_k and v_k, respectively, the steady-state probability that an arriving customer and a departing customer see exactly k customers in the system. For any system in which arrivals occur singly and services are offered to one at a time, then we have $u_k = v_k, k \geq 0$. However, note that π_k may differ from u_k (or v_k). The average arrival rate is given by:

$$\lambda_a = \lim_{t\to\infty} \frac{N(t)}{t} = \frac{1}{E(a_n)}. \qquad [1.1]$$

Note that in the case of an infinite buffer no customer will be lost and hence $\lambda_a = \lambda$, where λ is the average arrival rate. In the case of a finite buffer, it is possible for an arriving customer to be lost due to the buffer being full and hence $\lambda_a = \lambda(1 - P(loss))$.

A few words about the presentation of the materials in this book. The approach to the study of any queueing model can be classified into three categories: (1) embedded at departure epochs; (2) embedded at arrival epochs; and (3) at arbitrary times. The specific approach to be taken depends, most of the times, on the underlying queueing model. However, there are situations where one can have a choice in the approach.

One of the important aspects of any book is to provide different approaches for the same model so as to give a clear understanding of the underlying concepts to the reader. We take this into consideration and hence one will see the analysis of the same model from different perspectives. For example, the queueing model dealing with continuous time phase type (CPH) arrivals and CPH service times will be studied by looking at (a) the arrival epochs; (b) the service epochs; and (c) arbitrary time points. Similarly, the queueing model with CPH arrivals and exponential services will be discussed from both embedded at arrivals as well as at arbitrary time points of view.

In this chapter, our focus is on the study of queues by looking at departure epochs. That is, we focus on $M/G/1$-type queues. The chapter is organized as follows. In section 1.1, we present the most general single-server queueing model in which the arrivals occur according to a batch Markovian arrival process (BMAP) and the services are generally distributed. That is, we look at a $BMAP/G/1$ queue and its analysis. Illustrative examples covering a variety of scenarios follow the analysis. Special cases of the $BMAP/G/1$ queue are discussed very briefly in subsequent sections. Specifically, in sections 1.2, 1.3, 1.4, 1.5 and 1.6, respectively, we present the special cases involving $MAP/G/1$, $PH/G/1$, $M/G/1$, $PH/M/1$ and $M/M/1$ queues. However, more explanations will be provided when the special cases yield additional insights and/or explicit expressions.

1.1. *BMAP/G/1* queue

Consider a single-server system in which the customers arrive according to a BMAP with representation matrices $\{D_k\}$ of order m. Recall (see Chapter 6 in Volume 1) that a BMAP is characterized by the parameter matrices $\{D_k\}$ such that D_0 governs transitions corresponding to no arrivals and D_k governs transitions related to an arrival of a batch of k customers. The service times are generally distributed with the (cumulative) distribution function (CDF) given by $F(.)$ and assumed to have a finite mean, μ', and a finite variance, σ^2. The service rate is denoted by μ.

Let $\boldsymbol{\pi}$ denote the steady-state probability vector of the underlying (irreducible) generator, Q, of the BMAP. That is:

$$Q = \sum_{k=0}^{\infty} D_k = D_0 + D, \qquad [1.2]$$

and:

$$\boldsymbol{\pi} Q = \mathbf{0}, \quad \boldsymbol{\pi} \boldsymbol{e} = 1, \qquad [1.3]$$

Recall (see equation [6.4] in Volume 1) that the fundamental rate, λ, is defined as $\lambda = \boldsymbol{\pi} \sum_{k=1}^{\infty} k D_k \boldsymbol{e}$. The rate, λ_g, of group arrivals is given by $\lambda_g = \boldsymbol{\pi} D \boldsymbol{e}$ (see equation [6.5] in Volume 1).

We now look at the queueing system at departure epochs so as to study the system using Markov renewal process (MRP). Toward this end, let ϑ_n denote the epoch at which the nth departure (i.e. a service completion) occurs. Without loss of generality we take $\vartheta_0 = 0$. Suppose we define ζ_n and J_n, respectively, to be the number of customers in the system and the phase of the BMAP immediately after the nth departure. That is, we look at the system's state at epochs, ϑ_n^+. Note that the difference $\vartheta_{n+1} - \vartheta_n$ corresponds to the duration of the $(n+1)$st service.

We now define a number of auxiliary (matrix) probabilities for use in the following. Define $\hat{A}_n(t) = (\hat{a}_{i,j}(n,t))$ and $\hat{B}_n(t) = (\hat{b}_{i,j}(n,t))$ such that:

1) $\hat{a}_{i,j}(n,t)$ gives the conditional probability that the last departure left the server to be busy again with the BMAP in phase i and the next departure, which occurs no

later than time t, will see the BMAP in phase j with exactly n customers waiting in the system.

2) $\hat{b}_{i,j}(n,t)$ gives the conditional probability that the last departure left the server to be idle with the BMAP in phase i and the next departure, which occurs no later than time t, will see the BMAP in phase j with exactly n customers waiting in the system. Note that here a batch of one or more customers will have to arrive before time t and that a service completion must occur before time t. A pictorial description of this probability is given in Figure 1.1.

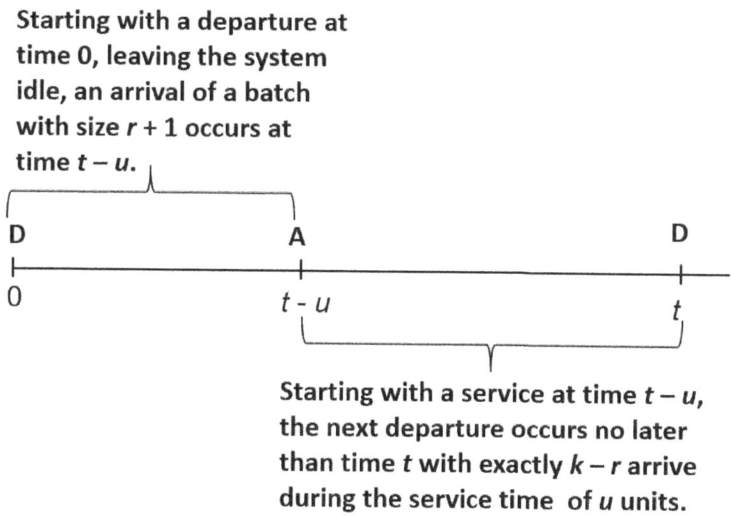

Figure 1.1. *Pictorial description for* $\hat{b}_{i,j}(n,t)$

Noting that $P(n,t)$ gives the (matrix) probability of the counting process associated with the BMAP (see equation [6.19] in Volume 1), we see that:

$$\hat{A}_n(t) = \int_0^t P(n,u)dF(u), \quad n \geq 0, \ t \geq 0,$$

$$\hat{B}_n(t) = \int_0^t e^{D_0(t-u)} \sum_{r=0}^{n} D_{r+1}\hat{A}_{n-r}(u)du, \quad n \geq 0, \ t \geq 0. \qquad [1.4]$$

While it is very clear how $\hat{A}_n(t)$ is obtained, a quick look at Figure 1.1 shows how one needs to split into two parts to get the expression for $\hat{B}_n(t)$.

The process, $\{(\zeta_n, J_n, \vartheta_{n+1} - \vartheta_n) : n \geq 0\}$, is an MRP on the state space $\{(i, j, t) : i \geq 0,\ 1 \leq j \leq m,\ t \geq 0\}$ and its TPM is given by:

$$\hat{P}(t) = \begin{bmatrix} \hat{B}_0(t) & \hat{B}_1(t) & \hat{B}_2(t) & \hat{B}_3(t) & \cdots \\ \hat{A}_0(t) & \hat{A}_1(t) & \hat{A}_2(t) & \hat{A}_3(t) & \cdots \\ & \hat{A}_0(t) & \hat{A}_1(t) & \hat{A}_2(t) & \cdots \\ & & \hat{A}_0(t) & \hat{A}_1(t) & \cdots \\ & & & \hat{A}_0(t) & \cdots \\ & & & & \ddots \end{bmatrix}, \qquad [1.5]$$

where the (matrix) entries are as given in equation [1.4]. For later use, we register a number of auxiliary matrices. The joint transform matrices, $A^*(z, s)$ and $B^*(z, s)$, respectively, of $\{\hat{A}_n(t)\}$ and $\{\hat{B}_n(t)\}$ are defined as:

$$A^*(z, s) = \sum_{n=0}^{\infty} z^n \int_0^{\infty} e^{-st} d\hat{A}_n(t),\ Re(s) \geq 0, |z| \leq 1, \qquad [1.6]$$

and:

$$B^*(z, s) = \sum_{n=0}^{\infty} z^n \int_0^{\infty} e^{-st} d\hat{B}_n(t),\ Re(s) \geq 0, |z| \leq 1. \qquad [1.7]$$

Let $D^*(z)$ denote the (matrix) probability generating function (PGF) of the BMAP. That is:

$$D^*(z) = \sum_{n=0}^{\infty} z^n D_n,\ |z| \leq 1. \qquad [1.8]$$

The following result (see Lucantoni (1993) and Neuts (1989)) establishes expressions for the transform matrices.

RESULT 1.1.– Suppose that $f^*(s)$ is the Laplace transform (LT) of $F(.)$. Then, the transform matrices $A^*(z, s)$ and $B^*(z, s)$ can be simplified as:

$$A^*(z, s) = \int_0^{\infty} e^{-st} e^{D(z)\,t} dF(t) = f^*(sI - D(z)),\ Re(s) \geq 0, |z| \leq 1, \qquad [1.9]$$

$$B^*(z, s) = z^{-1}(sI - D_0)^{-1}[D(z) - D_0]A^*(z, s),\ Re(s) \geq 0, |z| \leq 1. \qquad [1.10]$$

PROOF.– From the definition of $A^*(z, s)$ and using result 6.5 in Volume 1 (which says $P^*(z, t) = e^{D(z)t}$), we see:

$$A^*(z, s) = \int_0^{\infty} e^{-st} \sum_{n=0}^{\infty} z^n P(n, t) dF(t) = \int_0^{\infty} e^{-st} e^{D(z)\,t} dF(t), \qquad [1.11]$$

from which the stated result on $A^*(z,s)$ follows. From the definition of $B^*(z,s)$ and applying Leibnitz's rule, we get:

$$B^*(z,s) = \sum_{n=0}^{\infty} z^n \int_0^{\infty} e^{-st} d\hat{B}_n(t) = \sum_{n=0}^{\infty} z^n \int_0^{\infty} e^{-st} dt \sum_{r=0}^{n} D_{r+1}\hat{A}_{n-r}(t)$$

$$+ \sum_{n=0}^{\infty} z^n \int_0^{\infty} e^{-st} dt \int_0^t e^{D_0(t-u)} D_0 \sum_{r=0}^{n} D_{r+1}\hat{A}_{n-r}(u)du,$$

which upon simplification results in:

$$B^*(z,s) = \frac{1}{sz}\Big[(D(z)-D_0)A^*(z,s)+(sI-D_0)^{-1}D_0(D(z)-D_0)A^*(z,s)\Big],$$

from which the stated result follows immediately. Note that we used the fact that $\int_0^{\infty} e^{-st} \sum_{n=0}^{\infty} z^n \hat{A}_n(t)dt = \frac{1}{s}A^*(z,s)$. □

REMARK 1.1.– From equations [1.4], [1.9] and [1.10], it is easy to see:

$$A_n = \hat{A}_n(\infty) = \int_0^{\infty} P(n,t)dF(t), \; n \geq 0,$$

$$B_n = \hat{B}_n(\infty) = (-D_0)^{-1} \sum_{r=0}^{n} D_{r+1}A_{n-r}, \; n \geq 0,$$

$$A = A^*(1,0) = \sum_{n=0}^{\infty} A_n = \int_0^{\infty} e^{Qt}dF(t), \qquad [1.12]$$

$$B = B^*(1,0) = (-D_0)^{-1}(Q - D_0)A = (I - D_0^{-1}Q)A.$$

REMARK 1.2.– On noting that the (i,j)th entry of $(-D_0^{-1}D_k)$ gives the conditional probability that a BP starts with an arrival of a batch of size k with the BMAP in state j given that the idle period started with the BMAP in state i, the entries of B_n as given in equation [1.12] are intuitively obvious.

REMARK 1.3.– It is easy to verify that the vector $\boldsymbol{\pi}$ is also the steady-state vector of A. This follows from the fact that $\boldsymbol{\pi}e^{Qt} = \boldsymbol{\pi}$, for all $t \geq 0$.

The following result (see Neuts (1989)) plays a key role and gives an expression for the vector of the conditional mean number of arrivals during a service given the phase of the BMAP.

RESULT 1.2.– The vector, $\boldsymbol{\beta}^* = \sum_{k=1}^{\infty} kA_k\boldsymbol{e}$, is given by:

$$\boldsymbol{\beta}^* = \frac{\lambda}{\mu}\boldsymbol{e} + (I - A)(\boldsymbol{e\pi} - Q)^{-1}\sum_{k=1}^{\infty} kD_k\boldsymbol{e}. \quad [1.13]$$

PROOF.– Suppose that $M_k(t) = \sum_{n=k}^{\infty} \frac{n!}{(n-k)!}P(n,t)$. Then, (see Neuts (1989), Narayana and Neuts (1992) and Nielsen et al. (2007)), it is easy to verify:

$$\int_0^t dM_1(u) = \int_0^t M_1(x)Qdx + \int_0^t M_0(x)dx \sum_{r=0}^{\infty} rD_r, \quad [1.14]$$

which on post-multiplying by $\boldsymbol{e}$ along with the fact that $M_0(t) = e^{Q\,t}$, we get:

$$M_1(t)\boldsymbol{e} = \int_0^t e^{Qx}dx \sum_{r=0}^{\infty} rD_r\boldsymbol{e}. \quad [1.15]$$

Now applying result 4.25 in Volume 1, which gives an expression for the integral in equation [1.15], we get:

$$M_1(t)\boldsymbol{e} = t\lambda\boldsymbol{e} + [I - e^{Qt}](\boldsymbol{e\pi} - Q)^{-1}\sum_{r=0}^{\infty} rD_r\boldsymbol{e}. \quad [1.16]$$

From definition, we have:

$$\boldsymbol{\beta}^* = \sum_{k=1}^{\infty} kA_k\boldsymbol{e} = \int_0^t M_1(t)\boldsymbol{e}dF(t) = \frac{\lambda}{\mu}\boldsymbol{e} + (I - A)(\boldsymbol{e\pi} - Q)^{-1}\sum_{k=1}^{\infty} kD_k\boldsymbol{e}. \quad [1.17]$$

RESULT 1.3.– Neuts (1989) The $BMAP/G/1$ queue is stable if and only if $\lambda < \mu$.

PROOF.– Since $\boldsymbol{\pi\beta}^*$ gives the mean number of arrivals during a service time, the queue is stable if and only if this mean is less than one. Thus, using the expression for $\boldsymbol{\beta}^*$ as given in equation [1.17], we have:

$$\boldsymbol{\pi\beta}^* < 1 \;\Leftrightarrow\; \boldsymbol{\pi\beta}^* = \frac{\lambda}{\mu} < 1 \;\Leftrightarrow\; \lambda < \mu. \quad [1.18]$$

Thus, we get the stated result. □

Steady-state probabilities at departures: Suppose we look at the embedded Markov chain (EMC) corresponding to the MRP with TPM, $\hat{P}(t)$ given in equation [1.5]. That is, we look at the TPM, $P = \hat{P}(\infty)$:

$$P = \begin{bmatrix} B_0 & B_1 & B_2 & B_3 & \cdots \\ A_0 & A_1 & A_2 & A_3 & \cdots \\ & A_0 & A_1 & A_2 & \cdots \\ & & A_0 & A_1 & \cdots \\ & & & A_0 & \cdots \\ & & & & \ddots \end{bmatrix}, \qquad [1.19]$$

where the (matrix) entries are as given in equation [1.12]. We assume that $P = \hat{P}(\infty)$ is irreducible and that the condition $\lambda < \mu$ holds good so that P is Ergodic. Suppose we denote by $\boldsymbol{x} = (\boldsymbol{x}_0, \boldsymbol{x}_1, \boldsymbol{x}_2, \cdots)$ the steady-state probability vector of P. That is:

$$\boldsymbol{x}P = \boldsymbol{x}, \quad \boldsymbol{x}\boldsymbol{e} = 1. \qquad [1.20]$$

Note that the vectors, $\boldsymbol{x}_i,\ i \geq 0$, are all of dimension m.

The steady-state probability vector $\boldsymbol{x}$ of P satisfying equation [1.20] can be rewritten as:

$$\begin{aligned} \boldsymbol{x}_k &= \boldsymbol{x}_0 B_k + \sum_{j=1}^{k+1} \boldsymbol{x}_j\, A_{k-j+1},\ k \geq 0, \\ \sum_{j=0}^{\infty} \boldsymbol{x}_j\, \boldsymbol{e} &= 1. \end{aligned} \qquad [1.21]$$

The following result gives an expression for the (vector) PGF, $\boldsymbol{X}(z) = \sum_{k=0}^{\infty} z^k \boldsymbol{x}_k$.

RESULT 1.4.– The (vector) PGF, $\boldsymbol{X}(z)$, of $\{\boldsymbol{x}_k\}$ is given by:

$$\boldsymbol{X}(z) = \boldsymbol{x}_0(-D_0)^{-1}D(z)A(z)[zI - A(z)]^{-1},\ \ 0 < z < 1. \qquad [1.22]$$

PROOF.–In section 7.2 of Volume 1, we derived an expression for $\boldsymbol{X}(z)$ (see equation [7.36]), which is also valid here. Thus, we have:

$$\boldsymbol{X}(z) = \boldsymbol{x}_0[z\,B(z) - A(z)][z\,I - A(z)]^{-1},\ \ 0 < z < 1. \qquad [1.23]$$

From equation [1.10], by taking $s = 0$, we get:

$$B(z) = B^*(z, 0) = \sum_{k=0}^{\infty} z^k B_k = z^{-1}(-D_0)^{-1}[D(z) - D_0]A(z),\ |z| \leq 1, \qquad [1.24]$$

which can be rewritten as:

$$zB(z) - A(z) = (-D_0)^{-1}D(z)A(z), \; |z| \leq 1. \qquad [1.25]$$

In the above equations, $A(z) = A^*(z, 0) = \sum_{k=0}^{\infty} z^k A_k$ denotes the (matrix) PGF for $\{A_k\}$. The stated result follows by substituting the expression for $zB(z) - A(z)$ given in equation [1.25] into equation [1.23]. □

REMARK 1.4.– The PGF, $\boldsymbol{X}(z)$, is determined once $\boldsymbol{x}_0$ is computed. In section 7.2 of Volume 1, it was computed using the eigenvalue approach. Here, we take a different approach that has relevance to queueing. To compute $\boldsymbol{x}_0$, we first register the following key observations (see Neuts (1989)):

1) Note that the quantity x_{0j} gives the steady-state probability that a departure leaves the system empty with BMAP in phase j. Also, $\frac{1}{x_{0j}}$ gives the mean number of transitions in the EMC with TPM P between successive visits to $(0, j)$. Thus, to study the return time distributions to level $\mathbf{0}$, we need to study the first passage time distributions from level $\boldsymbol{i} + \mathbf{1}$ to level $\boldsymbol{i}$.

2) The EMC is skip-free to the left. That is, to reach level $\boldsymbol{j}$ from level $(\boldsymbol{i} + \mathbf{1})$, for $0 \leq j \leq i$, all the levels $\boldsymbol{j} + \mathbf{1}, \cdots, \boldsymbol{i}$ must be visited at least once.

3) Due to the homogeneity of the EMC, the first passage time distribution from $(\boldsymbol{i} + \mathbf{1})$ to level $\boldsymbol{i}$ is independent of i (away from the boundary state(s)).

The computation of $\boldsymbol{x}_0$ is done by first looking at the first passage time from level $(\boldsymbol{i} + \mathbf{1})$ to level $\boldsymbol{i}$, away from the boundary state $\mathbf{0}$, and then the recurrence time from level $\mathbf{1}$ to level $\mathbf{0}$. It should be pointed out that the first passage time from level $\boldsymbol{i}$ to level $(\boldsymbol{i} - \mathbf{1})$ is referred to as the fundamental period (see Neuts (1989)) and for the queueing model under study it is also the BP. In general, the BP may be different from the fundamental period. In general, studying the BPs is very involved. More on this will be seen later on.

First passage times from level $(\boldsymbol{i} + \mathbf{1})$ *to level* $\boldsymbol{i}$: we assume that the EMC is irreducible, which is the case in most applications. Let $T(i+r, j; i, k)$ denote the first passage time from state $(i + r, j)$ to the state (i, k), for $i, r \geq 1, 1 \leq j, k \leq m$. That is, $T(i+r, j; i, k)$ is the duration that the MRP, $\hat{P}(.)$, starting in state $(i+r, j)$ spends before visiting level $\boldsymbol{i}$ for the first time by entering the state (i, k). Let $V(i+r, j; i, k)$ denote the number of transitions involved during the first passage time $T(i+r, j; i, k)$.

The joint probability function of $T(i + r, j; i, k)$ and $V(i + r, j; i, k)$ plays a key role in $M/G/1$-type queues. We now define the matrix $G^{(r)}(n, t)$ whose (j, k)th entry gives the probability defined as (note that these matrices do not depend on i due to the structure), for $r \geq 0, 1 \leq j, k \leq m, \; n \geq 0$,

$$g_{j,k}^{(r)}(n, t) = P\{T(i + r, j; i, k) \leq t, V(i + r, j; i, k) = n\}, \qquad [1.26]$$

where we take:

$$g_{j,k}^{(0)}(n,t) = \begin{cases} 1, \ j=k, \ n=0, \ t=0, \\ \qquad 0, \ \text{elsewhere.} \end{cases} \qquad [1.27]$$

Define the joint transform matrix, $G(z,s), |z| \leq 1, Re(s) \geq 0$, of $G(n,t)$ as:

$$G(z,s) = \sum_{n=1}^{\infty} z^n \int_0^{\infty} e^{-st} dG(n,t). \qquad [1.28]$$

RESULT 1.5 (Neuts 1989).– The joint transform matrix $G(z,s)$ satisfies:

$$G(z,s) = z \sum_{r=0}^{\infty} A_r(s) G^r(z,s), \ |z| \leq 1, \ Re(s) \geq 0, \qquad [1.29]$$

where $A_r(s) = \int_0^{\infty} e^{-st} d\hat{A}_r(t), \ r \geq 0$.

PROOF.– First note that the Markov property of the MRP $\hat{P}(.)$ (see equation [1.5]) implies that the pairs $\{(T(i+r, \zeta_{i+r}; i+r-1, \zeta_{i+r-1}), V(i+r, \zeta_{i+r}; i+r-1, \zeta_{i+r-1}))\}$ are mutually conditionally independent given the random variables, $\zeta_i, \cdots, \zeta_{i+r}$, that keep track of the phase of the BMAP at those epochs when the corresponding levels are visited. This leads to:

$$G^{(r)}(n,t) = \sum_{S} G^{(1)}(r_1,.) * G^{(1)}(r_2,.) * \cdots * G^{(1)}(r_p,t), \qquad [1.30]$$

where the set S is such that $S = \{(r_1, \cdots, r_p) : r_q \geq 1, 1 \leq q \leq p, \ r_1 + \cdots + r_p = r\}$. The equation [1.30] implies that $G^{(r)}(n,t) = G^r(n,t)$. That is, the r-fold convolution of $G(.,.)$ is nothing but the rth power of $G(.,.)$. By the definition of $G(.,.)$ it is clear that $\hat{A}_0(.)$ governs the probabilities of the process going from level $(\boldsymbol{i}+1)$ to level $\boldsymbol{i}$ in one transition. To have exactly $r, \ r \geq 2$, transitions to go from level $(\boldsymbol{i}+1)$ to level $\boldsymbol{i}$, the process has to stay or go to some level to the right of level $(\boldsymbol{i}+1)$, say level $(\boldsymbol{i}+\boldsymbol{k}), \ k \geq 1$, and then return to level $\boldsymbol{i}$ but one level down at a time due to the skip-free to the left property. Thus, we have the following relationship between $\hat{A}_r(.)$ and $G(.,.)$:

$$\begin{aligned} G(1,t) &= \hat{A}_0(t), \\ G(r,t) &= \sum_{k=1}^{\infty} \hat{A}_k(.) * G^{(k)}(r-1,t), \ r \geq 2. \end{aligned} \qquad [1.31]$$

Multiplying the first equation in [1.31] by z, the second equation in [1.31] by z^r, adding them over r, and taking the LST, we get:

$$\int_0^{\infty} e^{-st} \sum_{r=1}^{\infty} z^r dG(r,t) = zA_0(s) + \sum_{r=2}^{\infty} z^r \sum_{k=1}^{\infty} A_k(s) G^k(r-1,s), \qquad [1.32]$$

from which the stated result follows. □

The following result (see Lucantoni (1991) and Lucantoni and Neuts (1994)) gives a different way to look at the joint transform matrix $G(.,.)$. Before seeing the result, we need to define:

$$D[G(z,s)] = \sum_{k=0}^{\infty} D_k G^k(z,s), \ Re(s) \geq 0, \ |z| \leq 1. \qquad [1.33]$$

RESULT 1.6.– The joint transform matrix $G(z,s)$ satisfies:

$$G(z,s) = z \int_0^{\infty} e^{-st} e^{D[G(z,s)]t} dF(t), \qquad [1.34]$$

from which we see that the matrix G satisfies:

$$G = \int_0^{\infty} e^{D[G]t} dF(t), \qquad [1.35]$$

where:

$$D[G] = \sum_{k=0}^{\infty} D_k G^k. \qquad [1.36]$$

REMARK 1.5.– The matrix $G(n,t) = G^{(1)}(n,t), n \geq 1, t \geq 0$, is such that its (j,k)th entry, $g_{j,k}(n,t)$, is the conditional probability that the first passage time from $(i+1,j)$ to (i,k), for $i \geq 1, 1 \leq j,k \leq m$, occurs in exactly n transitions and no later than t.

REMARK 1.6.– The substochastic matrix, $\tilde{G}(t) = \sum_{n=1}^{\infty} G(n,t)$, gives the matrix distribution for the fundamental period. That is, $\tilde{g}_{j,k}(t)$ is the conditional probability that the first passage time from $(i+1,j)$ to (i,k), for $1 \leq j,k \leq m$, occurs no later than t.

REMARK 1.7.– The sequence, $\{\hat{G}(n) = G(n,\infty)\}$, $n \geq 1$, of matrices, gives the (matrix) probability mass functions for the number of transitions during a fundamental period. That is, $\hat{g}_{j,k}(n), 1 \leq j,k \leq m, n \geq 1$, is the conditional probability that exactly n transitions occur during a first passage time from $(i+1,j)$ to (i,k).

REMARK 1.8.– The matrix $G = \sum_{n=1}^{\infty} \hat{G}(n)$ is such that its (j,k)th entry gives the conditional probability that the MRP eventually visits the state (i,k) for the first time starting in state $(i+1,j)$. This matrix plays an important role in the $M/G/1$-type queues (Neuts (1989)) like the matrix R in $GI/M/1$-type queues (Neuts (1981)).

REMARK 1.9.– If the MRP is irreducible, then G does not have zero rows. Furthermore, G has an eigenvalue of maximum modulus, which is positive.

REMARK 1.10.– The matrix G satisfies the following nonlinear matrix equation:

$$G = \sum_{i=0}^{\infty} A_i G^i, \qquad [1.37]$$

and is the minimal non-negative solution.

REMARK 1.11.– If the MRP is recurrent, then G is stochastic.

REMARK 1.12 (Structural properties of G (Neuts 1989)).– Some interesting structural properties are worth registering here. They are as follows:

1) To every zero column in A_0, the corresponding column in G is also zero.

2) If $(I - A_1)^{-1}A_0$ is irreducible, then G is also irreducible.

3) Suppose that A is reducible and is partitioned as:

$$A = \begin{bmatrix} A(1) & 0 & \cdots & 0 & 0 \\ 0 & A_1 & \cdots & 0 & 0 \\ \vdots & \vdots & \cdots & \vdots & \vdots \\ 0 & 0 & \cdots & A(c) & 0 \\ \tilde{A}_1 & \tilde{A}_2 & \cdots & \tilde{A}_c & \tilde{A}_0 \end{bmatrix}, \qquad [1.38]$$

where $A(r), 1 \leq r \leq c$, are irreducible stochastic matrices. The last (block) row, which may be absent in some applications, consists of substochastic matrices (which, of course, will add to get a stochastic matrix). In this case, G is also reducible and has the same structure as A.

4) Earlier results (when A is assumed to be irreducible) hold good by looking at c such conditions like $\boldsymbol{\pi}(r)\boldsymbol{\beta}(r) \leq 1,\ 1 \leq r \leq c$, and c such G matrices.

5) $D[G]$ has a nice probabilistic interpretation (see Lucantoni et al. (1990)). It is as follows. When the queue is stable, $D[G]$ is the generator of a continuous-time Markov chain (CTMC) in which all BPs are excised. That is, the CTMC looks at the phase of the arrival process during the times when the system is idle.

6) The matrix G and $D(G)$ commute. That is, $GD(G) = D(G)G$. This can be seen immediately by looking at equation [1.36] and the fact that $D[G]$ and $e^{D[G]}$ commute (due to the nature of the exponential matrix).

Recurrence time of level **0**

Here, we study the first passage (or return or recurrence) time from level **0** to level **0**. Toward this end, let $\vartheta_{i,j}$ denote the time duration the underlying MRP takes to return to state $(0, j)$ given that it started in state $(0, i)$. Let $U_{i,j}$ denote the number of transitions required to return to state $(0, j)$ from state $(0, i)$.

Let $K(n,t) = \{k_{i,j}(n,t)\}, n \geq 1, t \geq 0$, be such that $k_{i,j}(n,t)$ is the conditional probability that the first passage (return or recurrence) time from $(0,i)$ to $(0,j)$, for $1 \leq i,j \leq m$, occurs in exactly n transitions and no later than t.

$$k_{i,j}(n,t) = P\{\vartheta_{i,j} \leq t, U_{i,j} = n\},\ n \geq 0, 1 \leq i,j \leq m. \qquad [1.39]$$

The joint transform matrix, $K(z,s), |z| \leq 1, Re(s) \geq 0$, is defined as:

$$K(z,s) = \sum_{n=1}^{\infty} z^n \int_0^{\infty} e^{-sx} dK(n,x). \qquad [1.40]$$

The following result is similar to result 1.5 and the proof is left as an exercise.

RESULT 1.7 (Neuts 1989).– The joint transform matrix $K(z,s)$ satisfies:

$$K(z,s) = z\sum_{r=0}^{\infty} B_r(s)G^r(z,s),\ |z| \leq 1,\ Re(s) \geq 0, \qquad [1.41]$$

where $B_r(s) = \int_0^{\infty} e^{-st} d\hat{B}_r(t),\ r \geq 0$.

Suppose we define $K = K(1,0) = \{k_{i,j}\}$ such that $k_{i,j}$ gives the conditional probability that the MRP with TPM $P(.)$ of the BMAP at departure epochs will eventually hit level **0** by visiting $(0,j)$ given that it started in state $(0,i)$.

RESULT 1.8.– The matrix K is given by:

$$K = K(1,0) = I - D_0^{-1} D[G], \qquad [1.42]$$

where $D[G]$ is as given in equation [1.36].

PROOF.– Using the form of B_n as given in equation [1.12] in the definition of K, we get:

$$K = K(1,0) = \sum_{n=0}^{\infty} B_n G^n = \sum_{n=0}^{\infty} -(D_0)^{-1} \sum_{r=0}^{n} D_{r+1} A_{n-r} G^n$$

$$= -(D_0)^{-1} \sum_{r=0}^{\infty} D_{r+1} \sum_{n=r}^{\infty} A_{n-r} G^n = -(D_0)^{-1} \sum_{r=0}^{\infty} D_{r+1} \sum_{n=0}^{\infty} A_n G^{n+r},$$

$$= -(D_0)^{-1} \sum_{r=0}^{\infty} D_{r+1} G^{r+1} = -(D_0)^{-1}(D[G] - D_0) = I - (D_0)^{-1}[D[G]. \qquad \square$$

Suppose that $\tilde{\boldsymbol{\mu}}_G$ is the vector (of dimension m) of means such that the jth component gives the expected number of transitions required to reach level $\boldsymbol{i}$ from state $(i+1,j)$. Let $\hat{\boldsymbol{\mu}}_G$ denote the vector (of dimension m) of means such that the

jth component gives the expected duration to reach level i starting from state $(i+1, j)$. The following result (due to Neuts (1989)) gives the expressions for these mean vectors. For use in the following, we define $\boldsymbol{g}$ to be the steady-state probability vector of G. That is:

$$\boldsymbol{g}G = \boldsymbol{g}, \ \ \boldsymbol{g}\boldsymbol{e} = 1. \qquad [1.43]$$

RESULT 1.9.– The vectors $\tilde{\boldsymbol{\mu}}_G$ and $\hat{\boldsymbol{\mu}}_G$ are given by:

$$\begin{aligned} \tilde{\boldsymbol{\mu}}_G &= (I - G + \boldsymbol{e}\,\boldsymbol{g})[I - A + (\boldsymbol{e} - \boldsymbol{\beta}^*)\boldsymbol{g}]^{-1}\boldsymbol{e}, \\ \hat{\boldsymbol{\mu}}_G &= \frac{1}{\mu}(I - G + \boldsymbol{e}\,\boldsymbol{g})[I - A + (\boldsymbol{e} - \boldsymbol{\beta}^*)\boldsymbol{g}]^{-1}\boldsymbol{e}. \end{aligned} \qquad [1.44]$$

where $\boldsymbol{\beta}^*$ is as given in equation [1.17].

PROOF.– We prove the result $\tilde{\boldsymbol{\mu}}_G$ as it is similar for $\hat{\boldsymbol{\mu}}_G$. By definition:

$$\begin{aligned} \tilde{\boldsymbol{\mu}}_G &= \frac{\partial}{\partial z} G(z,s)\Big]_{s=0,z=1} \boldsymbol{e} = \sum_{r=0}^{\infty} A_r G^r \boldsymbol{e} \\ &+ \sum_{r=1}^{\infty} A_r \sum_{k=0}^{r-1} G^k \frac{\partial}{\partial z} G(z,s)\Big]_{s=0,z=1} G^{r-1-k}\boldsymbol{e}, \end{aligned} \qquad [1.45]$$

which reduces to (using the fact that G is stochastic under the assumption that the stability condition $\rho < 1$ holds good):

$$\Big[I - \sum_{r=1}^{\infty} A_r \sum_{k=0}^{r-1} G^k\Big]\tilde{\boldsymbol{\mu}}_G = \boldsymbol{e}. \qquad [1.46]$$

Note that:

$$\begin{aligned} &\Big[I - \sum_{r=1}^{\infty} A_r \sum_{k=0}^{r-1} G^k\Big][I - G + \boldsymbol{e}\,\boldsymbol{g}] \\ &= [I - G + \boldsymbol{e}\,\boldsymbol{g} - \sum_{r=1}^{\infty} A_r \sum_{k=0}^{r-1} G^k(I - G) - \sum_{r=1}^{\infty} rA_r\boldsymbol{e}\,\boldsymbol{g}] \\ &= [I - A + (\boldsymbol{e} - \boldsymbol{\beta}^*)\boldsymbol{g}]. \end{aligned} \qquad [1.47]$$

First, note that $[I - A + (\boldsymbol{e} - \boldsymbol{\beta}^*)\boldsymbol{g}]$ is non-singular. To see this, suppose that $\boldsymbol{u}$ is a non-zero vector such that:

$$[I - A + (\boldsymbol{e} - \boldsymbol{\beta}^*)\boldsymbol{g}]\boldsymbol{u} = \boldsymbol{0} \Rightarrow \ (I - A)\boldsymbol{u} + (\boldsymbol{e} - \boldsymbol{\beta}^*)\boldsymbol{g}\,\boldsymbol{u} = \boldsymbol{0}. \qquad [1.48]$$

Now pre-multiplying the equation [1.48] by $\boldsymbol{\pi}$, we get $\boldsymbol{g}\,\boldsymbol{u} = 0$, from which we see that $(I - A)\boldsymbol{u} = \boldsymbol{0}$. This implies that $\boldsymbol{u} = c\boldsymbol{e}$ with $c > 0$ due to A being a stochastic matrix. This is not possible since $\boldsymbol{g}\,\boldsymbol{u} = 0 \Rightarrow c = 0$. This indicates that $[I - A + (\boldsymbol{e} - \boldsymbol{\beta}^*)\boldsymbol{g}]$ has to be non-singular.

Since both $[I - A + (\boldsymbol{e} - \boldsymbol{\beta}^*\boldsymbol{g})]$ and $(I - G + \boldsymbol{e}\,\boldsymbol{g})$ are non-singular, we see that $\left[I - \sum_{r=1}^{\infty} A_r \sum_{k=0}^{r-1} G^k\right]$ is also non-singular.

Now combining equations [1.46] and [1.47] we get the stated result. □

REMARK 1.13.– The connection between $\tilde{\boldsymbol{\mu}}_G$ and $\hat{\boldsymbol{\mu}}_G$ is very clear.

REMARK 1.14.– It is easy to verify that the following equalities hold good and will serve as internal accuracy checks for computing G and other items related to the BP to be seen later.

$$\boldsymbol{g}\,\tilde{\boldsymbol{\mu}}_{\boldsymbol{G}} = \frac{1}{1-\rho} \text{ and } \boldsymbol{g}\,\hat{\boldsymbol{\mu}}_{\boldsymbol{G}} = \frac{1}{\mu(1-\rho)}. \qquad [1.49]$$

RESULT 1.10.– The vector $\tilde{\boldsymbol{\mu}}_K$ is given by:

$$\tilde{\boldsymbol{\mu}}_K = (-D_0^{-1})[Q - D[G] + \boldsymbol{d}\,\boldsymbol{g}][I - A + (\boldsymbol{e} - \boldsymbol{\beta}^*)\boldsymbol{g}]^{-1}\boldsymbol{e}, \qquad [1.50]$$

where:

$$\boldsymbol{d} = \sum_{k=0}^{\infty} k D_k \boldsymbol{e}. \qquad [1.51]$$

PROOF.– By definition:

$$\begin{aligned}\tilde{\boldsymbol{\mu}}_K &= \frac{\partial}{\partial z}K(z,s)\Big]_{s=0,z=1}\boldsymbol{e} = \sum_{r=0}^{\infty} B_r G^r \boldsymbol{e}\\ &+ \sum_{r=1}^{\infty} B_r \sum_{k=0}^{r-1} G^k \frac{\partial}{\partial z}G(z,s)\Big]_{s=0,z=1} G^{r-1-k}\boldsymbol{e},\end{aligned} \qquad [1.52]$$

which reduces to (using the fact that G is stochastic under the assumption that the stability condition $\rho < 1$ holds good):

$$\begin{aligned}\tilde{\boldsymbol{\mu}}_K &= \boldsymbol{e} + \sum_{r=1}^{\infty} B_r \sum_{k=0}^{r-1} G^k \tilde{\boldsymbol{\mu}}_G\\ &= \boldsymbol{e} + \sum_{r=0}^{\infty} B_r(I - G^r + r\boldsymbol{e}\,\boldsymbol{g})[I - A + (\boldsymbol{e} - \boldsymbol{\beta}^*)\boldsymbol{g}]^{-1}\boldsymbol{e}.\end{aligned} \qquad [1.53]$$

The stated result follows on noting (recall $D = Q - D_0$) the following:

$$\sum_{r=0}^{\infty} B_r = (-D_0^{-1})DA,$$

$$\sum_{r=0}^{\infty} B_r G^r = (-D_0^{-1})(D[G] - D_0), \qquad [1.54]$$

$$\sum_{r=0}^{\infty} r B_r \boldsymbol{e}\,\boldsymbol{g} = (-D_0^{-1})\boldsymbol{d}\,\boldsymbol{g} - (-D_0^{-1})D\boldsymbol{e}\,\boldsymbol{g}.$$

□

Once we get an expression for $\boldsymbol{x}_0$, the PGF, $\boldsymbol{X}(z)$, as given in equation [1.22] can be used to solve for the rest of the steady-state probability vector $\boldsymbol{x}$ at departures. We demonstrate this by first establishing the following result that gives an expression for the vector $\boldsymbol{x}_0$.

RESULT 1.11.– The vector $\boldsymbol{x}_0$ is given by:

$$\boldsymbol{x}_0 = \frac{(1-\rho)}{\lambda}\boldsymbol{g}(-D_0), \qquad [1.55]$$

where $\boldsymbol{g}$ is the steady-state vector of G (see equation [1.43]).

PROOF.– Let κ denote the steady-state vector of the stochastic matrix K as given in equation [1.42]. That is:

$$\boldsymbol{\kappa} K = \boldsymbol{\kappa}, \quad \boldsymbol{\kappa}\boldsymbol{e} = 1. \qquad [1.56]$$

Note that due to the form of K, we see $\boldsymbol{\kappa}$ can be obtained in terms of $\boldsymbol{g}$ as follows:

$$\boldsymbol{\kappa} = \frac{1}{\boldsymbol{g}D\boldsymbol{e}}\boldsymbol{g}(-D_0). \qquad [1.57]$$

To see this, note that $\boldsymbol{g}(-D_0)K = \boldsymbol{g}(-D_0) - \boldsymbol{g}D[G] = \boldsymbol{g}(-D_0)$, since $\boldsymbol{g}G = \boldsymbol{g} \Rightarrow \boldsymbol{g}D[G] = \mathbf{0}$. The uniqueness of $\boldsymbol{\kappa}$ yields the result. □

Applying result 2.60 from Volume 1, a classical result in MRP that relates the steady-state probability to the mean recurrence time, we note that:

$$\boldsymbol{x}_0 = \frac{1}{\boldsymbol{\kappa}\tilde{\boldsymbol{\mu}}_K}\boldsymbol{\kappa}. \qquad [1.58]$$

From equations [1.50] and [1.57], we note:

$$\boldsymbol{\kappa}\tilde{\boldsymbol{\mu}}_K = [\boldsymbol{g}D\boldsymbol{e}]^{-1}\boldsymbol{g}\Big[(Q - D[G] + \boldsymbol{d}\,\boldsymbol{g})(I - A + (\boldsymbol{e} - \boldsymbol{\beta}^*)\boldsymbol{g})^{-1}\boldsymbol{e}\Big]. \qquad [1.59]$$

Suppose we define $\boldsymbol{u} = [I - A + (\boldsymbol{e} - \boldsymbol{\beta}^*)\boldsymbol{g}]^{-1}\boldsymbol{e}$. Then, we see that:

$$[I - A + (\boldsymbol{e} - \boldsymbol{\beta}^*)\boldsymbol{g}]\boldsymbol{u} = \boldsymbol{e} \;\Rightarrow\; (I - A)\boldsymbol{u} + (\boldsymbol{e} - \boldsymbol{\beta}^*)\boldsymbol{g}\boldsymbol{u} = \boldsymbol{e}$$
$$\Rightarrow\; \boldsymbol{\pi}(\boldsymbol{e} - \boldsymbol{\beta}^*)\boldsymbol{g}\boldsymbol{u} = 1 \;\Rightarrow\; \boldsymbol{g}\boldsymbol{u} = \frac{1}{1-\rho}, \qquad [1.60]$$

which gives us:

$$(I - A)\boldsymbol{u} + \frac{1}{1-\rho}(\boldsymbol{e} - \boldsymbol{\beta}^*) = \boldsymbol{e}. \qquad [1.61]$$

Hence, noting that $\boldsymbol{g}D[G] = \mathbf{0}$, equation [1.59] can be simplified to:

$$\kappa\tilde{\boldsymbol{\mu}}_K = [\boldsymbol{g}D\boldsymbol{e}]^{-1}\Big[\boldsymbol{g}Q[I - A + (\boldsymbol{e} - \boldsymbol{\beta}^*)\boldsymbol{g}]^{-1}\boldsymbol{e} + (1-\rho)^{-1}\boldsymbol{g}\,\boldsymbol{d}\,\boldsymbol{g}\Big]. \qquad [1.62]$$

Using the expression for $\boldsymbol{\beta}^*$ (see equation [1.17]) in equation [1.61] and routine simplifications yield:

$$(I - A)\Big[\boldsymbol{u} - (1-\rho)^{-1}(\boldsymbol{e}\,\boldsymbol{\pi} - Q)^{-1}\boldsymbol{d}\Big] = \mathbf{0}, \qquad [1.63]$$

where $\boldsymbol{d}$ is as given in equation [1.51]. Now, using the uniqueness of the right eigenvector of $(I - A)$ corresponding to the eigenvalue 1, we get:

$$\boldsymbol{u} = c\boldsymbol{e} + (1-\rho)^{-1}(\boldsymbol{e}\,\boldsymbol{\pi} - Q)^{-1}\boldsymbol{d}. \qquad [1.64]$$

Using equation [1.64] in the first expression within parentheses of equation [1.62], we get:

$$\kappa\tilde{\boldsymbol{\mu}}_K = [(1-\rho)\boldsymbol{g}D\boldsymbol{e}]^{-1}\Big[\boldsymbol{g}Q(\boldsymbol{e}\,\boldsymbol{\pi} - Q)^{-1}\boldsymbol{d} + \boldsymbol{g}\,\boldsymbol{d}\,\boldsymbol{g}\Big]. \qquad [1.65]$$

Once we show that $\boldsymbol{g}Q(\boldsymbol{e}\,\boldsymbol{\pi} - Q)^{-1}\boldsymbol{d} = \lambda - \boldsymbol{g}\,\boldsymbol{d}$, the proof will be completed. Toward this end, suppose that $\boldsymbol{v} = (\boldsymbol{e}\,\boldsymbol{\pi} - Q)^{-1}\boldsymbol{d}$. Then, we have:

$$\boldsymbol{d} = \boldsymbol{e}\boldsymbol{\pi}\,\boldsymbol{v} - Q\boldsymbol{v} \;\Rightarrow\; \boldsymbol{\pi}\,\boldsymbol{d} = \boldsymbol{\pi}\,\boldsymbol{v} \;\Rightarrow\; \boldsymbol{\pi}\,\boldsymbol{v} = \lambda, \; (\because\; \boldsymbol{\pi}\,\boldsymbol{d} = \lambda)$$
$$\boldsymbol{g}\,\boldsymbol{d} = \boldsymbol{\pi}\,\boldsymbol{v} - \boldsymbol{g}Q\boldsymbol{v} \;\Rightarrow\; \boldsymbol{g}\,\boldsymbol{d} = \lambda - \boldsymbol{g}Q\boldsymbol{v} \;\Rightarrow\; \boldsymbol{g}Q\boldsymbol{v} = \lambda - \boldsymbol{g}\,\boldsymbol{d}. \qquad [1.66]$$

□

For the sake of completeness, we now summarize the results seen so far with regard to the steady-state vector and also recall the statements of theorem 7.2 in Volume 1 that pertain to our model here.

THEOREM 1.1.– When $\rho < 1$, the steady-state vector $\boldsymbol{x}$ is given by:

$$\boldsymbol{x}_0 = \frac{(1-\rho)}{\lambda}\boldsymbol{g}(-D_0),$$
$$\boldsymbol{x}_k = \Big[\boldsymbol{x}_0\tilde{B}_k + \sum_{j=1}^{k-1}\boldsymbol{x}_j\,\tilde{A}_{k-j+1}\Big](I - \tilde{A}_1)^{-1}, \; k \geq 1, \qquad [1.67]$$

where:

$$\tilde{A}_k = \sum_{j=k}^{\infty} A_j\, G^{j-k}, \quad \tilde{B}_k = \sum_{j=k}^{\infty} B_j\, G^{j-k}, \quad k \geq 1. \qquad [1.68]$$

REMARK 1.15.– When only the measures like the mean need to be computed, then we can use equation [1.35] to obtain the matrix G. The algorithmic steps for computing G are given below. This option does not require the knowledge of the matrices, $\{A_k\}$. However, if one needs to compute the steady-state vector $\boldsymbol{x}$, then we need to use equation [1.37]. The algorithmic steps in the computation of G as well as K matrices are spelled out in section 7.2 of Volume 1. Note that for this option we need to compute the matrices $\{P(n,t)\}$, which can be obtained using equation [6.36] in Volume 1 as follows (below we take $\theta = max_i\{-D_0\}_{ii}$):

$$A_n = \int_0^{\infty} P(n,t)dF(t) = \int_0^{\infty} \Big[\sum_{r=\lfloor \frac{n}{k^*} \rfloor k^*}^{\infty} e^{-\theta\, t}\, \frac{(\theta\, t)^r}{r!}\, \tilde{F}_{n,r} \Big] dF(t), \; n \geq 0. \qquad [1.69]$$

The computation of A_n depends on the nature of the service time distribution function $F(.)$. For example, if the services are deterministic or have masses only at a finite number of points, then we need to evaluate $P(n,t)$ at a finite number of time points. If the services are assumed to be CPH, then A_n is computed without any numerical integration. Suppose we consider four cases: (1) constant services with a mean of, say, d; (2) service times with masses at $\{d_1, \cdots, d_L\}$ with probabilities $\{p_1, \cdots, p_L\}$; (3) services are of CPH with representation $(\boldsymbol{\beta}, S)$ of order n; and (4) services follow a (continuous) uniform distribution on (u_1, u_2), for $0 < u_1 < u_2 < \infty$. Then, we have, for $n \geq 0$,

$$A_n = \begin{cases} P(n,d), & \text{constant services,} \\ \sum_{i=1}^{L} p_i P(n, d_i), & \text{discrete services,} \\ \sum_{r=\lfloor \frac{n}{k^*} \rfloor k^*}^{\infty} \gamma_r\, \tilde{F}_{n,r}, & \text{CPH services,} \\ \dfrac{1}{u_2 - u_1} \sum_{i=1}^{N} P(n, a_i), & \text{uniform services,} \end{cases} \qquad [1.70]$$

where $\{\gamma_k\}$ follows a discrete PH distribution with representation $(\theta\boldsymbol{\beta}(\theta I - S)^{-1}, \theta(\theta I - S)^{-1})$ of order n, and $\{a_i\}$ is a finite partition of the interval (u_1, u_2) such that $u_1 = a_1 < a_2 < a_3 < \cdots < a_{N-1} < a_N = u_2$. The choice of the partition is important so as to not miss out capturing the number of BMAP arrivals in that small interval. We refer the reader to Volume 1 for illustrative numerical examples dealing with the computation of $\{P(n,t)\}$ under a variety of scenarios in the context of MAM, and Neuts and Li (1996) for full details on the algorithm itself. Here, we present a number of examples in the context of queueing theory.

Algorithm 1.1. Algorithmic steps in computing the matrix G based on equation [1.35]

Here, we use the uniformization technique. Toward that end, let $\theta = max_i\{-D_0\}_{ii}$, $\gamma_r = \int_0^\infty e^{-\theta t}\frac{(\theta t)^r}{r!}dF(t), r \geq 0$.

Step 0: Set $k = 0, n = 0, G_k = 0, H_{0,k} = I$, for all k.

Step 1: Use Horner's method to get $D[G_k]$ (truncate $D[G]$ at some point, say, n_1, and the infinite sum of G at a point, say, n_2):

$$Y_0 = D_{n_1},\ Y_j = D_{n_1-j} + Y_{j-1}G_k,\ 1 \leq j \leq n_1,$$

and evaluate:

$$H_{r+1,k} = \left[I + \frac{1}{\theta}D[G_k]\right] H_{r,k},\ r = 0, 1, 2, \cdots, n_2.$$

Step 2: $k \leftarrow k+1,\ G_{k+1} = \sum_{r=0}^{n_2} \gamma_r H_{r,k}$.

Step 3: If $||G_{k+1} - G_k||_\infty < \epsilon$, where ϵ is a pre-determined small value, stop; otherwise go to step 1.

Steady-state probabilities at arbitrary times: Here, we look at the (joint) steady-state probability vector of the number in the system as well as the phase of the arrival process at arbitrary times. Let $\boldsymbol{y} = (\boldsymbol{y}_0, \boldsymbol{y}_1, \cdots)$ be the steady-state probability vector such that the jth component of $\boldsymbol{y}_i$, $i \geq 0$, gives the steady-state probability that there are i customers in the system and the phase of the arrival process is in state j, $1 \leq j \leq m$, at an arbitrary time.

Suppose that $\hat{M}(t) = \sum_{n=0}^{\infty} \hat{P}^{(n)}(t)$ is the Markov renewal function associated with the MRP whose TPM is given in equation [1.5]. Recall that $d\hat{M}_{i_0,j_0;i,j}(u)$ is the conditional probability that the MRP enters into state (i, j) during the interval $(u, u + du)$ given that at time 0 it was in state (i_0, j_0). In the context of $BMAP/G/1$ queue, this corresponds to the conditional probability that at a departure epoch the system has i customers and the phase of the BMAP is j given that at time 0 the system has i_0 customers and the phase of the BMAP is j_0.

The following result gives an expression for $\boldsymbol{y}$.

THEOREM 1.2.– When $\rho < 1$, the steady-state vector $\boldsymbol{y}$ is given by:

$$\boldsymbol{y}_0 = (1-\rho)\boldsymbol{g},$$
$$\boldsymbol{y}_k = \left[\sum_{i=0}^{k-1} \boldsymbol{y}_i D_{k-i} - \lambda(\boldsymbol{x}_{k-1} - \boldsymbol{x}_k)\right](-D_0)^{-1}, \ k \geq 1. \qquad [1.71]$$

PROOF.– Let $y_{i,j}(t)$, $i \geq 0, 1 \leq j \leq m$, $t \geq 0$, denote the probability that the queueing system under study has i customers in the system and the phase of the BMAP is in j at time t. Using the interpretation of $d\hat{M}_{i_0,j_0;i,j}(u)$ and the law of total probability by looking at all scenarios, it is easy to verify that:

$$y_{0,j}(t) = \sum_{r=1}^{m} \int_0^t d\hat{M}_{i_0,j_0;0,r}(u) \left(e^{D_0(t-u)}\right)_{r,j},$$

$$y_{k,j}(t) = \sum_{r=1}^{m} \int_0^t d\hat{M}_{i_0,j_0;0,r}(u) \int_0^{t-u} \left(e^{D_0 v} \sum_{i=1}^{k} D_i P(k-i, t-u-v)\right)_{r,j} \times$$
$$[1 - F(t-u-v)]dv$$
$$+ \sum_{i=1}^{k} \sum_{r=1}^{m} \int_0^t d\hat{M}_{i_0,j_0;i,r}(u) \left(P(k-i, t-u)\right)_{r,j} [1 - F(t-u)], \ k \geq 1.$$

From the classical key renewal theorem for MRP (see result 2.58 in Volume 1), we get (in matrix form):

$$\boldsymbol{y}_0 = \lambda \boldsymbol{x}_0 (-D_0)^{-1},$$
$$\boldsymbol{y}_k = \lambda \boldsymbol{x}_0 (-D_0)^{-1} \sum_{i=1}^{k} D_i \int_0^\infty P(k-i, t)[1 - F(t)]dt \qquad [1.72]$$
$$+ \lambda \sum_{i=1}^{k} \boldsymbol{x}_i \int_0^\infty P(k-i, t)[1 - F(t)]dt, \ k \geq 1.$$

Let $\boldsymbol{Y}(z)$ denote the (vector) PGF of $\boldsymbol{y}$. Then, from equation [1.72] through multiplying by z^k and adding over k, we get:

$$\boldsymbol{Y}(z) = \boldsymbol{y}_0 + \lambda \boldsymbol{x}_0 (-D_0)^{-1} \sum_{k=1}^{\infty} z^k \sum_{i=1}^{k} D_i \int_0^\infty P(k-i, t)[1 - F(t)]dt$$
$$+ \lambda \sum_{k=1}^{\infty} z^k \sum_{i=1}^{k} \boldsymbol{x}_i \int_0^\infty P(k-i, t)[1 - F(t)]dt,$$

which on simplification yields:

$$\begin{aligned}\boldsymbol{Y}(z) &= \boldsymbol{y}_0 + \lambda\boldsymbol{x}_0(-D_0)^{-1}[D(z) - D_0]\int\limits_0^\infty e^{D(z)\,t}[1 - F(t)]dt \\ &+\lambda[\boldsymbol{X}(z) - \boldsymbol{x}_0]\int\limits_0^\infty e^{D(z)\,t}[1 - F(t)]dt,\end{aligned} \tag{1.73}$$

where $\boldsymbol{X}(z)$ is the (vector) PGF of the steady-state vector $\boldsymbol{x}$ and is given in equation [1.22]. Recalling $A(z) = \sum\limits_{n=0}^{\infty} z^n A_n = \int\limits_0^\infty e^{D(z)\,t} d\,F(t)$, and using the fact that:

$$\begin{aligned}&\int\limits_0^\infty e^{D(z)\,t}[1 - F(t)]dt = \int\limits_0^\infty e^{D(z)\,t}\int\limits_t^\infty dF(u)dt = \int\limits_0^\infty dF(u)\int\limits_0^u e^{D(z)\,t}dt \\ &= [A(z) - I]D^{-1}(z),\ |z| \le 1,\end{aligned} \tag{1.74}$$

equation [1.73] can now be rewritten as:

$$\begin{aligned}&\boldsymbol{Y}(z)D(z) = \lambda\boldsymbol{x}_0(-D_0)^{-1}D(z) + \lambda\boldsymbol{x}_0(-D_0)^{-1}[D(z) - D_0][A(z) - I] \\ &+\lambda[\boldsymbol{X}(z) - \boldsymbol{x}_0][A(z) - I] = \lambda(z-1)\boldsymbol{X}(z),\ |z| \le 1.\end{aligned} \tag{1.75}$$

Using the expression for $\boldsymbol{x}_0$ and the identification of like coefficients of $z^k,\ k \ge 0$, in equation [1.75], we get the stated result. □

The following result will serve as an accuracy check in the computation of the probability vectors, $\boldsymbol{x}$ and $\boldsymbol{y}$.

RESULT 1.12.– We have:

$$\begin{aligned}\boldsymbol{X}(1) &= \sum_{k=0}^{\infty} \boldsymbol{x}_k = \boldsymbol{\pi} + \boldsymbol{x}_0(-D_0)^{-1}Q\,A(I - A + \boldsymbol{e}\,\boldsymbol{\pi})^{-1}, \\ \boldsymbol{Y}(1) &= \sum_{k=0}^{\infty} \boldsymbol{y}_k = \boldsymbol{\pi},\end{aligned} \tag{1.76}$$

where $\boldsymbol{\pi}$ is the steady-state probability vector of the generator $Q = \sum\limits_{k=0}^{\infty} D_k$ governing the BMAP.

PROOF.– The proof follows immediately by noting from equations [1.22] and [1.75] that $\boldsymbol{X}(1)[I - A] = \boldsymbol{x}_0(-D_0)^{-1}Q\,A$, and hence $\boldsymbol{Y}(1)Q = \boldsymbol{0}$. □

REMARK 1.16.– It is intuitively clear that $\boldsymbol{Y}(1) = \boldsymbol{\pi}$ due to the fact that the phase of BMAP in steady-state should be given by $\boldsymbol{\pi}$.

Stationary virtual waiting time: We assume that the FCFS rule is assumed. The virtual waiting time is the amount of time a virtual customer arriving at time t has to wait before receiving a service. That is, if an imaginary customer was to enter the system at time t, the virtual waiting time will be the time the customer has to wait before starting the service. The derivation of the Laplace-Stieltjes transform (LST) is quite involved and is due to Neuts (1989). This is the generalization of the well-known and classical Pollaczek-Khinchin formula for the $M/G/1$ queue to $BMAP/G/1$. We only state the result here and refer the reader to Neuts (1989) for full details.

RESULT 1.13.– Suppose that $\boldsymbol{w}(t)$ denote the vector whose jth component gives the joint probability that at an arbitrary time the arrival process is in phase j and that the virtual customer who arrives at that time waits at most t units of time before receiving a service. The LST, $\boldsymbol{w}^*(s)$, of $\boldsymbol{w}(t)$ is given by:

$$\boldsymbol{w}^*(s) = s(1-\rho)\boldsymbol{g}\Big(sI + D[f^*(s)]\Big)^{-1}, \ Re(s) \geq 0, \qquad [1.77]$$

where $f^*(s)$ is the LST of the service time distribution and $D[.]$ is as defined in equation [1.36].

REMARK 1.17.– Once various queue length densities and the waiting time distributions are obtained, we can compute the system performance measures like the mean, the median, the mode, the standard deviation and various percentiles, mostly through numerical computation. In the following examples, we present such measures.

Now we present a number of examples under various scenarios. These examples were generated using Fortran code (LF95 compiler), and a number of internal accuracy checks outlined earlier including the ones in result 1.12 were used to check the accuracy of the codes. Also, we used the exponential matrices, $e^{D_0 t}$ and $e^{Q t}$, and compared them with the computed matrices $P(n,t)$ (see remark 6.10 in Volume 1 for details on this).

In the following examples, we let μ_a denote the mean of the batch size.

EXAMPLE 1.1.– In this example, we look at BMAP arrivals in which the inter-arrival times are Erlang order 3, and the service times are assumed to be constant with a mean of 0.9. Specifically, we have:

$$D_0 = \begin{pmatrix} -1.2 & 1.2 & 0 \\ 0 & -1.2 & 1.2 \\ 0 & 0 & -1.2 \end{pmatrix}, \quad D_k = a_k \begin{pmatrix} 0 & 0 & 0 \\ 0 & 0 & 0 \\ 1.2 & 0 & 0 \end{pmatrix}, \ k \geq 1.$$

Here, $\{a_k\}$ is assumed to be uniform on $\{1, 2, 3, 4\}$ so that $D_k = 0$, for $k \geq 5$.

Verify that we have $\mu_a = 2.5$, $\lambda_g = \frac{1}{2.5}$, $\lambda = 1$, $\sigma_g = 1.44338$, and the 1-lag correlation coefficient, ρ_g, is zero due to the independence of the successive inter-arrival times. The matrices G and K along with their invariant vectors, $\boldsymbol{g}$ and $\boldsymbol{\kappa}$, are:

$$G = \begin{pmatrix} 0.3552 & 0.4006 & 0.2441 \\ 0.0461 & 0.443 & 0.5109 \\ 0.0965 & 0.2297 & 0.6738 \end{pmatrix}, \quad K = \begin{pmatrix} 0.1888 & 0.3595 & 0.4517 \\ 0.1888 & 0.3595 & 0.4517 \\ 0.1888 & 0.3595 & 0.4517 \end{pmatrix},$$

$$\boldsymbol{g} = (0.1087, 0.3156, 0.5757), \quad \boldsymbol{\kappa} = (0.1888, 0.3595, 0.4517).$$

Selected queue length statistics are listed in Table 1.1 and the steady-state probabilities of the number of customers in the system at departure, arbitrary and arrival epochs are plotted in Figure 1.2.

Epoch type	*Mean*	*Median*	*Mode*	*Modal value*	*SD*
Departure	6.974	5	3	0.1016	6.451
Arbitrary	6.794	5	0	0.1000	6.461
Arrival	8.479	7	4	0.1133	6.488

Table 1.1. *Selected queue length statistics for example 1.1*

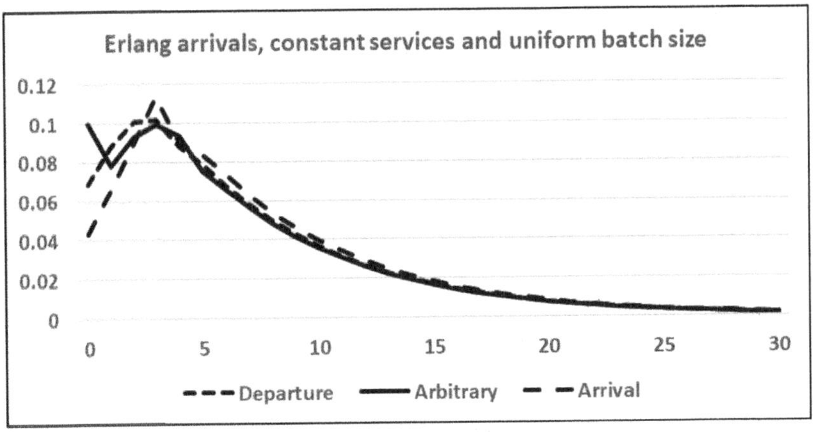

Figure 1.2. *Plots of steady-state probabilities for example 1.1*

EXAMPLE 1.2.– In this example, we look at BMAP arrivals in which the inter-arrival times are exponential, the service times are assumed to be constant with a mean of 0.9 and the batch sizes are uniformly distributed. That is, we have:

$$D_0 = (-0.4), \quad D_k = (0.1), \ 1 \leq k \leq 4, \quad D_k = (0), \ k \geq 5.$$

Verify that we have $\mu_a = 2.5$, $\lambda_g = \frac{1}{2.5}$, $\lambda = 1$, $\sigma_g = 2.5$, and the 1-lag correlation coefficient, ρ_g, is zero due to the independence of successive inter-arrival times. Obviously, G, K, $\boldsymbol{g}$ and $\boldsymbol{\kappa}$ are all scalars and given by 1. Selected queue length statistics are displayed in Table 1.2 and the steady-state probabilities of the number of customers in the system at departure, arbitrary and arrival epochs are plotted in Figure 1.3.

Epoch type	*Mean*	*Median*	*Mode*	*Modal value*	*SD*
Departure	14.835	11	3	0.0545	14.305
Arbitrary	13.837	9	0	0.1000	14.270
Arrival	16.334	12	4	0.0613	14.316

Table 1.2. *Selected queue length statistics for example 1.2*

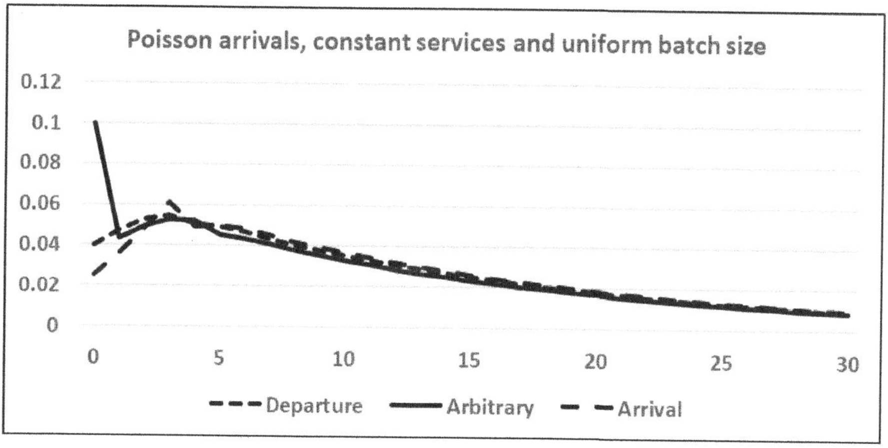

Figure 1.3. *Plots of steady-state probabilities for example 1.2*

EXAMPLE 1.3.– This is similar to the previous two examples, except that the inter-arrivals times are taken to be hyperexponential. Specifically, we have:

$$D_0 = \begin{pmatrix} -0.64 & 0 & 0 \\ 0 & -0.32 & 0 \\ 0 & 0 & -0.064 \end{pmatrix}, \quad D_k = \frac{1}{4}\begin{pmatrix} 0.5120 & 0.0960 & 0.0320 \\ 0.2560 & 0.0480 & 0.0160 \\ 0.0512 & 0.0096 & 0.0032 \end{pmatrix}, \quad k \geq 1,$$

and $D_k = 0,\ k \geq 5$.

Verify that we have $\mu_a = 2.5$, $\lambda_g = \frac{1}{2.5}$, $\lambda = 1$, $\sigma_g = 5.0$, and the 1-lag correlation coefficient, ρ_g, is zero due to the independence of successive inter-arrival times. The matrices, G and K, and their invariant vectors, $\boldsymbol{g}$ and $\boldsymbol{\kappa}$, are displayed below:

$$G = \begin{pmatrix} 0.7251 & 0.0935 & 0.1815 \\ 0.0936 & 0.8032 & 0.1032 \\ 0.0210 & 0.0119 & 0.9670 \end{pmatrix}, \quad K = \begin{pmatrix} 0.4235 & 0.2113 & 0.3652 \\ 0.4235 & 0.2113 & 0.3652 \\ 0.4235 & 0.2113 & 0.3652 \end{pmatrix},$$

$$\boldsymbol{g} = (0.0941, 0.094, 0.8119), \quad \boldsymbol{\kappa} = (0.4235, 0.2113, 0.3652).$$

Selected queue length statistics are displayed in Table 1.3 and the steady-state probabilities of the number of customers in the system at departure, arbitrary and arrival epochs are plotted in Figure 1.4.

Epoch type	*Mean*	*Median*	*Mode*	*Modal value*	*SD*
Departure	44.499	31	3	0.0201	43.482
Arbitrary	40.480	27	0	0.1000	43.298
Arrival	45.969	33	4	0.0222	43.417

Table 1.3. *Selected queue length statistics for example 1.3*

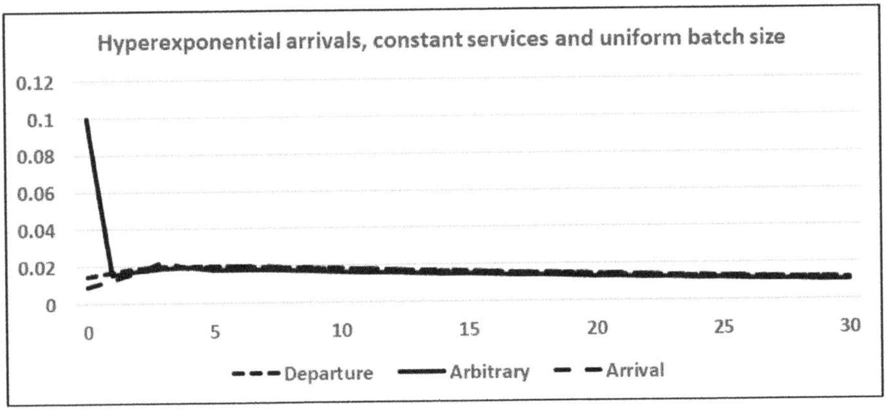

Figure 1.4. *Plots of steady-state probabilities for example 1.3*

EXAMPLE 1.4.– In this example, we look at BMAP arrivals and the service times are assumed to be constant with a mean of 0.9. We model the inter-arrival times with two distributions, one having a 1-lag negative correlation and the other having a 1-lag positive correlation. The BMAP representation for negatively correlated arrivals is given by:

$$D_0 = \begin{pmatrix} -0.5 & 0.5 & 0 \\ 0 & -0.5 & 0.5 \\ 0 & 0 & -1.0 \end{pmatrix}, \quad D_k = \frac{1}{4}\begin{pmatrix} 0 & 0 & 0 \\ 0.005 & 0 & 0.495 \\ 0.990 & 0 & 0.010 \end{pmatrix}, \quad 1 \le k \le 4,$$

and the representation for the positively correlated arrivals is given by:

$$D_0 = \begin{pmatrix} -0.5 & 0.5 & 0 \\ 0 & -0.5 & 0.5 \\ 0 & 0 & -1.0 \end{pmatrix}, \quad D_k = \frac{1}{4}\begin{pmatrix} 0 & 0 & 0 \\ 0.495 & 0 & 0.005 \\ 0.010 & 0 & 0.099 \end{pmatrix}, \quad 1 \le k \le 4,$$

and $D_k = 0, \ k \ge 5$, for both the cases. Note that both the BMAPs have the same mean and the same standard deviation for the inter-arrival times of successive groups. Like in those examples, we assume the batch sizes to follow a uniform distribution.

Verify that we have $\mu_a = 2.5, \ \lambda_g = \frac{1}{2.5}, \ \lambda = 1, \ \sigma_g = 2.3383$, and the 1-lag correlation coefficient, ρ_g, are, respectively, given by -0.32667 and 0.32667. We identify the nature of the correlated arrivals with superscripts NC and PC, respectively, for negative and positive 1-lag correlation. The matrices G and K along with their invariant vectors, $\boldsymbol{g}$ and $\boldsymbol{\kappa}$, are:

$$G^{(NC)} = \begin{pmatrix} 0.6588 & 0.3279 & 0.0132 \\ 0.1023 & 0.8396 & 0.0581 \\ 0.2336 & 0.3312 & 0.4352 \end{pmatrix}, \quad K^{(NC)} = \begin{pmatrix} 0.2755 & 0.5102 & 0.2143 \\ 0.2755 & 0.5102 & 0.2143 \\ 0.4513 & 0.5078 & 0.0409 \end{pmatrix},$$

$$\boldsymbol{g}^{(NC)} = (0.2529, 0.6720, 0.0751), \quad \boldsymbol{\kappa} = (0.3076, 0.5098, 0.1826).$$

$$G^{(PC)} = \begin{pmatrix} 0.6721 & 0.3277 & 0.0002 \\ 0.1607 & 0.8384 & 0.0009 \\ 0.1651 & 0.3327 & 0.5022 \end{pmatrix}, \quad K^{(PC)} = \begin{pmatrix} 0.4901 & 0.5069 & 0.0030 \\ 0.4901 & 0.5069 & 0.0030 \\ 0.2548 & 0.5109 & 0.2343 \end{pmatrix},$$

$$\boldsymbol{g}^{(PC)} = (0.3289, 0.6698, 0.0013), \quad \boldsymbol{\kappa}^{(PC)} = (0.4892, 0.5069, 0.0039).$$

Selected performance measures are listed in Table 1.4 and the steady-state probabilities of the number of customers in the system at departure, arbitrary and arrival epochs are plotted in Figure 1.5.

Type of BMAP	*Epoch Type*	*Mean*	*Median*	*Mode*	*Modal value*	*SD*
NC	*Departure*	11.798	9	3	0.0658	11.022
	Arbitrary	11.113	8	0	0.1000	11.044
	Arrival	13.304	10	4	0.0708	11.053
PC	*Departure*	420.718	268	0	0.0336	474.929
	Arbitrary	378.806	217	0	0.1000	467.027
	Arrival	421.867	270	4	0.0329	474.021

Table 1.4. *Selected queue length statistics for example 1.4*

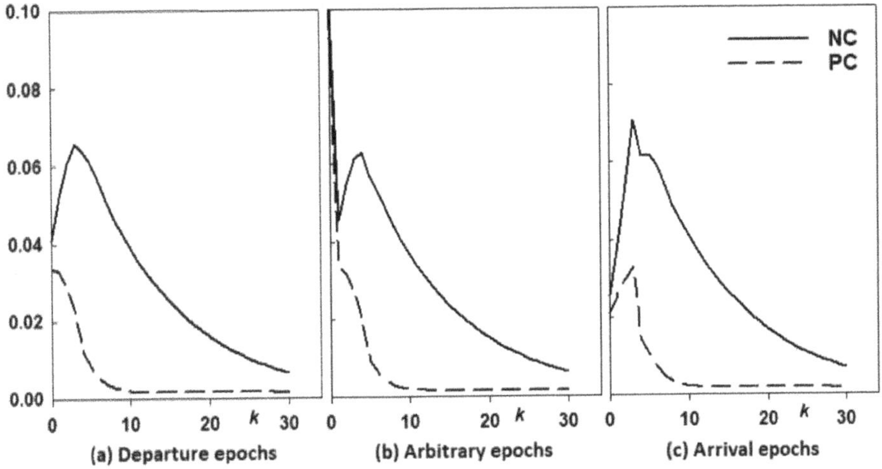

Figure 1.5. *Plots of steady-state probabilities for example 1.4*

EXAMPLE 1.5.– Referring to examples 1.1 through 1.4, suppose that the group arrivals are now modeled using a Poisson distribution with parameter 1.25 such that:

$$a_k = e^{-1.25}\frac{1.25^{k-1}}{(k-1)!}, \quad k \geq 1.$$

In order to fix $\lambda = 1$ like in the previous examples for proper comparisons, we need to normalize $\{D_k\}$ appropriately. The representations for the BMAPs for the five cases to be discussed in this example are as follows:

$$D_0^{(ER)} = \frac{1}{3}\begin{pmatrix} -4 & 4 & 0 \\ 0 & -4 & 4 \\ 0 & 0 & -4 \end{pmatrix}, D_k^{(ER)} = \frac{a_k}{3}\begin{pmatrix} 0 & 0 & 0 \\ 0 & 0 & 0 \\ 4 & 0 & 0 \end{pmatrix}, k \geq 1,$$

$$D_0^{(EX)} = \left(-\frac{1}{2.25}\right), \quad D_k^{(EX)} = \left(\frac{a_k}{2.25}\right), \quad k \geq 1,$$

$$D_0^{(HE)} = \frac{1}{45}\begin{pmatrix} -32 & 0 & 0 \\ 0 & -16 & 0 \\ 0 & 0 & -3.2 \end{pmatrix}, \; D_k^{(HE)} = \frac{a_k}{45}\begin{pmatrix} 25.60 & 4.80 & 1.60 \\ 12.80 & 2.40 & 0.80 \\ 2.56 & 0.48 & 0.16 \end{pmatrix}, \; k \geq 1,$$

$$D_0^{(NC)} = \frac{1}{9}\begin{pmatrix} -5 & 5 & 0 \\ 0 & -5 & 0 \\ 0 & 0 & -10 \end{pmatrix}, \; D_k^{(NC)} = \frac{a_k}{9}\begin{pmatrix} 0 & 0 & 0 \\ 0.05 & 0 & 4.95 \\ 9.90 & 0 & 0.10 \end{pmatrix}, \; k \geq 1,$$

$$D_0^{(PC)} = \frac{1}{9}\begin{pmatrix} -5 & 5 & 0 \\ 0 & -5 & 0 \\ 0 & 0 & -10 \end{pmatrix}, \; D_k^{(PC)} = \frac{a_k}{9}\begin{pmatrix} 0 & 0 & 0 \\ 4.95 & 0 & 0.95 \\ 0.10 & 0 & 9.90 \end{pmatrix}, \; k \geq 1,$$

Note that all these five BMAPs have the same mean batch size of 2.25 and the same arrival rate, $\lambda = 1.0$. However, the other parameters vary as follows. The standard deviations of these five arrival processes are, respectively, given by 1.2990, 2.25, 4.5, 2.3383 and 2.3383. Also, the 1-lag correlation, ρ_g, is as in example 1.4. For this example, G and K matrices along with their invariant vectors, $\boldsymbol{g}$ and $\boldsymbol{\kappa}$, are displayed along with their identifiers:

$$G^{(ER)} = \begin{pmatrix} 0.3203 & 0.4043 & 0.2754 \\ 0.0510 & 0.4204 & 0.5286 \\ 0.0979 & 0.2431 & 0.6589 \end{pmatrix}, \; K^{(ER)} = \begin{pmatrix} 0.1852 & 0.3635 & 0.4513 \\ 0.1852 & 0.3635 & 0.4513 \\ 0.1852 & 0.3635 & 0.4513 \end{pmatrix},$$

$$\boldsymbol{g}^{(ER)} = (0.1068, 0.3164, 0.5767), \; \boldsymbol{\kappa}^{(ER)} = (0.1852, 0.3635, 0.4513),$$

$$G^{(HE)} = \begin{pmatrix} 0.7033 & 0.1012 & 0.1955 \\ 0.1026 & 0.7848 & 0.1126 \\ 0.0234 & 0.0133 & 0.9634 \end{pmatrix}, \; K^{(HE)} = \begin{pmatrix} 0.4292 & 0.2117 & 0.3591 \\ 0.4292 & 0.2117 & 0.3591 \\ 0.4292 & 0.2117 & 0.3591 \end{pmatrix},$$

$$\boldsymbol{g}^{(HE)} = (0.0966, 0.0953, 0.8081), \; \boldsymbol{\kappa}^{(HE)} = (0.4292, 0.2117, 0.3591),$$

$$G^{(NC)} = \begin{pmatrix} 0.6322 & 0.3523 & 0.0155 \\ 0.1123 & 0.8265 & 0.0611 \\ 0.2468 & 0.3547 & 0.3986 \end{pmatrix}, \; K^{(NC)} = \begin{pmatrix} 0.2789 & 0.5072 & 0.2139 \\ 0.2789 & 0.5072 & 0.2139 \\ 0.4543 & 0.5055 & 0.0402 \end{pmatrix},$$

$$\boldsymbol{g}^{(NC)} = (0.2549, 0.6704, 0.0747), \; \boldsymbol{\kappa}^{(NC)} = (0.3108, 0.5069, 0.1822),$$

$$G^{(PC)} = \begin{pmatrix} 0.6476 & 0.3521 & 0.0002 \\ 0.1734 & 0.8257 & 0.0009 \\ 0.1773 & 0.3558 & 0.4669 \end{pmatrix}, \; K^{(PC)} = \begin{pmatrix} 0.4921 & 0.5049 & 0.003 \\ 0.4921 & 0.5049 & 0.003 \\ 0.2544 & 0.5078 & 0.2378 \end{pmatrix},$$

$$\boldsymbol{g}^{(PC)} = (0.3298, 0.6689, 0.0013), \; \boldsymbol{\kappa}^{(PC)} = (0.4912, 0.5049, 0.0039),$$

The plots of the steady-state probabilities at various epochs of the five arrival processes are displayed in Figure 1.6, and selected performance measures are listed in Table 1.5.

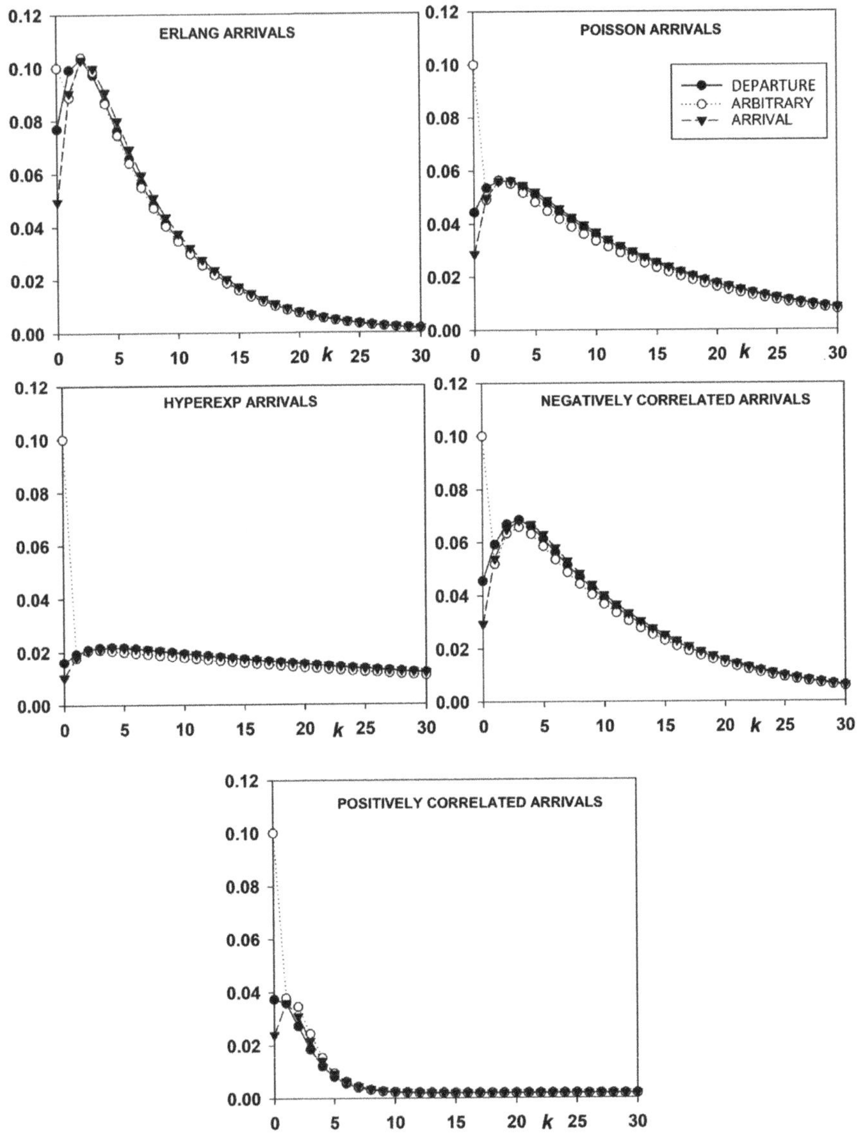

Figure 1.6. *Plots of steady-state probabilities at various epochs for example 1.5*

Type of BMAP	*Epoch Type*	*Mean*	*Median*	*Mode*	*Modal value*	*SD*
ER	*Departure*	6.797	5	2	0.1034	6.441
	Arbitrary	6.628	5	2	0.1040	6.419
	Arrival	8.143	6	3	0.1031	6.457
EX	*Departure*	13.877	10	2	0.0564	13.473
	Arbitrary	12.975	9	0	0.1000	13.431
	Arrival	15.223	11	4	0.0563	13.480
HE	*Departure*	40.618	28	4	0.0217	39.746
	Arbitrary	36.994	25	0	0.1000	39.584
	Arrival	41.943	30	5	0.0217	39.696
NC	*Departure*	11.136	8	3	0.0686	10.526
	Arbitrary	10.512	7	0	0.1000	10.518
	Arrival	12.482	9	4	0.0682	10.535
PC	*Departure*	379.179	242	0	0.0373	427.967
	Arbitrary	341.459	196	0	0.1	420.870
	Arrival	380.218	243	2	0.0360	427.168

Table 1.5. *Selected queue length statistics for example 1.5*

EXAMPLE 1.6.– This example is similar to example 1.5 except that the batch sizes are now modeled using a discrete PH distribution. The representation for this PH distribution is given by:

$$\boldsymbol{\alpha} = (0.7, 0.3), \quad T = \begin{pmatrix} 0.1 & 0.7 \\ 0.2 & 0.3 \end{pmatrix},$$

First, recall from section 3.1 (see result 3.1) from Volume 1 that the PMF, $\{a_k\}$, is obtained as $a_0 = 0$ and $a_k = \boldsymbol{\alpha} T^{k-1} \boldsymbol{T^0}, \ k \geq 1$. Verify first that the mean batch size, $\mu_a = 2.6735$, and hence the matrices $D_k, \ k \geq 0$, need to multiplied by an appropriate constant to get $\lambda = 1$. This constant, say, d, is obtained as follows. In example 1.5, the mean batch size is 2.25 and hence by taking $d = \dfrac{2.25}{2.6735} \simeq 0.8416$, we get the proper BMAP representations to guarantee $\lambda = 1$ for this example. For example, the representation for the Erlang arrivals for this example is given by:

$$D_0^{(ER)} = \frac{0.8416}{3} \begin{pmatrix} -4 & 4 & 0 \\ 0 & -4 & 4 \\ 0 & 0 & -4 \end{pmatrix}, D_k^{(ER)} = \frac{0.8416\, a_k}{3} \begin{pmatrix} 0 & 0 & 0 \\ 0 & 0 & 0 \\ 4 & 0 & 0 \end{pmatrix}, k \geq 1.$$

Thus, for this example, all these five BMAPs have the same mean batch size of 2.6735, the same arrival rate, $\lambda = 1.0$. However, the other parameters vary as follows. The standard deviations of these five arrival processes are, respectively, given by 1.5432, 2.6735, 5.3457, 2.7777 and 2.7777. Also, the 1-lag correlation, ρ_g, is the same as in example 1.4.

The matrices G and K matrices along with their invariant vectors, $\boldsymbol{g}$ and $\boldsymbol{\kappa}$, are:

$$G^{(ER)} = \begin{pmatrix} 0.3798 & 0.3981 & 0.2221 \\ 0.0488 & 0.4622 & 0.489 \\ 0.1075 & 0.2302 & 0.6623 \end{pmatrix}, \quad K^{(ER)} = \begin{pmatrix} 0.2198 & 0.371 & 0.4092 \\ 0.2198 & 0.371 & 0.4092 \\ 0.2198 & 0.371 & 0.4092 \end{pmatrix},$$

$$\boldsymbol{g}^{(ER)} = (0.1214, 0.3263, 0.5523), \quad \boldsymbol{\kappa}^{(ER)} = (0.2198, 0.371, 0.4092),$$

$$G^{(HE)} = \begin{pmatrix} 0.758 & 0.0885 & 0.1535 \\ 0.0994 & 0.814 & 0.0866 \\ 0.0222 & 0.0111 & 0.9667 \end{pmatrix}, \quad K^{(HE)} = \begin{pmatrix} 0.4677 & 0.2083 & 0.324 \\ 0.4677 & 0.2083 & 0.324 \\ 0.4677 & 0.2083 & 0.324 \end{pmatrix},$$

$$\boldsymbol{g}^{(HE)} = (0.1134, 0.101, 0.7856), \quad \boldsymbol{\kappa}^{(HE)} = (0.4677, 0.2083, 0.3240),$$

$$G^{(NC)} = \begin{pmatrix} 0.6762 & 0.3099 & 0.0139 \\ 0.1017 & 0.8334 & 0.0649 \\ 0.2407 & 0.299 & 0.4603 \end{pmatrix}, \quad K^{(NC)} = \begin{pmatrix} 0.2847 & 0.4649 & 0.2504 \\ 0.2847 & 0.4649 & 0.2504 \\ 0.4835 & 0.4732 & 0.0433 \end{pmatrix},$$

$$\boldsymbol{g}^{(NC)} = (0.2667, 0.6485, 0.0849), \quad \boldsymbol{\kappa}^{(NC)} = (0.3259, 0.4666, 0.2075),$$

$$G^{(PC)} = \begin{pmatrix} 0.6891 & 0.3107 & 0.0002 \\ 0.1617 & 0.8372 & 0.0011 \\ 0.1548 & 0.2929 & 0.5522 \end{pmatrix}, \quad K^{(PC)} = \begin{pmatrix} 0.5203 & 0.476 & 0.0037 \\ 0.5203 & 0.476 & 0.0037 \\ 0.2449 & 0.4609 & 0.2942 \end{pmatrix},$$

$$\boldsymbol{g}^{(PC)} = (0.3422, 0.6561, 0.0017), \quad \boldsymbol{\kappa}^{(PC)} = (0.5188, 0.4760, 0.0052).$$

The plots of the steady-state probabilities at various epochs of the five arrival processes are displayed in Figure 1.7, and selected performance measures are listed in Table 1.6.

EXAMPLE 1.7.– Here, we compare the impact of the batch size distribution when services are constant by looking at examples 1.1 through 1.6. A few selected measures are displayed in Table 1.7 and the steady-state probability vectors at departure epochs under various scenarios are plotted in Figure 1.8.

Type of BMAP	*Epoch Type*	*Mean*	*Median*	*Mode*	*Modal value*	*SD*
ER	*Departure*	11.354	8	1	0.0697	11.201
	Arbitrary	10.702	7	0	0.1000	11.100
	Arrival	12.501	9	2	0.0723	11.183
EX	*Departure*	19.795	14	1	0.0415	19.467
	Arbitrary	18.212	12	0	0.1000	19.260
	Arrival	20.878	15	2	0.0432	19.358
HE	*Departure*	51.933	36	4	0.0171	51.056
	Arbitrary	46.545	31	0	0.1000	49.892
	Arrival	52.446	38	5	0.0171	50.127
NC	*Departure*	16.507	12	3	0.0479	15.993
	Arbitrary	15.287	10	0	0.1000	15.869
	Arrival	17.623	13	4	0.0482	15.932
PC	*Departure*	453.714	290	0	0.0308	511.203
	Arbitrary	346.241	235	0	0.1000	385.220
	Arrival	392.159	292	2	0.0307	401.8610

Table 1.6. *Selected measures under DPH batch size for example 1.6*

Batch	*MEASURE*	*ERL*	*EXP*	*HEX*	*NCA*	*PCA*
UB	*MEAN*	6.974	14.836	44.499	11.798	420.718
	MEDIAN	5	11	31	9	268
	MODE	3	3	3	3	0
	MVALUE	0.1016	0.0545	0.0201	0.0658	0.0336
	SD	6.451	14.305	43.482	11.022	474.929
PB	*MEAN*	6.797	13.877	40.618	11.136	379.179
	MEDIAN	5	10	28	8	242
	MODE	2	2	4	3	0
	MVALUE	0.1034	0.0564	0.0217	0.0686	0.0373
	SD	6.441	13.473	39.746	10.526	427.967
DP	*MEAN*	11.354	19.796	51.933	16.507	453.714
	MEDIAN	8	14	36	12	290
	MODE	1	1	4	3	0
	MVALUE	0.0697	0.0415	0.0171	0.0479	0.0308
	SD	11.201	19.467	51.056	15.993	511.2030

Table 1.7. *Selected measures for example 1.7 under various batch sizes*

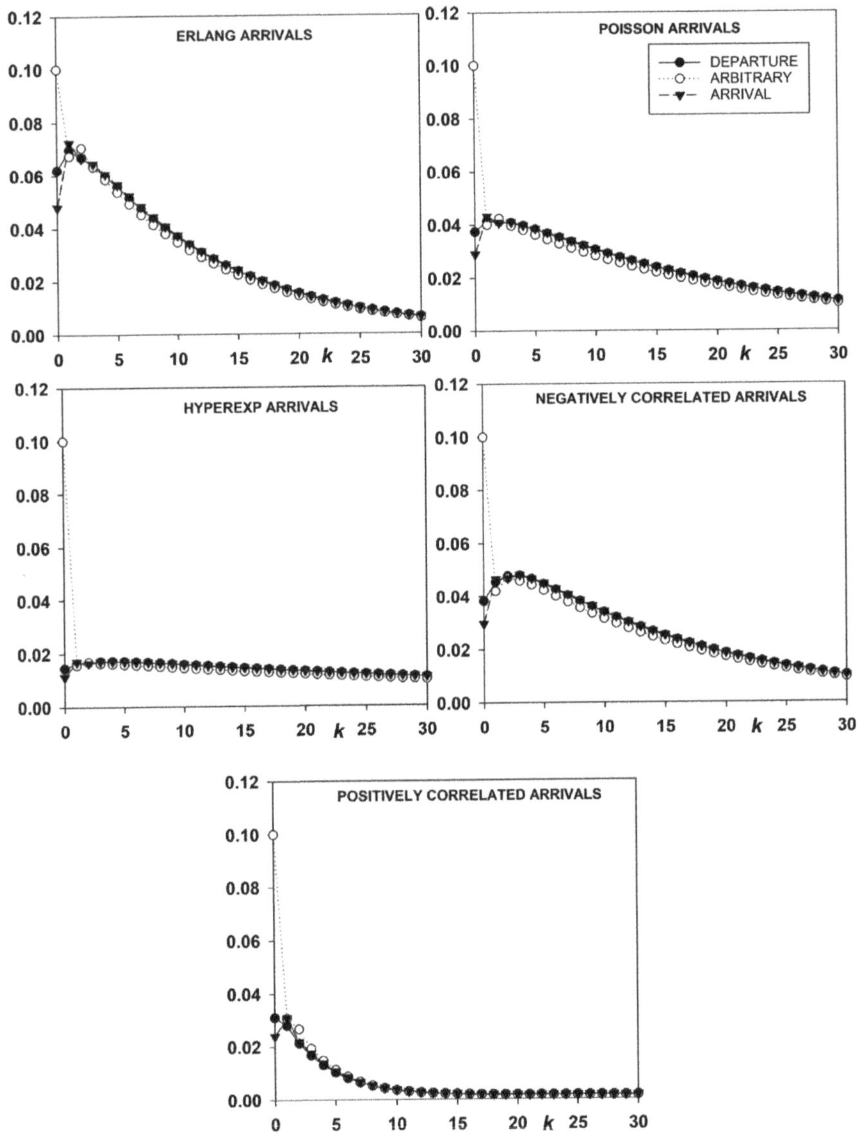

Figure 1.7. *Plots of steady-state probabilities at various epochs for example 1.6*

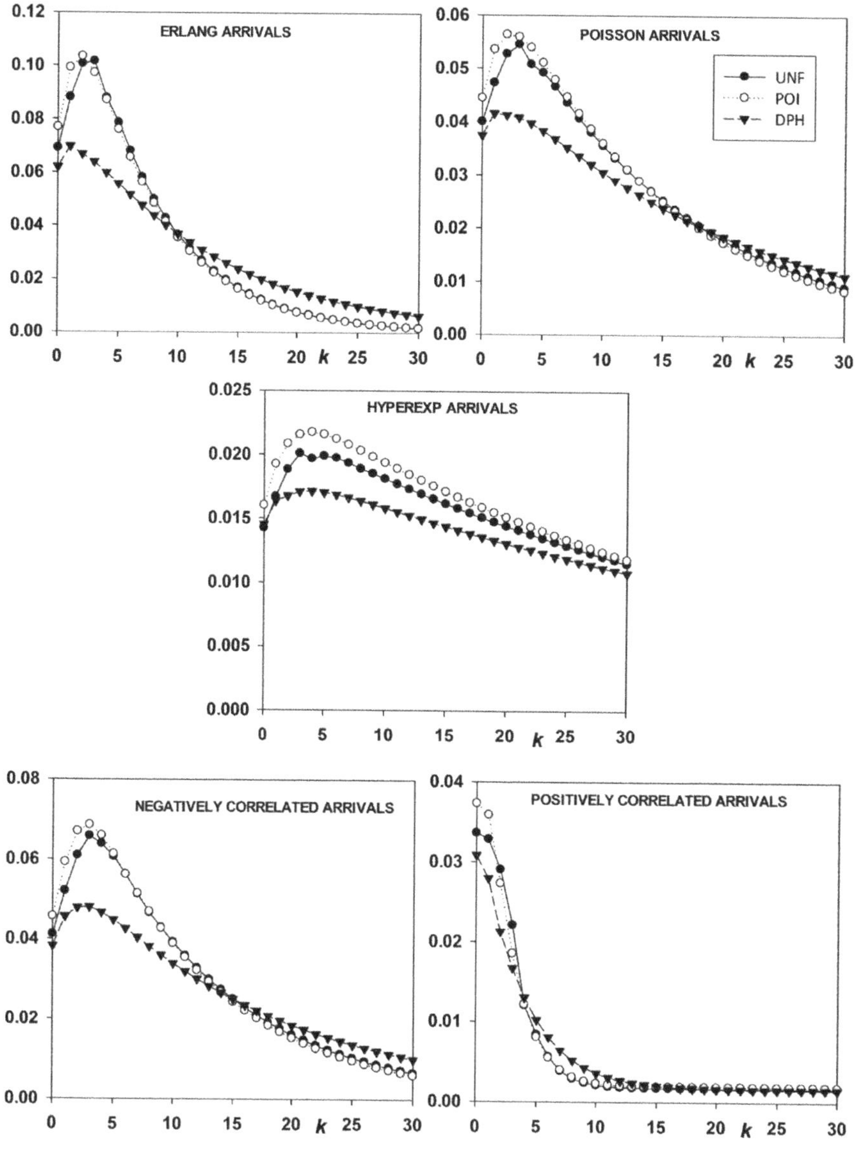

Figure 1.8. *Plots of steady-state probabilities at departure epochs for example 1.7*

EXAMPLE 1.8.– In this example, we look at BMAP arrivals as given in example 1.6 and consider two service time distributions. In the first one, we take the services to follow an arbitrary discrete probability function denoted as ARS for ease of reference in the table and figure below. Denoting by Y, the service time, we take $P(Y = 0.7) = 0.8, P(Y = 1.3) = 0.15$ and $P(Y = 2.9) = 0.05$. In the second one, we look at uniform services (UFS) on the interval $(0.4, 1.4)$.

Note that these two services have a mean of 0.9. However, the standard deviations are, respectively, given by 0.50596 and 0.28868. For batch sizes we look at uniform (UB) on {1, 2, 3, 4} as well as Poisson (PB) with parameter 1.25.

In Table 1.8, selected queue length statistics are displayed and in Figure 1.9 the plot of the steady-state vector under various scenarios is given.

$Service$	$Batch$	$MEASURE$	ERL	EXP	HEX	NCA	PCA
		$MEAN$	7.426	15.159	44.916	12.129	421.815
		$MEDIAN$	6	11	35	9	274
	UB	$MODE$	2	3	3	3	0
		$MVALUE$	0.0906	0.0516	0.02	0.0615	0.0332
		SD	7.828	15.885	45.791	12.557	476.394
ARS		$MEAN$	5.998	11.787	37.36881	9.421	379.494
		$MEDIAN$	6	11	29	9	264
	PB	$MODE$	2	2	3	3	0
		$MVALUE$	0.0916	0.0537	0.0213	0.0632	0.0368
		SD	6.618	14.376	43.781	11.115	430.142
		$MEAN$	7.428	15.362	45.298	12.292	424.522
		$MEDIAN$	5	11	32	8	269
	UB	$MODE$	3	3	3	3	0
		$MVALUE$	0.0971	0.0534	0.02	0.0641	0.0334
UFS		SD	7.031	15.076	44.930	11.719	485.652
		$MEAN$	7.253	14.391	41.391	11.624	382.641
		$MEDIAN$	5	10	29	8	242
	PB	$MODE$	2	2	4	3	0
		$MVALUE$	0.0988	0.0553	0.0215	0.0664	0.0371
		SD	7.0320	14.2090	41.1090	11.2050	437.6460

Table 1.8. *Selected queue length statistics at departure epochs for $BMAP/G/1$ queue of example 1.8*

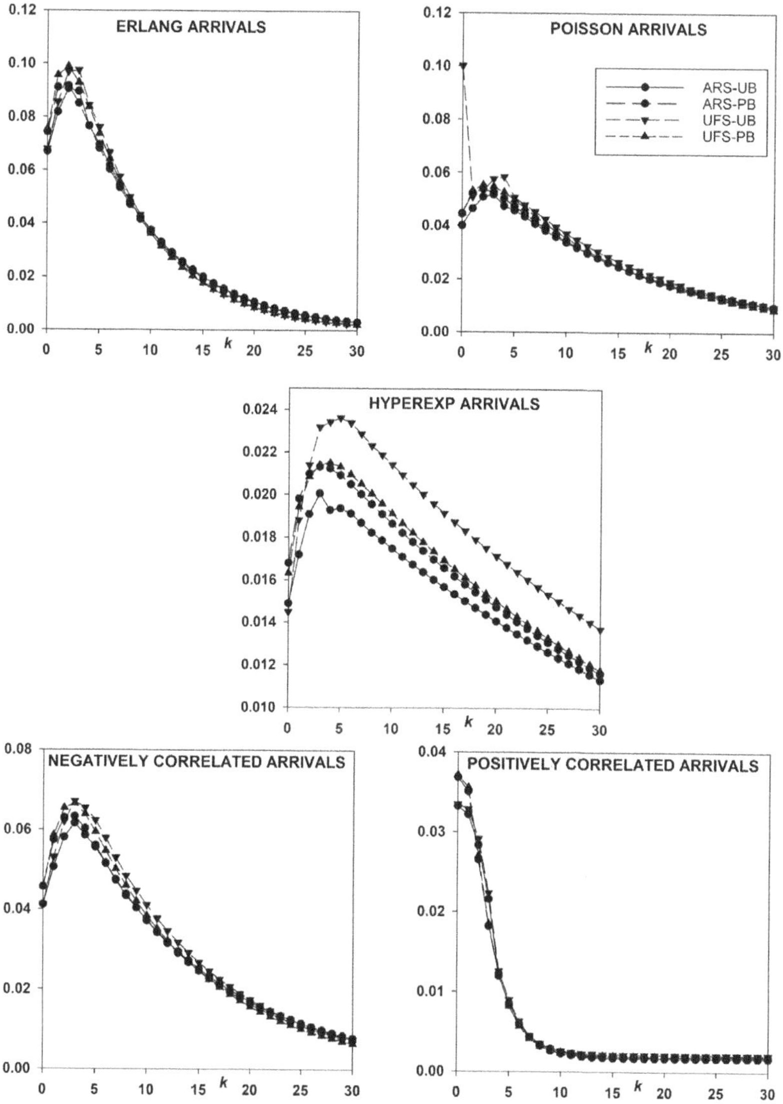

Figure 1.9. *Plots of steady-state probabilities at departure epochs for example 1.8*

EXAMPLE 1.9.– In this example, we look at $BMAP/PH/1$-type queues by referring to example 1.5, but here the service times are modeled using Erlang of order 3 with parameter $\frac{4}{3}$. The arrival rate, λ_g, of the groups will be normalized such that $\lambda = 1$. The service rate will be normalized so that the mean service time will be 0.9. Thus, in this example, arrivals occur in batches, which are modeled using Poisson distribution with parameter 1.25, and due to the shift in the masses the mean batch size is 2.25. We refer the reader to the representations given in example 1.5.

For this example, G and K matrices along with their invariant vectors, $\boldsymbol{g}$ and $\boldsymbol{\kappa}$, are:

$$G^{(ER)} = \begin{pmatrix} 0.3907 & 0.3636 & 0.2457 \\ 0.0574 & 0.4787 & 0.4639 \\ 0.1083 & 0.2235 & 0.6682 \end{pmatrix}, \quad K^{(ER)} = \begin{pmatrix} 0.2335 & 0.3580 & 0.4085 \\ 0.2335 & 0.3580 & 0.4085 \\ 0.2335 & 0.3580 & 0.4085 \end{pmatrix},$$

$$\boldsymbol{g}^{(ER)} = (0.1279, 0.3241, 0.5479), \quad \boldsymbol{\kappa}^{(ER)} = (0.2335, 0.358, 0.4085),$$

$$G^{(HE)} = \begin{pmatrix} 0.7321 & 0.0927 & 0.1752 \\ 0.1031 & 0.7930 & 0.1039 \\ 0.0242 & 0.0129 & 0.9629 \end{pmatrix}, \quad K^{(HE)} = \begin{pmatrix} 0.4601 & 0.2070 & 0.3329 \\ 0.4601 & 0.2070 & 0.3329 \\ 0.4601 & 0.2070 & 0.3329 \end{pmatrix},$$

$$\boldsymbol{g}^{(HE)} = (0.1095, 0.0985, 0.792), \quad \boldsymbol{\kappa}^{(HE)} = (0.4601, 0.207, 0.3329),$$

$$G^{(NC)} = \begin{pmatrix} 0.6588 & 0.3224 & 0.0188 \\ 0.1076 & 0.8268 & 0.0656 \\ 0.2324 & 0.3127 & 0.4549 \end{pmatrix}, \quad K^{(NC)} = \begin{pmatrix} 0.2777 & 0.4658 & 0.2565 \\ 0.2777 & 0.4658 & 0.2565 \\ 0.4808 & 0.4735 & 0.0457 \end{pmatrix},$$

$$\boldsymbol{g}^{(NC)} = (0.264, 0.6488, 0.0872), \quad \boldsymbol{\kappa}^{(NC)} = (0.3207, 0.4674, 0.2119),$$

$$G^{(PC)} = \begin{pmatrix} 0.6763 & 0.3234 & 0.0003 \\ 0.1683 & 0.8306 & 0.0011 \\ 0.1623 & 0.3079 & 0.5298 \end{pmatrix}, \quad K^{(PC)} = \begin{pmatrix} 0.5199 & 0.4764 & 0.0037 \\ 0.5199 & 0.4764 & 0.0037 \\ 0.2455 & 0.4626 & 0.2919 \end{pmatrix},$$

$$\boldsymbol{g}^{(PC)} = (0.3419, 0.6562, 0.0017), \quad \boldsymbol{\kappa}^{(PC)} = (0.5185, 0.4763, 0.0052),$$

The plots of the steady-state probabilities at various epochs of the five arrival processes are displayed in Figure 1.10, and selected performance measures are listed in Table 1.9.

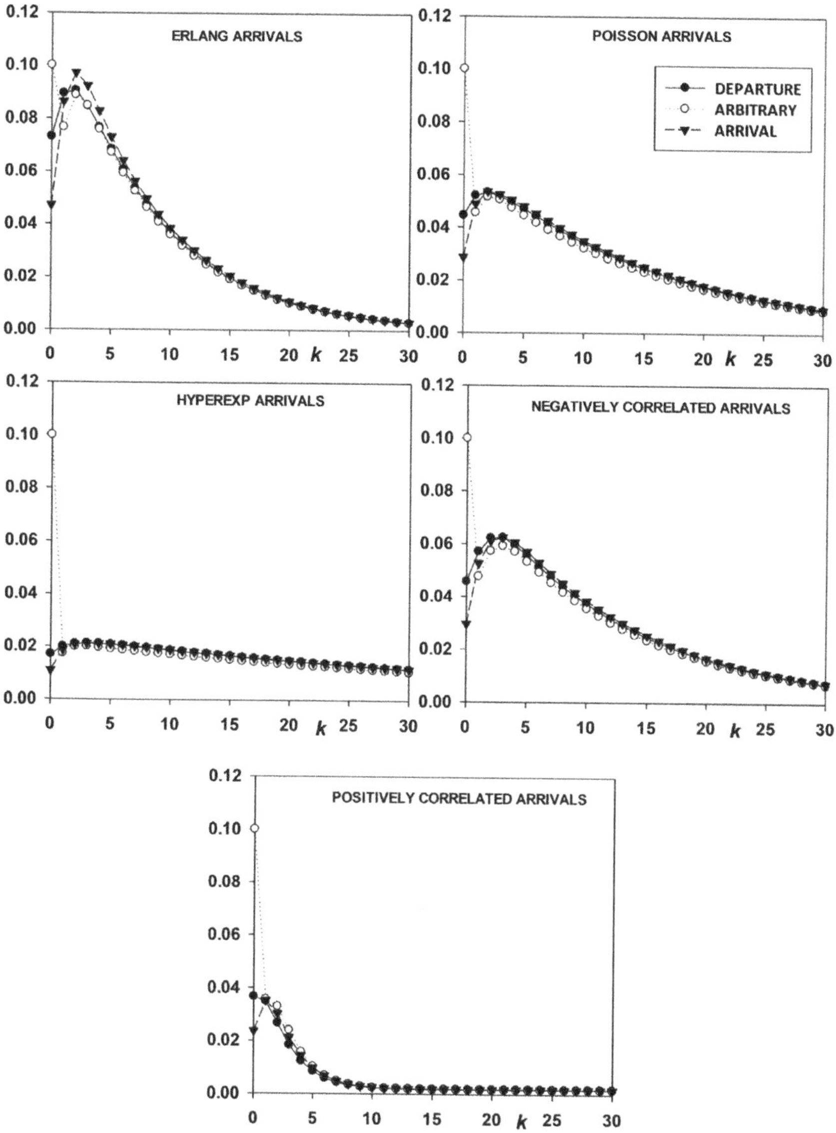

Figure 1.10. *Plots of steady-state probabilities at various epochs for example 1.9*

Type of BMAP	*Epoch Type*	*Mean*	*Median*	*Mode*	*Modal value*	*SD*
ER	*Departure*	8.104	6	2	0.09	7.885
	Arbitrary	7.934	6	0	0.1	7.889
	Arrival	9.456	7	3	0.091	7.913
EX	*Departure*	15.212	11	2	0.053	14.947
	Arbitrary	14.31	10	0	0.1	14.909
	Arrival	16.558	12	3	0.053	14.953
HE	*Departure*	42.164	29	3	0.021	41.454
	Arbitrary	38.536	26	0	0.1	41.303
	Arrival	43.495	31	4	0.021	41.421
NC	*Departure*	12.449	9	3	0.062	11.995
	Arbitrary	11.832	8	0	0.1	12.006
	Arrival	13.801	10	4	0.063	12.019
PC	*Departure*	380.543	243	0	0.037	429.336
	Arbitrary	342.823	197	0	0.1	422.251
	Arrival	381.582	244	2	0.035	428.5350

Table 1.9. *Selected quality statistics for* $BMAP/ER/1$ *queue of example 1.9*

EXAMPLE 1.10.– In this example, we look at $BMAP/PH/1$-type queues by referring to example 1.9, but here the service times are modeled using hyperexponential with parameters $\frac{19}{9}$ and $\frac{19}{90}$ with probabilities, respectively, given by 0.9 and 0.1. Note that the mean and the standard deviation of the services are 0.9 and 2.0203, respectively. The arrival rate, λ_g, of the groups will be normalized such that $\lambda = 1$. The service rate will be normalized so that the mean service time will be 0.9.

Note first that in this example the arrivals occur in batches, which are modeled using a Poisson distribution with parameter 1.25. We refer the reader to the representations given in example 1.5. Here, G and K matrices along with their invariant vectors, $\boldsymbol{g}$ and $\boldsymbol{\kappa}$, are:

$$G^{(ER)} = \begin{pmatrix} 0.5942 & 0.2669 & 0.1391 \\ 0.0556 & 0.6404 & 0.3040 \\ 0.1216 & 0.1567 & 0.7217 \end{pmatrix}, \quad K^{(ER)} = \begin{pmatrix} 0.3999 & 0.3327 & 0.2674 \\ 0.3999 & 0.3327 & 0.2674 \\ 0.3999 & 0.3327 & 0.2674 \end{pmatrix},$$

$$\boldsymbol{g}^{(ER)} = (0.1875, 0.3435, 0.4689), \quad \boldsymbol{\kappa}^{(ER)} = (0.3999, 0.3327, 0.2674),$$

$$G^{(HE)} = \begin{pmatrix} 0.8517 & 0.0614 & 0.0869 \\ 0.0961 & 0.8464 & 0.0575 \\ 0.0259 & 0.0110 & 0.9631 \end{pmatrix}, \quad K^{(HE)} = \begin{pmatrix} 0.6044 & 0.1930 & 0.2026 \\ 0.6044 & 0.1930 & 0.2026 \\ 0.6044 & 0.1930 & 0.2026 \end{pmatrix},$$

$$\boldsymbol{g}^{(HE)} = (0.2004, 0.1280, 0.6716), \quad \boldsymbol{\kappa}^{(HE)} = (0.6044, 0.1930, 0.2026),$$

$$G^{(NC)} = \begin{pmatrix} 0.7695 & 0.2098 & 0.0207 \\ 0.0832 & 0.8471 & 0.0697 \\ 0.1923 & 0.1731 & 0.6346 \end{pmatrix}, \quad K^{(NC)} = \begin{pmatrix} 0.2773 & 0.3024 & 0.4203 \\ 0.2773 & 0.3024 & 0.4203 \\ 0.6065 & 0.3425 & 0.0510 \end{pmatrix},$$

$$\boldsymbol{g}^{(NC)} = (0.3089, 0.5658, 0.1253), \quad \boldsymbol{\kappa}^{(NC)} = (0.3784, 0.3147, 0.3069),$$

$$G^{(PC)} = \begin{pmatrix} 0.7850 & 0.2145 & 0.0005 \\ 0.1373 & 0.8611 & 0.0015 \\ 0.0947 & 0.1443 & 0.7611 \end{pmatrix}, \quad K^{(PC)} = \begin{pmatrix} 0.6394 & 0.3537 & 0.0069 \\ 0.6394 & 0.3537 & 0.0069 \\ 0.1764 & 0.2642 & 0.5594 \end{pmatrix},$$

$$\boldsymbol{g}^{(PC)} = (0.3892, 0.6061, 0.0047), \quad \boldsymbol{\kappa}^{(PC)} = (0.6323, 0.3523, 0.0154).$$

Selected performance measures are listed in Table 1.10, and the plots of the steady-state probabilities at various epochs of the five arrival processes are displayed in Figure 1.11.

Type of BMAP	*Epoch Type*	*Mean*	*Median*	*Mode*	*Modal value*	*SD*
ER	*Departure*	26.909	17	0	0.063	29.763
	Arbitrary	26.719	17	0	0.100	29.794
	Arrival	28.26	19	2	0.060	29.782
EX	*Departure*	34.106	22	1	0.045	36.734
	Arbitrary	33.204	21	0	0.100	36.717
	Arrival	35.452	24	2	0.044	36.736
HE	*Departure*	62.614	42	1	0.024	64.93
	Arbitrary	58.909	38	0	0.100	64.687
	Arrival	63.94	44	2	0.024	64.882
NC	*Departure*	31.251	20	1	0.047	33.835
	Arbitrary	30.626	20	0	0.100	33.847
	Arrival	32.597	22	2	0.046	33.838
PC	*Departure*	399.891	256	0	0.034	449.547
	Arbitrary	362.204	210	0	0.100	442.603
	Arrival	400.95	257	2	0.031	448.791

Table 1.10. *Selected measures for* $BMAP/HE/1$ *queue*

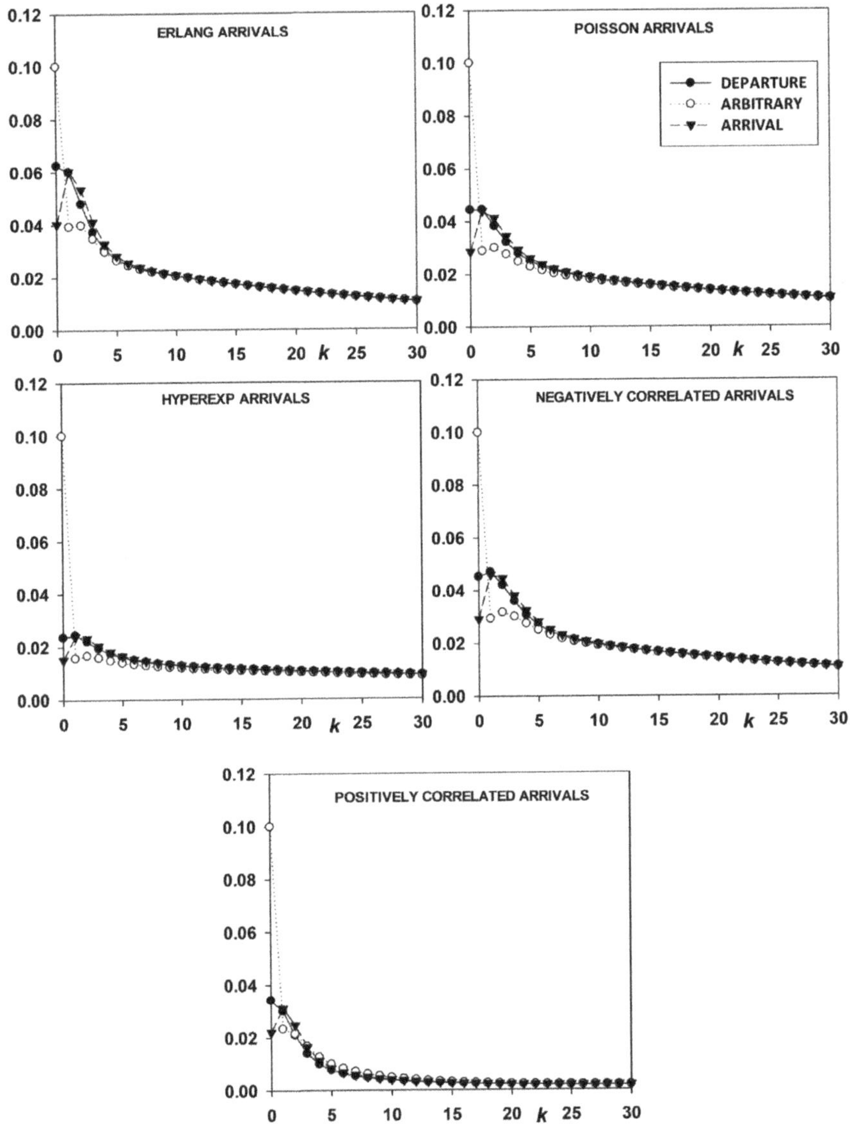

Figure 1.11. *Plots of steady-state probabilities at various epochs for example 1.10*

EXAMPLE 1.11.– In this example, we look at $BMAP/G/1$-type queues by looking at a batch MMPP and a batch MSPP for the arrivals and for the services we look at three cases: (1) CTS: constant services with a mean of 0.8; (2) ERS Erlang of order 3 with rate 3.75 in each state; (c) HES: hyperexponential with rates 2.625, 1.3125 and 0.2625 with mixing probabilities, respectively, given by 0.7, 0.2 and 0.1. Note that these three have the same mean of 0.8 and the standard deviations are 0, 0.46188 and 1.64251, respectively. The representations of the batch MMPP and the batch MSPP are as follows:

$$D_0 = \begin{pmatrix} -0.26345 & 0.03764 & 0.01882 & 0.01882 \\ 0.05645 & -0.56454 & 0.05645 & 0.07527 \\ 0 & 0.03764 & -0.65863 & 0.05645 \\ 0.01882 & 0.05645 & 0.07527 & -0.90326 \end{pmatrix},$$

$$D_k = a_k \begin{pmatrix} 0.18817 & 0 & 0 & 0 \\ 0 & 0.37637 & 0 & 0 \\ 0 & 0 & 0.56454 & 0 \\ 0 & 0 & 0 & 0.75272 \end{pmatrix},$$

$$D_0 = \begin{pmatrix} -1.20112 & 0 & 0 & 0 \\ 0 & -0.90085 & 0 & 0 \\ 0 & 0 & -0.60057 & 0 \\ 0 & 0 & 0 & -0.30029 \end{pmatrix},$$

$$D_k = a_k \begin{pmatrix} 0.30028 & 0.30028 & 0.30028 & 0.30028 \\ 0 & 0.30028 & 0 & 0.60057 \\ 0.06006 & 0.12011 & 0.06006 & 0.36034 \\ 0.09009 & 0.06006 & 0.12011 & 0.03003 \end{pmatrix}.$$

These two arrival processes have the same (batch) arrival rate of 0.5. However, MMPP process has a standard deviation of 2.37064 and its 1-lag correlation is 0.10657, whereas the standard deviation and the 1-lag correlation of MSPP are 2.49355 and -0.05798, respectively. For the batch size distribution, we look at four types: (1) Poisson (POI) with parameter 1, (2) Geometric (GEO) with parameter 0.5, (3) Uniform (UNF) on $\{1,2,3\}$ and (4) DPH with representation given by:

$$\boldsymbol{\alpha} = (0.1, 0.6, 0.3), \quad T = \begin{pmatrix} 0.2 & 0.2 & 0.1 \\ 0.1 & 0.1 & 0.2 \\ 0.1 & 0.1 & 0.4 \end{pmatrix}.$$

Note that the mean batch size for all four distributions is 2.0. In Figures 1.12 and 1.13, respectively, we display the steady-state probability vector at departure epochs for the batch MMPP and for the batch MSPP arrivals. Obviously, one can see the impact of the type of arrivals, services and the batch sizes on this steady-state vector.

In Figure 1.14, we display the steady-state vectors at various epochs for both arrival processes and for CTS so as to compare them.

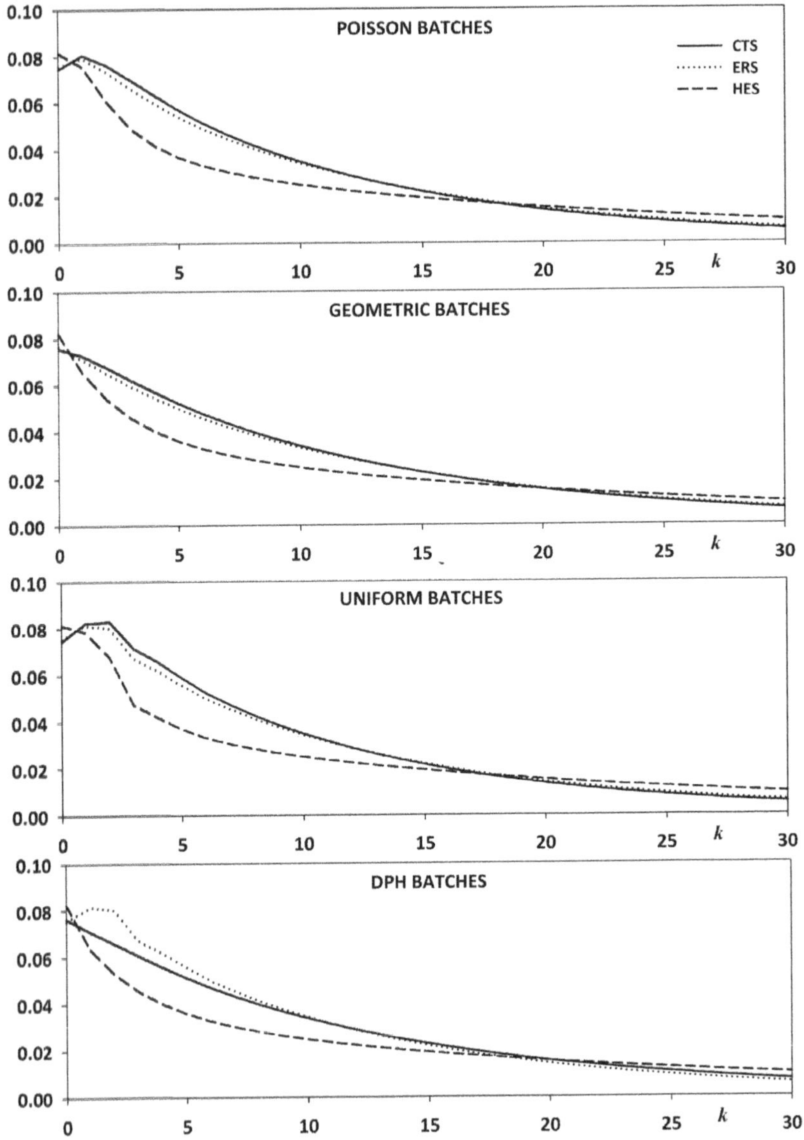

Figure 1.12. *Steady-state probabilities at departure epochs for batch MMPP for example 1.11*

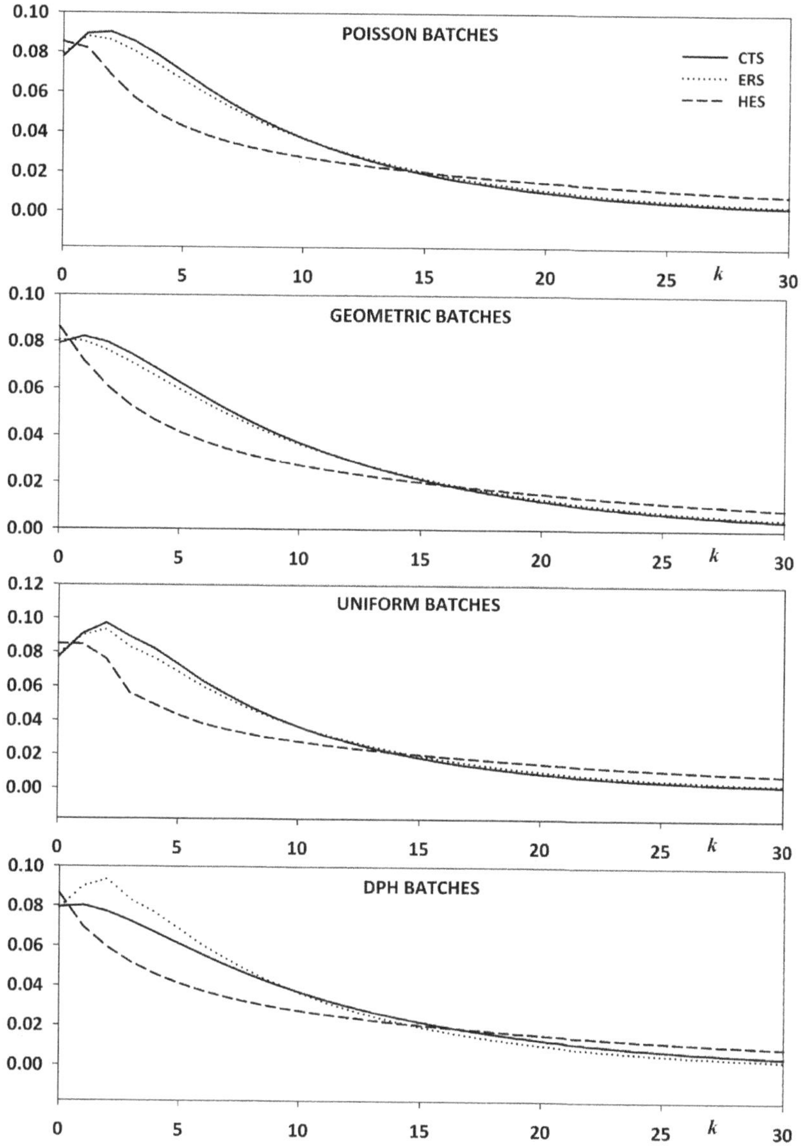

Figure 1.13. *Steady-state probabilities at departure epochs for batch MSPP for example 1.11*

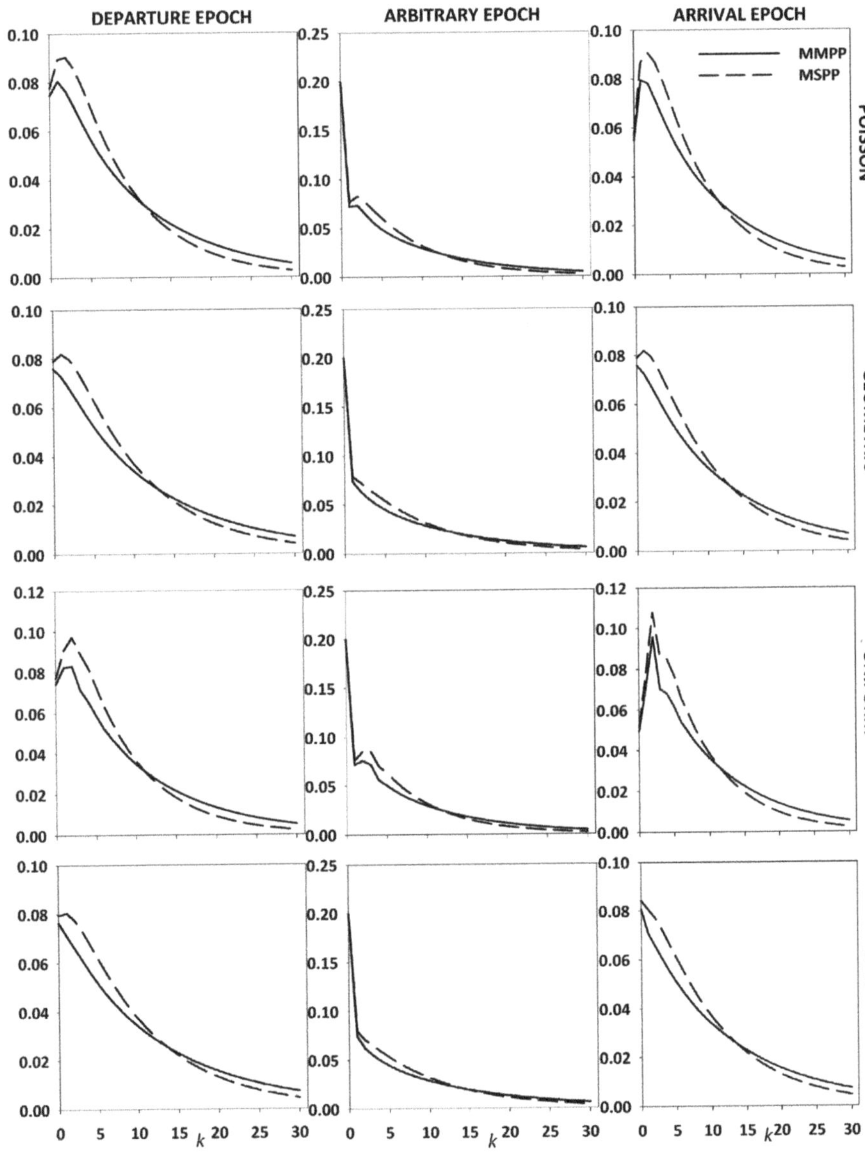

Figure 1.14. *Steady-state probabilities at various epochs for batch MMPP and MSPP and for CTS for example 1.11*

Using the notation $G^{(a,b)}$, where a stands for the arrival process abbreviated as MM for MMPP and MS for MSPP, and b for the type of batch size distribution abbreviated as PB for POI, GB for GEO, UB for UNF and DB for DPH, we display G and K matrices for CTS. Others are left as exercises.

$$G^{(\mathrm{MM,\,PB})} = \begin{pmatrix} 0.9256 & 0.0353 & 0.0214 & 0.0177 \\ 0.0785 & 0.7826 & 0.0714 & 0.0675 \\ 0.0370 & 0.0713 & 0.8217 & 0.0700 \\ 0.0899 & 0.1071 & 0.1370 & 0.6660 \end{pmatrix}, \quad G^{(\mathrm{MS,\,PB})} = \begin{pmatrix} 0.4140 & 0.0639 & 0.1126 & 0.4095 \\ 0.0088 & 0.5439 & 0.0452 & 0.4021 \\ 0.0120 & 0.0329 & 0.6667 & 0.2884 \\ 0.0128 & 0.0208 & 0.0504 & 0.9160 \end{pmatrix},$$

$$G^{(\mathrm{MM,\,GB})} = \begin{pmatrix} 0.9260 & 0.0350 & 0.0213 & 0.0177 \\ 0.0760 & 0.7864 & 0.0701 & 0.0675 \\ 0.0339 & 0.0680 & 0.8286 & 0.0695 \\ 0.0833 & 0.1026 & 0.1334 & 0.6807 \end{pmatrix}, \quad G^{(\mathrm{MS,\,GB})} = \begin{pmatrix} 0.4182 & 0.0698 & 0.1155 & 0.3965 \\ 0.0098 & 0.5491 & 0.0450 & 0.3961 \\ 0.0135 & 0.0358 & 0.6674 & 0.2833 \\ 0.0145 & 0.0226 & 0.0520 & 0.9109 \end{pmatrix},$$

$$G^{(\mathrm{MM,\,UB})} = \begin{pmatrix} 0.9255 & 0.0354 & 0.0215 & 0.0176 \\ 0.0794 & 0.7812 & 0.0719 & 0.0675 \\ 0.0382 & 0.0726 & 0.8190 & 0.0703 \\ 0.0925 & 0.1088 & 0.1384 & 0.6603 \end{pmatrix}, \quad G^{(\mathrm{MS,\,UB})} = \begin{pmatrix} 0.4123 & 0.0616 & 0.1114 & 0.4147 \\ 0.0085 & 0.5418 & 0.0452 & 0.4045 \\ 0.0114 & 0.0318 & 0.6664 & 0.2905 \\ 0.0121 & 0.0200 & 0.0497 & 0.9182 \end{pmatrix},$$

$$G^{(\mathrm{MM,\,DB})} = \begin{pmatrix} 0.9261 & 0.0349 & 0.0213 & 0.0178 \\ 0.0755 & 0.7873 & 0.0698 & 0.0675 \\ 0.0332 & 0.0673 & 0.8302 & 0.0693 \\ 0.0818 & 0.1016 & 0.1325 & 0.6840 \end{pmatrix}, \quad G^{(\mathrm{MS,\,DB})} = \begin{pmatrix} 0.4192 & 0.0711 & 0.1161 & 0.3936 \\ 0.0100 & 0.5503 & 0.0449 & 0.3948 \\ 0.0138 & 0.0365 & 0.6675 & 0.2822 \\ 0.0149 & 0.0230 & 0.0524 & 0.9097 \end{pmatrix},$$

$$K^{(\mathrm{MM,\,PB})} = \begin{pmatrix} 0.6632 & 0.1353 & 0.1183 & 0.0832 \\ 0.1878 & 0.4827 & 0.1855 & 0.1440 \\ 0.0896 & 0.1422 & 0.6355 & 0.1327 \\ 0.1575 & 0.1749 & 0.2319 & 0.4357 \end{pmatrix}, \quad K^{(\mathrm{MS,\,PB})} = \begin{pmatrix} 0.0705 & 0.1241 & 0.1948 & 0.6106 \\ 0.0154 & 0.1387 & 0.0750 & 0.7709 \\ 0.0378 & 0.1004 & 0.1235 & 0.7383 \\ 0.0816 & 0.1116 & 0.2609 & 0.5459 \end{pmatrix},$$

$$K^{(\mathrm{MM,\,GB})} = \begin{pmatrix} 0.6633 & 0.1337 & 0.1177 & 0.0853 \\ 0.1796 & 0.4927 & 0.1817 & 0.1460 \\ 0.0823 & 0.1340 & 0.6519 & 0.1318 \\ 0.1445 & 0.1648 & 0.2230 & 0.4676 \end{pmatrix}, \quad K^{(\mathrm{MS,\,GB})} = \begin{pmatrix} 0.0800 & 0.1346 & 0.1994 & 0.5860 \\ 0.0167 & 0.1510 & 0.0735 & 0.7589 \\ 0.0424 & 0.1087 & 0.1246 & 0.7243 \\ 0.0927 & 0.1207 & 0.2693 & 0.5173 \end{pmatrix},$$

$$K^{(\mathrm{MM,\,UB})} = \begin{pmatrix} 0.6633 & 0.1359 & 0.1185 & 0.0823 \\ 0.1910 & 0.4789 & 0.1869 & 0.1432 \\ 0.0924 & 0.1456 & 0.6290 & 0.1330 \\ 0.1628 & 0.1790 & 0.2352 & 0.4230 \end{pmatrix}, \quad K^{(\mathrm{MS,\,UB})} = \begin{pmatrix} 0.0669 & 0.1199 & 0.1929 & 0.6203 \\ 0.0148 & 0.1338 & 0.0755 & 0.7759 \\ 0.0360 & 0.0971 & 0.1229 & 0.7440 \\ 0.0774 & 0.1080 & 0.2576 & 0.5570 \end{pmatrix},$$

$$K^{(\mathrm{MM,\,DB})} = \begin{pmatrix} 0.6633 & 0.1333 & 0.1176 & 0.0858 \\ 0.1778 & 0.4950 & 0.1809 & 0.1463 \\ 0.0807 & 0.1322 & 0.6556 & 0.1315 \\ 0.1416 & 0.1626 & 0.2210 & 0.4748 \end{pmatrix}, \quad K^{(\mathrm{MS,\,DB})} = \begin{pmatrix} 0.0821 & 0.1369 & 0.2004 & 0.5806 \\ 0.0170 & 0.1537 & 0.0730 & 0.7563 \\ 0.0434 & 0.1106 & 0.1247 & 0.7213 \\ 0.0951 & 0.1227 & 0.2712 & 0.5110 \end{pmatrix}.$$

While we see the differences in the two arrival processes under various scenarios with regard to the steady-state probabilities at various epochs, the picture is even more clear when looking at the G and K matrices. For example, the entries of G matrices indicate a vast difference when going from MMPP to MSPP; however, they appear to be similar when looking at different batch size distributions when the arrival process is fixed. This is an interesting case and worth further investigation.

A number of queueing models can be seen as special cases of $BMAP/G/1$ and in some cases simplifications occur. We explore them in the next few sections and point out where additional simplifications are possible analytically. Where such simplifications are not possible, we add more illustrative examples that are not provided in the previous sections.

1.2. *MAP/G/1* queue

Here, the arrivals occur singly resulting in many simplifications mainly for numerical implementation. That is, here we have $D_k = 0, \ k \geq 2$, and hence $B_n = (-D_0)^{-1} D_1 A_n, \ n \geq 0$. The single arrivals make it easy to obtain the measures numerically as compared to the group arrivals. Also, the vector $\boldsymbol{y}$ is simplified as:

$$\boldsymbol{y}_0 = (1-\rho)\boldsymbol{g}, \ \ \boldsymbol{y}_k = [\boldsymbol{y}_{k-1} D_1 - \lambda(\boldsymbol{x}_{k-1} - \boldsymbol{x}_k)](-D_0)^{-1}, \ k \geq 1. \quad [1.78]$$

REMARK 1.18.– It is easy to verify that the steady-state probabilities at arrival epochs are identical to the ones at departures. To see this, suppose that $\boldsymbol{z} = (z_1, z_2, \cdots)$ denotes the steady-state probability vector at arrivals such that z_i denotes the probability that an arriving customer will see $i, \ i \geq 1$, customers (including the arrived one). Then, we have $z_i = \frac{1}{\lambda} \boldsymbol{y}_{i-1} D_1 \boldsymbol{e}, \ i \geq 1$. Now using equations [1.55] and [1.78] (in the case of single arrivals), we get $z_1 = \boldsymbol{x}_0 \boldsymbol{e}$. Using the recursive scheme given in equation [1.78], we get the stated result.

EXAMPLE 1.12.– In this example, we look at different negatively as well as positively correlated processes as input to a $MAP/G/1$ queue with service times assumed to be constant (CTS) with a value of 0.9. These MAPs are selected by first setting $\lambda_1 = 1.25 + 0.5(k-2)$ and $\lambda_2 = 2\lambda_1$ and vary k from 2 to 9 (see section 6.3 in Volume 1). Specifically, the matrices D_0 and D_1 are such that

$$D_0 = \begin{pmatrix} E_k & 0 \\ 0 & -\lambda_2 \end{pmatrix}, \ \ D_1 = \begin{pmatrix} \lambda_1 p_1 \boldsymbol{e}_1(\boldsymbol{k}) \boldsymbol{e}_1^T(\boldsymbol{k}) & \lambda_1 q_1 \boldsymbol{e}_1^T(\boldsymbol{k}) \\ & \\ q_2 \lambda_2 \boldsymbol{e}_1^T(\boldsymbol{k}) & p_2 \lambda_2 \end{pmatrix},$$

where p_i and q_i are such that $0 < p_i < 1, q_i = 1 - p_i, \ i = 1, 2$, and E_k denotes Erlang of order k with parameter λ_1 in each stage. First, verify Table 1.11 containing the values of the standard deviation and the 1-lag correlation coefficient. Also, note that these MAPs are qualitatively different even though they all have the same mean of 1.

k	TaP	$p_1 = p_2$	ρ	σ	TaP	$p_1 = p_2$	ρ	σ
2	NC_1	0.01	-0.3267	1.0392	PC_1	0.99	0.3267	1.0392
3	NC_2	0.01	-0.4804	1.0202	PC_2	0.99	0.4804	1.0202
4	NC_3	0.01	-0.5786	1.0123	PC_3	0.99	0.5786	1.0123
5	NC_4	0.01	-0.6454	1.0082	PC_4	0.99	0.6454	1.0082
6	NC_5	0.01	-0.6935	1.0059	PC_5	0.99	0.6935	1.0059
7	NC_6	0.01	-0.7296	1.0044	PC_6	0.99	0.7296	1.0044
8	NC_7	0.01	-0.7577	1.0035	PC_7	0.99	0.7577	1.0035
9	NC_8	0.01	-0.7802	1.0028	PC_8	0.99	0.7802	1.0028

Table 1.11. *Selected MAPs with* $\lambda_1 = 1.25 + 0.5(k-2)$ *and* $\lambda_2 = 2\lambda_1$

The graphs of the steady-state probability vector at departure epochs are plotted in Figure 1.15.

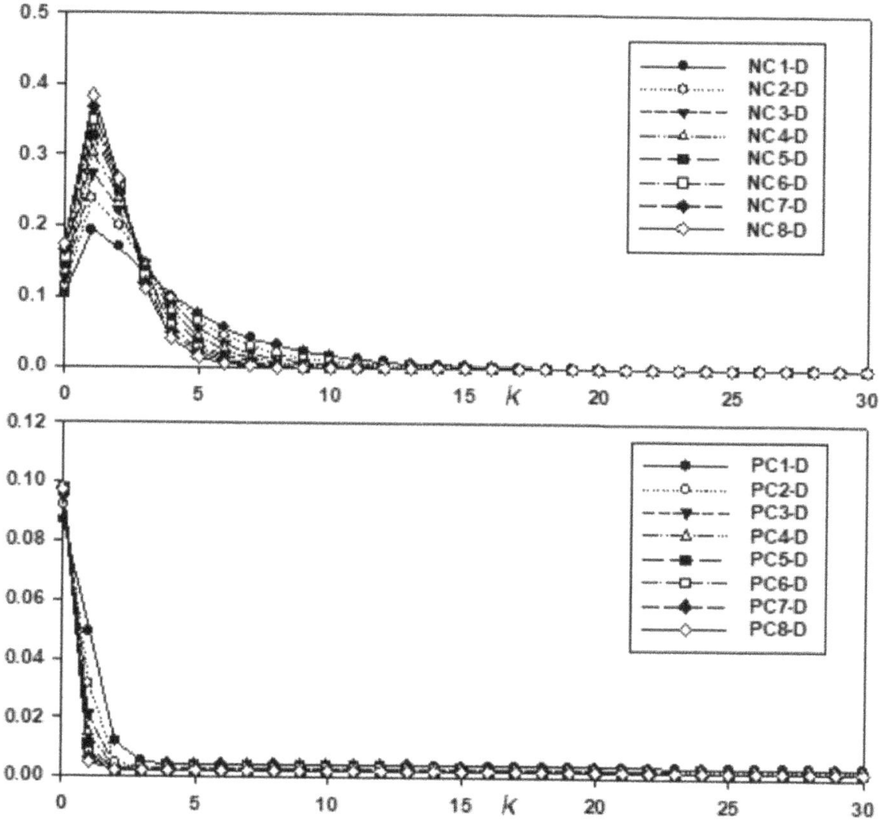

Figure 1.15. *Plots of steady-state probabilities at departure epochs for example 1.12*

These graphs clearly illustrate the significant role played by the (1-lag) correlation. For example, the tail probabilities for the negatively correlated arrivals tend to decay faster to zero, whereas that is not the case for the positively correlated ones.

EXAMPLE 1.13.– This example is similar to example 1.12 except now the services are modeled using an arbitrary distribution (ABS). Denoting by Y, the service time, we take $P(Y = 0.7) = 0.8, P(Y = 1.3) = 0.15$ and $P(Y = 2.9) = 0.05$. Verify that the mean service time is 0.9 and the standard deviation is 0.50596. The graphs of the steady-state probability vector at departure epochs are plotted in Figure 1.16.

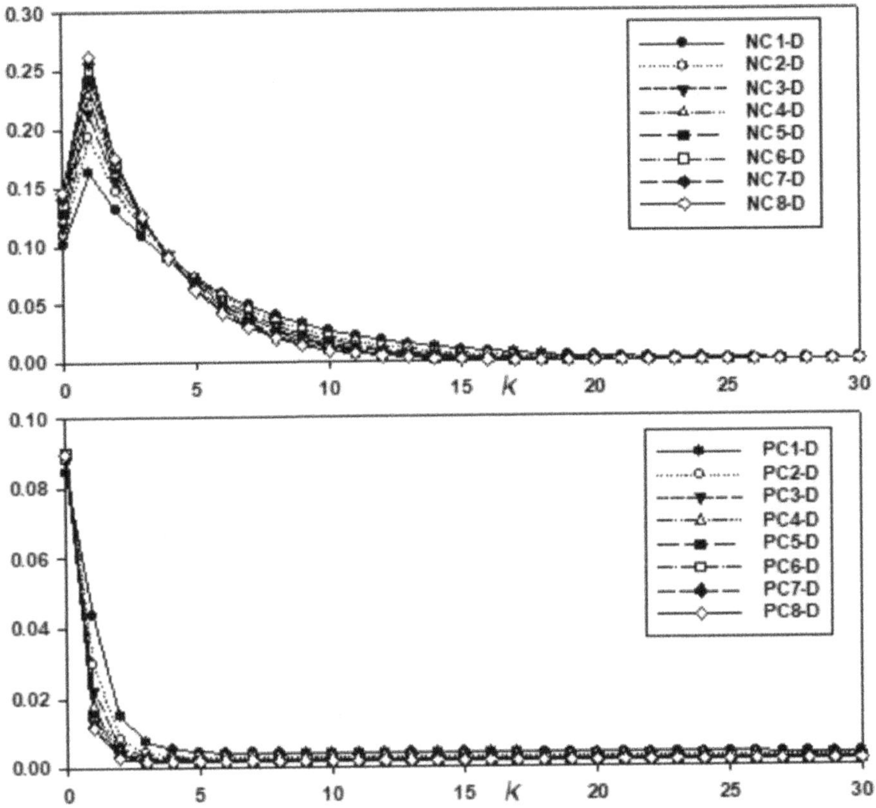

Figure 1.16. *Plots of steady-state probabilities at departure epochs for example 1.13*

Like in example 1.12, we see a similar trend indicating the significant role played by the (1-lag) correlation.

EXAMPLE 1.14.– Here again the example is similar to example 1.13 except that the service times are now uniform (UFS) on the interval $(0.4, 1.4)$ services. Verify that the mean and the standard deviation of the services are, respectively, given by 0.9 and 0.28868. The graphs of the steady-state probability vector at departure epochs are plotted in Figure 1.17.

It is obvious to infer from these figures that the positively correlated arrivals have a larger mean number and a higher median values as compared to those of the negatively correlated arrivals. Furthermore, in the case of positively correlated arrivals, the median values appear to be insensitive to the type of services; however, the median values depend on the magnitude of the correlation.

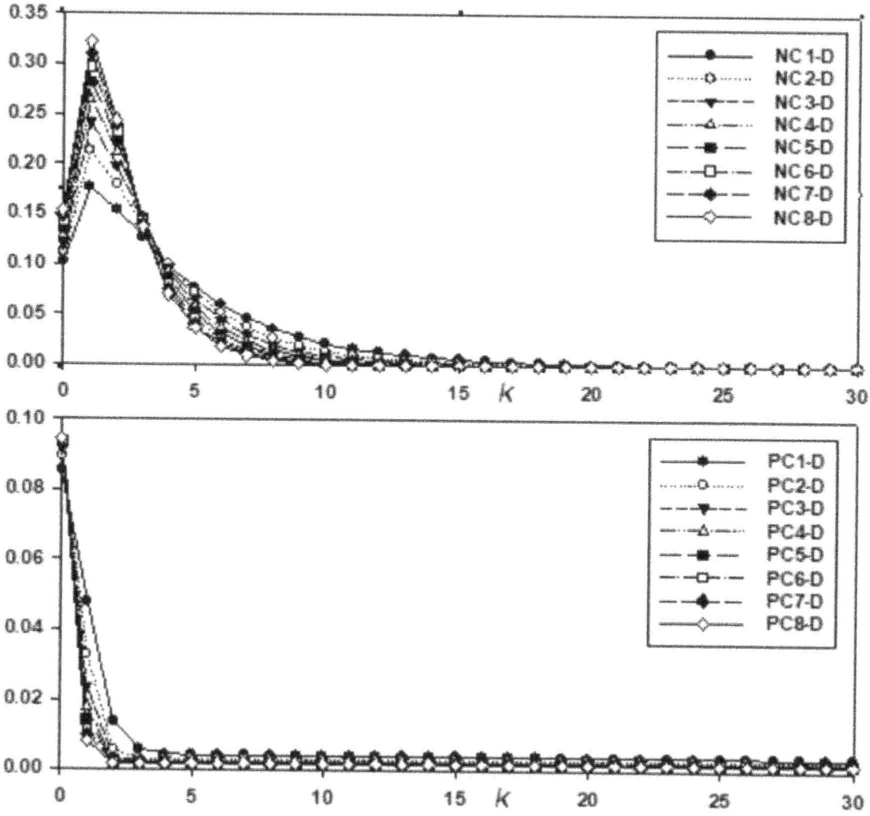

Figure 1.17. *Plots of steady-state probabilities at departure epochs for example 1.14*

EXAMPLE 1.15.– In this example, we compare the impact of the service time distributions, namely, CTS, ABS and UFS, considered in examples 1.12 through 1.14. Note that these services all have a mean of 0.9 but with different coefficient of variation. The graphs of the natural log of the mean number in system and the median of the steady-state vector at departure epochs are plotted in Figures 1.18 and 1.19, respectively.

EXAMPLE 1.16.– In this example, we look at the case of single arrivals with inter-arrivals times that are either negatively or positively correlated. The services are modeled using the minimum of three distributions: exponential with parameter 1, Erlang of order 2 with parameter 2 and a hyperexponential distribution with parameters 10 and 1 with mixing probabilities 0.9 and 0.1. Recall from section 4.1 (see result 4.8) in Volume 1 that this minimum is a CPH with:

$$\beta = (0.9, 0.1, 0, 0), \quad S = \begin{pmatrix} -1.68584 & 0 & 0.68810 & 0 \\ 0 & -1.09752 & 0 & 0.68810 \\ 0 & 0 & -1.68584 & 0 \\ 0 & & 0 & -1.09752 \end{pmatrix}.$$

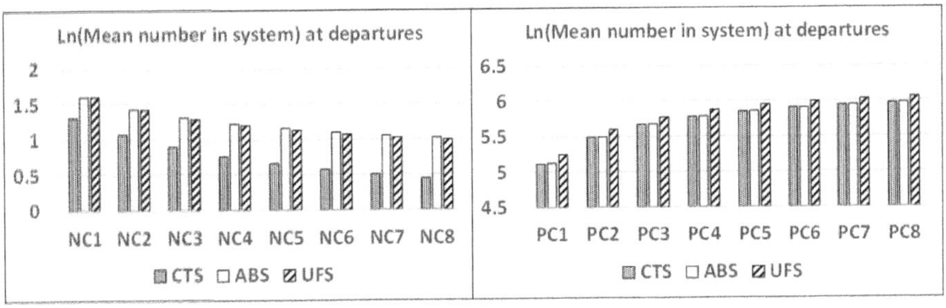

Figure 1.18. *Ln(mean number in system) at departure epochs for example 1.15*

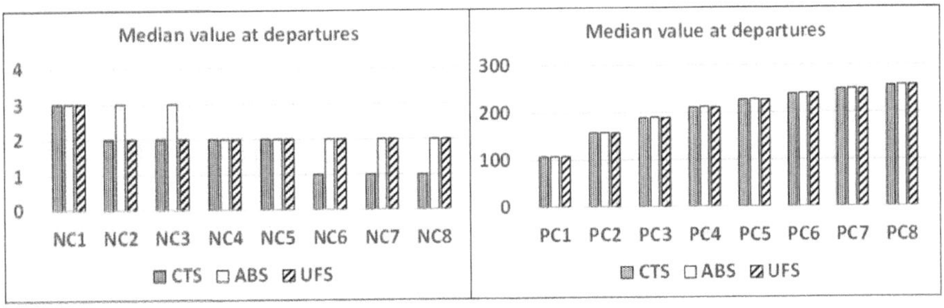

Figure 1.19. *Median of the number in system at departure epochs for example 1.15*

Verify that the above CPH has a mean of 0.9 and a standard deviation of 0.84534. The plots of the steady-state probabilities at departure epochs for the eight pairs of negatively and positively correlated MAPs (see example 1.12 for the details on these MAPs) are given in Figure 1.20. Also, in Table 1.12 we list the various measures under various scenarios.

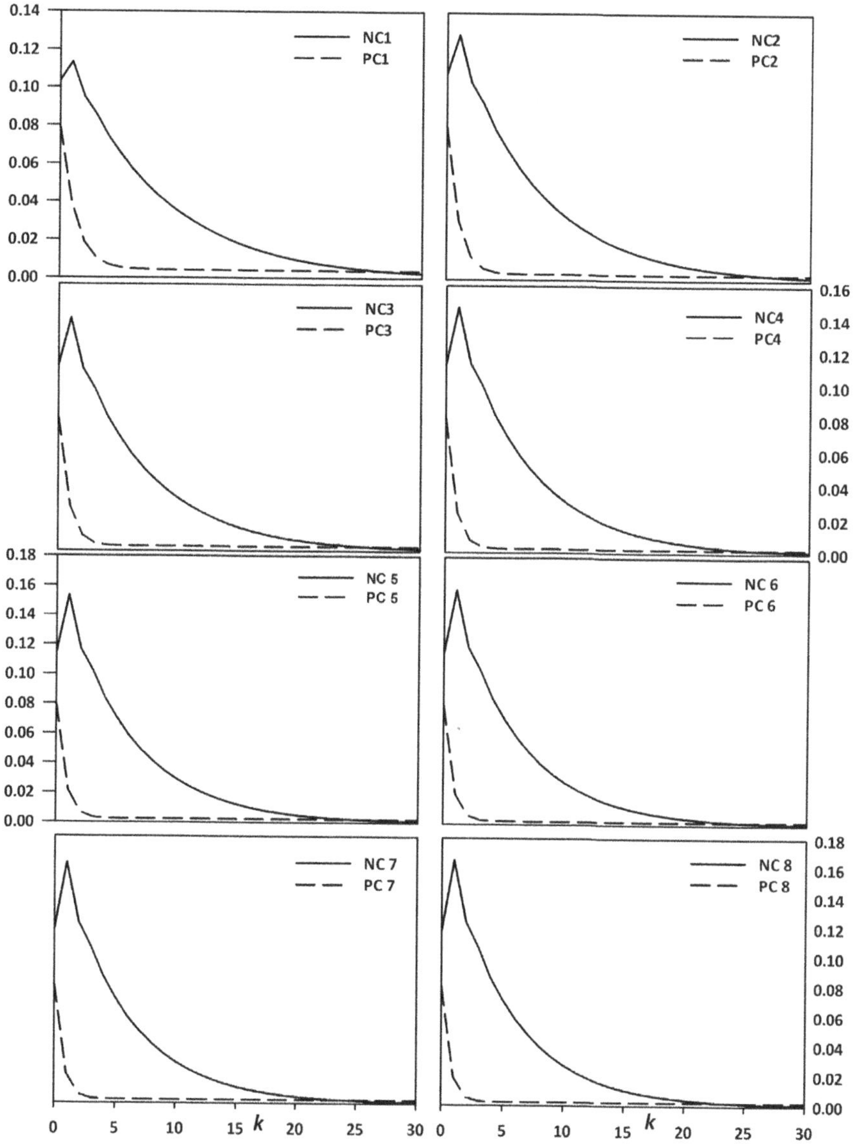

Figure 1.20. *Plots of the steady-state probabilities under various scenarios for example 1.16*

$Type\ of\ BMAP$	CC	$Mean$	$Median$	$Mode$	$Modal\ value$	SD
NCA	-0.7802	4.923	3	1	0.165	5.035
	-0.7577	5.027	3	1	0.162	5.148
	-0.7296	5.156	4	1	0.158	5.285
	-0.6935	5.321	4	1	0.153	5.458
	-0.6454	5.549	4	1	0.147	5.701
	-0.5786	5.873	4	1	0.139	6.048
	-0.4804	6.36	4	1	0.129	6.552
	-0.3267	7.204	5	1	0.113	7.426
PCA	0.3267	170.874	109	0	0.079	192.741
	0.4804	245.593	159	0	0.08	272.422
	0.5786	293.251	191	0	0.08	322.734
	0.6454	325.864	213	0	0.08	356.996
	0.6935	349.487	230	0	0.08	381.754
	0.7296	367.345	242	0	0.08	400.424
	0.7577	381.315	251	0	0.08	415.016
	0.7802	392.524	259	0	0.08	426.695

Table 1.12. *Selected measures for $MAP/PH/1$ queue for example 1.16*

1.3. *PH/G/1* queue

Here, the arrivals occur singly and according to a CPH distribution with representation given by $(\boldsymbol{\alpha}, T)$ of order m. We assume that there is no mass at the origin so that the inter-arrival times are positive with probability 1. Verify that $D_0 = T$ and $D_1 = \boldsymbol{T^0}\boldsymbol{\alpha}$. The matrix G is obtained as solution to:

$$G = \int_0^\infty exp\{(T + \boldsymbol{T^0}\boldsymbol{\alpha}G)t\}dF(t).$$

The steady-state vector $\boldsymbol{g}$ of G is of the form: $\boldsymbol{g} = [\boldsymbol{\alpha}G(-T)^{-1}\boldsymbol{e}]^{-1}\,\boldsymbol{\alpha}G(-T)^{-1}$. The matrix K is given by $K = \boldsymbol{e}\,\boldsymbol{\alpha}G$ and hence $\boldsymbol{\kappa} = \boldsymbol{\alpha}G$.

Unlike for the $MAP/G/1$ queue, which required an iterative process involving a matrix, here the iterative process is on the vector $\boldsymbol{u}$, which is defined as $\boldsymbol{u} = \boldsymbol{\alpha}G$. With the knowledge of $\boldsymbol{u}$, should there be a need for obtaining G, we can compute it directly. The details are given below.

Algorithm 1.2. Algorithmic steps in computing the matrix G

From the definition of $\boldsymbol{u}$ and the form of G, we get:

$$\boldsymbol{u} = \sum_{n=0}^{\infty} \gamma_n \boldsymbol{\alpha} \left[I + \frac{1}{\theta}(T + \boldsymbol{T^0 u}) \right]^n, \text{ where } \theta = -\min\{T_{i,i}\}. \qquad [1.79]$$

The vector $\boldsymbol{u}$ is first computed using the uniformization method as follows: For a given $\epsilon > 0$, compute γ_n, where $\gamma_n = \int_0^\infty e^{-\theta t} \frac{(\theta t)^n}{k!} dF(t)$, for $0 \leq n \leq n^*$. The truncation point, n^*, is identified as $\sum_{n=0}^{n^*} \gamma_n > 1 - \epsilon$. Define $\boldsymbol{h}_n = \boldsymbol{\alpha} \left[I + \frac{1}{\theta}(T + \boldsymbol{T^0 u}) \right]^n$, $n \geq 0$.

Step 0: Set $k = 0, \boldsymbol{u}^{(0)} = \boldsymbol{0}$.

Step 1: Set $k \leftarrow k + 1$.

Step 1A: Set $n = 0, \boldsymbol{h}_0 = \boldsymbol{\alpha}.\ \boldsymbol{u}^{(k)} = \gamma_0 \boldsymbol{h}_0$.

Step 1B: $n \leftarrow n + 1; \boldsymbol{h}_n = \boldsymbol{h}_{n-1}[I + \frac{1}{\theta}(T + \boldsymbol{T^0 u})].\ \boldsymbol{u}^{(k)} \rightarrow \boldsymbol{u}^{(k)} + \gamma_n \boldsymbol{h}_n$.

Step 1C: If $n < n^*$ go to step 1B. Otherwise go to step 2.

Step 2: If $||\boldsymbol{u}^{(k)} - \boldsymbol{u}^{(k-1)}||_\infty < \epsilon$ (note one can use the same ϵ used for finding n^* or a different one), go to step 3. If not, go to step 1.

Step 3: If G is required, then it is computed as $G = \sum_{n=0}^{n^*} \gamma_n \left[I + \frac{1}{\theta}(T + \boldsymbol{T^0 u}) \right]^n$. One can employ Horner's rule to compute this efficiently by first computing and storing the matrix $I + \frac{1}{\theta}(T + \boldsymbol{T^0 u})$.

EXAMPLE 1.17.– In this example, we look at a $PH/G/1$-type queue such that we consider Erlangs of order 1 through 10 for arrivals, all of which are normalized to have the arrival rate be 1. We look at three different service times (see examples 1.12–1.14), namely, constant (CTS), arbitrary (ABS) and uniform (UFS). These service time distributions will be parameterized to have a mean of 0.9. However, the standard deviations of these are 0, 0.50596 and 0.28868, respectively. The steady-state probability vectors at departure epochs are plotted in Figure 1.21. The mean and the median of the number in the system at departure epochs under various scenarios are plotted in Figure 1.22.

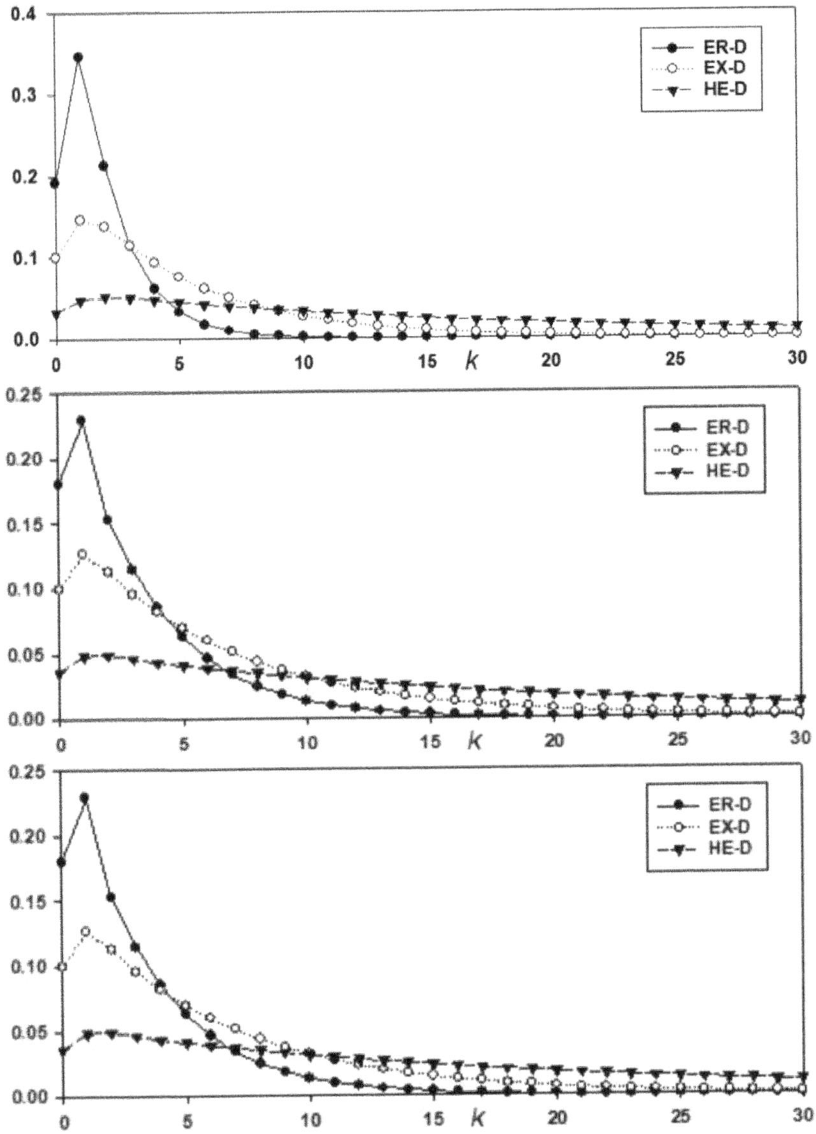

Figure 1.21. *Steady-state vectors at departure epochs for CTS (top), ABS (middle) and UFS (bottom) – example 1.17*

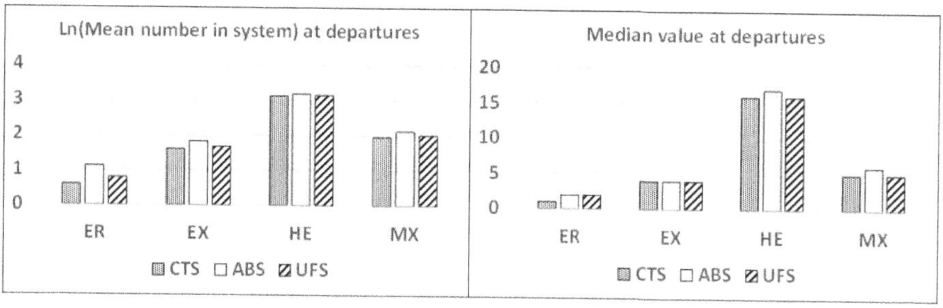

Figure 1.22. *Mean and median of the number in system at departure epochs for example 1.17*

EXAMPLE 1.18.– In this example, we look at four different arrival processes in the context of $PH/G/1$ and with three different service times: constant (CTS), arbitrary (ABS) and uniform (UFS). The four PH distributions considered are Erlang (ER) of order 3; exponential (EX); HE: hyperexponential (HE) with rates 10 and 1 and mixing probabilities 0.9 and 0.1; and a mixture of (EH) of Erlang of order 2 and HE with mixing equal mixing probabilities. These will be normalized to have a mean of 1. For services we look at constant (CTS) with a value of 0.95; arbitrary (ABS) distribution, which has masses at 0.3, 1.02, 2.22 and 3.22 with probabilities, respectively, given by 0.5, 0.3, 0.15 and 0.05; and uniform (UFS) on the interval $(0.45, 1.45)$. Note that all these three service times have a mean of 0.95 and the standard deviations are, respectively, given by 0, 0.28868 and 0.84398. Specifically, the CPH representations for the arrivals are as follows:

$$\mathbf{ER}: \ \boldsymbol{\alpha} = (1,0,0), \ T = \begin{pmatrix} -3 & 3 & 0 \\ 0 & -3 & 3 \\ 0 & 0 & -3 \end{pmatrix}, \mathbf{HE}: \ \boldsymbol{\alpha} = (0.9, 0.1), \ \ T = \begin{pmatrix} -10 & 0 \\ 0 & -1 \end{pmatrix},$$

$$\mathbf{EX}: \ \boldsymbol{\alpha} = 1, \ \ T = (-1), \ \mathbf{EH}: \ \boldsymbol{\alpha} = (0.5, 0, 0.45, 0.05), \ T = \begin{pmatrix} -2 & 2 & 0 & 0 \\ 0 & -2 & 0 & 0 \\ 0 & 0 & -10 & 0 \\ 0 & 0 & 0 & -1 \end{pmatrix}$$

The steady-state probability vectors at departure and at arbitrary epochs are plotted in Figures 1.23 and 1.24, respectively.

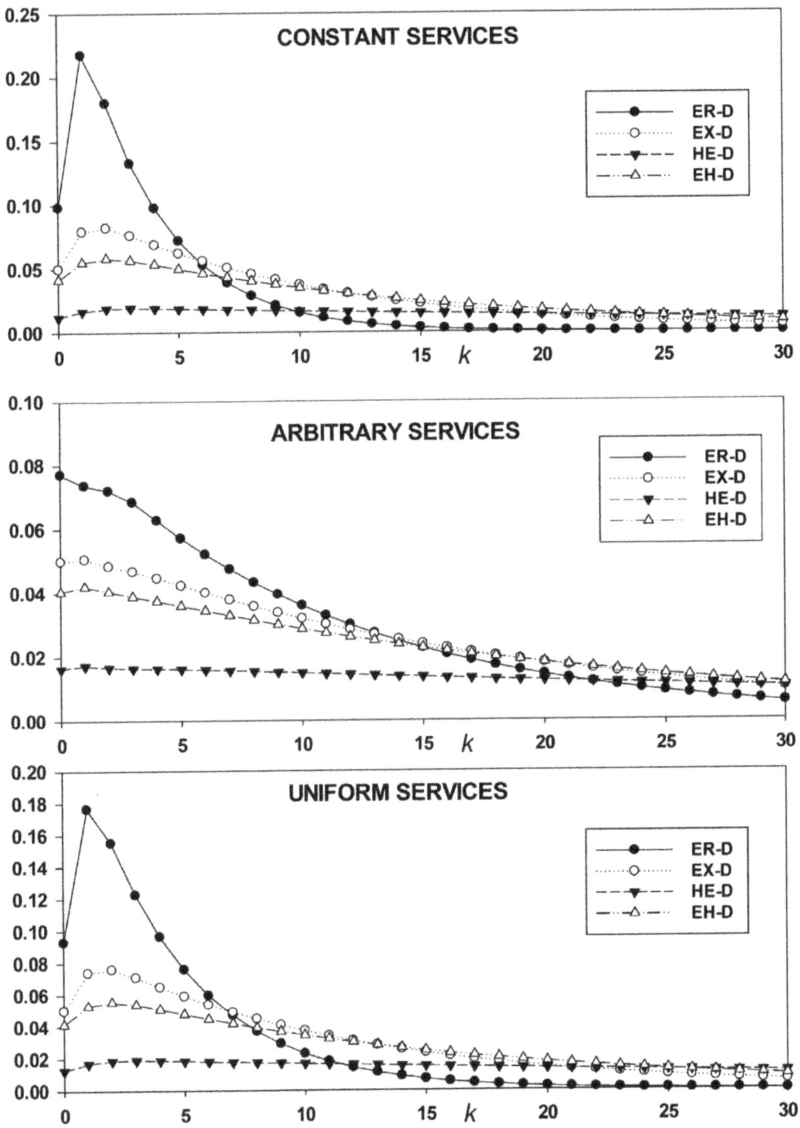

Figure 1.23. *The steady-state vector at departure epochs under various scenarios for example 1.18*

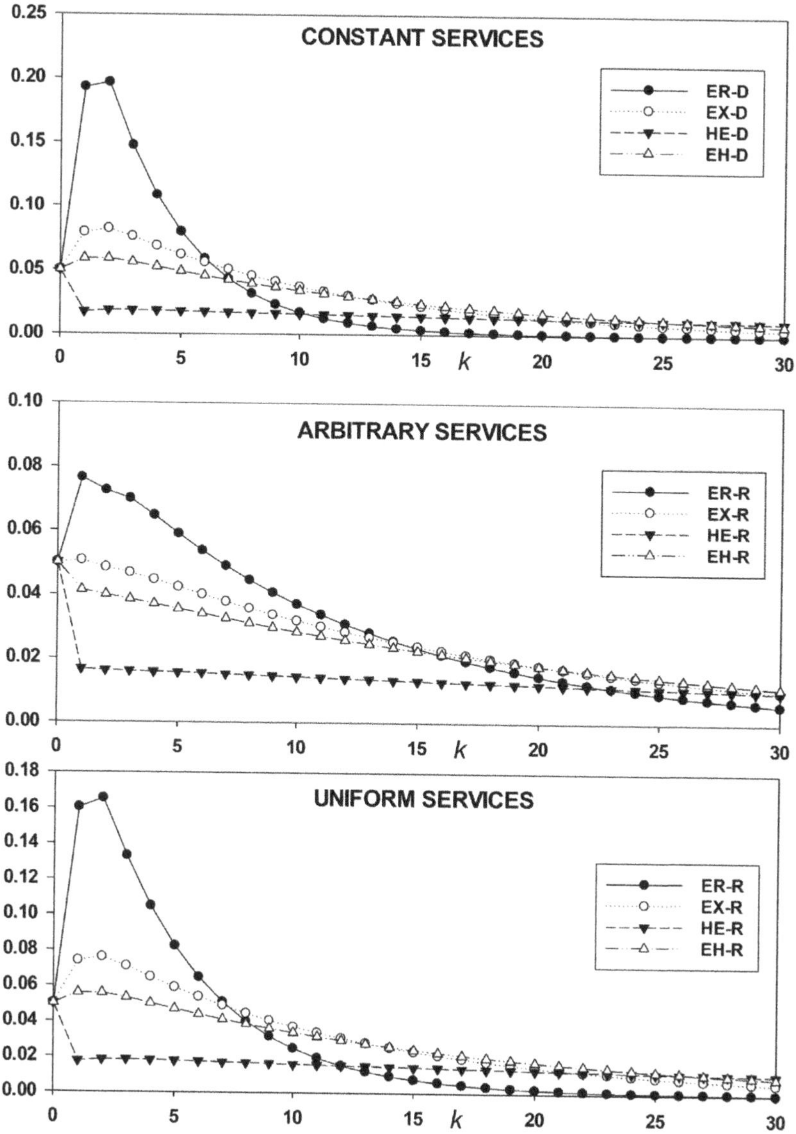

Figure 1.24. *The steady-state vector at arbitrary epochs under various scenarios for example 1.18*

1.4. *M/G/1* queue

The assumptions here are that (1) the arrivals occur singly according to a Poisson process with rate λ and (2) the service times are generally distributed with a finite mean and a finite variance. Let $F(.)$ denote the CDF of the service times and let $\rho = \frac{\lambda}{\mu}$, where μ is the service rate.

It is easy to verify that, for this model, we have (1) $D_0 = -\lambda$ and $D_1 = \lambda$; (2) G is a scalar and obviously equal to 1; (3) $A^*(z,s)$, which becomes a scalar, is given by $A^*(z,s) = f^*(s + \lambda(1-z))$; and (d) $A(z)$, which is the (scalar) PGF of A_n, is $f^*(\lambda(1-z))$.

The above observations result in the scalar probabilities:

$$A_n = \int_0^\infty e^{-\lambda t} \frac{(\lambda t)^n}{n!} dF(t), \; n \geq 0, \qquad [1.80]$$

and the PGFs of $\{x_k\}$ and $\{y_k\}$ are identical and are given by:

$$X(z) = Y(z) = \frac{(1-\rho)(1-z)A(z)}{A(z)-z}. \qquad [1.81]$$

The LT of the stationary waiting time distribution in the queue of an arriving (same as at an arbitrary time due to Poisson arrivals) customer is given by:

$$w_q^*(s) = s(1-\rho)[s - \lambda(1 - f^*(s))]^{-1}. \qquad [1.82]$$

Note that $w_q^*(s)$ can be rewritten as:

$$w_q^*(s) = (1-\rho)\left[1 - \frac{\lambda(1-f^*(s))}{s}\right]^{-1} = (1-\rho)\sum_{k=0}^{\infty} \rho^k \left[\frac{\mu(1-f^*(s))}{s}\right]^k. \qquad [1.83]$$

Thus, the stationary waiting time distribution in the queue of an arrival in the $M/G/1$ queue is a geometric mixture of successive convolutions of the CDF given by:

$$\mu \int_0^t [1 - F(x)]dx.$$

The mean number, say, μ_{NS}, of customers at departures (as well as at an arbitrary time) in the system is given by (see Bhat (2015)):

$$\mu_{NS} = \rho + \frac{\rho^2(\mu^2\sigma^2 + 1)}{2(1-\rho)}, \qquad [1.84]$$

where σ^2 is the variance of the service times.

EXAMPLE 1.19.– Suppose that the arrivals occur according to a Poisson process with rate $\lambda = 1$. We consider three service times: (1) constant services (CTS) with a value of 0.95; (2) arbitrary services (ABS) with masses at 0.3, 1.02, 2.22 and 3.22 with probabilities, respectively, given by 0.5, 0.3, 0.15 and 0.05; and (3) uniform services (UFS) on the interval (0.45, 1.45). In all these three cases, the mean is 0.95. The standard deviations of CTS, ABS and UFS are 0, 0.843978 and 0.288675, respectively.

The steady-state probability vector at departure epochs (note that here this vector is also the vector at arbitrary epochs) for the three services is plotted in Figure 1.25.

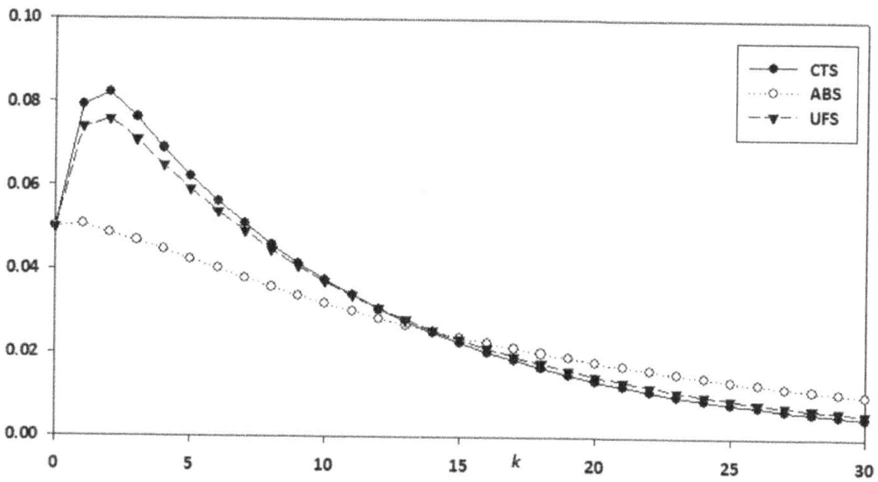

Figure 1.25. *The steady-state vector at arbitrary epochs under various three services for example 1.19*

The selected queue length statistics for these three cases are displayed in Table 1.13.

Type of services	*Mean*	*Median*	*Mode*	*Modal value*	*SD*
CTS	9.900	7	2	0.0822	9.626
ABS	16.964	12	1	0.0506	16.993
UFS	10.741	8	2	0.0759	10.504

Table 1.13. *Selected queue length statistics for the $M/G/1$ queue of example 1.19 under three scenarios*

Using the explicit expression for the mean as given in equation [1.84], verify the means for CTS, ABS, and UFS are, respectively, given by 9.975, 17.098 and 10.808.

EXAMPLE 1.20.– The purpose of this example is to see the behavior of the steady-state probabilities at departure epochs and other measures in the context of $M/D/1$ queue as ρ is varied. Toward this end, we first plot the graph of the steady-state probability as a function of k and ρ in Figure 1.26, and in Figure 1.27 we plot various measures at departure epochs.

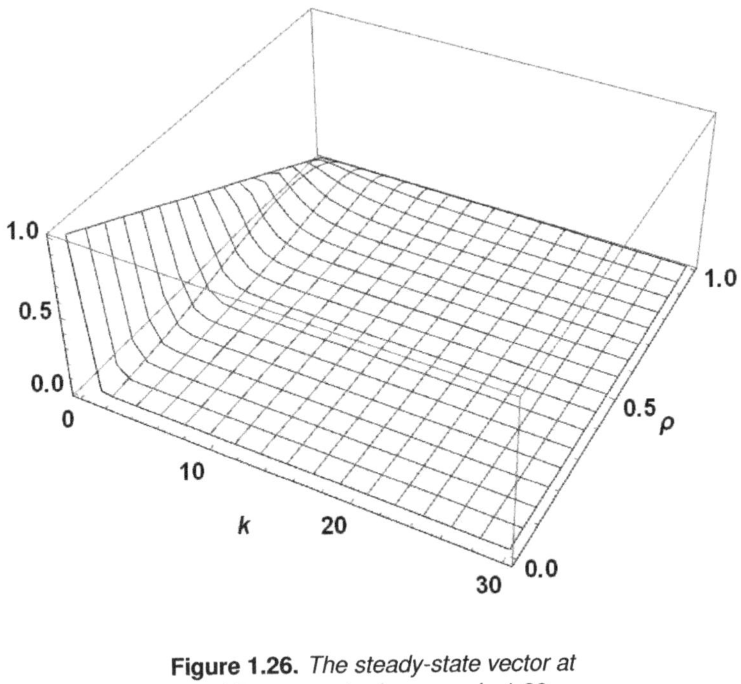

Figure 1.26. *The steady-state vector at arbitrary epochs for example 1.20*

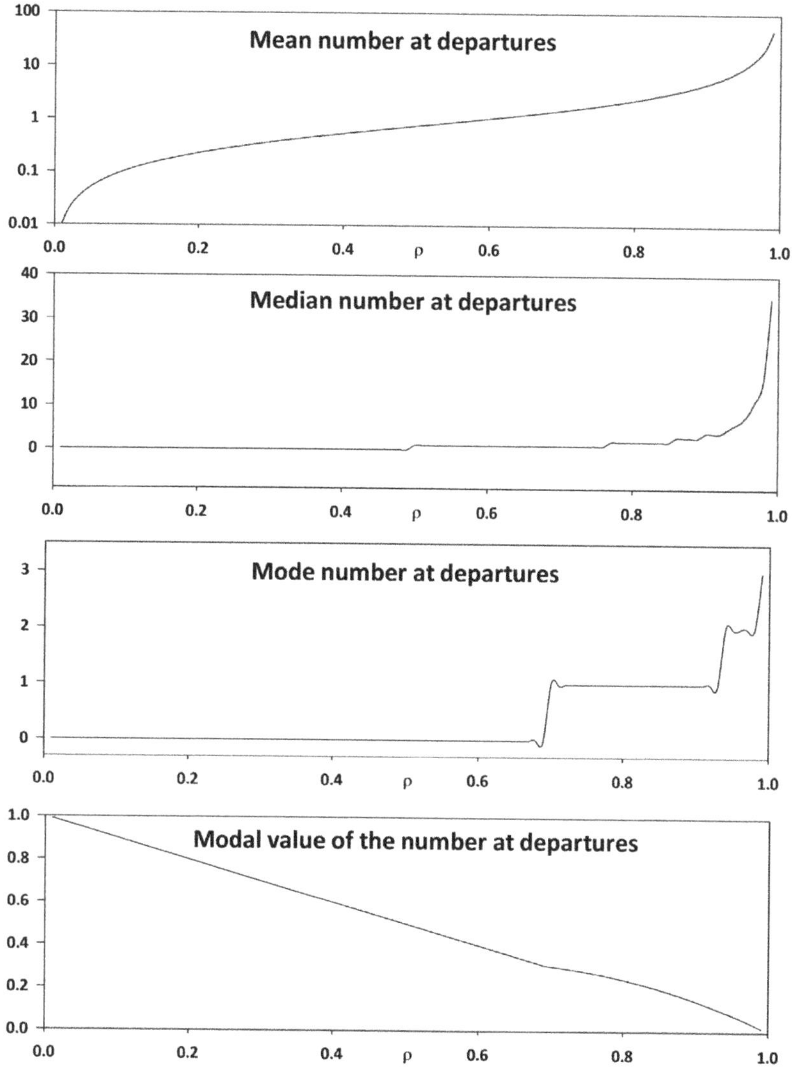

Figure 1.27. *Selected measures for $M/D/1$ queue for example 1.20*

1.5. *PH/M/1* queue

Here, the arrivals occur singly and according to a CPH distribution with representation given by $(\boldsymbol{\alpha}, T)$ of order m. Note that the arrival rate, λ, is $\lambda = [\boldsymbol{\alpha}(-T)^{-1}\mathbf{e}]^{-1}$. The service times are exponential with rate μ. It is easy to verify that:

$$A_n = \mu \int_0^\infty P(n,t)e^{-\mu t}dt, \; n \geq 0.$$

Suppose that $A(z) = \sum_{n=0}^{\infty} z^n A_n$ is the matrix PGF. Then, verify that (see equation [4.36] in Volume 1) $P^*(z,t) = e^{[T+z\boldsymbol{T}^0\boldsymbol{\alpha}]t}, \; t \geq 0, \; |z| \leq 1$:

$$A(z) = (\mu I - T - z\boldsymbol{T}^0\boldsymbol{\alpha})^{-1}, \; |z| \leq 1.$$

The matrix G is now obtained as a solution to:

$$G = \mu[\mu I - T - \boldsymbol{T}^0\boldsymbol{\alpha}G]^{-1}, \quad [1.85]$$

from which we get the following matrix-quadratic equation:

$$\boldsymbol{T}^0\boldsymbol{\alpha}G^2 + (T - \mu I)G + \mu I = 0.$$

EXAMPLE 1.21.– Suppose that the inter-arrival times of customers entering into a single-server system are of phase type with representation given by:

$$\boldsymbol{\alpha} = (0.9, 0.1, 0, 0), \; T = \begin{pmatrix} -1.51726 & 0. & 0.61929 & 0. \\ 0. & -0.98776 & 0. & 0.61929 \\ 0. & 0. & -1.51726 & 0. \\ 0. & 0. & 0. & -0.98776 \end{pmatrix}.$$

The service times are exponential with mean 0.95. Compute the G matrix and its invariant vector, $\boldsymbol{g}$. Also, compute various other measures.

Using the matrix-quadratic equation for G, verify that:

$$G = \begin{pmatrix} 0.6512 & 0.0510 & 0.2521 & 0.0457 \\ 0.1610 & 0.5498 & 0.1022 & 0.1870 \\ 0.2902 & 0.0612 & 0.5937 & 0.0549 \\ 0.2379 & 0.0502 & 0.1510 & 0.5609 \end{pmatrix}, \; \boldsymbol{g} = (0.4214, 0.1084, 0.3379, 0.1323).$$

The mean, the median, the mode, the modal values and the standard deviation of the queue length in steady-state at departure epochs are 17.824, 12, 0, 0.053 and 18.277, respectively.

The mean, the median, the mode, the modal values and the standard deviation of the queue length in steady-state at an arbitrary epoch are 17.882, 12, 1, 0.050 and 18.279, respectively.

1.6. *M/M/1* queue

The simplest single-server queueing model in continuous-time is the one with Poisson arrivals and exponential services. Here, all the results are explicitly known. The approach taken here uses the embedded points at departure epochs. We briefly summarize them here as they are obtained by taking the simplest form for D_0 and D_1 matrices. In the current setup, $M/M/1$ queue is obtained by taking $D_0 = -\lambda$, $D_1 = -\lambda$ and $F(t) = 1 - e^{-\mu t}$. This results in:

$$A_n = B_n = \mu \int_0^{\infty} e^{-\lambda t} \frac{(\lambda t)^n}{n!} e^{-\mu t} dt = \frac{\mu}{\lambda + \mu}\left(\frac{\mu}{\lambda + \mu}\right)^n, \ n \geq 0.$$

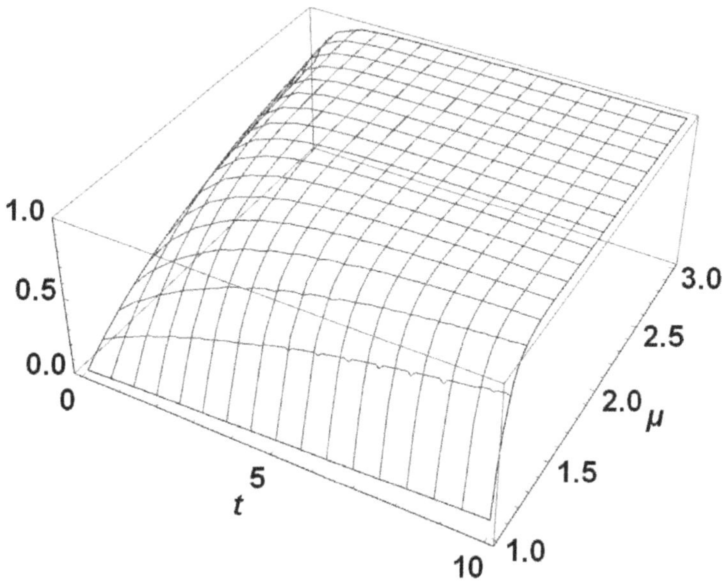

Figure 1.28. *The CDF of the waiting time in queue for $M/M/1$ queue*

Obviously, $G = K = 1$, and the steady-state probabilities are: $x_k = y_k = (1-\rho)\rho^k, \ k \geq 0$. Using equation [1.82] by taking $f^*(s) = \dfrac{\mu}{s+\mu}$ for the current case

$\left(\text{note that here } \frac{\mu(1-f^*(s))}{s} = f^*(s)\right)$, we get $w_q^*(s) = (1-\rho)\sum_{k=0}^{\infty}\rho^k\left(\frac{\mu}{s+\mu}\right)^k$, which can be inverted to obtain $F_q(t) = P(T_q \le t) = 1 - \rho e^{-\mu(1-\rho)t}$, $t \ge 0$. In Figure 1.28, we display the graph of $F_q(t)$ by fixing $\lambda = 1$ and varying μ and t.

1.7. Exercises

Exercise 1.1. Referring to examples 1.1 through 1.4, develop and implement a code for computing various quantities and point out some key observations based on the measures displayed in the tables and the plots seen in the figures in those examples.

Exercise 1.2. Referring to example 1.5, develop and implement a code for computing various quantities and point out some key observations based on the measures displayed in the table and the plots seen in the figure.

Exercise 1.3. Referring to example 1.6, develop and implement a code for computing various quantities and point out some key observations based on the measures displayed in the table and the plots seen in the figure.

Exercise 1.4. Referring to example 1.7, develop and implement a code for computing various quantities and point out some key observations based on the measures displayed in the table and the plots seen in the figure.

Exercise 1.5. Referring to example 1.8, develop and implement a code for computing various quantities and point out some key observations based on the measures displayed in the table and the plots seen in the figure.

Exercise 1.6. Referring to example 1.9, develop and implement a code for computing various quantities and point out some key observations based on the measures displayed in the table and the plots seen in the figure.

Exercise 1.7. Referring to example 1.10, develop and implement a code for computing various quantities and point out some key observations based on the measures displayed in the table and the plots seen in the figure.

Exercise 1.8. Referring to example 1.11, develop and implement a code for computing various quantities and point out some key observations based on the measures displayed in the figures. Verify the G and K matrices given for constant services. Obtain these matrices for Erlang and hyperexponential services. Investigate further the role of these two BMAP under various scenarios.

Exercise 1.9. Referring to examples 1.12 through 1.16, develop and implement a code for computing various quantities and point out some key observations based on the measures displayed in the tables and the plots seen in the figures for these examples.

Exercise 1.10. Referring to examples 1.17 and 1.18, develop and implement a code for computing various quantities and point out some key observations based on the measures displayed in the tables and the plots seen in the figures for these examples.

Exercise 1.11. Referring to examples 1.19 and 1.20, develop and implement a code for computing various quantities and point out some key observations based on the measures displayed in the tables and the plots seen in the figures for these examples.

Exercise 1.12. Referring to example 1.21, verify the results for G and its invariant vector.

Note: For exercises dealing with computational ones, your answers may differ based on the error tolerance levels chosen.

Exercise 1.13. Suppose that the customers arrive in batches that are modeled using a Poisson process with parameter 0.5 and the inter-arrival times of the batches are Erlang of order 4 with parameter $7/3$. That is:

$$D_0 = \frac{7}{3}\begin{pmatrix} -1 & 1 & 0 & 0 \\ 0 & -1 & 1 & 0 \\ 0 & 0 & -1 & 1 \\ 0 & 0 & 0 & -1 \end{pmatrix}, \quad D_k = \frac{7}{3}e^{-0.5}\frac{0.5^{k-1}}{(k-1)!}\begin{pmatrix} 0 & 0 & 0 & 0 \\ 0 & 0 & 0 & 0 \\ 0 & 0 & 0 & 0 \\ 1 & 0 & 0 & 0 \end{pmatrix}, \quad k \geq 1.$$

There is a single server to serve the customers on an FCFS basis and the service times are constant with a value of 0.8 unit. Answer the following questions:

a) Find G, K and their respective invariant vectors, $\boldsymbol{g}$ and $\boldsymbol{\kappa}$.

b) Compute the selected queue length statistics: the mean, the median, the mode, the modal value and the standard deviation at various epochs.

c) Compute the first few steady-state probability vectors.

Exercise 1.14. Suppose that the customers arrive in batches that are modeled using a geometric distribution with parameter $2/3$ and the inter-arrival times of the batches are Erlang of order 4 with parameter $7/3$. That is:

$$D_0 = \frac{7}{3}\begin{pmatrix} -1 & 1 & 0 & 0 \\ 0 & -1 & 1 & 0 \\ 0 & 0 & -1 & 1 \\ 0 & 0 & 0 & -1 \end{pmatrix}, \quad D_k = \frac{14}{9}\left(\frac{1}{3}\right)^{k-1}\begin{pmatrix} 0 & 0 & 0 & 0 \\ 0 & 0 & 0 & 0 \\ 0 & 0 & 0 & 0 \\ 1 & 0 & 0 & 0 \end{pmatrix}, \quad k \geq 1.$$

There is a single server to serve the customers on an FCFS basis and the service times are constant with a value of 0.8 unit. Answer the following questions:

a) Find G, K and their respective invariant vectors, $\boldsymbol{g}$ and $\boldsymbol{\kappa}$.

b) Compute the selected queue length statistics: the mean, the median, the mode, the modal value and the standard deviation at various epochs.

c) Compute the first few steady-state probability vectors.

Exercise 1.15. Suppose that the customers arrive in batches that are modeled using a (discrete) uniform on $\{1, 2\}$ and the inter-arrival times of the batches are Erlang of order 4 with parameter $7/3$. That is:

$$D_0 = \frac{7}{3}\begin{pmatrix} -1 & 1 & 0 & 0 \\ 0 & -1 & 1 & 0 \\ 0 & 0 & -1 & 1 \\ 0 & 0 & 0 & -1 \end{pmatrix}, \quad D_k = 0.5\begin{pmatrix} 0 & 0 & 0 & 0 \\ 0 & 0 & 0 & 0 \\ 0 & 0 & 0 & 0 \\ 1 & 0 & 0 & 0 \end{pmatrix},$$
$$k = 1, 2, \; D_k = 0, \; k \geq 3.$$

There is a single server to serve the customers on an FCFS basis and the service times are constant with a value of 0.8 unit. Answer the following questions:

a) Find G, K and their respective invariant vectors, $\boldsymbol{g}$ and $\boldsymbol{\kappa}$.

b) Compute the selected queue length statistics: the mean, the median, the mode, the modal value and the standard deviation at various epochs.

c) Compute the first few steady-state probability vectors.

Exercise 1.16. Suppose that the customers arrive according to a BMAP with representation:

$$D_0 = \frac{7}{3}\begin{pmatrix} -1 & 1 & 0 & 0 \\ 0 & -1 & 1 & 0 \\ 0 & 0 & -1 & 1 \\ 0 & 0 & 0 & -1 \end{pmatrix}, \quad D_k = (\boldsymbol{\alpha} T^{k-1} \boldsymbol{T^0})\begin{pmatrix} 0 & 0 & 0 & 0 \\ 0 & 0 & 0 & 0 \\ 0 & 0 & 0 & 0 \\ 1 & 0 & 0 & 0 \end{pmatrix}, \; k \geq 1,$$

where:

$$\boldsymbol{\alpha} = (0.7397, 0.2603), \quad T = \begin{pmatrix} 0.1000 & 0.2000 \\ 0.1512 & 0.2341 \end{pmatrix}, \quad \boldsymbol{T^0} = \boldsymbol{e} - T\boldsymbol{e}.$$

There is a single server to serve the customers on an FCFS basis and the service times are constant with a value of 0.8 unit. Answer the following questions:

a) Find G, K and their respective invariant vectors, $\boldsymbol{g}$ and $\boldsymbol{\kappa}$.

b) Compute the selected queue length statistics: the mean, the median, the mode, the modal value and the standard deviation at various epochs.

c) Compute the first few steady-state probability vectors.

Exercise 1.17. Suppose that the customers arrive in batches that are modeled using a Poisson process with parameter 0.5 and the inter-arrival times of the batches are assumed to be hyperexponential. That is, set $a_k = \frac{7}{3}e^{-0.5}\frac{0.5^{k-1}}{(k-1)!}$, $k \geq 1$,

$$D_0 = \begin{pmatrix} -1.4 & 0 & 0 & 0 \\ 0 & -0.7 & 0 & 0 \\ 0 & 0 & -0.35 & 0 \\ 0 & 0 & 0 & -0.175 \end{pmatrix},$$

$$D_k = a_k \begin{pmatrix} 0.7 & 0.42 & 0.21 & 0.07 \\ 0.35 & 0.21 & 0.105 & 0.035 \\ 0.175 & 0.105 & 0.0525 & 0.0175 \\ 0.0875 & 0.0525 & 0.02625 & 0.00875 \end{pmatrix}.$$

There is a single server to serve the customers on an FCFS basis and the service times are constant with a value of 0.8 unit. Answer the following questions:

a) Find G, K and their respective invariant vectors, $\boldsymbol{g}$ and $\boldsymbol{\kappa}$.

b) Compute the selected queue length statistics: the mean, the median, the mode, the modal value and the standard deviation at various epochs.

c) Compute the first few steady-state probability vectors.

Exercise 1.18. Suppose that the customers arrive in batches that are modeled using a geometric distribution with parameter 2/3 and the inter-arrival times of the batches are assumed to be hyperexponential. That is, set $a_k = (2/3)(1/3)^{k-1}$, $k \geq 1$,

$$D_0 = \begin{pmatrix} -1.4 & 0 & 0 & 0 \\ 0 & -0.7 & 0 & 0 \\ 0 & 0 & -0.35 & 0 \\ 0 & 0 & 0 & -0.175 \end{pmatrix},$$

$$D_k = a_k \begin{pmatrix} 0.7 & 0.42 & 0.21 & 0.07 \\ 0.35 & 0.21 & 0.105 & 0.035 \\ 0.175 & 0.105 & 0.0525 & 0.0175 \\ 0.0875 & 0.0525 & 0.02625 & 0.00875 \end{pmatrix}.$$

There is a single server to serve the customers on an FCFS basis and the service times are constant with a value of 0.8 unit. Answer the following questions:

a) Find G, K and their respective invariant vectors, $\boldsymbol{g}$ and $\boldsymbol{\kappa}$.

b) Compute the selected queue length statistics: the mean, the median, the mode, the modal value and the standard deviation at various epochs.

c) Compute the first few steady-state probability vectors.

Exercise 1.19. Suppose that the customers arrive in batches that are modeled using a (discrete) uniform on $\{1, 2\}$ and the inter-arrival times of the batches are assumed to be hyperexponential. That is, set $a_k = 0.5, \ k = 1, 2, \ a_k = 0, \ k \geq 3$,

$$D_0 = \begin{pmatrix} -1.4 & 0 & 0 & 0 \\ 0 & -0.7 & 0 & 0 \\ 0 & 0 & -0.35 & 0 \\ 0 & 0 & 0 & -0.175 \end{pmatrix},$$

$$D_k = a_k \begin{pmatrix} 0.7 & 0.42 & 0.21 & 0.07 \\ 0.35 & 0.21 & 0.105 & 0.035 \\ 0.175 & 0.105 & 0.0525 & 0.0175 \\ 0.0875 & 0.0525 & 0.02625 & 0.00875 \end{pmatrix}.$$

There is a single server to serve the customers on an FCFS basis and the service times are constant with a value of 0.8 unit. Answer the following questions:

a) Find G, K and their respective invariant vectors, $\boldsymbol{g}$ and $\boldsymbol{\kappa}$.

b) Compute the selected queue length statistics: the mean, the median, the mode, the modal value and the standard deviation at various epochs.

c) Compute the first few steady-state probability vectors.

Exercise 1.20. Suppose that the customers arrive according to a BMAP with representation:

$$D_0 = \begin{pmatrix} -1.4 & 0 & 0 & 0 \\ 0 & -0.7 & 0 & 0 \\ 0 & 0 & -0.35 & 0 \\ 0 & 0 & 0 & -0.175 \end{pmatrix},$$

$$D_k = a_k \begin{pmatrix} 0.7 & 0.42 & 0.21 & 0.07 \\ 0.35 & 0.21 & 0.105 & 0.035 \\ 0.175 & 0.105 & 0.0525 & 0.0175 \\ 0.0875 & 0.0525 & 0.02625 & 0.00875 \end{pmatrix},$$

where:

$$\boldsymbol{\alpha} = (0.7397, 0.2603), \quad T = \begin{pmatrix} 0.1000 & 0.2000 \\ 0.1512 & 0.2341 \end{pmatrix}, \quad \boldsymbol{T^0} = \boldsymbol{e} - T\boldsymbol{e}.$$

There is a single server to serve the customers on an FCFS basis and the service times are constant with a value of 0.8 unit. Answer the following questions:

a) Find G, K and their respective invariant vectors, $\boldsymbol{g}$ and $\boldsymbol{\kappa}$.

b) Compute the selected queue length statistics: the mean, the median, the mode, the modal value and the standard deviation at various epochs.

c) Compute the first few steady-state probability vectors.

Exercise 1.21. Suppose that the customers arrive in batches according to a BMAP with representation given by:

$$D_0 = \frac{1}{3}\begin{pmatrix} -5 & 5 & 0 & 0 \\ 0 & -5 & 5 & 0 \\ 0 & 0 & -5 & 0 \\ 0 & 0 & 0 & -7 \end{pmatrix}, \quad D_k = a_k \begin{pmatrix} 0 & 0 & 0 & 0 \\ 0 & 0 & 0 & 0 \\ 0.01167 & 0 & 0 & 1.155 \\ 2.31 & 0 & 0 & 0.02333 \end{pmatrix}.$$

Consider the following four distributions for the batch size:

$$a_k^{(r)} = \begin{cases} e^{-0.5}\dfrac{0.5^{k-1}}{(k-1)!}, \; k \geq 1, \; r = 1, \\[2ex] (2/3)(1/3)^{k-1}, \; k \geq 1, \; r = 2, \\[2ex] 0.5, \; k = 1, 2, \; r = 3, \\[2ex] \boldsymbol{\alpha} T^{k-1} \boldsymbol{T^0}, \; k \geq 1, \; r = 4, \end{cases}$$

where:

$$\boldsymbol{\alpha} = (0.7397, 0.2603), \quad T = \begin{pmatrix} 0.1000 & 0.2000 \\ 0.1512 & 0.2341 \end{pmatrix},$$

with $Te + T^0 = e$. There is a single server to serve the customers on an FCFS basis and the service times are constant with a value of 0.8 unit. Answer the following questions:

a) Find the 1-lag correlation of two successive inter-arrival times.

b) Find G, K and their respective invariant vectors, $\boldsymbol{g}$ and $\boldsymbol{\kappa}$, for these four types of group sizes.

c) Compute the selected queue length statistics: the mean, the median, the model, the modal value and the standard deviation at various epochs, for these four types of group sizes.

d) Compute the first few steady-state probability vectors, for these four types of group sizes.

Exercise 1.22. Suppose that the customers arrive in batches according to a BMAP with representation given by:

$$D_0 = \frac{1}{3}\begin{pmatrix} -5 & 5 & 0 & 0 \\ 0 & -5 & 5 & 0 \\ 0 & 0 & -5 & 0 \\ 0 & 0 & 0 & -7 \end{pmatrix}, D_k = a_k \begin{pmatrix} 0 & 0 & 0 & 0 \\ 0 & 0 & 0 & 0 \\ 1.155 & 0 & 0 & 0.01167 \\ 0.02333 & 0 & 0 & 2.31 \end{pmatrix}.$$

Consider the following four distributions for the batch size:

$$a_k^{(r)} = \begin{cases} e^{-0.5}\dfrac{0.5^{k-1}}{(k-1)!}, \; k \geq 1, \; r = 1, \\ (2/3)(1/3)^{k-1}, \; k \geq 1, \; r = 2, \\ 0.5, \; k = 1, 2, \; r = 3, \\ \boldsymbol{\alpha} T^{k-1} \boldsymbol{T}^0, \; k \geq 1, \; r = 4, \end{cases}$$

where:

$$\boldsymbol{\alpha} = (0.7397, 0.2603), \;\; T = \begin{pmatrix} 0.1000 & 0.2000 \\ 0.1512 & 0.2341 \end{pmatrix},$$

with $Te + T^0 = e$. There is a single server to serve the customers on an FCFS basis and the service times are constant with a value of 0.8 unit. Answer the following questions:

a) Find the 1-lag correlation of two successive inter-arrival times.

b) Find G, K and their respective invariant vectors, $\boldsymbol{g}$ and $\boldsymbol{\kappa}$, for these four types of group sizes.

c) Compute the selected queue length statistics: the mean, the median, the model, the modal value and the standard deviation at various epochs, for these four types of group sizes.

d) Compute the first few steady-state probability vectors, for these four types of group sizes.

Exercise 1.23. Suppose that the customers arrive in batches according to a BMAP with representation given by:

$$D_0 = \frac{7}{3}\begin{pmatrix} -1 & 1 & 0 & 0 \\ 0 & -1 & 1 & 0 \\ 0 & 0 & -1 & 1 \\ 0 & 0 & 0 & -1 \end{pmatrix}, \quad D_k = \frac{7}{3}a_k \begin{pmatrix} 0 & 0 & 0 & 0 \\ 0 & 0 & 0 & 0 \\ 0 & 0 & 0 & 0 \\ 1 & 0 & 0 & 0 \end{pmatrix}, \quad k \geq 1.$$

Consider the following four distributions for the batch size:

$$a_k^{(r)} = \begin{cases} e^{-0.5}\dfrac{0.5^{k-1}}{(k-1)!}, \; k \geq 1, \; r = 1, \\[2ex] (2/3)(1/3)^{k-1}, \; k \geq 1, \; r = 2, \\[2ex] 0.5, \; k = 1, 2, \; r = 3, \\[2ex] \boldsymbol{\alpha} T^{k-1} \boldsymbol{T^0}, \; k \geq 1, \; r = 4, \end{cases}$$

where:

$$\boldsymbol{\alpha} = (0.7397, 0.2603), \quad T = \begin{pmatrix} 0.1000 & 0.2000 \\ 0.1512 & 0.2341 \end{pmatrix},$$

with $T\boldsymbol{e} + \boldsymbol{T^0} = \boldsymbol{e}$. There is a single server to serve the customers on an FCFS basis and the service times are uniform on the interval $(0.3, 1.3)$. Answer the following questions:

a) Find the 1-lag correlation of two successive inter-arrival times.

b) Find G, K and their respective invariant vectors, $\boldsymbol{g}$ and $\boldsymbol{\kappa}$, for these four types of group sizes.

c) Compute the selected queue length statistics: the mean, the median, the model, the modal value and the standard deviation at various epochs, for these four types of group sizes.

d) Compute the first few steady-state probability vectors, for these four types of group sizes.

Exercise 1.24. Referring to exercise 1.23, suppose that the customers arrive in batches according to a BMAP with representation given by:

$$D_0 = \begin{pmatrix} -1.4 & 0 & 0 & 0 \\ 0 & -0.7 & 0 & 0 \\ 0 & 0 & -0.35 & 0 \\ 0 & 0 & 0 & -0.175 \end{pmatrix},$$

$$D_k = a_k \begin{pmatrix} 0.7 & 0.42 & 0.21 & 0.07 \\ 0.35 & 0.21 & 0.105 & 0.035 \\ 0.175 & 0.105 & 0.0525 & 0.0175 \\ 0.0875 & 0.0525 & 0.02625 & 0.00875 \end{pmatrix}.$$

All other parameters remain the same as in exercise 1.23. Answer the following questions:

a) Find the 1-lag correlation of two successive inter-arrival times.

b) Find G, K and their respective invariant vectors, $\boldsymbol{g}$ and $\boldsymbol{\kappa}$, for these four types of group sizes.

c) Compute the selected queue length statistics: the mean, the median, the model, the modal value and the standard deviation at various epochs, for these four types of group sizes.

d) Compute the first few steady-state probability vectors, for these four types of group sizes.

Exercise 1.25. Referring to exercise 1.23, suppose that the customers arrive in batches according to a BMAP with representation given by:

$$D_0 = \frac{1}{3}\begin{pmatrix} -5 & 5 & 0 & 0 \\ 0 & -5 & 5 & 0 \\ 0 & 0 & -5 & 0 \\ 0 & 0 & 0 & -7 \end{pmatrix}, D_k = a_k \begin{pmatrix} 0 & 0 & 0 & 0 \\ 0 & 0 & 0 & 0 \\ 0.01167 & 0 & 0 & 1.155 \\ 2.31 & 0 & 0 & 0.02333 \end{pmatrix}.$$

All other parameters remain the same as in exercise 1.23. Answer the following questions:

a) Find G, K and their respective invariant vectors, $\boldsymbol{g}$ and $\boldsymbol{\kappa}$, for these four types of group sizes.

b) Compute the selected queue length statistics: the mean, the median, the model, the modal value and the standard deviation at various epochs, for these four types of group sizes.

c) Compute the first few steady-state probability vectors, for these four types of group sizes.

Exercise 1.26. Referring to exercise 1.23, suppose that the customers arrive in batches according to a BMAP with representation given by:

$$D_0 = \frac{1}{3}\begin{pmatrix} -5 & 5 & 0 & 0 \\ 0 & -5 & 5 & 0 \\ 0 & 0 & -5 & 0 \\ 0 & 0 & 0 & -7 \end{pmatrix}, D_k = a_k \begin{pmatrix} 0 & 0 & 0 & 0 \\ 0 & 0 & 0 & 0 \\ 1.155 & 0 & 0 & 0.01167 \\ 0.02333 & 0 & 0 & 2.31 \end{pmatrix}.$$

All other parameters remain the same as in exercise 1.23. Answer the following questions:

a) Find G, K and their respective invariant vectors, $\boldsymbol{g}$ and $\boldsymbol{\kappa}$, for these four types of group sizes.

b) Compute the selected queue length statistics: the mean, the median, the model, the modal value and the standard deviation at various epochs, for these four types of group sizes.

c) Compute the first few steady-state probability vectors, for these four types of group sizes.

Exercise 1.27. Suppose that the customers arrive in batches according to a BMAP with representation given by:

$$D_0 = \frac{7}{3}\begin{pmatrix} -1 & 1 & 0 & 0 \\ 0 & -1 & 1 & 0 \\ 0 & 0 & -1 & 1 \\ 0 & 0 & 0 & -1 \end{pmatrix}, \quad D_k = a_k \begin{pmatrix} 0 & 0 & 0 & 0 \\ 0 & 0 & 0 & 0 \\ 0 & 0 & 0 & 0 \\ 1 & 0 & 0 & 0 \end{pmatrix}, \quad k \geq 1,$$

where $\{a_k\}$ has one of the four distributions for the batch size as described in exercise 1.23. Suppose that the service times are modeled using an Erlang of order 2 with parameter 2.5 in each stage so that the mean service time is 0.8. Answer the following questions:

a) Find G, K and their respective invariant vectors, $\boldsymbol{g}$ and $\boldsymbol{\kappa}$, for these four types of group sizes.

b) Compute the selected queue length statistics: the mean, the median, the model, the modal value and the standard deviation at various epochs, for these four types of group sizes.

c) Compute the first few steady-state probability vectors, for these four types of group sizes.

Exercise 1.28. Referring to exercise 1.27, suppose that the customers arrive in batches according to a BMAP with representation given by:

$$D_0 = \begin{pmatrix} -1.4 & 0 & 0 & 0 \\ 0 & -0.7 & 0 & 0 \\ 0 & 0 & -0.35 & 0 \\ 0 & 0 & 0 & -0.175 \end{pmatrix},$$

$$D_k = a_k \begin{pmatrix} 0.7 & 0.42 & 0.21 & 0.07 \\ 0.35 & 0.21 & 0.105 & 0.035 \\ 0.175 & 0.105 & 0.0525 & 0.0175 \\ 0.0875 & 0.0525 & 0.02625 & 0.00875 \end{pmatrix},$$

where $\{a_k\}$ has one of the four distributions for the batch size as described in exercise 1.23. All other parameters remain the same as in exercise 1.27. Answer the following questions:

a) Find G, K and their respective invariant vectors, $\boldsymbol{g}$ and $\boldsymbol{\kappa}$, for these four types of group sizes.

b) Compute the selected queue length statistics: the mean, the median, the model, the modal value and the standard deviation at various epochs, for these four types of group sizes.

c) Compute the first few steady-state probability vectors, for these four types of group sizes.

Exercise 1.29. Referring to exercise 1.27, suppose that the customers arrive in batches according to a BMAP with representation given by:

$$D_0 = \frac{1}{3}\begin{pmatrix} -5 & 5 & 0 & 0 \\ 0 & -5 & 5 & 0 \\ 0 & 0 & -5 & 0 \\ 0 & 0 & 0 & -7 \end{pmatrix}, D_k = a_k \begin{pmatrix} 0 & 0 & 0 & 0 \\ 0 & 0 & 0 & 0 \\ 0.01167 & 0 & 0 & 1.155 \\ 2.31 & 0 & 0 & 0.02333 \end{pmatrix},$$

where $\{a_k\}$ has one of the four distributions for the batch size as described in exercise 1.23. All other parameters remain the same as in exercise 1.27. Answer the following questions:

a) Find G, K and their respective invariant vectors, $\boldsymbol{g}$ and $\boldsymbol{\kappa}$, for these four types of group sizes.

b) Compute the selected queue length statistics: the mean, the median, the model, the modal value and the standard deviation at various epochs, for these four types of group sizes.

c) Compute the first few steady-state probability vectors, for these four types of group sizes.

Exercise 1.30. Referring to exercise 1.27, suppose that the customers arrive in batches according to a BMAP with representation given by:

$$D_0 = \frac{1}{3}\begin{pmatrix} -5 & 5 & 0 & 0 \\ 0 & -5 & 5 & 0 \\ 0 & 0 & -5 & 0 \\ 0 & 0 & 0 & -7 \end{pmatrix}, D_k = a_k \begin{pmatrix} 0 & 0 & 0 & 0 \\ 0 & 0 & 0 & 0 \\ 1.155 & 0 & 0 & 0.01167 \\ 0.02333 & 0 & 0 & 2.31 \end{pmatrix},$$

where $\{a_k\}$ has one of the four distributions for the batch size as described in exercise 1.23. All other parameters remain the same as in exercise 1.27. Answer the following questions:

a) Find G, K and their respective invariant vectors, $\boldsymbol{g}$ and $\boldsymbol{\kappa}$, for these four types of group sizes.

b) Compute the selected queue length statistics: the mean, the median, the model, the modal value and the standard deviation at various epochs, for these four types of group sizes.

c) Compute the first few steady-state probability vectors, for these four types of group sizes.

Exercise 1.31. Suppose that arrivals in batches and the batch arrivals follow a Poisson process with a mean of $2/3$ per unit of time. The batch size PMF, $\{a_k\}$, is modeled using one of four types: (1) PoB: $a_k = e^{-0.5}\frac{0.5^{k-1}}{(k-1)!}$, $k \geq 1$; (2) GeB: $a_k = (2/3)(1/3)^{k-1}$, $k \geq 1$; (3) $UfB : a_k = 0.5$, $k = 1, 2; = 0$, elsewhere; and (4) $DpB : a_k = \boldsymbol{\alpha} T^{k-1}\boldsymbol{T^0}$, $k \geq 1$, where:

$$\boldsymbol{\alpha} = (0.7397, 0.2603), \quad T = \begin{pmatrix} 0.1000 & 0.2000 \\ 0.1512 & 0.2341 \end{pmatrix}.$$

There is a single server and the service times are modeled using one of four types: (1) CTS: constant with a value of 0.8 unit; (2) UFS: uniform on the interval $(0.3, 1.3)$; (3) ERS: Erlang of order 2 with parameter 2.5 in each state; (4) HES: hyperexponential with parameters 2.375 and 0.2375, with mixing probabilities, respectively, taken as 0.9 and 0.1. Compute the mean and the standard deviation of various steady-state probabilities under different scenarios.

Exercise 1.32. Suppose that the customers arrive singly according to a MAP (corresponding to Erlang inter-arrival times) with representation given by:

$$D_0 = \begin{pmatrix} -4 & 4 & 0 & 0 \\ 0 & -4 & 4 & 0 \\ 0 & 0 & -4 & 4 \\ 0 & 0 & 0 & -4 \end{pmatrix}, D_1 = \begin{pmatrix} 0 & 0 & 0 & 0 \\ 0 & 0 & 0 & 0 \\ 0 & 0 & 0 & 0 \\ 4 & 0 & 0 & 0 \end{pmatrix}.$$

There is a single server in the system and for services we look at four scenarios as displayed below:

A: CTS: Constant services with a mean of 0.8 unit and a standard deviation of 0.

B: UFS: Uniform services on the interval $(0.3, 1.3)$ with a mean of 0.8 unit and a standard deviation of 0.28868 unit.

C: ERS: Erlang services with two states with rate 2.5 in each state. The mean and the standard deviation of 0.8 unit and a standard deviation of 0.56569 unit.

D: HES: Hyperexponential services with parameters 2.375 and 0.2375 with mixing probabilities, respectively, given by 0.8 unit and a standard deviation of 1.79578 units.

Answer the following questions, where the superscripts contain the type of services used:

a) Find G, K and their respective invariant vectors, $\boldsymbol{g}$ and $\boldsymbol{\kappa}$, for these four types of services.

c) Compute the selected queue length statistics: the mean, the median, the model, the modal value and the standard deviation at various epochs, for these four types of services.

d) Compute the first few steady-state probability vectors, for these four types of services.

Exercise 1.33. Referring to exercise 1.32, we keep all parameters the same except the arrival process, which is taken to be a hyperexponential one with representation given by:

$$D_0 = \begin{pmatrix} -2.1 & 0 & 0 & 0 \\ 0 & -1.05 & 0 & 0 \\ 0 & 0 & -0.525 & 0 \\ 0 & 0 & 0 & -0.2625 \end{pmatrix},$$

$$D_1 = \begin{pmatrix} 1.05 & 0.63 & 0.315 & 0.105 \\ 0.525 & 0.315 & 0.1575 & 0.0525 \\ 0.2625 & 0.1575 & 0.07875 & 0.02625 \\ 0.13125 & 0.07875 & 0.03938 & 0.01312 \end{pmatrix}.$$

Answer the following questions, where the superscripts contain the type of services used:

a) Find G, K and their respective invariant vectors, $\boldsymbol{g}$ and $\boldsymbol{\kappa}$, for these four types of services.

c) Compute the selected queue length statistics: the mean, the median, the model, the modal value and the standard deviation at various epochs, for these four types of services.

d) Compute the first few steady-state probability vectors, for these four types of services.

Exercise 1.34. Referring to exercise 1.32, we keep all parameters the same except the arrival process, which is taken to be a MAP with a negative correlation, with representation given by:

$$D_0 = \begin{pmatrix} -1.75 & 1.75 & 0 & 0 \\ 0 & -1.75 & 1.75 & 0 \\ 0 & 0 & -1.75 & 0 \\ 0 & 0 & 0 & -3.5 \end{pmatrix}, D_1 = \begin{pmatrix} 0 & 0 & 0 & 0 \\ 0 & 0 & 0 & 0 \\ 0.0175 & 0 & 0 & 1.7325 \\ 3.465 & 0 & 0 & 0.035 \end{pmatrix}.$$

Answer the following questions, where the superscripts contain the type of services used:

a) Find G, K and their respective invariant vectors, $\boldsymbol{g}$ and $\boldsymbol{\kappa}$, for these four types of services.

c) Compute the selected queue length statistics: the mean, the median, the model, the modal value and the standard deviation at various epochs, for these four types of services.

d) Compute the first few steady-state probability vectors, for these four types of services.

Exercise 1.35. Referring to exercise 1.32, we keep all parameters the same except the arrival process, which is taken to be a MAP with a negative correlation, with representation given by:

$$D_0 = \begin{pmatrix} -1.75 & 1.75 & 0 & 0 \\ 0 & -1.75 & 1.75 & 0 \\ 0 & 0 & -1.75 & 0 \\ 0 & 0 & 0 & -3.5 \end{pmatrix}, D_1 = \begin{pmatrix} 0 & 0 & 0 & 0 \\ 0 & 0 & 0 & 0 \\ 1.7325 & 0 & 0 & 0.0175 \\ 0.035 & 0 & 0 & 3.465 \end{pmatrix}.$$

Answer the following questions, where the superscripts contain the type of services used:

a) Find G, K and their respective invariant vectors, $\boldsymbol{g}$ and $\boldsymbol{\kappa}$, for these four types of services.

c) Compute the selected queue length statistics: the mean, the median, the model, the modal value and the standard deviation at various epochs, for these four types of services.

d) Compute the first few steady-state probability vectors, for these four types of services.

Exercise 1.36. Show that G and $D[G]$ (see equation [1.36]) commute.

Exercise 1.37. Show that for $MAP/G/1$ queueing model (assuming the system is stable) that the steady-state probability vectors at arrival epochs and departure epochs are identical.

Exercise 1.38. Show that β^* is as given in equation [1.17].

Exercise 1.39. Prove the result for $\hat{\mu}_G$ as given in equation [1.44].

Exercise 1.40. Prove the following result (see equation [1.53]):

$$\sum_{r=1}^{\infty} B_r \sum_{k=0}^{r-1} G^k \tilde{\mu}_G = \sum_{r=0}^{\infty} B_r(I - G^r + re\,g)[I - A + (e - \beta^*)g]^{-1}e.$$

Exercise 1.41. Prove result 1.7.

Exercise 1.42. Prove result 1.14.

Exercise 1.43. Prove equation [1.59] from equations [1.50] and [1.57].

Exercise 1.44. Prove the simplification mentioned in equation [1.62].

Exercise 1.45. Prove result given in equation [1.65].

Exercise 1.46. Prove the expression given in equation [1.69].

Exercise 1.47. Prove result 1.13.

Exercise 1.48. Using equation [1.75] and with the help of the identification of the like coefficients, prove the expressions given in equation [1.71].

Exercise 1.49. Prove the expressions given in equation [1.78].

Exercise 1.50. Use the expressions given in equation [1.78] to show that the steady-state probabilities at arrivals and at departures are identical.

Exercise 1.51. For the $PH/G/1$ queue, show that the steady-state vector g of G is of the form: $g = [\alpha G(-T)^{-1}e]^{-1}\, \alpha G(-T)^{-1}$.

Exercise 1.52. For the $PH/G/1$ queue, show that the matrix K is given by $K = e\,\alpha G$ and hence $\kappa = \alpha G$.

2

Single-Server Queues Embedded at Arrival Epochs

In this chapter, we look at queueing systems embedded at arrival epochs and study them under $GI/M/1$-paradigm. Recall the basics of this paradigm from Chapter 7 of Volume 1. This chapter is organized as follows. In section 2.1, we look at the $GI/PH/1$ queue and in sections 2.2, 2.3 and 2.4, respectively, we look at the special cases, namely, $GI/M/1$, $M/PH/1$ and $M/M/1$ queues.

2.1. *GI/PH/1* queue

In this section, we briefly discuss a queueing model in which the inter-arrival times are $i.i.d.$ random variables with a common distribution function, say, $F(x)$, and with probability density function, $f(x)$, and its Laplace transform (LT) is denoted by $f^*(s)$. It is assumed that the mean and variance are finite. Let λ denote the average rate of arrivals per unit of time. The service times are assumed to follow a continuous phase type (CPH) distribution with representation given by $(\boldsymbol{\beta}, S)$ of order ν, and with the mean given by $\boldsymbol{\beta}(-S)^{-1}\mathbf{e}$. For use in the following, let $\mu = [\boldsymbol{\beta}(-S)^{-1}\mathbf{e}]^{-1}$ denote the average service rate. We need to look at the system at points of arrivals.

It should be pointed out that most of the fundamental ideas needed in the analysis of $GI/M/1$-queues were presented in sections 7.3 and 7.4 of Volume 1 and only the items pertaining to the queues are discussed here. Thus, the results needed for getting the expressions in the queueing context will be stated without repeating the details from those sections.

Embedded process at arrivals: Let $\{t_k : k \geq 0\}$ with $t_0 = 0$ denote the time points at which the kth arrival occurs. Thus, for $k \geq 0$, $t_{k+1} - t_k$ denotes the duration of the time between the kth and the $(k+1)$st arrivals, and this inter-arrival time has the distribution function given by $F(x)$, which is independent of k. We assume that the distribution function, $F(x)$, of the non-negative random variable $\tau_{k+1} = t_{k+1} - t_k$ has a finite mean given by $\lambda^{-1} = \int_0^\infty t\, dF(t)$.

Let N_i denote the number of customers in the queue just *prior* to the ith arrival, and J_i denote the phase of the service just after the ith arrival. Choosing (without loss of generality) $\tau_0 = 0$, the sequence $\{(N_i, J_i, \tau_i) : i \geq 0\}$ forms a Markov renewal sequence on the state space $\Omega = \{(n, j, t) : n \geq 0, 1 \leq j \leq n, t \geq 0\}$.

Recall the probability matrices, $P(n,t) = \{p_{i,j}(n,t)\}$, associated with the counting process of the CPH from section 4.2 (see equation [4.31] of Volume 1). These correspond to the number of services completed during $(0,t]$ in the context of the study here.

Suppose we define the matrices $A_n(t)$ and $B_n(t)$ as follows:

1) The (i,j)th element of $A_n(t)$ is defined as follows: Given that an arriving customer finds n customers in the system with the phase of the service right after that arrival is in state i, the probability that the next arrival to occur by time t will see $(k+1-n)$, $n \leq k$, customers with the phase of the service process to be in j. Note that due to the level of independence, the conditional probabilities are independent of n. Basically, these conditional probabilities correspond to completing n services during the time between two successive arrivals.

2) The (i,j)th element of $B_n(t)$ is defined as follows: Given that an arriving customer finds n customers in the system with the phase of the service after that arrival to be in state i, the probability that the next arrival to occur by time t will see an empty system with the phase of the new service to start in phase j. Basically, these conditional probabilities correspond to completing at least $n+1$ services between two successive arrivals that occur within t units of time. That is, the system should be cleared in order to take the next arrival into service.

It is easy to verify that:

$$A_n(t) = \int_0^t P(n,u)dF(u),\ t \geq 0,\ \ A_n = A_n(\infty) = \int_0^\infty P(n,t)dF(t),\ n \geq 0, \qquad [2.1]$$

and on noting that requiring all customers to be cleared from the system and to take the next arrival into service results in:

$$B_n(t) = \sum_{k=n+1}^{\infty} \int_0^t P(k,u)\boldsymbol{e}\boldsymbol{\beta}dF(u),\ n \geq 0,\ t \geq 0,$$

$$B_n = B_n(\infty) = \sum_{k=n+1}^{\infty} \int_0^\infty P(k,t)dF(t)\boldsymbol{e}\boldsymbol{\beta} = \sum_{k=n+1}^{\infty} A_k\boldsymbol{e}\boldsymbol{\beta},\ n \geq 0. \qquad [2.2]$$

It should be pointed out (see Neuts (1981)) that if any entry of $\boldsymbol{\beta}$ is zero, then the corresponding column of all B_n matrices will also be zero. So, in order to make $\tilde{P}$ irreducible it is understood that the corresponding states are deleted. If that is not carried out, the steady-state probability vector will accordingly assign a zero value to such states.

The transition probability matrix (TPM), $\tilde{P}(t)$, of the Markov renewal process (MRP) is given by:

$$\tilde{P}(t) = \begin{pmatrix} B_0(t) & A_0(t) & 0 & 0 & \cdots \\ B_1(t) & A_1(t) & A_0(t) & 0 & \cdots \\ B_2(t) & A_2(t) & A_1(t) & A_0(t) & \cdots \\ B_3(t) & A_3(t) & A_2(t) & A_1(t) & \cdots \\ \vdots & \vdots & \vdots & \vdots & \ddots \end{pmatrix}, \qquad [2.3]$$

and the TPM of the embedded discrete-time Markov chain (DTMC) for the process $\{(N_k, J_k) : k \geq 0\}$ is given by:

$$\tilde{P} = \tilde{P}(\infty) = \begin{pmatrix} B_0 & A_0 & 0 & 0 & \cdots \\ B_1 & A_1 & A_0 & 0 & \cdots \\ B_2 & A_2 & A_1 & A_0 & \cdots \\ B_3 & A_3 & A_2 & A_1 & \cdots \\ \vdots & \vdots & \vdots & \vdots & \ddots \end{pmatrix}. \qquad [2.4]$$

REMARK 2.1.– The embedded Markov chain with TPM given in equation [2.4] is skip-free to the right since to reach level $\boldsymbol{j}$ from level $\boldsymbol{i}$, for $j \geq i+1$, all the levels $\boldsymbol{i+1}, \cdots, \boldsymbol{j}$ must be visited at least once.

Let $\boldsymbol{\pi}$ denote the steady-state probability vector of the renewal process associated with the service times' CPH. Recall from section 4.2 (see result [4.20] in Volume 1 but by using the CPH representation given here) that the vector $\boldsymbol{\pi}$ is:

$$\boldsymbol{\pi} = \mu\boldsymbol{\beta}(-S)^{-1}. \qquad [2.5]$$

Since we have:

$$A = \sum_{n=0}^{\infty} A_n = \int_0^{\infty} e^{[S+\boldsymbol{S}^0\boldsymbol{\beta}]\,t} dF(t), \qquad [2.6]$$

and since A is stochastic, it is easy to see that:

$$\boldsymbol{\pi} A = \boldsymbol{\pi}. \qquad [2.7]$$

Now applying result 4.27 in Volume 1 to the current model, we get (with $Q = S + \boldsymbol{S}^0\boldsymbol{\beta}$):

$$\boldsymbol{\beta}^* = \sum_{n=1}^{\infty} nA_n\boldsymbol{e} = \mu\int_0^{\infty} tdF(t)\boldsymbol{e}+\mu\int_0^{\infty}[I-e^{Qt}]S^{-1}\boldsymbol{e} = \frac{1}{\lambda}\mu\boldsymbol{e}+\mu[I-A]S^{-1}\boldsymbol{e}. \quad [2.8]$$

The DTMC is positive recurrent if and only if $\boldsymbol{\pi}\,\boldsymbol{\beta}^* > 1$ (see Neuts (1981)) holds good. Note that this results in the familiar requirement that $\lambda < \mu$. That is:

$$\boldsymbol{\pi}\,\boldsymbol{\beta}^* > 1 \;\Rightarrow\; \lambda < \mu. \quad [2.9]$$

Matrix-geometric solution: Suppose that $\boldsymbol{x} = (\boldsymbol{x}_0, \boldsymbol{x}_1, \cdots)$ denotes the steady-state system size at arrival epochs. Then, we have (Neuts 1981):

$$\boldsymbol{x}\tilde{P} = \boldsymbol{x}, \;\; \boldsymbol{x}\,\boldsymbol{e} = 1. \quad [2.10]$$

The following key result follows from theorem 7.5 in Volume 1. This is Neuts's famous matrix-geometric solution for the steady-state probability vector for $GI/M/1$-type queues.

THEOREM 2.1.– Under the stability condition that $\lambda < \mu$, the steady-state probabilities of the system size (not including the one just arrived) at arrivals are given by:

$$\boldsymbol{x}_k = c\boldsymbol{\beta}R^k, \; k \geq 0, \quad [2.11]$$

where $c = [\boldsymbol{\beta}(I - R)^{-1}\mathbf{e}]^{-1}$ and R is the minimal non-negative solution to:

$$R = \sum_{n=0}^{\infty} R^n A_n. \quad [2.12]$$

REMARK 2.2.– The key to getting the above result is due to the simple expression of

$$B[R] = \sum_{k=0}^{\infty} R^k B_k = \boldsymbol{e}\boldsymbol{\beta}, \quad [2.13]$$

which gives its steady-state vector to be $\boldsymbol{\beta}$, resulting in $\boldsymbol{x}_0 = c\boldsymbol{\beta}$.

REMARK 2.3.– Using the probabilistic interpretation of the rate matrix R (recall that here the (i, j)th element of R gives the expected number of visits to the state $(1, j)$ before returning to level **0** given that the DTMC under study starts in state $(0, i)$), it is easy to see that the ith component of $(I - R)^{-1}\boldsymbol{e}$ gives the conditional expected number of transitions before returning to level **0** starting from state $(0, i)$. Thus, the quantity $\boldsymbol{\beta}(I - R)^{-1}\boldsymbol{e}$ gives the mean number of customers served during a busy period. This is due to the fact that the process starting in state $(0, i)$ indicates that an arrival occurs to an empty system and so the service begins in phase i with probability β_i.

REMARK 2.4.– It is easy to verify that the mean, $L_s^{(A)}$, number in the system at arrival epochs is given by:

$$L_s^{(A)} = \sum_{k=1}^{\infty} k\boldsymbol{x}_k\boldsymbol{e} = \frac{1}{\boldsymbol{\beta}(I-R)^{-1}\boldsymbol{e}}\Big[\boldsymbol{\beta}(I-R)^{-2}\boldsymbol{e}\Big] - 1. \qquad [2.14]$$

REMARK 2.5.– The rate matrix R can be obtained without the knowledge of the individual matrices, A_n. This is due to Sengupta (1989). Suppose that:

$$A(z) = \sum_{n=0}^{\infty} z^n A_n. \qquad [2.15]$$

Then, the nonlinear equation given in equation [2.12] can be written as:

$$R = \int_0^{\infty} e^{(S+R\boldsymbol{S}^{\mathbf{0}}\boldsymbol{\beta})t} dF(t). \qquad [2.16]$$

The above equation provides an efficient iterative scheme for computing R using the uniformization technique (see section 2.2.2 in Volume 1 for the uniformization method). The key steps involved are summarized as follows:

Algorithm 2.1. Algorithmic steps in computing the matrix R for $GI/PH/1$-type models using equation [2.16]

Step 0: $k = 0, R_0 = 0, k \leftarrow k+1$. Let $\theta = \max_i |(S)_{i,i}|$.

Step 1: Compute the next iterate value as follows. Define $K = I + \frac{1}{\theta}[S + R_{k-1}\boldsymbol{S}^{\mathbf{0}}\boldsymbol{\beta}]$ and calculate $R_k = \sum_{n=0}^{\infty} \gamma_n K^n$, where $\gamma_n = \int_{n=0}^{\infty} e^{-\theta t}\frac{(\theta t)^n}{n!} dF(t), \quad n \geq 0$. The infinite sum is truncated based on a given small positive number, ϵ.

Step 2: Continue step 1 until $||R_{k+1} - R_k||_\infty < \epsilon$, where ϵ is a very small positive number like 10^{-8}.

REMARK 2.6.– The computation of R depends on the nature of the inter-arrival time distribution function $F(.)$. For example, if the arrivals are deterministic or have masses only at finite number of points, then we need to evaluate the exponential matrix, which appears as an integrand only at a finite number of time points. For example, if the inter-arrival times are $i.i.d$ random variables with a common discrete probability mass function (PMF), with masses at $a_1, \cdots, a_N$ with probabilities given by $p_1, \cdots, p_N$, respectively, then equation [2.16] is written as:

$$R = \sum_{k=1}^{N} p_k e^{(S+R\boldsymbol{S}^{\mathbf{0}}\boldsymbol{\beta})a_k}.$$

One can use the uniformization method (described and shown through many examples in earlier sections and in Volume 1) to compute R iteratively.

In the case when the inter-arrival times are of CPH with representation $(\boldsymbol{\alpha}, T)$ of order m, the probabilities, γ_n can be computed without integration since $\{\gamma_k\}$ follows a discrete PH distribution with representation $(\theta\boldsymbol{\alpha}(\theta I - T)^{-1}, \theta(\theta I - T)^{-1})$ of order m. This is similar (see remark 1.15) to the $M/G/1$-type model seen earlier.

In the case when the distribution of the inter-arrival times is (continuous) uniform on the interval (a, b), we can solve R as follows. Suppose we denote by U the matrix $U = S + R\boldsymbol{S^0}\boldsymbol{\beta}$. Then, we have:

$$R = \frac{1}{b-a}(-U)^{-1}\left[e^{U\,a} - e^{U\,b}\right].$$

It is easy to verify that U is non-singular. For example, suppose that $\boldsymbol{u}$ is a vector satisfying $\boldsymbol{u}U = \mathbf{0}$. If we show that the only solution to this equation is $\boldsymbol{u} = 0$, then U has to be non-singular. Under the condition that $\lambda < \mu$ (i.e. the underlying queueing model is stable), we know that $(I - R)$ is non-singular. Thus:

$$\boldsymbol{u}R = \boldsymbol{u}\int_0^\infty e^{U\,t}dF(t) = \boldsymbol{u} \Rightarrow \boldsymbol{u}R = \boldsymbol{u} \Rightarrow \boldsymbol{u}(I - R) = \mathbf{0} \Rightarrow \boldsymbol{u} = \mathbf{0}.$$

As pointed out in Volume 1 (see remark 7.15), Neuts has established a number of interesting properties dealing with the rate matrix R, its spectral radius and the corresponding left and right eigenvectors. Here, we briefly summarize them and refer the reader to Neuts (1981, 1986) for details.

REMARK 2.7.– The spectral radius, η, of the rate matrix R plays an important role. As pointed out earlier, η can either be computed from the knowledge of R or from solving $\eta = \chi(\eta)$, where (recall) $\chi(z)$ is the maximal eigenvalue of $A(z) = \sum_{k=0}^{\infty} z^k A_k$. The latter does not require the knowledge of R (see remark 7.9 in Volume 1). Assume that R is irreducible. If it is not, then we can find the submatrix of R that is irreducible and contains the spectral radius. Let $\boldsymbol{u}$ and $\boldsymbol{v}$, respectively, denote the left and right eigenvectors of η. That is:

$$\boldsymbol{u}R = \eta\boldsymbol{u}, R\boldsymbol{v} = \eta\boldsymbol{v}.$$

Furthermore, we can normalize $\boldsymbol{u}$ and $\boldsymbol{v}$, such that $\boldsymbol{u}\boldsymbol{e} = 1 = \boldsymbol{u}\boldsymbol{v}$.

1) Based on a classical theorem on irreducible non-negative matrices, we have $R^n = \eta^n\boldsymbol{v}\boldsymbol{u} + o(\eta^n)$, as $n \to \infty$.

2) Item 1 above makes the tail probabilities to be computed easily as $\boldsymbol{x}_{*+n-1} = \eta^n(\boldsymbol{x}_*\boldsymbol{v})\boldsymbol{u} + o(\eta^n)$ as $n \to \infty$. Here, $\boldsymbol{x}_*$ denotes the probability vector with whose knowledge along with R results in knowing the rest of the non-boundary probability vectors.

3) Neuts coined the term caudal characteristic curve ("Cauda" means "tail" in Latin) for the graph of η as a function of the traffic intensity of the model (possessing matrix-geometric solution) under study.

4) Whether R is irreducible or not, it is true that $R^n = O(\eta^n)$ as $n \to \infty$.

5) If X denotes the queue length of the model under study, then $P(X > n) = N\eta^n + o(\eta^n)$, as $n \to \infty$. While the known quantity, η, is indicative of the tail behavior (i.e. the decay rate), the unknown constant N (indicating the magnitude) is generally computable only for the simplest models.

6) The ratio of the expected time spent in level $\boldsymbol{k+1}$ to that spent in level $\boldsymbol{k}$ is approximately equal to η. That is:

$$\lim_{k\to\infty} \frac{\boldsymbol{x}_* R^{k+1}\boldsymbol{e}}{\boldsymbol{x}_* R^k \boldsymbol{e}} = \eta.$$

7) The ratio of the expected time spent in $(k+1, j)$ to that spent in state (k, j) for large k is η. That is:

$$\lim_{k\to\infty} \frac{(\boldsymbol{x}_* R^{k+1})_j}{(\boldsymbol{x}_* R^k)_j} = \eta,\ 1 \le j \le m.$$

8) The conditional probability of being in state (k, j) given that the chain is in level $\boldsymbol{k}$, for large k, is approximately given by:

$$\lim_{k\to\infty} \frac{(\boldsymbol{x}_* R^{k+1})_j}{(\boldsymbol{x}_* R^k)\boldsymbol{e}} = u_j,\ 1 \le j \le m.$$

9) A high value of η for a specific ρ indicates the underlying queue will exhibit large and long-lived excursions. Such excursions (though rare) affect the queue length as well as the waiting time.

10) When η is large for small ρ, it is possible that the paths of the underlying queue exhibit (rare) upsurges (i.e. rare visits to longer queue lengths and staying for a longer period of time) which may take a long time to clear. Upsurges occur when the caudal characteristic curve increases rapidly near 0. This will remain true for such queues for all ρ. If η is large (and close to 1) then we will see upsurges. For queues that have upsurges the overall mean queue length will be an unreliable descriptor.

11) When the curve has an S-shape, a phenomenon known as "coalescing of excursions" is said to occur. In this case, η stays much smaller than ρ initially. Then η steeply increases over the rest of the values of ρ, probably with a slower rate toward

1. Thus, the queue may have significant but not extreme upsurges (and a fast recovery from such upsurges) at low values of ρ. However, as ρ increases a full recovery from upsurges are infrequent, and the queue will occasionally go into long-lived excursions.

12) Small values of ρ indicate the system will have more mass on the low values of the queue length. Still the system can visit (at least occasionally) higher levels of the queue length. However, if η is significantly smaller (than 1), then the queue size will quickly get back to low levels. Thus, ρ and η provide complementary information on the model under study.

Steady-state probabilities at arbitrary times: Suppose that $\boldsymbol{y} = (y_0, \boldsymbol{y}_1, \boldsymbol{y}_2, \cdots)$ denotes the steady-state vector giving the number of customers in the system at arbitrary times. Note that (1) y_0 gives the steady-state probability that at an arbitrary time the system is idle, and (2) $\boldsymbol{y}_n$ gives the steady-state probability vector that at an arbitrary time there will be n, $n \geq 1$, customers in the system with the service phase in one of ν phases. In order to apply the key renewal theorem to get an expression for $\boldsymbol{y}$, we denote by $\tilde{M}(t)$ the Markov renewal function associated with the MRP, whose TPM is given in equation [2.3]. That is, $\tilde{M}(t) = \sum\limits_{n=0}^{\infty} \tilde{P}^{(n)}(t)$.

THEOREM 2.2 (Neuts 1981).– Under the stability condition that $\lambda < \mu$, the steady-state probabilities of the system size at arbitrary times are given by:

$$y_0 = 1 - \rho, \quad \boldsymbol{y}_n = c\lambda\boldsymbol{\beta}R^{n-1}[I - R + R\boldsymbol{e}\,\boldsymbol{\beta}](-S)^{-1}, \; n \geq 1. \qquad [2.17]$$

PROOF.– Suppose we define $y_0(t)$ to be the probability that at an arbitrary time the system is idle and let $y_{n,j}(t)$, $n \geq 1, 1 \leq j \leq m$, $t \geq 0$, denote the probability that the queueing system under study has n customers in the system and the phase of the service is in j at time t. Using the interpretation of $d\tilde{M}_{i_0,j_0;i,j}(u)$ and the law of total probability by looking at all scenarios, it is easy to verify that:

$$y_0(t) = \sum_{i=0}^{\infty}\sum_{j=1}^{\nu}\int_0^t d\tilde{M}_{i_0,j_0;i,j}(u) \sum_{r=i+1}^{\infty} P(r, t-u)\boldsymbol{e}\Big)_j [1 - F(t-u)],$$

$$y_{n,k}(t) = \sum_{i=n-1}^{\infty}\sum_{j=1}^{\nu}\int_0^t d\tilde{M}_{i_0,j_0;i,j}(u)\Big(P(i-n+1, t-u)\Big)_{j,k}[1 - F(t-u)],$$
$$n \geq 1, \; 1 \leq k \leq \nu.$$

From the classical key renewal theorem for MRP (see result 2.58 in Volume 1), we get (in matrix form):

$$y_0 = \lambda \sum_{i=0}^{\infty} \boldsymbol{x}_i \sum_{r=i+1}^{\infty} \int_0^{\infty} P(r, t)\boldsymbol{e}[1 - F(t)]dt,$$
$$\boldsymbol{y}_n = \lambda \sum_{i=n-1}^{\infty} \boldsymbol{x}_i \int_0^{\infty} P(i-n+1, t)[1 - F(t)]dt, \; n \geq 1. \qquad [2.18]$$

To find an expression for the integrals appearing in equation [2.18], we first define:

$$U_i = \lambda \int_0^\infty P(i,t)[1 - F(t)]dt, \ i \geq 0. \qquad [2.19]$$

From the definition of $\{A_i\}$ (see equation [2.1]), we get:

$$A_0 = \int_0^\infty P(0,t)dF(t) = -\int_0^\infty P(0,t)d[1 - F(t)]$$

$$= -P(0,t)[1 - F(t)]\Big)_0^\infty + \int_0^\infty P'(0,t)[1 - F(t)]dt = I + \int_0^\infty P'(0,t)[1 - F(t)]dt,$$

and similarly, for $i \geq 1$,

$$A_i = -P(i,t)[1-F(t)]\Big)_0^\infty + \int_0^\infty P'(i,t)[1-F(t)]dt = \int_0^\infty P'(i,t)[1-F(t)]dt. \quad [2.20]$$

In the above equation, we used the facts that $P(0,0) = I$ and $P(i,0) = 0, i \geq 1$, and the property of the (cumulative) distribution function (CDF). Now using the differential-difference equation for $\{P(i,t)\}$ (see equation [4.32] in Volume 1) of the current CPH with representation $(\boldsymbol{\beta}, S)$), we get:

$$A_0 = I + \frac{1}{\lambda}U_0 S, \text{ and } A_i = \frac{1}{\lambda}U_i S + \frac{1}{\lambda}U_{i-1}\boldsymbol{S}^0\boldsymbol{\beta}, \ \ i \geq 1. \qquad [2.21]$$

Pre-multiplying A_i by R^i and adding over i along with A_0 in the above equation, we get:

$$\sum_{i=0}^\infty R^i A_i = I + \frac{1}{\lambda}U_0 S + \frac{1}{\lambda}\sum_{i=1}^\infty R^i U_i S + \frac{1}{\lambda}\sum_{i=1}^\infty R^i U_{i-1}\boldsymbol{S}^0\boldsymbol{\beta}, \qquad [2.22]$$

which reduces to:

$$R = I + \frac{1}{\lambda}\sum_{i=0}^\infty R^i U_i S + R\frac{1}{\lambda}\sum_{i=0}^\infty R^i U_i \boldsymbol{S}^0\boldsymbol{\beta}. \qquad [2.23]$$

Post-multiplying equation [2.23] by $\boldsymbol{e}$ and using the fact that $(I - R)^{-1}$ exists, we get:

$$\sum_{i=0}^\infty R^i U_i \boldsymbol{S}^0 = \lambda \boldsymbol{e}, \qquad [2.24]$$

which upon substitution in equation [2.23] and simplifying yields:

$$\sum_{i=0}^{\infty} R^i U_i = \lambda[I - R + Re\boldsymbol{\beta}](-S)^{-1}. \qquad [2.25]$$

The stated result for $\boldsymbol{y}_n$, $n \geq 1$, follows immediately by using theorem 2.1 and the expression obtained in equation [2.25]. The expression for y_0 is obtained in a similar fashion but by noting that:

$$\frac{1}{\boldsymbol{\beta}(I-R)^{-1}\boldsymbol{e}}\lambda\boldsymbol{\beta}(I-R)^{-1}[I-R+Re\,\boldsymbol{\beta}](-S)^{-1}\boldsymbol{e}$$

$$= \frac{1}{\boldsymbol{\beta}(I-R)^{-1}\boldsymbol{e}}\lambda\boldsymbol{\beta}[I+(I-R)^{-1}\boldsymbol{e}\,\boldsymbol{\beta} - \boldsymbol{e}\,\boldsymbol{\beta}](-S)^{-1}]\boldsymbol{e} = \rho. \qquad \square$$

REMARK 2.8.– Noting that the marginal probabilities of the number in the system at an arbitrary time are given by:

$$y_0 = 1-\rho, \quad \boldsymbol{y}_k\boldsymbol{e} = \frac{\lambda}{\boldsymbol{\beta}(I-R)^{-1}\boldsymbol{e}}\Big[\rho\boldsymbol{\beta}R^k\boldsymbol{e} + \boldsymbol{\beta}R^{k-1}(I-R)(-S)^{-1}\boldsymbol{e}\Big],\ k \geq 1, \qquad [2.26]$$

verify that the mean, L_s, number of customers in the system at an arbitrary time is given by:

$$L_s = \sum_{k=1}^{\infty} k\boldsymbol{y}_k\boldsymbol{e} = \rho L_s^{(A)} + \frac{\lambda}{\boldsymbol{\beta}(I-R)^{-1}\boldsymbol{e}}\Big[\boldsymbol{\beta}(I-R)^{-1}(-S)^{-1}\boldsymbol{e}\Big]. \qquad [2.27]$$

REMARK 2.9.– It is easy to verify that the marginal stationary probabilities at departure epochs, z_n, $n \geq 0$, are obtained as:

$$z_n = d\boldsymbol{y}_{n+1}\boldsymbol{S}^0,\ n \geq 0 \Rightarrow z_n = [\boldsymbol{\beta}(I-R)^{-1}\boldsymbol{e}]^{-1}\boldsymbol{\beta}R^n\boldsymbol{e},\ n \geq 0, \qquad [2.28]$$

where the normalizing constant d is calculated to be $\frac{1}{\lambda}$.

REMARK 2.10.– Recall from section 7.4 in Volume 1 that the (j,k)th entry of R (obtained by looking at the embedded process) gives the expected number of visits to state $(1,k)$ before returning to level **0** given that the process started in state $(0,j)$. Note that the probability of starting in state $(0,j)$ is β_j (this is due to the fact that an arrival seeing the system empty will begin the service in one of ν phases according to the probability vector $\boldsymbol{\beta}$) and combining this with the fact that the jth component of the vector $(I-R)^{-1}\boldsymbol{e} = \boldsymbol{e}+R\boldsymbol{e}+R^2\boldsymbol{e}+\cdots$ gives the expected number of transitions before returning for the first time to level **0**, we note that the quantity $\boldsymbol{\beta}(I-R)^{-1}\boldsymbol{e}$ gives the mean number of service completions during a busy period.

Stationary waiting times: The stationary waiting times, such as (1) the virtual waiting times in the system (or in the queue), (2) the waiting times in the system (or in the queue) at arrivals and (3) the waiting times in the system (or in the queue) at an arbitrary time, can be obtained under the first come, first served (FCFS) rule. Depending on the need, one can choose to study any or all of them. In this section, we give the expressions for the LT of these waiting times and refer the reader to Neuts (1981) for full details.

In order to get the expressions for each of the waiting times, we need the residual service time (if the server is busy at the epoch of interest) and the number of service completions of those present ahead. Once these are obtained, the derivations of the waiting times are straightforward. We recall the LT of the service times from section 4.1 (see equation [4.5] in Volume 1 modified for the current model here):

$$f^*(s) = \boldsymbol{\beta}(sI - S)^{-1}\boldsymbol{S}^0, \quad Re(s) \geq 0. \qquad [2.29]$$

Stationary waiting time in the queue (at arrivals): Note that an arriving customer can have a zero waiting time in the queue when at that moment the server is idle. This occurs with probability $\boldsymbol{x}_0\boldsymbol{e} = [\boldsymbol{\beta}(I - R)^{-1}\boldsymbol{e}]^{-1}$; otherwise the arriving customer finding k customers (not including the arrived one) has to wait for the duration of the current service and k services, before entering into service. Thus, we have the following result:

RESULT 2.1.– The LT, $w^*_{AQ}(s)$, of the waiting time in the queue of an arriving customer is obtained as:

$$w^*_{AQ}(s) = [\boldsymbol{\beta}(I - R)^{-1}\boldsymbol{e}]^{-1} + \sum_{k=1}^{\infty} \boldsymbol{x}_k(sI - S)^{-1}\boldsymbol{S}^0[f^*(s)]^k, \ Re(s) \geq 0,$$

which is simplified using the expression for $\boldsymbol{x}_k$ given in equation [2.11] as:

$$w^*_{AQ}(s) = \frac{1}{\boldsymbol{\beta}(I - R)^{-1}\boldsymbol{e}}\boldsymbol{\beta}\sum_{k=0}^{\infty} R^k[(sI - S)^{-1}\boldsymbol{S}^0\boldsymbol{\beta}]^k\boldsymbol{e}, \ Re(s) \geq 0. \qquad [2.30]$$

The mean waiting time in the queue of an arriving customer is:

$$W_{AQ} = -\frac{d}{ds}w^*_{AQ}(s)\Big]_{s=0} = \frac{1}{\lambda}L_s - \frac{1}{\mu}. \qquad [2.31]$$

Stationary waiting time in the system (at arrivals): Note that an arriving customer can have a 0 waiting time in the queue when at that moment the server is idle. This customer's waiting time in the system is just the service time. This occurs with probability $\boldsymbol{x}_0\boldsymbol{e} = (\boldsymbol{\beta}(I - R)^{-1}\boldsymbol{e})^{-1}$; otherwise the arriving customer finding k customers (not including the arrived one) has to wait for the duration of the current service and $k + 1$ services before the system. Thus, we have the following result:

RESULT 2.2.– The LT, $w_{AS}^*(s)$, of the waiting time in the system of an arriving customer is obtained as:

$$w_{AS}^*(s) = (\boldsymbol{\beta}(I-R)^{-1}\mathbf{e})^{-1} f^*(s) + \sum_{k=1}^{\infty} \boldsymbol{x}_k (sI - S)^{-1} \boldsymbol{S}^0 [f^*(s)]^{k+1}, \; Re(s) \geq 0,$$

which is simplified using the expression for $\boldsymbol{x}_k$ given in equation [2.11] as:

$$w_{AS}^*(s) = \frac{1}{\boldsymbol{\beta}(I-R)^{-1}\mathbf{e}} \boldsymbol{\beta} \sum_{k=0}^{\infty} R^k [(sI - S)^{-1} \boldsymbol{S}^0 \boldsymbol{\beta}]^{k+1} \mathbf{e}, \; Re(s) \geq 0. \quad [2.32]$$

The mean waiting time in the system of an arriving customer is:

$$W_{AS} = -\frac{d}{ds} w_{AS}^*(s) \Big]_{s=0} = \frac{1}{\lambda} L_s. \quad [2.33]$$

REMARK 2.11.– As is to be expected, equation [2.33] is the famous Little's result.

EXAMPLE 2.1.– In this example, we look at constant arrivals with the inter-arrival times having a value of 1 and the services are assumed to be Erlang of order 3. The parameter of this Erlang is chosen to be $\frac{3}{0.95}$ so that the mean service time is 0.95. This will result in the traffic intensity being 0.95 in the $GI/PH/1$ queue under study. We use the algorithmic steps outlined earlier to compute the rate matrix R as given in equation [2.16]. Due to constant arrivals assumed in this example, equation [2.16] reduces to:

$$R = e^{(S + R\boldsymbol{S}^0 \boldsymbol{\beta})},$$

which requires us to compute the exponential matrix for each iterate of R . This is accomplished using the uniformization method, which was discussed in more detail in the previous sections. Very briefly, here are the key steps. Once we obtain $\theta = \max_i |(S)_{i,i}| = \frac{3}{0.95}$, to calculate the kth iterate of R, we need to truncate the infinite sum (see step 1 in Algorithm 2.1) and here the truncated value, say, n^* of n for which $\sum_{n=n^*+1}^{\infty} \gamma_n < 10^{-8}$ is obtained as $n^* = 17$. Also, verify that:

$$\boldsymbol{\gamma} = (0.0425, 0.1343, 0.2120, 0.2231, 0.1762, 0.1113, 0.0586, 0.0264, 0.0104, \\ 0.0037, 0.0012, 0.0003, 0.0001, 0.0000, 0.0000, 0.0000, 0.0000, 0.0000).$$

Starting with $R_0 = 0$, successive iterates converge to:

$$R = \begin{pmatrix} 0.4856 & 0.5329 & 0.5158 \\ 0.1799 & 0.2108 & 0.2668 \\ 0.0390 & 0.0377 & 0.0732 \end{pmatrix}$$

The plots of the steady-state probability vector at arrival and at arbitrary epochs are given in Figure 2.1.

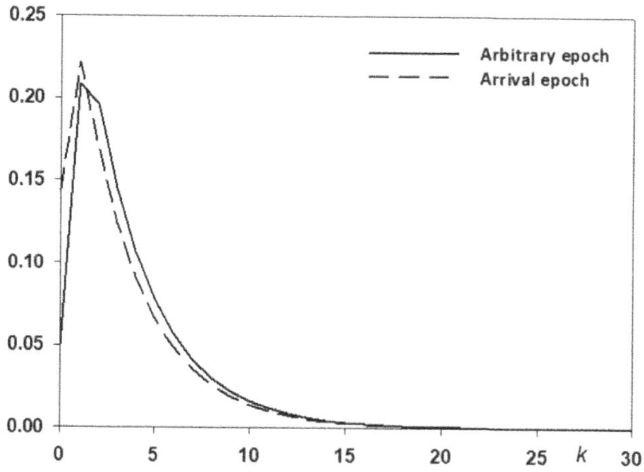

Figure 2.1. *Plots of the steady-state probability vectors for example 2.1*

It is easy to verify that (1) the spectral radius of R is 0.7331; (2) the mean, the median and the mode of $\{\boldsymbol{x}_k\}$ are, respectively, given by 3.2341, 2 and 2; and (3) the mean, the median and the mode of $\{\boldsymbol{y}_k\}$ are, respectively, given by 3.7363, 3 and 0.

EXAMPLE 2.2.– This example is similar to example 2.1 but with the inter-arrival times modeled using an arbitrary discrete probability function. Suppose that X denotes the time between ith and $(i+1)$st arrivals. Then, the PMF we use here is $P(X = 0.5) = 0.3, P(X = 1.0) = 0.6$ and $P(X = 2.5) = 0.1$. All other parameters are identical to the ones seen in example 2.1. Verify that the mean of X is 1.0 and the standard deviation of X is 0.54772. The traffic intensity is also 0.95 here. The rate matrix R here is solved as follows:

$$R = 0.3e^{0.5(S+RS^0\boldsymbol{\beta})} + 0.6e^{(S+RS^0\boldsymbol{\beta})} + 0.1e^{2.5(S+RS^0\boldsymbol{\beta})},$$

Once again we use the uniformization method to compute R. Once we obtain $\theta = \max_i |(S)_{i,i}| = \frac{3}{0.95}$, we can use the same θ to compute all three exponential matrices. The n^* values to truncate the infinite sum with an error of 10^{-8} (see step 1) are 13, 17 and 28, respectively. Starting with $R_0 = 0$, successive iterates converge to:

$$R = \begin{pmatrix} 0.5249 & 0.5315 & 0.4593 \\ 0.2318 & 0.2808 & 0.3206 \\ 0.0710 & 0.0613 & 0.1336 \end{pmatrix}$$

The plots of the steady-state probability vector at arrival and at arbitrary epochs are given in Figure 2.2.

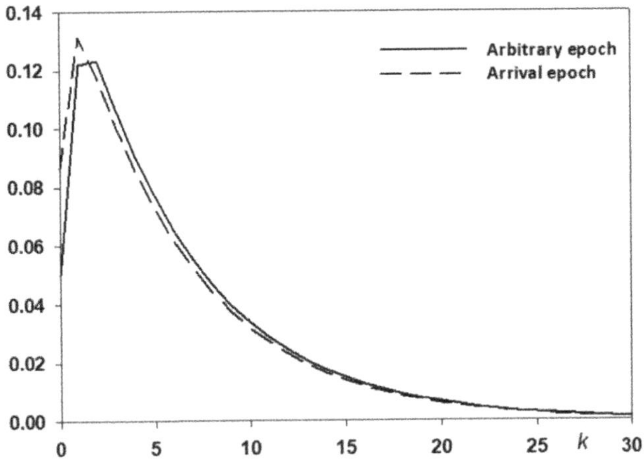

Figure 2.2. *Plots of the steady-state probability vectors for example 2.2*

It is easy to verify that (1) the spectral radius of R is 0.8489; (2) the mean, the median and the mode of the number in the system at arrival epochs are, respectively, given by 6.0921, 4 and 4; and (3) the mean, the median and the mode of the number in the system at arbitrary epochs are, respectively, given by 6.4413, 5 and 0.

EXAMPLE 2.3.– In this example, we assume that the inter-arrival times follow a (continuous) uniform on $(0.1, 1.9)$ so that the mean and the standard deviation of the inter-arrival times are, respectively, given by 1 and 0.51962. All other parameters are identical to the ones seen in example 2.1. The rate matrix R here is solved as follows:

$$R = \frac{1}{1.8}(-U)^{-1}[e^{0.1(S+R\boldsymbol{S}^{\mathbf{o}}\boldsymbol{\beta})} - e^{1.9(S+R\boldsymbol{S}^{\mathbf{o}}\boldsymbol{\beta})}].$$

Once again we use the uniformization method to compute R. Once we obtain $\theta = \max_i |(S)_{i,i}| = \frac{3}{0.95}$, we can use the same θ to compute the three exponential matrices. The n^* values to truncate the infinite sum with an error of 10^{-8} (see step 1) are 7 and 24, respectively. Starting with $R_0 = 0$, successive iterates converge to:

$$R = \begin{pmatrix} 0.5074 & 0.4735 & 0.4012 \\ 0.2250 & 0.3174 & 0.3125 \\ 0.0909 & 0.0770 & 0.1920 \end{pmatrix}$$

The plots of the steady-state probability vector at arrival and at arbitrary epochs are given in Figure 2.3.

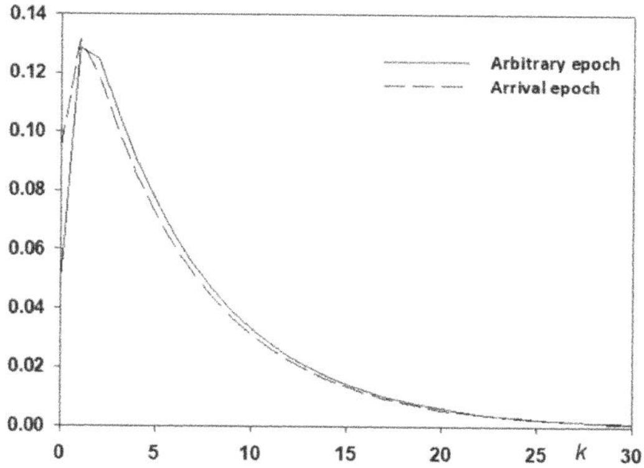

Figure 2.3. *Plots of the steady-state probability vectors for example 2.3*

It is easy to verify that (1) the spectral radius of R is 0.8455; (2) the mean, the median and the mode of the number in the system at arrival epochs are, respectively, given by 5.9202, 4 and 4; and (3) the mean, median and the mode of the number in the system at arbitrary epochs are, respectively, given by 6.2821, 4 and 0.

EXAMPLE 2.4.– In this example, we assume that the inter-arrival times follow an Erlang of order 3 with parameter 3 so that the mean and the standard deviation of the inter-arrival times are, respectively, given by 1 and 0.57735. All other parameters are identical to the ones seen in example 2.1. We use the uniformization method along the lines outlined in remark 2.6 of Volume 1. For the tolerance error set at 10^{-8}, the number of points needed to compute the associated DPH (recall that when dealing with $PH/PH/1$ queue embedded at arrivals, we need to calculate the number of PH renewals occurring during an arrival, which follows a discrete phase type (DPH) – see result 4.31 in Volume 1) is 41 and the probabilities (which are non-zero up to five decimal places) are displayed below:

$$(0.11563, 0.17789, 0.18245, 0.15594, 0.11996, 0.08612, 0.05889, 0.03883,\\ 0.02489, 0.01560, 0.00960, 0.00582, 0.00348, 0.00206, 0.00121, 0.00070,\\ 0.00040, 0.00023, 0.00013, 0.00007, 0.00004, 0.00002, 0.00001, 0.00001)$$

Starting with $R_0 = 0$, successive iterates converge to:

$$R = \begin{pmatrix} 0.5174 & 0.4997 & 0.4214 \\ 0.2330 & 0.3068 & 0.3221 \\ 0.0855 & 0.0721 & 0.1710 \end{pmatrix}$$

The plots of the steady-state probability vector at arrival and at arbitrary epochs are given in Figure 2.4.

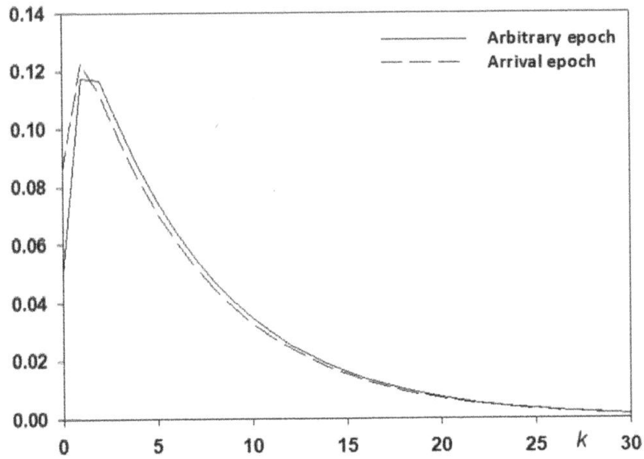

Figure 2.4. *Plots of the steady-state probability vectors for example 2.4*

It is easy to verify that (1) the spectral radius of R is 0.8574; (2) the mean, the median and the mode of the number in the system at arrival epochs are, respectively, given by 6.4707, 4 and 5; and (3) the mean, the median and the mode of the number in the system at arbitrary epochs are, respectively, given by 6.8020, 5 and 0.

EXAMPLE 2.5.– In examples 2.1 through 2.4, we looked at various arrival processes but all of them had the same type of services. Thus, in this example, we compare the measures reported there to bring out the qualitative nature of the models under study.

While the steady-state probability vectors (at arbitrary as well as at arrival epochs) show similar patterns for all four types of arrivals (constant to arbitrary to uniform to Erlang of order 3), we do see the queueing systems exhibit different patterns in other measures. For example, looking at the entries of the rate matrix, R, we see that the expected number of visits to, say, state $(i+1,2)$ before returning to level $\boldsymbol{i}$, starting in state $(i,1)$, is largest in the case of constant arrivals as well as for the arrivals with a discrete PMF; however, for the other two arrivals we notice the largest value for this expected number of visits occur by visiting the state $(i+1,1)$ before returning to $\boldsymbol{i}$, starting in state $(i,1)$. For all four types of arrivals, we notice the smallest expected number of visits is to the state $(i+1,2)$ before returning to $\boldsymbol{i}$, starting in state $(i,3)$.

EXAMPLE 2.6.– In this example, we discuss the behavior of the spectral radius as a function of the traffic intensity under various scenarios. Toward this end, we look at four types of arrival processes: (1) Constant (CTA) arrivals; (2) Uniform (UFA) arrivals on $(0.1,1.9)$; (3) Erlang of order (ER5) arrivals; and (4) Hyperexponential

(HE3) arrivals with parameters 3.8, 0.95 and 0.19 with probabilities 0.6, 0.3 and 0.1, respectively. These four arrival processes, covering a wide range of scenarios, will be normalized to have a mean of 1. The standard deviations (after normalizing to have a mean of 1) are 0, 0.5196, 0.4472 and 2.2996, respectively. For services we look at two CPH distributions, namely, Erlang of order 5 and hyperexponential with parameters 2.18000, 1.09, 0.545, 0.2725 and 0.13625 with probabilities, respectively, given by 0.5, 0.3, 0.15, 0.04 and 0.01. Since we vary the traffic intensity ρ from 0.01 to 0.99, the service rate will be taken as $\mu = \frac{1}{\rho}$ (since $\lambda = 1$), and hence the service distributions will be accordingly normalized. Note that the ratio of the standard deviation of hyperexponential services to that of the corresponding Erlang services remains the same and is given by 3.79483.

The graphs of the spectral radius as well as the mean number in system at arrival epochs as functions of ρ under various scenarios are, respectively, given in Figures 2.5 and 2.6. It should be pointed out that the line segment (labeled as 45°) is plotted to indicate the level by which the spectral radius deviates from ρ.

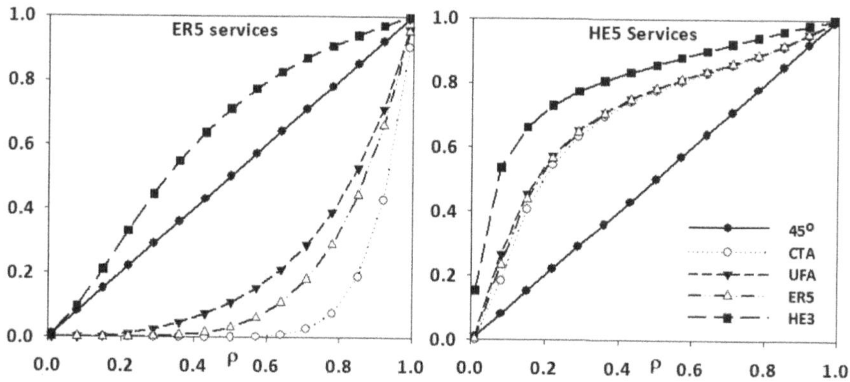

Figure 2.5. *Caudal characteristic curves under various scenarios for example 2.6*

As mentioned earlier (see remark 2.7), the caudal characteristic curve shows that for the hyperexponential services we notice the underlying queue exhibiting upsurges that take a longer time to clear. This is the case even for the hyperexponential arrivals. This also results in having a larger mean for the number in the system when dealing with hyperexponential arrivals or for hyperexponential services.

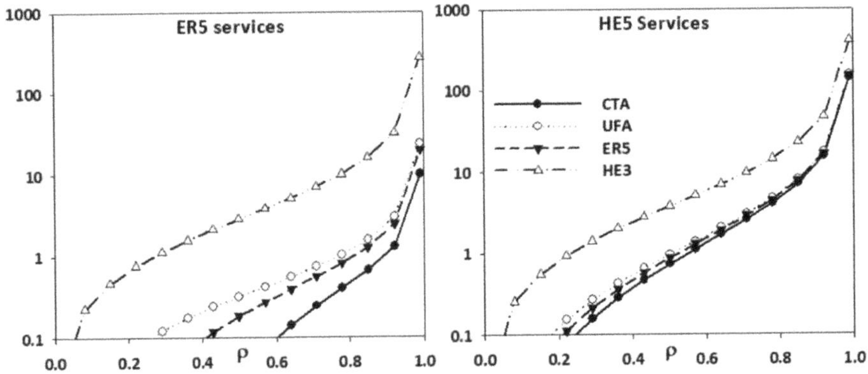

Figure 2.6. *Mean number in system at arrival epochs under various scenarios for example 2.6*

EXAMPLE 2.7.– In this example, we look at various scenarios by considering two types of arrivals and four types of services, which all follow CPH distribution. For arrivals, we look at Erlang of order 5, referred to as ER5, and hyperexponential with parameters 2.18, 1.09, 0.545, 0.2725 and 0.13625 with mixing probabilities, respectively, to be 0.5, 0.3, 0.15, 0.04 and 0.01. For services we look at four CPH distributions obtained: (1) (CNV): convolution of Erlang of order 2 with parameter 2 and hyperexponential of order 2 with parameters 10 and 1 with probabilities, respectively, given by 0.9 and 0.1; (2) (MXT): mixture (with equal probabilities) of Erlang of order 2 with parameter 2 and hyperexponential of order 2 with parameters 10 and 1 with probabilities, respectively, given by 0.9 and 0.1; (3) (MIN): minimum of exponential with parameter 1, Erlang of order 2 with parameter 2 and hyperexponential of order 2 with parameters 10 and 1 with probabilities, respectively, given by 0.9 and 0.1; and (4) (MAX): maximum of Erlang of order 2 with parameter 2 and hyperexponential of order 2 with parameters 10 and 1 with probabilities, respectively, given by 0.9 and 0.1. These four CPH distributions used for service times will be normalized to have a mean of 0.95. Note that in all scenarios, the traffic intensity is fixed at $\rho = 0.95$.

The following R matrices correspond to the arrivals that follow an Erlang of order 5:

$$R^{(CNV)} = \begin{pmatrix} 0.233 & 0.283 & 0.282 & 0.111 \\ 0.112 & 0.154 & 0.190 & 0.098 \\ 0.021 & 0.022 & 0.072 & 0.006 \\ 0.455 & 0.392 & 0.296 & 0.770 \end{pmatrix}, \quad R^{(MXT)} = \begin{pmatrix} 0.690 & 0.513 & 0.096 & 0.048 \\ 0.172 & 0.419 & 0.043 & 0.023 \\ 0.007 & 0.005 & 0.019 & 0.001 \\ 0.309 & 0.157 & 0.082 & 0.595 \end{pmatrix},$$

$$R^{(MIN)} = \begin{pmatrix} 0.635 & 0.054 & 0.219 & 0.016 \\ 0.659 & 0.480 & 0.156 & 0.235 \\ 0.247 & 0.035 & 0.313 & 0.010 \\ 0.408 & 0.057 & 0.100 & 0.405 \end{pmatrix},$$

$$R^{(MAX)} = \begin{pmatrix} 0.087 & 0.014 & 0.062 & 0.012 & 0.133 & 0.229 & 0.085 & 0.027 \\ 0.298 & 0.122 & 0.143 & 0.169 & 0.256 & 0.327 & 0.117 & 0.621 \\ 0.034 & 0.007 & 0.035 & 0.006 & 0.035 & 0.112 & 0.091 & 0.014 \\ 0.288 & 0.053 & 0.138 & 0.107 & 0.234 & 0.298 & 0.113 & 0.745 \\ 0.058 & 0.012 & 0.032 & 0.01 & 0.137 & 0.236 & 0.03 & 0.022 \\ 0.017 & 0.004 & 0.01 & 0.003 & 0.018 & 0.105 & 0.01 & 0.007 \\ 0.019 & 0.004 & 0.011 & 0.004 & 0.02 & 0.028 & 0.099 & 0.008 \\ 0.285 & 0.053 & 0.136 & 0.04 & 0.23 & 0.28 & 0.111 & 0.81 \end{pmatrix}.$$

The following R matrices correspond to the hyperexponential arrivals:

$$R^{(CNV)} = \begin{pmatrix} 0.459 & 0.336 & 0.233 & 0.086 \\ 0.173 & 0.382 & 0.260 & 0.092 \\ 0.106 & 0.078 & 0.317 & 0.021 \\ 0.214 & 0.165 & 0.122 & 0.8 \end{pmatrix}, \quad R^{(MXT)} = \begin{pmatrix} 0.743 & 0.353 & 0.069 & 0.032 \\ 0.174 & 0.601 & 0.053 & 0.024 \\ 0.073 & 0.036 & 0.210 & 0.010 \\ 0.207 & 0.111 & 0.062 & 0.691 \end{pmatrix},$$

$$R^{(MIN)} = \begin{pmatrix} 0.764 & 0.043 & 0.166 & 0.013 \\ 0.359 & 0.604 & 0.087 & 0.158 \\ 0.268 & 0.037 & 0.517 & 0.011 \\ 0.308 & 0.043 & 0.073 & 0.566 \end{pmatrix},$$

$$R^{(MAX)} = \begin{pmatrix} 0.312 & 0.022 & 0.128 & 0.015 & 0.201 & 0.223 & 0.09 & 0.036 \\ 0.147 & 0.307 & 0.063 & 0.192 & 0.121 & 0.145 & 0.048 & 0.381 \\ 0.104 & 0.018 & 0.227 & 0.012 & 0.070 & 0.195 & 0.157 & 0.028 \\ 0.143 & 0.025 & 0.061 & 0.298 & 0.100 & 0.136 & 0.047 & 0.554 \\ 0.118 & 0.020 & 0.049 & 0.013 & 0.378 & 0.288 & 0.036 & 0.032 \\ 0.076 & 0.013 & 0.032 & 0.009 & 0.051 & 0.356 & 0.023 & 0.020 \\ 0.078 & 0.013 & 0.032 & 0.009 & 0.052 & 0.059 & 0.332 & 0.021 \\ 0.139 & 0.024 & 0.059 & 0.017 & 0.098 & 0.116 & 0.046 & 0.834 \end{pmatrix}.$$

The graphs of the steady-state probability vectors at arrival and at arbitrary epochs are plotted in Figures 2.7 and 2.8, respectively

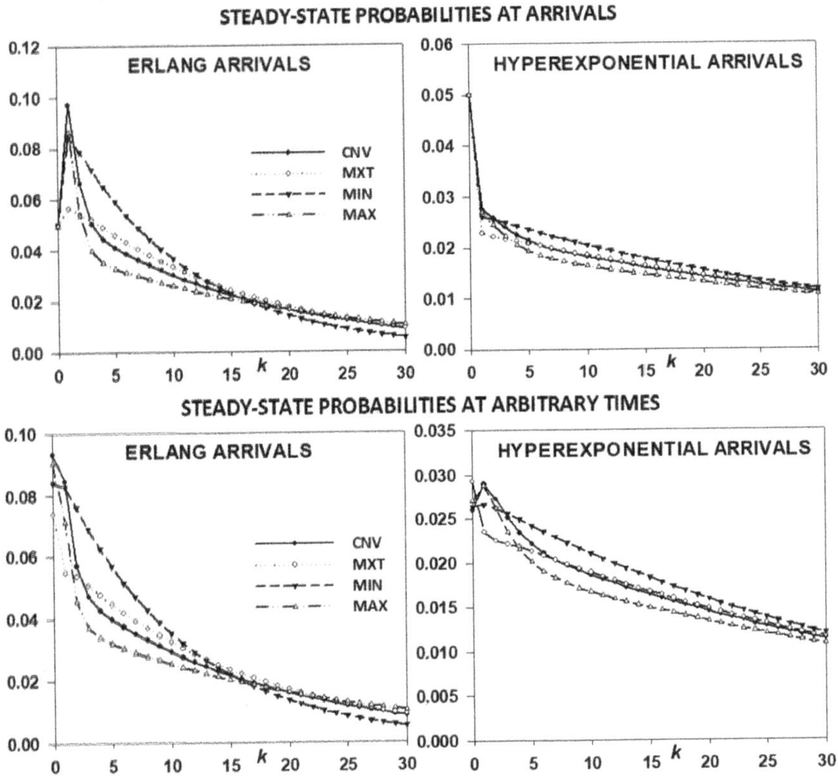

Figure 2.7. *Steady-state vectors at arrival (top) and arbitrary epochs for example 2.7*

The graphs of key measures under various scenarios are plotted in Figure 2.8.

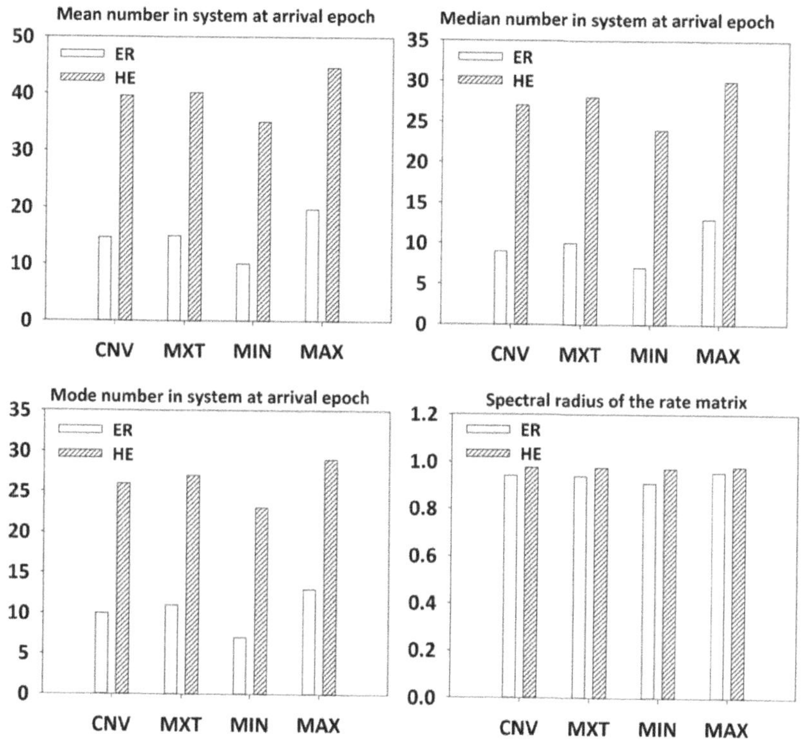

Figure 2.8. *Selected measures under various scenarios for example 2.7*

EXAMPLE 2.8.– In example 2.7, we looked at $PH/PH/1$-type queues embedded at arrivals and among other things and calculated the rate matrices. In this example, we look at one scenario and approach the analysis via embedded at departures, namely, using $M/G/1$-type tools. Using the tools of section 7.2 of Volume 1, verify that for Erlang of order 5 arrivals and with service times assumed to be the convolution of Erlang of order 2 having parameter 2 and hyperexponential of order 2 with parameters 10 and 1 with probabilities 0.9 and 0.1, respectively, the G and K matrices are given by:

$$G = \begin{pmatrix} 0.1037 & 0.1878 & 0.2312 & 0.2425 & 0.2348 \\ 0.0244 & 0.1478 & 0.2421 & 0.2881 & 0.2976 \\ 0.0309 & 0.0802 & 0.2166 & 0.3143 & 0.3580 \\ 0.0372 & 0.0981 & 0.1630 & 0.3034 & 0.3983 \\ 0.0414 & 0.1119 & 0.1902 & 0.2596 & 0.3969 \end{pmatrix},$$

$$K = \begin{pmatrix} 0.1037 & 0.1878 & 0.2312 & 0.2425 & 0.2348 \\ 0.1037 & 0.1878 & 0.2312 & 0.2425 & 0.2348 \\ 0.1037 & 0.1878 & 0.2312 & 0.2425 & 0.2348 \\ 0.1037 & 0.1878 & 0.2312 & 0.2425 & 0.2348 \\ 0.1037 & 0.1878 & 0.2312 & 0.2425 & 0.2348 \end{pmatrix}.$$

The invariant probability vectors of G and K are, respectively, given by:

$$\boldsymbol{g} = (0.0387, 0.1086, 0.1948, 0.2852, 0.3727),$$

$$\boldsymbol{\kappa} = (0.1037, 0.1878, 0.2312, 0.2425, 0.2348).$$

Note that $K = \boldsymbol{e}\,\boldsymbol{G}$ as is to be the case, and this results in $\boldsymbol{\kappa} = \boldsymbol{\alpha}\,\boldsymbol{G}$.

Now, we look at some special cases of $GI/PH/1$ queue.

2.2. *GI/M/1* queue

Here, the services are exponential and so $\boldsymbol{\beta} = 1$ and $S = (-\mu)$. It is easy to verify that (note that $f^*(s)$ is the LT of the PDF of the inter-arrival times):

$$a_k = \int_0^\infty e^{-\mu t}\frac{(\mu t)^k}{k!}dF(t),\ k \geq 0, \ \text{ and } b_k = b_k(\infty) = 1 - \sum_{i=0}^{k} a_i, k \geq 0. \quad [2.34]$$

In this case, R is a scalar and is given by ξ, where ξ is obtained as a unique solution in $(0, 1)$ of the equation:

$$\xi = f^*(\mu(1-\xi)). \quad [2.35]$$

To see this, R is the minimal non-negative solution to:

$$R = \sum_{n=0}^{\infty} R^n a_n \ \Rightarrow \ R = \int_0^\infty e^{-\mu t}\frac{(\mu R t)^k}{k!}dF(t) \ \Rightarrow \ R = f^*(\mu(1-R)). \quad [2.36]$$

Under the stability condition that $\lambda < \mu$, the steady-state probabilities at arrivals, $\{x_k\}$, and at arbitrary times, $\{y_k\}$, are given by:

$$x_k = (1-\xi)\xi^k, \ \ k \geq 0. \quad [2.37]$$

$$y_k = \begin{cases} 1-\rho, \ \ k = 0, \\ \rho(1-\xi)\xi^{k-1}, \ \ k \geq 1. \end{cases} \quad [2.38]$$

Verify that the various means are as follows:

$$L_s^{(A)} = \frac{\xi}{1-\xi}, \quad L_s = \frac{\rho}{1-\xi}, \quad W_{AQ} = \frac{\xi}{\mu(1-\xi)}, \quad W_s = \frac{1}{\mu(1-\xi)}.$$

The stationary waiting time in the queue (at arrivals) can be shown to be exponential with parameter $\mu(1-\xi)$. Toward this end, we first note that the LT of the waiting time in the queue is reduced to:

$$w_{AQ}^*(s) = (1-\xi)\sum_{k=0}^{\infty} \xi^k \left(\frac{\mu}{(s+\mu)}\right)^k, \quad Re(s) \geq 0. \qquad [2.39]$$

On noting that the waiting time in the queue (at arrivals), say, T_q, takes the value 0 with probability $1-\xi$, and applying result 4.12 in Volume 1 since we have an infinite (geometric) mixture of exponentials, we see that the probability density function (PDF) and the CDF of the waiting time distribution at arrivals are of the form:

$$f_q(t) = \begin{cases} (1-\xi), & t = 0, \\ \mu\xi(1-\xi)e^{-\mu(1-\xi)\,t}, & t > 0, \end{cases} \qquad [2.40]$$

$$F_q(t) = P(T_q \leq t) = 1 - \xi e^{-\mu(1-\xi)\,t}, \quad t \geq 0.$$

REMARK 2.12.– It is worth pointing out that in the special case when the underlying matrices are of dimension 2, Garimella and Rumyantsev (2018) provide an explicit expression for the rate matrix R using the Cayley-Hamilton theorem.

EXAMPLE 2.9.– In this example, we take the service times to be exponential with parameter μ. For arrivals we look at five different types: (1) Constant (CTA) arrivals; (2) Arbitrary (discrete) arrivals (ABA) taking values 0.3, 1.0, 2.5 and 3.5, with probabilities, respectively, given by 0.50, 0.30, 0.15 and 0.05; (3) Uniform (UFA) arrivals on $(0.1, 1.9)$; (4) Erlang of order 5 (ER5) with parameter 5; and (5) Hyperexponential (HE3) arrivals with parameters 3.8, 0.95 and 0.19 with probabilities, respectively, given by 0.6, 0.3 and 0.1. These five arrival processes, covering a wide range of scenarios, have a mean of 1. The standard deviations are, respectively, given by 0, 0.9460, 0.5196, 0.4472 and 2.2996. We vary the traffic intensity ρ from 0.01 to 0.99, and hence the service rate will be taken as $\mu = \frac{1}{\rho}$.

The graphs of the spectral radius as well as the mean number in system at arrival epochs as functions of ρ under various scenarios are, respectively, given in Figures 2.9 and 2.10. Also, the plots of the PDF of the waiting time (in the queue) at arrivals (see equation [2.40]) are plotted in Figure 2.11. In the plot, we also display the mean, the median and the mode (in that order) of the number in the system at arrival epochs.

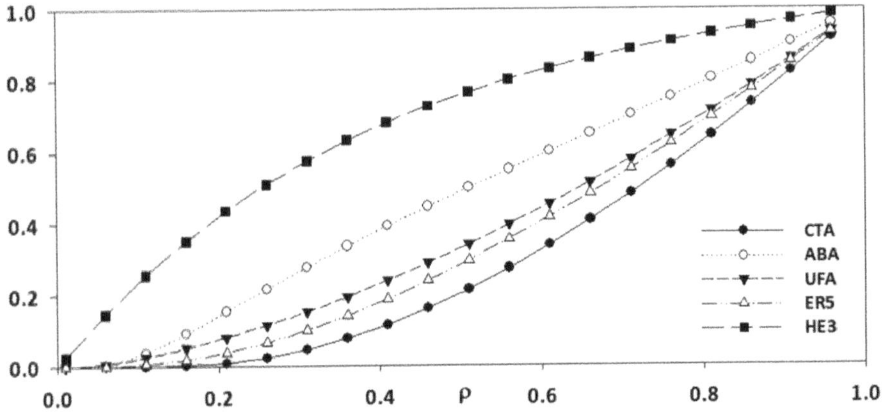

Figure 2.9. *Caudal characteristic curves under various scenarios for example 2.9*

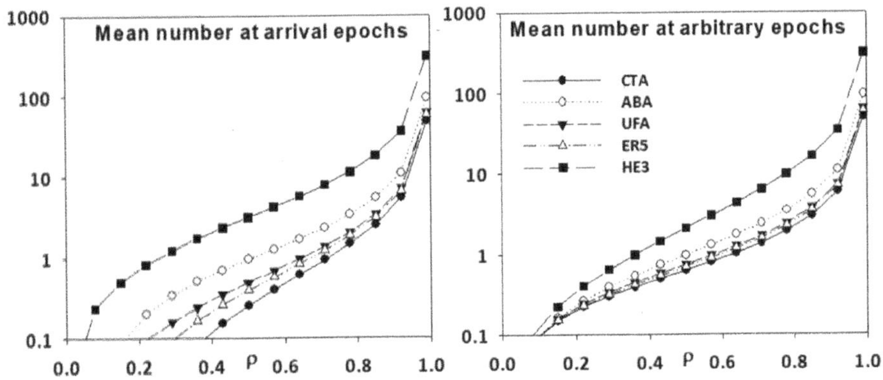

Figure 2.10. *Mean number in system at arrival epochs under various scenarios for example 2.9*

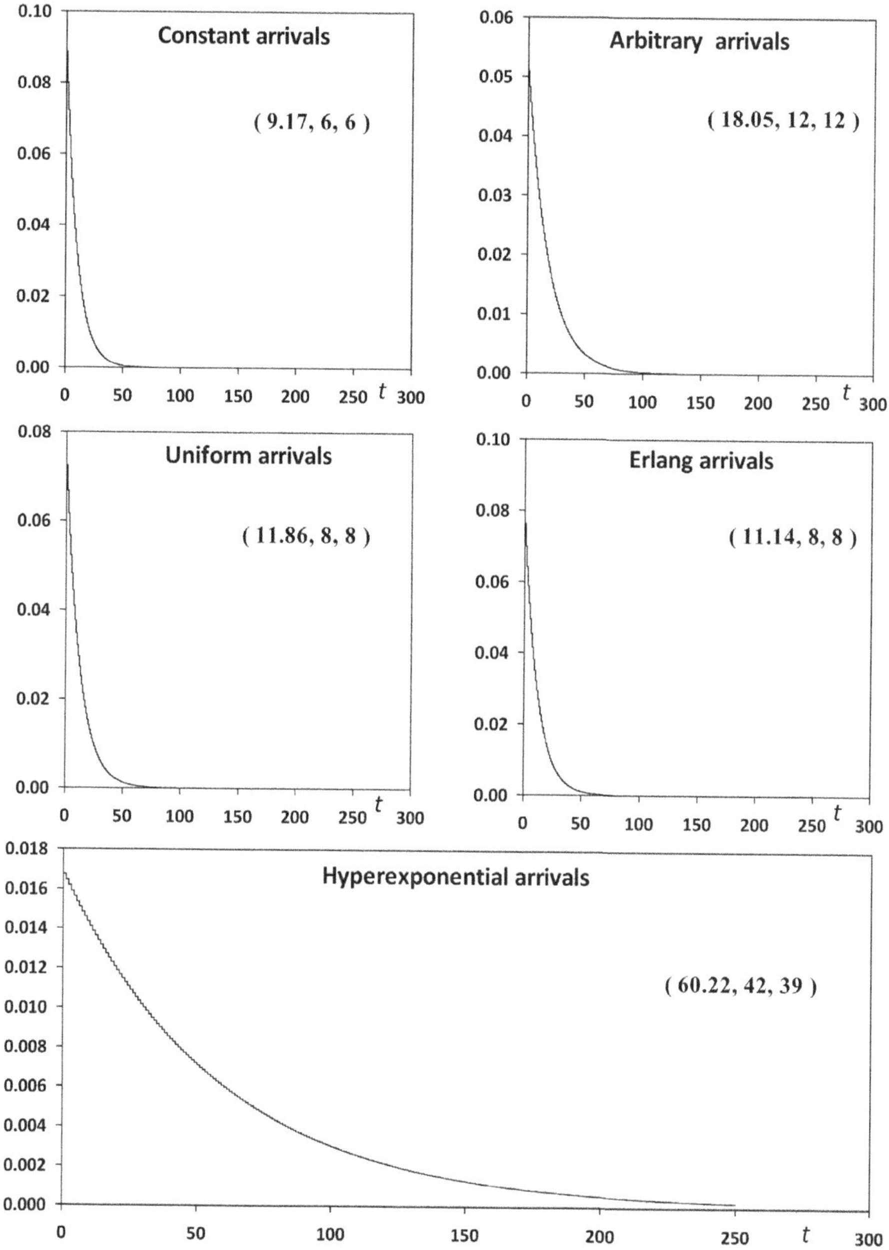

Figure 2.11. *Probability density functions of the waiting times at arrival epochs under various scenarios for example 2.9*

2.3. *M/PH/1* queue

In the case of Poisson arrivals and phase type services with representation $(\boldsymbol{\beta}, S)$ of order ν, we have:

$$A_n = \lambda \int_0^\infty P(n,t)e^{-\lambda t}dt, \quad n \geq 0. \tag{2.41}$$

THEOREM 2.3.– The matrix R is obtained explicitly as follows:

$$R = \lambda[\lambda I - \lambda \boldsymbol{e}\boldsymbol{\beta} - S]^{-1}. \tag{2.42}$$

PROOF.– Suppose we define the generating function $A(z) = \sum\limits_{n=0}^{\infty} z^n A_n$. Then, applying equation [4.36] in Volume 1 for the current model, we get:

$$A(z) = \lambda[\lambda I - S - z\boldsymbol{S^0}\boldsymbol{\beta}]^{-1} = \lambda(\lambda I - S)^{-1}\big[I - z\boldsymbol{S^0}\boldsymbol{\beta}(\lambda I - S)^{-1}\big]^{-1}. \tag{2.43}$$

Using the identification of coefficients of z^n in equation [2.43], it is obvious that:

$$A_n = \lambda(\lambda I - S)^{-1}\big[\boldsymbol{S^0}\boldsymbol{\beta}(\lambda I - S)^{-1}\big]^n, \quad n \geq 0. \tag{2.44}$$

Defining $f^*(\lambda) = \boldsymbol{\beta}(\lambda I - S)^{-1}\boldsymbol{S^0}$, it is easy to see that A_n can be rewritten as:

$$A_n = \begin{cases} \lambda(\lambda I - S)^{-1}, & n = 0, \\ \big[f^*(\lambda)\big]^{n-1} A_1, & n \geq 1. \end{cases} \tag{2.45}$$

It is worth pointing out that A_1 can also be written in terms of A_0 as:

$$A_1 = \frac{1}{\lambda} A_0 \boldsymbol{S^0}\boldsymbol{\beta} A_0. \tag{2.46}$$

First, verify that $R = \sum\limits_{n=0}^{\infty} R^n A_n$ implies that:

$$R = A_0 + \frac{1}{\lambda}[I - f^*(\lambda)R]^{-1} R A_0 \boldsymbol{S^0}\boldsymbol{\beta} A_0. \tag{2.47}$$

Simple matrix manipulations on equation [2.47] with the help of the fact that $\boldsymbol{\beta} A_0 \boldsymbol{S^0} = \lambda f^*(\lambda)$, we get:

$$R\boldsymbol{S^0} = [I - f^*(\lambda)R]^{-1} A_0 \boldsymbol{S^0}. \tag{2.48}$$

Now post-multiplying equation [2.47] by $(\lambda I - S)\boldsymbol{e}$ and using equation [2.48], we obtain:

$$R\boldsymbol{S^0} = \lambda\boldsymbol{e}. \qquad [2.49]$$

Using the fact that R and $[I - f^*(\lambda)R]^{-1}$ commute, it is easy to verify that equation [2.47], with the help of equations [2.48] and [2.49], yields:

$$R = A_0 + R\boldsymbol{e}\boldsymbol{\beta}A_0, \qquad [2.50]$$

from which the stated result follows immediately. □

REMARK 2.13.– The explicit expression for R, as given in equation [2.42], was obtained using the quasi-birth-and-death (QBD) approach in Neuts (1981), which we will see in the next chapter.

The explicit expression for R enables us to get explicit results for the stationary vectors. Note that the steady-state vectors $\{x_k\}$, at arrival epochs, and $\{y_k\}$, at arbitrary epochs, are equal to each other and are given by:

$$\boldsymbol{x}_k = \boldsymbol{y}_k = (1-\rho)\boldsymbol{\beta}R^k, \ k \geq 0, \qquad [2.51]$$

In the above equation, we used the fact $\boldsymbol{\beta}(I - R)^{-1}\boldsymbol{e} = \dfrac{1}{1-\rho}$.

Suppose that T_q stands for the waiting time in the queue of a customer at an arrival epoch (or at an arbitrary time due to Poisson arrivals). Since T_q is expressed as the infinite mixture of (self-)convolution of distributions (see equation [1.83]), using remark 4.7 of Volume 1, we get that T_q to follow a CPH with representation $(\lambda\boldsymbol{\beta}(-S)^{-1}, S + \lambda\boldsymbol{S^0}\boldsymbol{\beta}(-S)^{-1})$ of order ν (see theorem 2.2.7 in Neuts (1981)). Thus, we have:

$$W_q(t) = P(T_q \leq t) = 1 - \lambda\boldsymbol{\beta}(-S)^{-1}e^{[S+\lambda\boldsymbol{S^0}\boldsymbol{\beta}(-S)^{-1}]t}\boldsymbol{e}, \ t \geq 0. \qquad [2.52]$$

It can be verified that the mean waiting time, μ'_{W_q}, in the queue is obtained as follows. From the definition of the mean of a CPH distribution, we have:

$$\mu'_{W_q} = \lambda\boldsymbol{\beta}(-S)^{-1}[-(S + \lambda\boldsymbol{S^0}\boldsymbol{\beta}(-S)^{-1})]^{-1}\boldsymbol{e}.$$

Noting that:

$$[-(S + \lambda\boldsymbol{S^0}\boldsymbol{\beta}(-S)^{-1})]^{-1} = (-S)^{-1}\Big[I + \frac{\lambda}{1-\rho}\boldsymbol{S^0}\boldsymbol{\beta}(-S)^{-2}\Big],$$

we get:

$$\mu'_{W_q} = \frac{\lambda}{(1-\rho)}\boldsymbol{\beta}(-S)^{-2}\boldsymbol{e}.$$

Using Little's law (or with direct calculations), we get the mean number, L_q, of customers in the queue to be:

$$L_q = \frac{\lambda^2}{(1-\rho)}\boldsymbol{\beta}(-S)^{-2}\boldsymbol{e}.$$

With the knowledge of μ'_{W_q} and L_q, one can get the mean waiting time, μ'_{W_s}, and the mean number, L_s, in the system to be:

$$\mu'_{W_s} = \mu'_{W_q} + \frac{1}{\mu}, \quad L_s = L_q + \rho.$$

The above expressions agree with the known results (see Bhat (2015)) for the $M/G/1$ queue available in the literature.

REMARK 2.14.– The mean waiting times and the mean numbers are explicitly known in terms of input parameters and hence they can serve as internal accuracy checks in the numerical computation of the waiting time distributions and the steady-state probability vector.

EXAMPLE 2.10.– Suppose that the customers arrive according to a Poisson process with a rate of 10 per hour to a single-server system. If the service times are Erlang of order 3 such that the rate of service is 12 per hour, then the rate matrix R for this model is given by:

$$R = \begin{pmatrix} 0.454 & 0.355 & 0.278 \\ 0.176 & 0.355 & 0.278 \\ 0.099 & 0.077 & 0.278 \end{pmatrix}.$$

To see this, we list some key facts:

1) Using the facts that $\lambda = 10$, $\boldsymbol{\beta} = (1, 0, 0)$ and

$$S = \begin{pmatrix} -36 & 36 & 0 \\ 0 & -36 & 36 \\ 0 & 0 & -36 \end{pmatrix},$$

verify that:

$$(\lambda I - S - \lambda \boldsymbol{e}\boldsymbol{\beta}) = \begin{pmatrix} 36 & -36 & 0 \\ -10 & 46 & -36 \\ -10 & 0 & 46 \end{pmatrix},$$

and hence:

$$R = \begin{pmatrix} \frac{2645}{5832} & \frac{115}{324} & \frac{5}{18} \\ \frac{1025}{5832} & \frac{115}{324} & \frac{5}{18} \\ \frac{575}{5832} & \frac{25}{324} & \frac{5}{18} \end{pmatrix} = \begin{pmatrix} 0.454 & 0.355 & 0.278 \\ 0.176 & 0.355 & 0.278 \\ 0.099 & 0.077 & 0.278 \end{pmatrix}.$$

EXAMPLE 2.11.– Suppose that the arrivals occur according to a Poisson process with a rate of 1 per unit of time. We look at several service times that are modeled using CPH to see how the waiting time distribution in the queue as well as in the system behave as certain parameters of the $M/PH/1$ queue are varied. Specifically, we consider the following distributions for the services:

1) $ER\ r$: Erlang of order r, for $r = 1, 2$ and 5.

2) $HE\ 1$: Hyperexponential with parameters 20, 5 and 1 with mixing probabilities, respectively, given by 0.6, 0.3 and 0.1.

3) $HE\ 2$: Hyperexponential with parameters 20, 10, 5 and 1 with mixing probabilities, respectively, given by 0.6, 0.2, 0.12 and 0.08.

4) $HE\ 3$: Hyperexponential with parameters 50, 25, 12.5, 5 and 1 with mixing probabilities, respectively, given by 0.6, 0.25, 0.1, 0.04 and 0.01.

5) (MXT): Mixture (with equal probabilities) of Erlang of order 2 with parameter 2 and hyperexponential of order 2 with parameters 10 and 1 with probabilities, respectively, given by 0.9 and 0.1.

6) (MIN): Minimum of exponential with parameter 1, Erlang of order 2 with parameter 2 and hyperexponential of order 2 with parameters 10 and 1 with probabilities, respectively, given by 0.9 and 0.1.

7) (MAX): Maximum of Erlang of order 2 with parameter 2 and hyperexponential of order 2 with parameters 10 and 1 with probabilities, respectively, given by 0.9 and 0.1.

We look at four values for the traffic intensity, $\rho = \frac{\lambda}{\mu}$, namely, $\rho = 0.5, 0.8, 0.95$ and 0.99. In order to achieve these and since λ is fixed at 1, we normalize the service times CPH so that the service mean = ρ.

The plots of the PDF of each of the nine PH distributions considered for the services are displayed in Figure 2.12. Selected measures of the waiting times in the queue and in the system are, respectively, displayed in Tables 2.1 and 2.2.

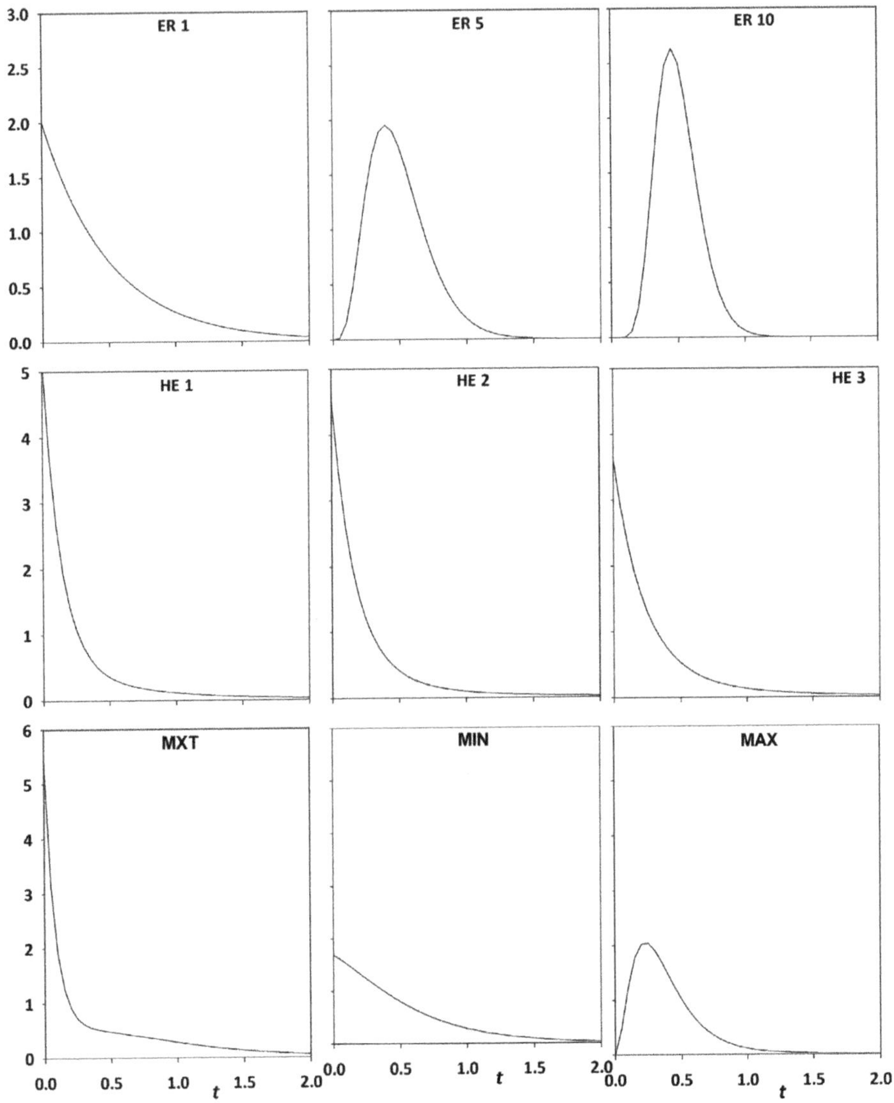

Figure 2.12. *Probability density functions of the CPH for services for example 2.11*

ρ	PH	σ_S	μ'_{W_q}	σ_{W_q}	25th	50th	75th	95th
0.5	$ER\,1$	0.500	0.500	0.866	0	0	0.693	2.301
	$ER\,5$	0.224	0.300	0.480	0	0	0.457	1.287
	$ER\,10$	0.158	0.275	0.431	0	0	0.428	1.158
	$HE\,1$	1.150	1.572	3.153	0	0	1.691	8.172
	$HE\,2$	1.269	1.862	3.819	0	0	1.902	9.859
	$HE\,3$	1.595	2.795	7.402	0	0	1.469	16.938
	MXT	0.597	0.607	1.040	0	0	0.862	2.762
	MIN	0.470	0.471	0.811	0	0	0.657	2.152
	MAX	0.696	0.734	1.587	0	0	0.649	3.992
0.8	$ER\,1$	0.800	3.200	3.919	0.258	1.880	4.652	11.094
	$ER\,5$	0.358	1.920	2.263	0.258	1.196	2.779	6.453
	$ER\,10$	0.253	1.760	2.055	0.258	1.108	2.544	5.875
	$HE\,1$	1.840	10.061	13.331	0.258	5.020	14.672	37.148
	$HE\,2$	2.031	11.914	15.995	0.258	5.773	17.406	44.453
	$HE\,3$	2.553	17.889	28.304	0.258	5.039	24.156	76.641
	MXT	0.955	3.882	4.730	0.258	2.309	5.641	13.406
	MIN	0.751	3.012	3.679	0.253	1.776	4.375	10.422
	MAX	1.114	4.700	6.509	0.253	2.018	6.734	18.031
0.95	$ER\,1$	0.950	18.05	18.976	4.492	12.195	25.359	55.938
	$ER\,5$	0.425	10.830	11.265	4.492	7.363	15.180	33.320
	$ER\,10$	0.300	9.928	10.300	4.492	6.762	13.906	30.469
	$HE\,1$	2.185	56.750	61.098	4.492	37.734	80.234	178.906
	$HE\,2$	2.412	67.204	72.658	4.492	44.570	95.078	212.500
	$HE\,3$	3.031	100.905	115.777	16.266	63.125	144.434	333.203
	MXT	1.135	21.898	22.988	16.266	14.805	30.762	67.812
	MIN	0.892	16.987	17.846	4.234	11.480	23.859	52.656
	MAX	1.322	26.511	28.960	4.234	17.398	37.578	84.453
0.99	$ER\,1$	0.990	98.010	98.995	27.484	67.617	136.230	295.703
	$ER\,5$	0.443	58.806	59.266	27.484	40.625	81.680	177.148
	$ER\,10$	0.313	53.905	54.300	27.484	37.246	74.883	162.305
	$HE\,1$	2.277	308.148	312.817	27.484	212.109	428.906	932.812
	$HE\,2$	2.514	364.913	370.780	27.484	250.977	508.203	1105.469
	$HE\,3$	3.159	547.903	564.299	145.312	374.219	765.625	1673.828
	MXT	1.182	118.905	120.063	145.312	82.031	165.234	358.594
	MIN	0.930	92.238	93.151	25.875	63.633	128.223	278.125
	MAX	1.378	143.955	146.600	25.875	98.906	200.586	436.719

Table 2.1. *Selected measures for the waiting time in queue under various scenarios for example 2.11*

ρ	PH	σ_S	μ'_{W_s}	σ_{W_s}	25th	50th	75th	95th
	$ER\,1$	0.500	1	1	0.288	0.693	1.386	2.996
	$ER\,5$	0.224	0.800	0.529	0.288	0.658	1.013	1.842
	$ER\,10$	0.158	0.775	0.459	0.288	0.640	0.952	1.688
	$HE\,1$	1.150	2.072	3.356	0.288	0.581	2.551	9.062
0.5	$HE\,2$	1.269	2.362	4.025	0.288	0.541	2.789	10.781
	$HE\,3$	1.595	3.295	7.572	0.288	0.566	2.266	17.875
	MXT	0.597	1.107	1.199	0.288	0.751	1.590	3.484
	MIN	0.470	0.971	0.937	0.288	0.694	1.334	2.832
	MAX	0.696	1.234	1.733	0.288	0.578	1.307	4.828
	$ER\,1$	0.800	4	4	1.15	2.772	5.545	11.984
	$ER\,5$	0.358	2.720	2.291	1.150	2.025	3.609	7.281
	$ER\,10$	0.253	2.560	2.071	1.150	1.925	3.359	6.695
	$HE\,1$	1.840	10.861	13.457	1.150	5.98	15.641	38.125
0.8	$HE\,2$	2.031	12.714	16.123	1.150	6.742	18.375	45.391
	$HE\,3$	2.553	18.689	28.419	1.150	5.977	25.156	77.656
	MXT	0.955	4.682	4.826	1.150	3.217	6.551	14.312
	MIN	0.751	3.812	3.755	1.147	2.662	5.262	11.297
	MAX	1.114	5.500	6.604	1.147	2.906	7.672	18.969
	$ER\,1$	0.950	19	19	5.465	13.172	26.344	56.953
	$ER\,5$	0.425	11.780	11.273	5.465	8.324	16.133	34.297
	$ER\,10$	0.300	10.877	10.305	5.465	7.715	14.859	31.445
	$HE\,1$	2.185	57.700	61.137	5.465	38.711	81.211	179.883
0.95	$HE\,2$	2.412	68.154	72.698	5.465	45.547	96.094	213.477
	$HE\,3$	3.031	101.855	115.817	5.465	64.141	145.508	334.375
	MXT	1.135	22.848	23.016	5.465	15.789	31.738	68.750
	MIN	0.892	17.937	17.868	5.211	12.453	24.844	53.594
	MAX	1.322	27.461	28.990	5.211	18.383	38.574	85.469
	$ER\,1$	0.990	99	99	28.469	68.633	137.207	296.484
	$ER\,5$	0.443	59.796	59.268	28.469	41.602	82.695	178.125
	$ER\,10$	0.313	54.895	54.301	28.469	38.223	75.859	163.281
	$HE\,1$	2.277	309.138	312.825	28.469	213.086	429.883	933.594
0.99	$HE\,2$	2.514	365.903	370.789	28.469	252.051	508.984	1105.469
	$HE\,3$	3.159	548.893	564.308	28.469	375.195	766.406	1675.781
	MXT	1.182	119.895	120.069	28.469	83.047	166.309	359.375
	MIN	0.930	93.228	93.156	26.875	64.648	129.199	279.297
	MAX	1.378	144.945	146.607	26.875	99.922	201.562	437.500

Table 2.2. *Selected measures for the waiting time in system under various scenarios for example 2.11*

In Figures 2.13 and 2.14, respectively, we display the CDFs of the waiting time in the queue and the waiting time in the system of a customer at an arrival epoch.

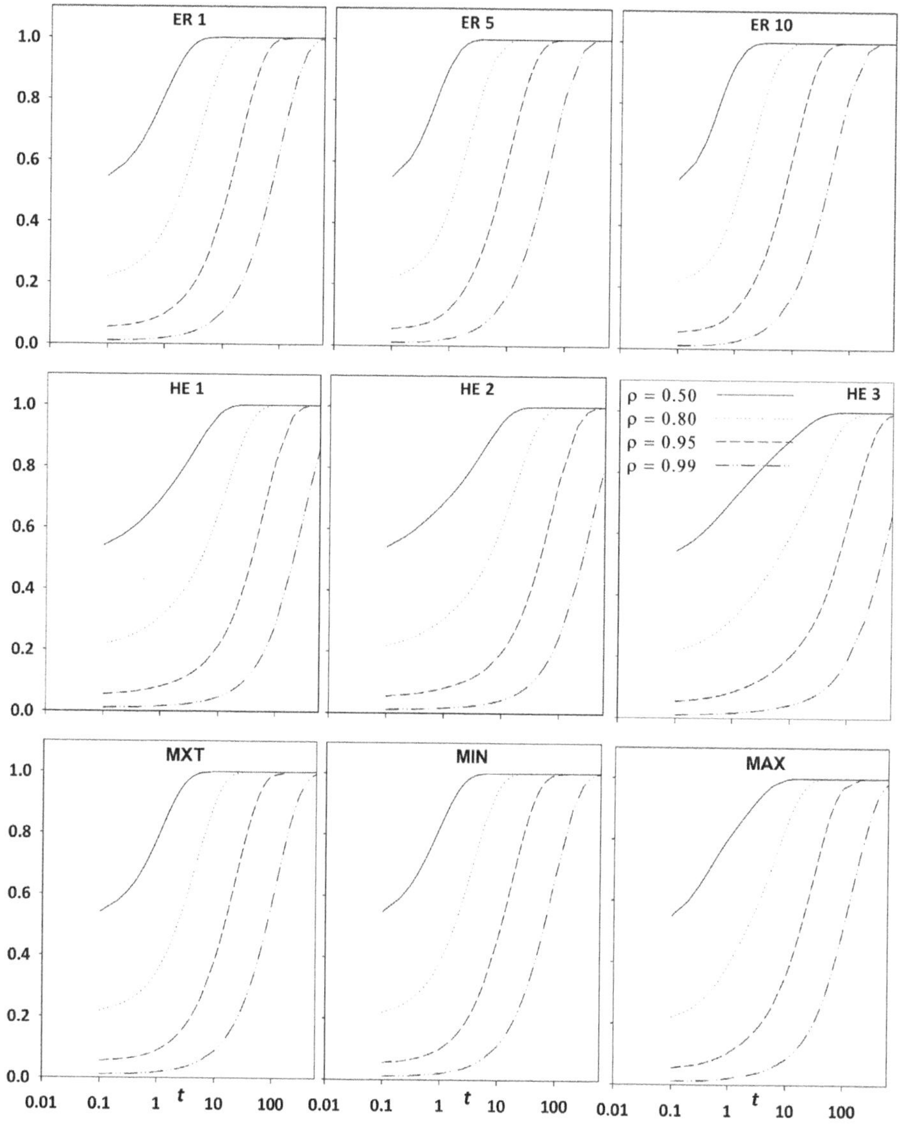

Figure 2.13. *Probability distribution function of the waiting time in the queue for example 2.11*

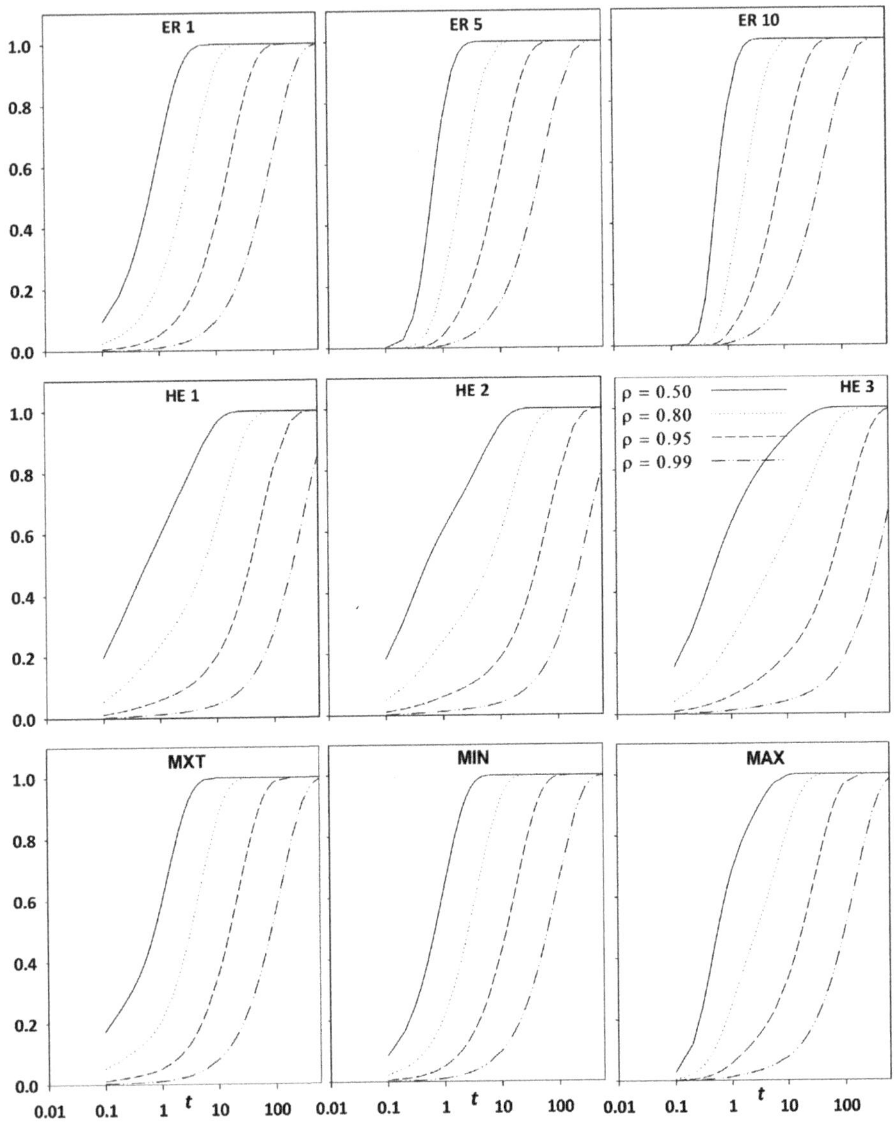

Figure 2.14. *Probability distribution function of the waiting time in the system for example 2.11*

2.4. *M/M/1* queue

Revisiting the simplest single-server queueing model in continuous-time in the context of embedded at arrival epochs, we briefly summarize the key items here. Note that all the quantities we are dealing with are scalars.

1) $A_n = \lambda \int\limits_0^\infty e^{-\mu t} \frac{(\mu t)^n}{n!} e^{-\lambda t} dt = \frac{\lambda}{\lambda+\mu}\Big(\frac{\mu}{\lambda+\mu}\Big)^n, \; n \geq 0.$

2) $B_n = 1 - \sum\limits_{i=0}^{n} A_i = \Big(\frac{\mu}{\lambda+\mu}\Big)^{n+1}, \; n \geq 0.$

3) $R = \rho.$

4) The steady-state probabilities are: $x_k = y_k = (1-\rho)\rho^k, \; k \geq 0.$

5) The stationary waiting time distribution in queue of an arrival (see equation [2.52]) is given by $W_q(t) = P(T_q \leq t) = 1 - \rho e^{(\mu-\lambda)t}, \; t \geq 0.$

2.5. Exercises

Exercise 2.1. Prove the expression given in equation [2.7].

Exercise 2.2. Prove the expression given in equation [2.8].

Exercise 2.3. Prove the expression given in equation [2.9].

Exercise 2.4. Prove the expression given in equation [2.14].

Exercise 2.5. Prove the expression given in equation [2.27].

Exercise 2.6. Prove the expression given in equation [2.28].

Exercise 2.7. Prove the expressions mentioned in result 2.1.

Exercise 2.8. Prove the expressions mentioned in result 2.2.

Exercise 2.9. Prove the expression given in equation [2.31].

Exercise 2.10. Prove the expression given in equation [2.33].

Exercise 2.11. Prove the expression given in equation [2.39].

Exercise 2.12. Prove the expression given in equation [2.44].

Exercise 2.13. Develop a code for $GI/PH/1$ queue and verify the results of example 2.9.

Exercise 2.14. Prove the expression given in equation [2.47].

Exercise 2.15. Prove the expression given in equation [2.48].

Exercise 2.16. Prove the expression given in equation [2.52].

Exercise 2.17. For the $M/PH/1$ model, show that:

$$\mu'_{W_q} = \frac{\lambda}{(1-\rho)}\beta(-S)^{-2}e.$$

Exercise 2.18. Develop a code for $M/PH/1$ queue and verify the results of example 2.11.

Exercise 2.19. Prove the expression given in equation [2.40].

Exercise 2.20. Referring to example 2.1, show that (a) the spectral radius of R is 0.7331; (b) the mean, the median and the mode of $\{x_k\}$ are, respectively, given by 3.2341, 2 and 2; and (c) the mean, the median and the mode of $\{y_k\}$ are, respectively, given by 3.7363, 3 and 0.

Exercise 2.21. Referring to example 2.2, show that (a) the spectral radius of R is 0.8489; (b) the mean, the median and the mode of the number in the system at arrival epochs are, respectively, given by 6.0921, 4 and 4; and (c) the mean, the median and the mode of the number in the system at arbitrary epochs are, respectively, given by 6.4413, 5 and 0.

Exercise 2.22. Referring to example 2.3, show that (a) the spectral radius of R is 0.8455; (b) the mean, the median and the mode of the number in the system at arrival epochs are, respectively, given by 5.9202, 4 and 4; and (c) the mean, median and the mode of the number in the system at arbitrary epochs are, respectively, given by 6.2821, 4 and 0.

Exercise 2.23. Referring to example 2.4, show that (a) the spectral radius of R is 0.8574; (b) the mean, the median and the mode of the number in the system at arrival epochs are, respectively, given by 6.4707, 4 and 5; and (c) the mean, the median and the mode of the number in the system at arbitrary epochs are, respectively, given by 6.8020, 5 and 0.

Exercise 2.24. Referring to example 2.8, verify the matrices $G, K, \boldsymbol{g}$ and $\boldsymbol{\kappa}$.

Exercise 2.25. Suppose that the customers arrive singly every minute to a single-server system and get services that are exponentially distributed with a mean of 0.5 minute.

a) Find the steady-state probability, x_k, that an arriving customer will see exactly $k, k \geq 0$, customers (not including the arrival).

b) Find the steady-state probability, y_k, of seeing exactly $k, k \geq 0$, customers in the system at an arbitrary time.

Exercise 2.26. Suppose that the customers arrive singly every minute to a single-server system and get services that are exponentially distributed with a mean of 0.8 minute.

a) Find the steady-state probability, x_k, that an arriving customer will see exactly $k, k \geq 0$, customers (not including the arrival).

b) Find the steady-state probability, y_k, of seeing exactly $k, k \geq 0$, customers in the system at an arbitrary time.

Exercise 2.27. Suppose that the customers arrive singly every minute to a single-server system and get services that are exponentially distributed with a mean of 0.96 minute.

a) Find the steady-state probability, x_k, that an arriving customer will see exactly $k, k \geq 0$, customers (not including the arrival).

b) Find the steady-state probability, y_k, of seeing exactly $k, k \geq 0$, customers in the system at an arbitrary time.

Exercise 2.28. Suppose that the customers arrive singly and the inter-arrival times are $i.i.d$ with a common PMF with masses at 0.5, 1 and 2.5, with probabilities 0.3, 0.6 and 0.1, respectively. There is a single server in the system offering services that have an exponential distribution with a mean of 0.5 minute.

a) Find the steady-state probability, x_k, that an arriving customer will see exactly $k, k \geq 0$, customers (not including the arrival).

b) Find the steady-state probability, y_k, of seeing exactly $k, k \geq 0$, customers in the system at an arbitrary time.

Exercise 2.29. Suppose that the customers arrive singly and the inter-arrival times are $i.i.d$ with a common PMF with masses at 0.5, 1 and 2.5, with probabilities 0.3, 0.6 and 0.1, respectively. There is a single server in the system offering services that have an exponential distribution with a mean of 0.8 minute.

a) Find the steady-state probability, x_k, that an arriving customer will see exactly $k, k \geq 0$, customers (not including the arrival).

b) Find the steady-state probability, y_k, of seeing exactly $k, k \geq 0$, customers in the system at an arbitrary time.

Exercise 2.30. Suppose that the customers arrive singly and the inter-arrival times are $i.i.d$ with a common PMF with masses at 0.5, 1 and 2.5, with probabilities 0.3, 0.6 and 0.1, respectively. There is a single server in the system offering services that have an exponential distribution with a mean of 0.96 minute.

a) Find the steady-state probability, x_k, that an arriving customer will see exactly $k, k \geq 0$, customers (not including the arrival).

b) Find the steady-state probability, y_k, of seeing exactly $k, k \geq 0$, customers in the system at an arbitrary time.

Exercise 2.31. Suppose that the customers arrive singly and the inter-arrival times are uniformly distributed on the interval $(0.1, 1.9)$. There is a single server in the system offering services that have an exponential distribution with a mean of 0.5 minute.

a) Find the steady-state probability, x_k, that an arriving customer will see exactly $k, k \geq 0$, customers (not including the arrival).

b) Find the steady-state probability, y_k, of seeing exactly $k, k \geq 0$, customers in the system at an arbitrary time.

Exercise 2.32. Suppose that the customers arrive singly and the inter-arrival times are uniformly distributed on the interval $(0.1, 1.9)$. There is a single server in the system offering services that have an exponential distribution with a mean of 0.8 minute.

a) Find the steady-state probability, x_k, that an arriving customer will see exactly $k, k \geq 0$, customers (not including the arrival).

b) Find the steady-state probability, y_k, of seeing exactly $k, k \geq 0$, customers in the system at an arbitrary time.

Exercise 2.33. Suppose that the customers arrive singly and the inter-arrival times are uniformly distributed on the interval $(0.1, 1.9)$. There is a single server in the system offering services that have an exponential distribution with a mean of 0.96 minute.

a) Find the steady-state probability, x_k, that an arriving customer will see exactly $k, k \geq 0$, customers (not including the arrival).

b) Find the steady-state probability, y_k, of seeing exactly $k, k \geq 0$, customers in the system at an arbitrary time.

Exercise 2.34. Suppose that the customers arrive singly every minute to a single server system and get services that are distributed as Erlang of order 4 with parameter $25/6$ in each of the four states.

a) Find the mean, the standard deviation of the services and the traffic intensity.

b) Find the rate matrix R and its spectral radius.

c) Find the mean and the standard deviation of the number of customers in the system at the time of an arrival (not including the one arrived).

d) Find the mean and the standard deviation of the number of customers in the system at an arbitrary time.

e) Find the first six (marginal) steady-state probabilities of the number in the system that an arrival (not including the arrived one) sees.

f) Find the first six (marginal) steady-state probabilities of the number of customers in the system.

Exercise 2.35. Suppose that the customers arrive singly every minute to a single-server system and get services that are distributed as hyperexponential with parameters 2.18750, 1.09375, 0.43750 and 0.21875, respectively, with mixing probabilities 0.6, 0.25, 0.10 and 0.05.

a) Find the mean, the standard deviation of the services and the traffic intensity.

b) Find the rate matrix R and its spectral radius.

c) Find the mean and the standard deviation of the number of customers in the system at the time of an arrival (not including the one arrived).

d) Find the mean and the standard deviation of the number of customers in the system at an arbitrary time.

e) Find the first six (marginal) steady-state probabilities of the number in the system that an arrival (not including the arrived one) sees.

f) Find the first six (marginal) steady-state probabilities of the number of customers in the system.

Exercise 2.36. Suppose that the customers arrive singly every minute to a single-server system and get services that are modeled using a mixture of two Erlang and hyperexponential. The CPH representation is given by:

$$\boldsymbol{\beta} = (0.5, 0, 0.45, 0.05), \quad S = \begin{pmatrix} -1.23958 & 1.23958 & 0 & 0 \\ 0 & -1.23958 & 0 & 0 \\ 0 & 0 & -6.19792 & 0 \\ 0 & 0 & 0 & -0.61979 \end{pmatrix}.$$

a) Find the mean, the standard deviation of the services and the traffic intensity.

b) Find the rate matrix R and its spectral radius.

c) Find the mean and the standard deviation of the number of customers in the system at the time of an arrival (not including the one arrived).

d) Find the mean and the standard deviation of the number of customers in the system at an arbitrary time.

e) Find the first six (marginal) steady-state probabilities of the number in the system that an arrival (not including the arrived one) sees.

f) Find the first six (marginal) steady-state probabilities of the number of customers in the system.

Exercise 2.37. Suppose that the customers arrive singly every minute to a single-server system and get services that are modeled using a CPH obtained as a minimum of exponential with parameter 1, Erlang of order 2 with parameter 2 in each state and a hyperexponential with parameters 1.9 and 0.19 with mixing probabilities 0.9 and 0.1, respectively. The CPH representation (after normalized to get a specific mean of 0.96) is given by:

$$\boldsymbol{\beta} = (0.9, 0.1, 0, 0), \quad S = \begin{pmatrix} -1.58048 & 0 & 0.64509 & 0 \\ 0 & -1.02892 & 0 & 0.64509 \\ 0 & 0 & -1.58048 & 0 \\ 0 & 0 & 0 & -1.02892 \end{pmatrix}.$$

a) Find the mean, the standard deviation of the services and the traffic intensity.

b) Find the rate matrix R and its spectral radius.

c) Find the mean and the standard deviation of the number of customers in the system at the time of an arrival (not including the one arrived).

d) Find the mean and the standard deviation of the number of customers in the system at an arbitrary time.

e) Find the first six (marginal) steady-state probabilities of the number in the system that an arrival (not including the arrived one) sees.

f) Find the first six (marginal) steady-state probabilities of the number of customers in the system.

Exercise 2.38. Referring to exercise 2.34, suppose that the customers arrive singly and the inter-arrival times are $i.i.d$ with a common PMF with masses at 0.5, 1 and 2.5, with probabilities 0.3, 0.6 and 0.1, respectively. All other parameters remain the same.

a) Find the mean, the standard deviation of the services and the traffic intensity.

b) Find the rate matrix R and its spectral radius.

c) Find the mean and the standard deviation of the number of customers in the system at the time of an arrival (not including the one arrived).

d) Find the mean and the standard deviation of the number of customers in the system at an arbitrary time.

e) Find the first six (marginal) steady-state probabilities of the number in the system that an arrival (not including the arrived one) sees.

f) Find the first six (marginal) steady-state probabilities of the number of customers in the system.

Exercise 2.39. Referring to exercise 2.35, suppose that the customers arrive singly and the inter-arrival times are $i.i.d$ with a common PMF with masses at 0.5, 1 and 2.5, with probabilities 0.3, 0.6 and 0.1, respectively. All other parameters remain the same.

a) Show that the mean, the standard deviation of the services and the traffic intensity are, respectively, given by 0.96, 1.69759 and 0.96.

b) Find the mean, the standard deviation of the services and the traffic intensity.

c) Find the rate matrix R and its spectral radius.

d) Find the mean and the standard deviation of the number of customers in the system at the time of an arrival (not including the one arrived).

e) Find the mean and the standard deviation of the number of customers in the system at an arbitrary time.

f) Find the first six (marginal) steady-state probabilities of the number in the system that an arrival (not including the arrived one) sees.

g) Find the first six (marginal) steady-state probabilities of the number of customers in the system.

Exercise 2.40. Referring to exercise 2.36, suppose that the customers arrive singly and the inter-arrival times are $i.i.d$ with a common PMF with masses at 0.5, 1 and 2.5, with probabilities 0.3, 0.6 and 0.1, respectively. All other parameters remain the same.

a) Find the mean, the standard deviation of the services and the traffic intensity.

b) Find the rate matrix R and its spectral radius.

c) Find the mean and the standard deviation of the number of customers in the system at the time of an arrival (not including the one arrived).

d) Find the mean and the standard deviation of the number of customers in the system at an arbitrary time.

e) Find the first six (marginal) steady-state probabilities of the number in the system that an arrival (not including the arrived one) sees.

f) Find the first six (marginal) steady-state probabilities of the number of customers in the system.

Exercise 2.41. Referring to exercise 2.37, suppose that the customers arrive singly and the inter-arrival times are $i.i.d$ with a common PMF with masses at 0.5, 1 and 2.5, with probabilities 0.3, 0.6 and 0.1, respectively. All other parameters remain the same.

a) Find the mean, the standard deviation of the services and the traffic intensity.

b) Find the rate matrix R and its spectral radius.

c) Find the mean and the standard deviation of the number of customers in the system at the time of an arrival (not including the one arrived).

d) Find the mean and the standard deviation of the number of customers in the system at an arbitrary time.

e) Find the first six (marginal) steady-state probabilities of the number in the system that an arrival (not including the arrived one) sees.

f) Find the first six (marginal) steady-state probabilities of the number of customers in the system.

Exercise 2.42. Referring to exercise 2.34, suppose that the customers arrive singly and the inter-arrival times are uniformly distributed on the interval $(0.1, 1.9)$. All other parameters remain the same.

a) Find the mean, the standard deviation of the services and the traffic intensity.

b) Find the rate matrix R and its spectral radius.

c) Find the mean and the standard deviation of the number of customers in the system at the time of an arrival (not including the one arrived).

d) Find the mean and the standard deviation of the number of customers in the system at an arbitrary time.

e) Find the first six (marginal) steady-state probabilities of the number in the system that an arrival (not including the arrived one) sees.

f) Find the first six (marginal) steady-state probabilities of the number of customers in the system.

Exercise 2.43. Referring to exercise 2.35, suppose that the customers arrive singly and the inter-arrival times are uniformly distributed on the interval $(0.1, 1.9)$. All other parameters remain the same.

a) Find the mean, the standard deviation of the services and the traffic intensity.

b) Find the rate matrix R and its spectral radius.

c) Find the mean and the standard deviation of the number of customers in the system at the time of an arrival (not including the one arrived).

d) Find the mean and the standard deviation of the number of customers in the system at an arbitrary time.

e) Find the first six (marginal) steady-state probabilities of the number in the system that an arrival (not including the arrived one) sees.

f) Find the first six (marginal) steady-state probabilities of the number of customers in the system.

Exercise 2.44. Referring to exercise 2.36, suppose that the customers arrive singly and the inter-arrival times are uniformly distributed on the interval $(0.1, 1.9)$. All other parameters remain the same.

a) Find the mean, the standard deviation of the services and the traffic intensity.

b) Find the rate matrix R and its spectral radius.

c) Find the mean and the standard deviation of the number of customers in the system at the time of an arrival (not including the one arrived).

d) Find the mean and the standard deviation of the number of customers in the system at an arbitrary time.

e) Find the first six (marginal) steady-state probabilities of the number in the system that an arrival (not including the arrived one) sees.

f) Find the first six (marginal) steady-state probabilities of the number of customers in the system.

Exercise 2.45. Referring to exercise 2.37, suppose that the customers arrive singly and the inter-arrival times are uniformly distributed on the interval $(0.1, 1.9)$. All other parameters remain the same.

a) Find the mean, the standard deviation of the services and the traffic intensity.

b) Find the rate matrix R and its spectral radius.

c) Find the mean and the standard deviation of the number of customers in the system at the time of an arrival (not including the one arrived).

d) Find the mean and the standard deviation of the number of customers in the system at an arbitrary time.

e) Find the first six (marginal) steady-state probabilities of the number in the system that an arrival (not including the arrived one) sees.

f) Find the first six (marginal) steady-state probabilities of the number of customers in the system.

Exercise 2.46. Referring to exercise 2.37, suppose that the customers arrive singly and the inter-arrival times are Erlang of order 3 with parameter 3 in each stage. All other parameters remain the same.

a) Find the mean, the standard deviation of the services and the traffic intensity.

b) Find the rate matrix R and its spectral radius.

c) Find the mean and the standard deviation of the number of customers in the system at the time of an arrival (not including the one arrived).

d) Find the mean and the standard deviation of the number of customers in the system at an arbitrary time.

e) Find the first six (marginal) steady-state probabilities of the number in the system that an arrival (not including the arrived one) sees.

f) Find the first six (marginal) steady-state probabilities of the number of customers in the system.

Exercise 2.47. Referring to exercise 2.38, suppose that the customers arrive singly and the inter-arrival times are Erlang of order 3 with parameter 3 in each stage. All other parameters remain the same.

a) Find the mean, the standard deviation of the services and the traffic intensity.

b) Find the rate matrix R and its spectral radius.

c) Find the mean and the standard deviation of the number of customers in the system at the time of an arrival (not including the one arrived).

d) Find the mean and the standard deviation of the number of customers in the system at an arbitrary time.

e) Find the first six (marginal) steady-state probabilities of the number in the system that an arrival (not including the arrived one) sees.

f) Find the first six (marginal) steady-state probabilities of the number of customers in the system.

Exercise 2.48. Referring to exercise 2.39, suppose that the customers arrive singly and the inter-arrival times are Erlang of order 3 with parameter 3 in each stage. All other parameters remain the same.

a) Find the mean, the standard deviation of the services and the traffic intensity.

b) Find the rate matrix R and its spectral radius.

c) Find the mean and the standard deviation of the number of customers in the system at the time of an arrival (not including the one arrived).

d) Find the mean and the standard deviation of the number of customers in the system at an arbitrary time.

e) Find the first six (marginal) steady-state probabilities of the number in the system that an arrival (not including the arrived one) sees.

f) Find the first six (marginal) steady-state probabilities of the number of customers in the system.

Exercise 2.49. Referring to exercise 2.40, suppose that the customers arrive singly and the inter-arrival times are Erlang of order 3 with parameter 3 in each stage. All other parameters remain the same.

a) Find the mean, the standard deviation of the services and the traffic intensity.

b) Find the rate matrix R and its spectral radius.

c) Find the mean and the standard deviation of the number of customers in the system at the time of an arrival (not including the one arrived).

d) Find the mean and the standard deviation of the number of customers in the system at an arbitrary time.

e) Find the first six (marginal) steady-state probabilities of the number in the system that an arrival (not including the arrived one) sees.

f) Find the first six (marginal) steady-state probabilities of the number of customers in the system.

Exercise 2.50. Referring to exercise 2.37, suppose that the customers arrive singly and the inter-arrival times are hyperexponential with parameters 2.1, 1.05 and 0.42, respectively, with mixing probabilities 0.5, 0.3 and 0.2. All other parameters remain the same.

a) Find the mean, the standard deviation of the services and the traffic intensity.

b) Find the rate matrix R and its spectral radius.

c) Find the mean and the standard deviation of the number of customers in the system at the time of an arrival (not including the one arrived).

d) Find the mean and the standard deviation of the number of customers in the system at an arbitrary time.

e) Find the first six (marginal) steady-state probabilities of the number in the system that an arrival (not including the arrived one) sees.

f) Find the first six (marginal) steady-state probabilities of the number of customers in the system.

Exercise 2.51. Referring to exercise 2.38, suppose that the customers arrive singly and the inter-arrival times are hyperexponential with parameters 2.1, 1.05 and 0.42, respectively, with mixing probabilities 0.5, 0.3 and 0.2. All other parameters remain the same.

a) Find the mean, the standard deviation of the services and the traffic intensity.

b) Find the rate matrix R and its spectral radius.

c) Find the mean and the standard deviation of the number of customers in the system at the time of an arrival (not including the one arrived).

d) Find the mean and the standard deviation of the number of customers in the system at an arbitrary time.

e) Find the first six (marginal) steady-state probabilities of the number in the system that an arrival (not including the arrived one) sees.

f) Find the first six (marginal) steady-state probabilities of the number of customers in the system.

Exercise 2.52. Referring to exercise 2.39, suppose that the customers arrive singly and the inter-arrival times are hyperexponential with parameters 2.1, 1.05 and 0.42, respectively, with mixing probabilities 0.5, 0.3 and 0.2. All other parameters remain the same.

a) Find the mean, the standard deviation of the services and the traffic intensity.

b) Find the rate matrix R and its spectral radius.

c) Find the mean and the standard deviation of the number of customers in the system at the time of an arrival (not including the one arrived).

d) Find the mean and the standard deviation of the number of customers in the system at an arbitrary time.

e) Find the first six (marginal) steady-state probabilities of the number in the system that an arrival (not including the arrived one) sees.

f) Find the first six (marginal) steady-state probabilities of the number of customers in the system.

Exercise 2.53. Referring to exercise 2.40, suppose that the customers arrive singly and the inter-arrival times are hyperexponential with parameters 2.1, 1.05 and 0.42, respectively, with mixing probabilities 0.5, 0.3 and 0.2. All other parameters remain the same.

a) Find the mean, the standard deviation of the services and the traffic intensity.

b) Find the rate matrix R and its spectral radius.

c) Find the mean and the standard deviation of the number of customers in the system at the time of an arrival (not including the one arrived).

d) Find the mean and the standard deviation of the number of customers in the system at an arbitrary time.

e) Find the first six (marginal) steady-state probabilities of the number in the system that an arrival (not including the arrived one) sees.

f) Find the first six (marginal) steady-state probabilities of the number of customers in the system.

Exercise 2.54. For the $M/PH/1$ queueing model, using the embedded (at arrivals) approach (see theorem 2.3), answer the following questions:

a) Verify $R = A_0 + \frac{1}{\lambda}[I - f^*(\lambda)R]^{-1}RA_0\boldsymbol{S^0}\boldsymbol{\beta}A_0$.

b) Show that $R\boldsymbol{S^0} = [I - f^*(\lambda)R]^{-1}A_0\boldsymbol{S^0} = \lambda\boldsymbol{e}$.

c) Show that $[-(S + \lambda\boldsymbol{S^0}\boldsymbol{\beta}(-S)^{-1})]^{-1} = (-S)^{-1}[I + \frac{\lambda}{1-\rho}\boldsymbol{S^0}\boldsymbol{\beta}(-S)^{-2}]$.

Exercise 2.55. For the $M/PH/1$ queueing model show that Little's law holds good by using the expressions for L_q directly.

Exercise 2.56. Prove equation [2.16] giving an expression for computing R without the knowledge of the matrices $\{A_n\}$.

Exercise 2.57. In the case when the distribution of the inter-arrival times is (continuous) uniform on the interval (a, b), suppose that $U = S + R\boldsymbol{S^0}\boldsymbol{\beta}$. Show that

$$R = \frac{1}{b-a}(-U)^{-1}\left[e^{U\,a} - e^{U\,b}\right].$$

Exercise 2.58. Prove the expressions given in equation [2.17].

Exercise 2.59. Prove the expressions given in equation [2.18].

Exercise 2.60. Prove the expressions given in equation [2.21].

3

Single-Server Queues Based on Arbitrary Epochs

In this chapter, we present ways to study queueing models by looking at arbitrary times. This approach is convenient and easy in some cases even though the dimensionality of the problem may increase. The type of queueing models to be studied via arbitrary time approach depends on whether one is dealing with $M/G/1$, $GI/M/1$ or quasi-birth-and-death (QBD) in continuous-time. It is worth pointing out that the QBD processes can be thought as the intersection of $M/G/1$-type and $GI/M/1$-type queues. Since this section relies heavily on Chapter 8 of Volume 1, readers who are not familiar with them are highly encouraged to review the materials from that chapter before pursuing further. The chapter is organized as follows. In section 3.1, we study $BMAP/PH/1$ queue, and in sections 3.2, 3.3, 3.4, 3.5 and 3.6, respectively, we look at the special cases of $BMAP/PH/1$, namely, $MAP/PH/1$, $PH/PH/1$, $PH/M/1$, $M/PH/1$ and $M/M/1$ queues.

3.1. *BMAP/PH/1* queue

Consider a single-server queueing model in which the customers arrive according to a batch Markovian arrival process (BMAP) with representation given by $\{D_k\}$ of order m. Let $D = \sum_{k=1}^{\infty} D_k$ and denote $\boldsymbol{\delta}$ to be the invariant vector of the irreducible generator $\tilde{Q} = D_0 + D$. That is, $\boldsymbol{\delta}$ satisfies:

$$\boldsymbol{\delta}\tilde{Q} = \mathbf{0} \text{ and } \boldsymbol{\delta}\, \boldsymbol{e} = 1.$$

The batch arrival rate, λ_g, and the arrival rate, λ, are then given by:

$$\lambda_g = \boldsymbol{\delta} D \boldsymbol{e} \text{ and } \lambda = \boldsymbol{\delta} \sum_{k=1}^{\infty} k D_k \boldsymbol{e}.$$

The service times are assumed to be of phase type (PH) with representation $(\boldsymbol{\beta}, S)$ of order n with the average service rate given by $\mu = [\boldsymbol{\beta}(-S)^{-1}\boldsymbol{e}]^{-1}$. Denote $N(t)$, $J_1(t)$ and $J_2(t)$, respectively, to be the number of customers in the system, the phase of the service and the phase of the arrival process at time t. The process $\{(N(t), J_1(t), J_2(t)) : t \geq 0\}$ completely describes the queueing model under study and this process is a continuous-time Markov chain (CTMC) on state space given by $\Omega = \{(0,k),\ 1 \leq k \leq m\} \bigcup \{(i,j,k) : 1 \leq j \leq n, 1 \leq k \leq m,\ i \geq 1\}$.

Note that here the level $\mathbf{0} = \{(0,1), \cdots, (0,m)\}$ corresponds to the server being idle and the arrival process is in phase k; the level $\boldsymbol{i} = \{((\boldsymbol{i}, \mathbf{1}), \cdots, (\boldsymbol{i}, \boldsymbol{n}))\}$, consisting of of mn states, corresponds to the system having i, $i \geq 1$, customers with the service and the arrival processes in various phases. The process is of $M/G/1$-type in continuous-time and its generator is obtained by first noting the following:

1) When the process is in level $\mathbf{0}$ (i.e. the system is idle), in an infinitesimal time interval, either (1) an arrival (of batch size k) occurs and so the system gets into level $\boldsymbol{k}$ with a new service starting in one of n phases or (2) the underlying BMAP is in one of the m states without an arrival occurring. The rates for these two events are $C_k = \boldsymbol{\beta} \otimes D_k$, $k \geq 1$, and $C_0 = D_0$, respectively, and thus we see the first (block) row of Q (see equation [3.1] below).

2) When the system is in level $\mathbf{1}$, in an infinitesimal interval, either (1) an arrival (of batch size k) occurs or (2) a service completion occurs with the server becoming idle; or (3) the underlying BMAP is in one of the m states without an arrival occurring and without a service completion. The rates of these three events are $H_k = I \otimes D_{k-1}$, $k \geq 2$, $F = \boldsymbol{S^0} \otimes I$ and $H_1 = S \oplus D_0$, respectively.

3) When the system is in level $\boldsymbol{i}$, in an infinitesimal interval, either (1) an arrival (of batch size k) occurs or (2) a service completion occurs and the server starts a new service; or (3) the underlying BMAP is in one of the m states without an arrival occurring and without a service completion. The rates of these three events are $H_k = I \otimes D_{k-1}$, $k \geq 2$, $H_0 = \boldsymbol{S^0}\boldsymbol{\beta} \otimes I$ and $H_1 = S \oplus D_0$, respectively.

Thus, the generator Q is of the form:

$$Q = \begin{pmatrix} C_0 & C_1 & C_2 & C_3 & C_4 & \cdots \\ F & H_1 & H_2 & H_3 & H_4 & \cdots \\ & H_0 & H_1 & H_2 & H_3 & \cdots \\ & & H_0 & H_1 & H_2 & \cdots \\ & & & \ddots & \ddots & \ddots \end{pmatrix}. \quad [3.1]$$

We can appeal to the results of section 8.2 of Volume 1 for the current model. Here, we briefly summarize the key points. Let $\boldsymbol{y} = (\boldsymbol{y}_0, \boldsymbol{y}_1, \boldsymbol{y}_2, \cdots)$ be the steady-state probability vector of Q such that $\boldsymbol{y}_0$ is of dimension m giving the steady-state probability that the system is empty with the server being idle; $\boldsymbol{y}_i$, $i \geq 1$, of dimension mn gives the steady-state probability vector of seeing i customers in the system with the service and the arrival processes being in various phases.

1) Let $\theta = \max_i\{|(S \oplus D_0)_{i,i}|\}$.

2) Define $B_0 = I + \frac{1}{\theta}C_0, \;\; B_k = \frac{1}{\theta}C_k, \; k \geq 1$, and $E = \frac{1}{\theta}F$.

3) Define $A_0 = \frac{1}{\theta}H_0, \;\; A_1 = I + \frac{1}{\theta}H_1, \;\; A_k = \frac{1}{\theta}H_k, \; k \geq 2$.

4) The transition probability matrix (TPM), P, obtained as $P = I + \frac{1}{\theta}Q$ is then given by:

$$P = \begin{pmatrix} B_0 & B_1 & B_2 & B_3 & B_4 & \cdots \\ E & A_1 & A_2 & A_3 & A_4 & \cdots \\ & A_0 & A_1 & A_2 & A_3 & \cdots \\ & & A_0 & A_1 & A_2 & \cdots \\ & & & \ddots & \ddots & \ddots \end{pmatrix}. \qquad [3.2]$$

It is important to point out that the matrix P is the TPM of the discrete-time Markov chain (DTMC) embedded by looking at the transitions corresponding to the associated CTMC with generator Q (see equation [3.1]). These transitions correspond to either the arrivals of customers or the service completions. The reader is encouraged to compare such embedded processes with the ones corresponding to only arrival epochs ($M/G/1$-paradigm) or only at service epochs ($GI/M/1$-paradigm).

5) Note that $\boldsymbol{y}$ is also the steady-state probability vector of P.

6) G is obtained by solving: $\sum_{n=0}^{\infty} A_n G^n = G$.

7) Let $\tilde{A}_k = \sum_{j=k}^{\infty} A_j \, G^{j-k}, \;\; \tilde{B}_k = \sum_{j=k}^{\infty} B_j \, G^{j-k}, \;\; k \geq 1$.

8) Define $\hat{K} = B_0 + \tilde{B}_1(I - \tilde{A}_1)^{-1}E$ and let $\hat{\boldsymbol{\kappa}}$ be the invariant vector of $\hat{K}$.

9) The vectors are computed as $\boldsymbol{y}_0 = \hat{c}\,\hat{\boldsymbol{\kappa}}$, where:

$$\hat{c} = \left(1 + \frac{1}{1 - \boldsymbol{\pi}\,\boldsymbol{a}}\hat{\boldsymbol{\kappa}}\left[\boldsymbol{b} + \Big(\sum_{i=1}^{\infty} B_i - \tilde{B}_1\,G\Big)(I - A + \boldsymbol{e}\,\boldsymbol{\pi})^{-1}\boldsymbol{a}\right]\right)^{-1},$$

$$\boldsymbol{a} = \sum_{k=1}^{\infty} kA_k\boldsymbol{e}, \boldsymbol{b} = \sum_{k=1}^{\infty} kB_k\boldsymbol{e}$$

and $\boldsymbol{y}_k = \left[\boldsymbol{y}_0\tilde{B}_k + \sum_{j=1}^{k-1} \boldsymbol{y}_j\,\tilde{A}_{k-j+1}\right](I - \tilde{A}_1)^{-1}, \; k \geq 1.$

10) A number of internal accuracy checks can be built in the code. Some of these are as follows. First, verify that $\boldsymbol{\pi}$, the steady-state probability vector of $A = \sum_{n=0}^{\infty} A_n$, is given by $\boldsymbol{\pi} = \mu\boldsymbol{\beta}(-S)^{-1} \otimes \boldsymbol{\delta}$.

11) The vector $\boldsymbol{y}$ should satisfy the following intuitively obvious relations:

$$a)\ \boldsymbol{y}_0 + \sum_{n=1}^{\infty} \boldsymbol{y}_n(\boldsymbol{e} \otimes I) = \boldsymbol{\delta}, \quad b)\ \frac{1}{(1 - \boldsymbol{y}_0\boldsymbol{e})} \sum_{n=1}^{\infty} \boldsymbol{y}_n(I \otimes \boldsymbol{e}) = \mu\boldsymbol{\beta}(-S)^{-1},$$

$$c)\ \sum_{n=1}^{\infty} \boldsymbol{y}_n(\boldsymbol{S^0} \otimes \boldsymbol{e}) = \lambda.$$

12) Suppose that we denote $\boldsymbol{x} = (\boldsymbol{x}_0, \boldsymbol{x}_1, \cdots)$ such that $\boldsymbol{x}_k,\ k \geq 0$, of dimension m gives the steady-state probability vector that at a departure epoch there are k customers left in the system with the arrival process in various phases. It is easy to verify that:

$$\boldsymbol{x}_{k-1} = \frac{1}{\lambda} \frac{1}{\mu(1 - \boldsymbol{y}_0\boldsymbol{e})} \boldsymbol{y}_k(\boldsymbol{S^0} \otimes I),\ k \geq 1.$$

13) Suppose that we denote $\boldsymbol{z} = (\boldsymbol{z}_1, \boldsymbol{z}_2, \cdots)$ such that $\boldsymbol{z}_k,\ k \geq 1$, of dimension mn gives the steady-state probability vector that just after an arrival epoch there are k customers in the system with arrival and services are in various phases. It is easy to verify that:

$$\boldsymbol{z}_k = \frac{1}{\lambda} \boldsymbol{y}_0(\boldsymbol{\beta} \otimes D_k) + \frac{1}{\lambda} \sum_{j=1}^{k-1} \boldsymbol{y}_{k-j}(I \otimes D_j),\ k \geq 1.$$

Another important point worthy to mention here is the following. Looking at the structure given in equation [3.2], the matrices G corresponding to the first passage time from level $\boldsymbol{i+1}$ to level $\boldsymbol{i}$ and K corresponding to the recurrence time of level $\boldsymbol{0}$ need to be supplemented with another to describe the first passage time from level $\boldsymbol{1}$ to level $\boldsymbol{0}$. The following discussion (see Neuts (1989)) is along the lines of the discussion of the first passage time from level $\boldsymbol{i+1}$ to level $\boldsymbol{i}$ and we present only briefly the relevant points and refer the reader to Neuts (1989) for further details.

Suppose that $L = \{l_{i,j}\}$ of dimension $mn \times m$ is such that $l_{i,j}$ gives the conditional probability that the Markov renewal process (MRP) with TPM given in equation [3.2] eventually reaches level $\boldsymbol{0}$ by visiting the state $(0, j)$ starting from state $(1, i)$. It should be pointed out that i refers to one of mn states since level $\boldsymbol{1}$ has the set of states $(1, j_1, j_2) : 1 \leq j_1 \leq n, 1 \leq j_2 \leq m$. In Neuts (1989), it is shown (using the transform analysis similar to the ones used for G and K matrices) that L is given by:

$$L = (I - \tilde{A}_1)^{-1}E, \qquad [3.3]$$

where $\tilde{A}_1$ is as defined in item 7 above.

REMARK 3.1.– Here's an important question that an astute reader may be thinking at this point: Which approach – embedded or arbitrary – should such an option be available? While in the embedded approach one needs more calculations to be done in the beginning, and also one deals with smaller (comparatively speaking) dimensions of the underlying matrices, in the arbitrary approach one needs to spend time to exploit the structure of the coefficient matrices appearing in G matrix or in the steady-state equations in later parts of the analysis, especially in coding to get the numerical results. Thus, the choice depends on the model under study. However, it is a very good idea to try both the methods to get a deeper understanding. This is the main reason the book shows various approaches to study a queueing model.

REMARK 3.2.– How is the G matrix obtained in the embedded $BMAP/G/1$-approach with phase type services related to the G matrix obtained in this section? Suppose that the dimension of the parameter matrices describing the BMAP process is m and the order of the phase type (PH) distribution for the service process is n. Let the service time representation be denoted by $(\boldsymbol{\beta}, S)$ of order n. In the following, we denote $G^{(e)}$ of dimension m and $G^{(a)}$ of dimension mn to be the G matrices obtained using the embedded and arbitrary time approaches, respectively. Using the interpretation of the G matrix and from a probabilistic argument, it is easily verified that the following relationship holds good:

$$G^{(e)} = (\boldsymbol{\beta} \otimes I)G^{(a)}(\mathbf{e} \otimes I).$$

Denoting $G_1^{(a)}, \cdots, G_n^{(a)}$ to be the block matrices of dimension m appearing in $G^{(a)}(\mathbf{e} \otimes I)$, the above equation can be written as:

$$G^{(e)} = \sum_{k=1}^{n} \beta_k G_k^{(a)}.$$

REMARK 3.3.– Earlier we saw how the G matrices of the embedded and arbitrary methods are connected. However, such a connection involving K in the embedded approach and $\hat{K}$ (see item 8 listed above) in the arbitrary time approach is not available. But one can obtain the K matrix of the embedded approach using the G and L matrices. This is possible using the probabilistic approach with the help of the interpretations of the underlying matrices. Note that in both the cases the dimensions of K and $\hat{K}$ are m. Suppose that the services are of phase type with representation $(\boldsymbol{\beta}, S)$ of order n. Using the interpretation of the G matrix and from a probabilistic argument, it is easily verified that the following relationship holds good:

$$K = \sum_{k=1}^{\infty} (\boldsymbol{\beta} \otimes (-D_0)^{-1} D_k)[G^{(a)}]^{k-1} L, \qquad [3.4]$$

where L is as given in equation [3.3].

EXAMPLE 3.1.– In this example, we revisit example 1.9 by using the tools of this section. Since the steady-state vector and the measures reported are almost identical (subject to the tolerance error used) to the ones given in example 1.9, we only list the quantities (corresponding to Erlang arrivals) that are unique to this example. The rest are assigned as exercises. First, verify that the solution to the equation listed in item 6 (using the uniformization technique) is given by:

$$G^{(a)} = \begin{pmatrix} 0.3907 & 0.3636 & 0.2457 & 0 & 0 & 0 & 0 & 0 & 0 \\ 0.0574 & 0.4787 & 0.4639 & 0 & 0 & 0 & 0 & 0 & 0 \\ 0.1083 & 0.2235 & 0.6682 & 0 & 0 & 0 & 0 & 0 & 0 \\ 0.5240 & 0.3176 & 0.1584 & 0 & 0 & 0 & 0 & 0 & 0 \\ 0.0370 & 0.5807 & 0.3823 & 0 & 0 & 0 & 0 & 0 & 0 \\ 0.0893 & 0.1738 & 0.7369 & 0 & 0 & 0 & 0 & 0 & 0 \\ 0.7189 & 0.2124 & 0.0687 & 0 & 0 & 0 & 0 & 0 & 0 \\ 0.0160 & 0.7435 & 0.2405 & 0 & 0 & 0 & 0 & 0 & 0 \\ 0.0562 & 0.1021 & 0.8417 & 0 & 0 & 0 & 0 & 0 & 0 \end{pmatrix},$$

and hence the block matrices in $G^{(a)}(\boldsymbol{e} \otimes I)$ are given by:

$$G_1^{(a)} = \begin{pmatrix} 0.3907 & 0.3636 & 0.2457 \\ 0.0574 & 0.4787 & 0.4639 \\ 0.1083 & 0.2235 & 0.6682 \end{pmatrix}, \quad G_2^{(a)} = \begin{pmatrix} 0.5240 & 0.3176 & 0.1584 \\ 0.0370 & 0.5807 & 0.3823 \\ 0.0893 & 0.1738 & 0.7369 \end{pmatrix},$$

$$G_3^{(a)} = \begin{pmatrix} 0.7189 & 0.2124 & 0.0687 \\ 0.0160 & 0.7435 & 0.2405 \\ 0.0562 & 0.1021 & 0.8417 \end{pmatrix}.$$

Now $G^{(e)}$ for Erlang is nothing but $G_1^{(a)}$, which is in agreement with the G matrix obtained in example 1.9.

Also, verify that L for our example (see equation [3.3]) is given by:

$$L = \begin{pmatrix} 0.3907 & 0.3636 & 0.2457 \\ 0.0574 & 0.4787 & 0.4639 \\ 0.1083 & 0.2235 & 0.6682 \\ 0.5240 & 0.3176 & 0.1584 \\ 0.0370 & 0.5807 & 0.3823 \\ 0.0893 & 0.1738 & 0.7369 \\ 0.7189 & 0.2124 & 0.0687 \\ 0.0160 & 0.7435 & 0.2405 \\ 0.0562 & 0.1021 & 0.8417 \end{pmatrix},$$

and using the expression for $\hat{K}$ (see item 8 above), verify that:

$$\hat{K} = \begin{pmatrix} 0.7143 & 0.2857 & 0 \\ 0 & 0.7143 & 0.2857 \\ 0.0667 & 0.1023 & 0.8310 \end{pmatrix}.$$

Now we compute K seen in the embedded approach using equation [3.4]. Toward this end, recall first that we assumed $D_k = a_k D_1$ with $\{a_k\}$ following a Poisson distribution. In order to compute K, the infinite sum given in equation [3.4] should be truncated. The truncation value, say, k^*, when $\epsilon = 10^{-12}$, is 16. So, K is obtained as:

$$K = (\boldsymbol{\beta} \otimes (-D_0)^{-1} D_1) \sum_{k=1}^{16} a_k [G^{(a)}]^{k-1} L.$$

and is given by:

$$K = \begin{pmatrix} 0.2335 & 0.3580 & 0.4085 \\ 0.2335 & 0.3580 & 0.4085 \\ 0.2335 & 0.3580 & 0.4085 \end{pmatrix},$$

which agrees with the one given in example 1.9.

EXAMPLE 3.2.– In this example, we revisit example 1.10 by using the tools of this section. Since the steady-state vector and the measures reported are almost identical (subject to the tolerance error used) to the ones given in example 1.10, we only list some quantities of interest (corresponding to the Erlang arrivals under consideration) that are unique to this setup, namely, a continuous-time approach to the model. The other items of interest are assigned as exercises. First, using the uniformization technique, verify that the solution to the equation listed in item 6 above is given by:

$$G^{(a)} = \begin{pmatrix} 0.5672 & 0.2325 & 0.1004 & 0.0630 & 0.0258 & 0.0112 \\ 0.0401 & 0.6005 & 0.2593 & 0.0045 & 0.0667 & 0.0288 \\ 0.1037 & 0.1264 & 0.6699 & 0.0115 & 0.0140 & 0.0744 \\ 0.2430 & 0.3096 & 0.3475 & 0.0270 & 0.0344 & 0.0386 \\ 0.1390 & 0.3586 & 0.4025 & 0.0154 & 0.0398 & 0.0447 \\ 0.1610 & 0.2728 & 0.4662 & 0.0179 & 0.0303 & 0.0518 \end{pmatrix},$$

and hence the block matrices in $G^{(a)}(\boldsymbol{e} \otimes I)$ are given by:

$$G_1^{(a)} = \begin{pmatrix} 0.6302 & 0.2583 & 0.1115 \\ 0.0446 & 0.6673 & 0.2881 \\ 0.1152 & 0.1405 & 0.7443 \end{pmatrix}, \quad G_2^{(a)} = \begin{pmatrix} 0.2700 & 0.3440 & 0.3861 \\ 0.1544 & 0.3984 & 0.4472 \\ 0.1788 & 0.3032 & 0.5180 \end{pmatrix},$$

Now $G^{(e)}$ is obtained as $G^{(e)} = 0.9G_1^{(a)} + 0.1G_2^{(a)}$, and since $\boldsymbol{\beta} = (0.9, 0.1)$ it is given as:

$$G^{(e)} = \begin{pmatrix} 0.5942 & 0.2669 & 0.1391 \\ 0.0556 & 0.6404 & 0.3040 \\ 0.1216 & 0.1567 & 0.7217 \end{pmatrix},$$

which is in agreement with the G matrix obtained in example 1.10.

Also, verify that L for our example (see equation [3.3]) is given by:

$$L = \begin{pmatrix} 0.6302 & 0.2583 & 0.1115 \\ 0.0446 & 0.6673 & 0.2881 \\ 0.1152 & 0.1405 & 0.7443 \\ 0.2700 & 0.3440 & 0.3860 \\ 0.1544 & 0.3984 & 0.4472 \\ 0.1788 & 0.3032 & 0.5180 \end{pmatrix},$$

and using the expression for $\hat{K}$ (see item 8 above), verify that:

$$\hat{K} = \begin{pmatrix} 0.6129 & 0.3871 & 0 \\ 0 & 0.6129 & 0.3871 \\ 0.1548 & 0.1288 & 0.7164 \end{pmatrix}.$$

Now we compute K seen in the embedded approach using equation [3.4]. Toward this end, recall first that we assumed $D_k = a_k D_1$ with $\{a_k\}$ following a Poisson distribution. The infinite sum given in equation [3.4] is truncated by taking $\epsilon = 10^{-12}$ and we identified the truncated value, say, k^*, to be 16. So, K is obtained as:

$$K = (\boldsymbol{\beta} \otimes (-D_0)^{-1} D_1) \sum_{k=1}^{16} a_k [G^{(a)}]^{k-1} L.$$

and is given by:

$$K = \begin{pmatrix} 0.3999 & 0.3327 & 0.2674 \\ 0.3999 & 0.3327 & 0.2674 \\ 0.3999 & 0.3327 & 0.2674 \end{pmatrix},$$

which agrees with the one given in example 1.10.

REMARK 3.4.– Having seen the analysis of $BMAP/PH/1$ queues via both embedded at departures and at arbitrary time approaches, the choice of which approach to take is up to the model under study. As pointed out earlier, the embedded approach takes more effort initially to get the block matrices appearing in the transition probability matrix, whereas in the arbitrary time approach the initial construction (to get the generator) may be relatively easy but a little more effort is probably needed during the implementation of the algorithmic procedures in the computation (through exploitation of any special structure of the coefficient matrices) of key quantities. It should be pointed out that in the embedded approach the computation of the $P(n,t)$ matrices appearing in the block matrices requires a careful planning and with a high precision for the entries of the TPM; otherwise, the round-off errors will accumulate, leading to inflating/deflating the measures. In any case, it is worthy for readers to try out both the approaches.

3.2. *MAP/PH/1* queueing model

In this special case, the arrivals occur singly. This results in some simplifications and also we obtain the waiting time distribution. Suppose that the arrival process MAP (Markovian arrival process) has the representation (D_0, D_1) of order m. Note here that $\lambda = \lambda_g = \boldsymbol{\delta} D_1 \boldsymbol{e}$, where $\boldsymbol{\delta}$ is the invariant vector of $D_0 + D_1$. The single server offers services to the customers on a first come, first served (FCFS) basis and the service times are assumed to follow a PH distribution with an irreducible representation $(\boldsymbol{\beta}, S)$ of order n. The mean service time is given by $\mu^{-1} = \boldsymbol{\beta}(-S)^{-1}\boldsymbol{e}$. Let $\boldsymbol{S^0}$ be such that $S\boldsymbol{e} + \boldsymbol{S^0} = \mathbf{0}$.

By keeping track of the number of customers, say, $N(t)$, in the system, the phase of the arrival process, say, $J_1(t)$, and the phase of the service process, say, $J_2(t)$ (when the server is busy) at time t, the model can be studied as a quasi-birth-and-death (QBD) process with the state space given by $\Omega = \{(0,k), 1 \le k \le m\} \bigcup \{(i,j,k) : i \ge 1, 1 \le j \le n, 1 \le k \le m\}$. Let the level $\mathbf{0}$, consisting of m states, correspond to the server being idle with the arrival process in one of m phases. Define the level $\boldsymbol{i}$, of dimension mn, to correspond to the case when the system has i customers and the service and the arrival processes are in various phases.

The generator of the Markov chain governing the QBD process is given by:

$$Q = \begin{pmatrix} D_0 & \boldsymbol{\beta} \otimes D_1 & & & \\ \boldsymbol{S^0} \otimes I & A_1 & A_0 & & \\ & A_2 & A_1 & A_0 & \\ & & A_2 & A_1 & A_0 \\ & & & \ddots & \ddots & \ddots \end{pmatrix}, \qquad [3.5]$$

where:

$$A_0 = I \otimes D_1, \quad A_1 = S \oplus D_0, \quad A_2 = \boldsymbol{S^0}\boldsymbol{\beta} \otimes I. \qquad [3.6]$$

Now we can apply the results of section 8.6 of Volume 1 to the model under study here and obtain the following result. Suppose that $\boldsymbol{y} = (\boldsymbol{y}_0, \boldsymbol{y}_1, \cdots)$ is the steady-state probability vector of Q. Note that $\boldsymbol{y}_0$ is of dimension m and $\boldsymbol{y}_i$, $i \ge 1$, is of dimension mn.

THEOREM 3.1.– For the $MAP/PH/1$ queueing model, under the stability condition that $\lambda < \mu$, we have:

$$\boldsymbol{y}_i = \boldsymbol{y}_0(\boldsymbol{\beta} \otimes I)R^i, \; i \ge 1, \qquad [3.7]$$

where R is the minimal non-negative solution to the matrix-quadratic equation:

$$R^2 A_2 + RA_1 + A_0 = 0, \qquad [3.8]$$

and the vector $\boldsymbol{y}_0$ is obtained by solving:

$$\begin{aligned} &\boldsymbol{y}_0[D_0 + (\boldsymbol{\beta} \otimes I)\, R(\boldsymbol{S^0} \otimes I)] = \mathbf{0}, \\ &\boldsymbol{y}_0(\boldsymbol{\beta} \otimes I)\,(I - R)^{-1}\boldsymbol{e} = 1. \end{aligned} \qquad [3.9]$$

PROOF.– Using Neuts's matrix-geometric approach, which is applicable for the current model, we get:

$$\boldsymbol{y}_i = \boldsymbol{y}_1 R^{i-1},\ i \geq 1, \qquad [3.10]$$

where R is solved using equation [3.8], and $\boldsymbol{y}_0$ and $\boldsymbol{y}_1$ are obtained by solving:

$$\begin{aligned} &\boldsymbol{y}_0 D_0 + \boldsymbol{y}_1(\boldsymbol{S^0} \otimes I) = \mathbf{0}, \\ &\boldsymbol{y}_0(\boldsymbol{\beta} \otimes D_1) + \boldsymbol{y}_1[A_1 + RA_2] = \mathbf{0}, \end{aligned} \qquad [3.11]$$

subject to the normalizing condition:

$$\boldsymbol{y}_0\boldsymbol{e} + \boldsymbol{y}_1(I - R)^{-1}\boldsymbol{e} = 1. \qquad [3.12]$$

Noting from equation [3.8] that $R = -A_0(A_1 + RA_2)^{-1}$, we get from equation [3.11] that:

$$\boldsymbol{y}_1 = \boldsymbol{y}_0(\boldsymbol{\beta} \otimes I)R \ \text{ and } \ \boldsymbol{y}_0[D_0 + (\boldsymbol{\beta} \otimes I)R(\boldsymbol{S^0} \otimes I)] = \mathbf{0}, \qquad [3.13]$$

and hence the stated result follows once we substitute the expression for $\boldsymbol{y}_1$ in the normalizing condition given in equation [3.12]. □

REMARK 3.5.– The computation of R is discussed in section 8.6 of Volume 1. Recall that the (j, k)th entry of R matrix gives the mean time spent in state $(i + 1, k)$ before the first return to level $\boldsymbol{i}$, given that the process starts in state (i, j) expressed in the time unit measured in terms of the mean time spent in state (i, j). This probabilistic interpretation enables one to identify the special structure, if any, present in a given model and can be exploited in numerical computation, especially when dealing with large values for m and n.

REMARK 3.6.– A number of internal accuracy checks are available. Some of these include $RA_2\boldsymbol{e} = A_0\boldsymbol{e}$, and the other one is the intuitively obvious one involving the phase of the arrival process in steady-state: $\boldsymbol{y}_0 + \sum_{i=1}^{\infty} \boldsymbol{y}_i(\boldsymbol{e} \otimes I) = \boldsymbol{\delta}$, where $\boldsymbol{\delta}$ is the invariant vector of the generator $D_0 + D_1$ of the CTMC governing the MAP.

It is easy to verify the following expression:

$$\boldsymbol{y}_0(\boldsymbol{\beta} \otimes I)(I - R)^{-1}(\boldsymbol{e} \otimes I) = \boldsymbol{\delta}. \qquad [3.14]$$

Steady-state vector at arrivals: The steady-state vector at arrival epochs, $\boldsymbol{z} = (\boldsymbol{z}_1, \boldsymbol{z}_2, \cdots)$, is computed as:

$$\boldsymbol{z}_1 = \frac{1}{\lambda}\boldsymbol{y}_0(\boldsymbol{\beta} \otimes D_1) \text{ and } \boldsymbol{z}_{i+1} = \frac{1}{\lambda}\boldsymbol{y}_1 R^{i-1}(I \otimes D_1),\ i \geq 1.$$

Defining $\hat{R} = -(A_1 + RA_2)^{-1}A_0$ and recalling that $R = -A_0(A_1 + RA_2)^{-1}$, the steady-state vector $\boldsymbol{z}$ can be rewritten as:

$$\boldsymbol{z}_i = \frac{1}{\lambda}\boldsymbol{y}_0(\boldsymbol{\beta} \otimes D_1)\hat{R}^{i-1},\ i \geq 1. \qquad [3.15]$$

Stationary waiting time in the system at arrival epochs: Using the technique due to Ozawa (2006), the stationary distribution of the waiting time in the system of an arriving customer can be obtained. Suppose that Υ is the waiting time in the system of an arriving customer. Verify that:

$$P(\Upsilon > t) = (\boldsymbol{e^T} \otimes \boldsymbol{\zeta})exp\{[(A_0 + A_1)^T \otimes I + A_2^T \otimes \hat{R}]t\}\boldsymbol{a}, \qquad [3.16]$$

where $\boldsymbol{\zeta} = \sum_{k=1}^{\infty} \boldsymbol{z}_k = \frac{1}{\lambda}\boldsymbol{y}_0(\boldsymbol{\beta} \otimes D_1)(\grave{I} - \hat{R})^{-1}$ and $\boldsymbol{a}$ is a column vector of dimension $(mn)^2$ defined as:

$$\boldsymbol{a} = \begin{bmatrix} \boldsymbol{e}_1(mn) \\ \boldsymbol{e}_2(mn) \\ \vdots \\ \boldsymbol{e}_{mn}(mn) \end{bmatrix}. \qquad [3.17]$$

REMARK 3.7.– The mean, μ_Υ, waiting time in the system of an arriving customer is calculated as $\mu_\Upsilon = (\boldsymbol{e^T} \otimes \boldsymbol{\zeta})\left[-\left((A_0 + A_1)^T \otimes I + A_2^T \otimes \hat{R}\right)^{-1}\right]\boldsymbol{a}$. One can use Little's law to obtain this mean and it will also serve as a good accuracy check in numerical calculations.

System performance measures: Qualitative analysis of any queueing model requires looking at one or more key system performance measures. A few of them are listed here. Others can be developed depending on the need in practice.

1) The probability, P_{idle}, that the system is idle is $P_{idle} = \boldsymbol{y}_0\boldsymbol{e} = 1 - \rho$.

2) The mean, L_q, queue length is calculated as $L_q = \boldsymbol{y}_1 R(I - R)^{-2}\boldsymbol{e}$.

3) The mean, L_s, number in the system is $L_s = \boldsymbol{y}_1(I - R)^{-2}\boldsymbol{e} = L_q + \rho$.

4) The probability, P_{wait}, that an arrival has to wait to get service is $P_{wait} = \frac{1}{\lambda}\boldsymbol{y}_1(I - R)^{-1}D_1\boldsymbol{e}$.

EXAMPLE 3.3.– Suppose we look at four types of arrivals in which the inter-arrival times have a non-zero 1-lag correlation. For services we look at two CPH distributions. The representations along with the notation (for plotting purposes) are as follows. The four arrival processes representations are:

$$\textbf{NC 1}:\ D_0 = \begin{pmatrix} -1.25 & 1.25 & 0 \\ 0 & -1.25 & 0 \\ 0 & 0 & -2.50 \end{pmatrix},\ D_1 = \begin{pmatrix} 0 & 0 & 0 \\ 0.0125 & 0 & 1.2375 \\ 2.4750 & 0 & 0.0250 \end{pmatrix}.$$

$$\textbf{NC 4}:\ D_0 = \begin{pmatrix} -2.75 & 2.75 & 0 & 0 & 0 & 0 \\ 0 & -2.75 & 2.75 & 0 & 0 & 0 \\ 0 & 0 & -2.75 & 2.75 & 0 & 0 \\ 0 & 0 & 0 & -2.75 & 2.75 & 0 \\ 0 & 0 & 0 & 0 & -2.75 & 0 \\ 0 & 0 & 0 & 0 & 0 & -5.5 \end{pmatrix},$$

$$D_1 = \begin{pmatrix} 0 & 0 & 0 & 0 & 0 & 0 \\ 0 & 0 & 0 & 0 & 0 & 0 \\ 0 & 0 & 0 & 0 & 0 & 0 \\ 0 & 0 & 0 & 0 & 0 & 0 \\ 0.0275 & 0 & 0 & 0 & 0 & 2.7225 \\ 5.445 & 0 & 0 & 0 & 0 & 0.055 \end{pmatrix}.$$

$$\textbf{PC 1}:\ D_0 = \begin{pmatrix} -1.25 & 1.25 & 0 \\ 0 & -1.25 & 0 \\ 0 & 0 & -2.50 \end{pmatrix},\ D_1 = \begin{pmatrix} 0 & 0 & 0 \\ 1.2375 & 0 & 0.0125 \\ 0.025 & 0 & 2.475 \end{pmatrix}.$$

$$\textbf{PC 4}:\ D_0 = \begin{pmatrix} -2.75 & 2.75 & 0 & 0 & 0 & 0 \\ 0 & -2.75 & 2.75 & 0 & 0 & 0 \\ 0 & 0 & -2.75 & 2.75 & 0 & 0 \\ 0 & 0 & 0 & -2.75 & 2.75 & 0 \\ 0 & 0 & 0 & 0 & -2.75 & 0 \\ 0 & 0 & 0 & 0 & 0 & -5.5 \end{pmatrix},$$

$$D_1 = \begin{pmatrix} 0 & 0 & 0 & 0 & 0 & 0 \\ 0 & 0 & 0 & 0 & 0 & 0 \\ 0 & 0 & 0 & 0 & 0 & 0 \\ 0 & 0 & 0 & 0 & 0 & 0 \\ 2.7225 & 0 & 0 & 0 & 0 & 0.0275 \\ 0.055 & 0 & 0 & 0 & 0 & 5.445 \end{pmatrix}.$$

These four arrival processes will be normalized to have an arrival rate $\lambda = 1$. The 1-lag correlation of these four arrival processes are, respectively, -0.3267, -0.6454, 0.3267, and 0.6454. The representations of the service times are:

$$\textbf{ERS}:\ \boldsymbol{\beta} = (1, 0),\ S = \begin{pmatrix} -2 & 2 \\ 0 & -2 \end{pmatrix},\ \textbf{HES}:\ \boldsymbol{\beta} = (0.9, 0.1),\ S = \begin{pmatrix} -10 & 0 \\ 0 & -1 \end{pmatrix}.$$

The service times will be normalized so that the mean will be equal to the traffic intensity (since $\lambda = 1$). We look at two possible scenarios: $\rho = 0.8$ and $\rho = 0.95$.

In Figures 3.1 and 3.2, respectively, we display the plots of the probability distribution of the sojourn time in the system of a customer at an arrival epoch for $\rho = 0.80$ and $\rho = 0.95$.

In Table 3.1, we display some selected measures under various scenarios of arrival/service and traffic intensity.

ρ	ToS	ToA	μ'_{W_s}	P_{wait}	25th	50th	75th	95th
0.8	ERS	$NC\,1$	2.762	0.7957	1.048	2.055	3.723	7.594
		$PC\,1$	62.256	0.8388	2.695	34.258	90.674	221.68
		$NC\,4$	2.205	0.7754	0.936	1.727	2.949	5.734
		$PC\,4$	127.289	0.8361	11.750	75.537	184.57	437.891
	HES	$NC\,1$	9.980	0.7975	0.956	5.262	14.391	35.586
		$PC\,1$	69.838	0.8494	5.719	38.477	101.270	247.168
		$NC\,4$	9.351	0.7994	0.877	4.781	13.477	33.672
		$PC\,4$	134.520	0.8587	13.812	79.883	194.629	461.328
0.95	ERS	$NC\,1$	11.993	0.9484	3.756	8.449	16.469	35.117
		$PC\,1$	338.716	0.9588	77.637	225.635	478.613	1065.918
		$NC\,4$	8.679	0.9404	77.637	6.203	11.840	24.922
		$PC\,4$	630.744	0.9569	153.711	424.707	887.695	1962.891
	HES	$NC\,1$	52.904	0.9491	12.375	35.332	74.609	165.625
		$PC\,1$	379.892	0.9620	88.867	253.76	535.889	1190.723
		$NC\,4$	49.513	0.9485	88.867	32.969	69.873	155.664
		$PC\,4$	671.658	0.9637	165.039	452.539	944.336	2087.5

Table 3.1. *Selected measures for the waiting time in system under various scenarios for example 3.3*

Let $F(t; A, B)$ be the distribution function of the waiting time, Υ, in the system of an arriving customer, when the arrival process is A and the service time distribution is B. In the case of the negatively correlated arrivals, we notice that $F(t; NC_4, ERS) \leq F(t; NC_1, ERS), \ \forall \ t \geq 0$, as well as $F(t; NC_4, HES) \leq F(t; NC_1, HES), \ \forall \ t \geq 0$. In the case of the positively correlated arrivals, we have $F(t; PC_4, ERS) \leq F(t; PC_1, ERS)$ as well as $F(t; PC_4, HES) \leq F(t; PC_1, HES)$, for all $t \geq 0$. This type of stochastic dominance has been recently mentioned in Chakravarthy (2021b) through several illustrative examples.

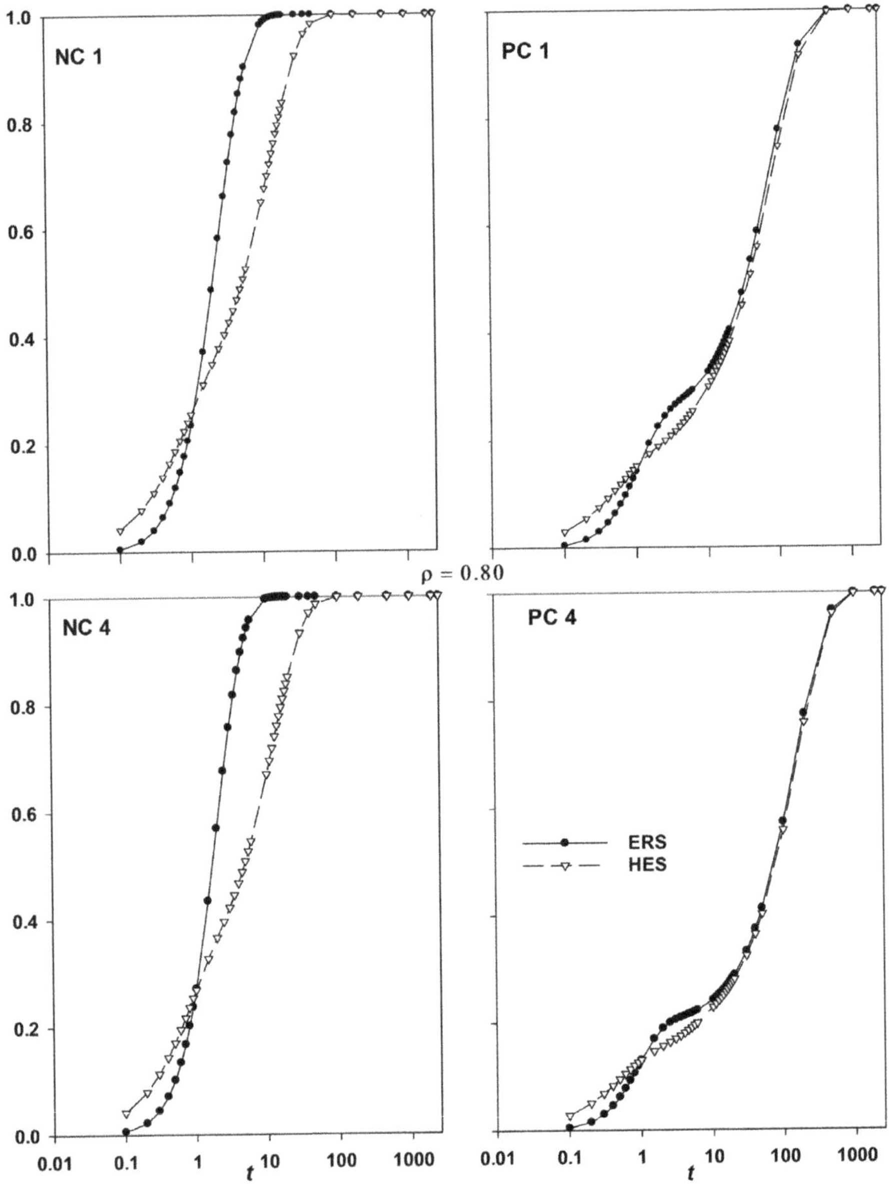

Figure 3.1. *Probability distribution of the waiting time in system when* $\rho = 0.80$ *for example 3.3*

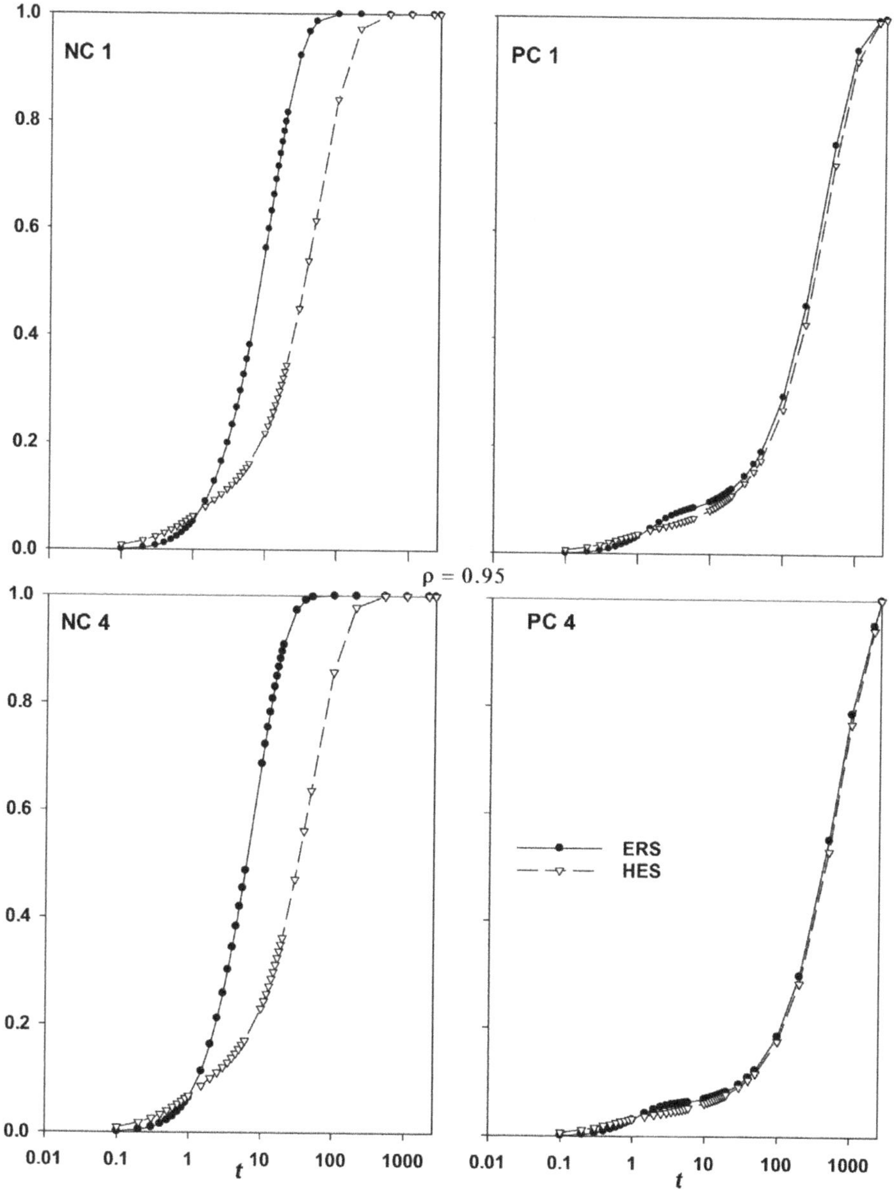

Figure 3.2. *Probability distribution of the waiting time in system when $\rho = 0.95$ for example 3.3*

Now comparing the type of services used by fixing the arrival process, it can be verified that there exists a cut-off point, say, $t^* > 0$, such that $F(t; NC_1, ERS) \leq F(t; NC_1, HES)$, $t \leq t^*$, and $F(t; NC_1, HES) < F(t; NC_1, ERS)$, $t > t^*$, for negatively correlated arrivals. Similarly, for positively correlated arrivals there exists, say, $\tilde{t}^* > 0$, such that $F(t; PC_1, ERS) \leq F(t; PC_1, HES)$, $t \leq \tilde{t}^*$, and $F(t; PC_1, HES) < F(t; PC_1, ERS)$, $t > \tilde{t}^*$.

3.3. *PH/PH/1* queue

Consider a single-server queueing model in which the customers arrive singly according to a CPH process with PH representation given by $(\boldsymbol{\alpha}, T)$ of order m. The rate, λ, of arrivals is then given by $\lambda = [\boldsymbol{\alpha}(-T)^{-1}\boldsymbol{e}]^{-1}$. The service times are of phase type with representation $(\boldsymbol{\beta}, S)$ of order n. Denote $N(t), J_1(t)$ and $J_2(t)$, respectively, to be the number of customers in the system, the phase of the service and the phase of the arrival process at time t. The process $\{(N(t), J_1(t), J_2(t)) : t \geq 0\}$ completely describes the queueing model under study and this process is a CTMC on state space given by $\Omega = \{(0, k), 1 \leq k \leq m\} \bigcup \{(i, j, k): 1 \leq j \leq n, 1 \leq k \leq m, \ i \geq 0\}$.

Note that here the level $\mathbf{0} = \{(0, 1), \cdots, (0, m)\}$ corresponds to the server being idle and the arrival process is in phase k; the level $\boldsymbol{i} = \{((\boldsymbol{i}, \mathbf{1}), \cdots, (\boldsymbol{i}, \boldsymbol{n}))\}$, consisting of of mn states, corresponds to the system having i, $i \geq 1$, customers with the service and the arrival processes in various phases. The process is a QBD process and its generator is obtained by first noting the following:

1) When the process is in level $\mathbf{0}$ (i.e. the system is idle), in an infinitesimal time interval, an arrival may occur and so the system gets into level $\mathbf{1}$ whereupon the service will start in one of n phases, and the arrival process will start in one of m phases to generate the next arrival. This results in the entry $\boldsymbol{T^0}\boldsymbol{\alpha} \otimes \boldsymbol{\beta}$ in the first (block) row of Q. The entry T in the first block corresponds to the transitions within level $\mathbf{0}$ without having an arrival.

2) When the system is in level $\boldsymbol{i}$, in an infinitesimal interval, either (1) an arrival occurs or (2) a service completion occurs with the server becoming idle; or (3) a service completion occurs and the server starts the next service. The rates of these three events are, respectively, given by $\boldsymbol{T^0}\boldsymbol{\alpha} \otimes I$, $I \otimes \boldsymbol{S^0}$ and $I \otimes \boldsymbol{S^0}\boldsymbol{\beta}$.

Thus, the generator Q is of the form:

$$Q = \begin{bmatrix} T & \boldsymbol{T^0}\boldsymbol{\alpha} \otimes \boldsymbol{\beta} & & & \\ I \otimes \boldsymbol{S^0} & T \oplus S & \boldsymbol{T^0}\boldsymbol{\alpha} \otimes I & & \\ & I \otimes \boldsymbol{S^0}\boldsymbol{\beta} & T \oplus S & \boldsymbol{T^0}\boldsymbol{\alpha} \otimes I & \\ & & \ddots & \ddots & \ddots \end{bmatrix}. \quad [3.18]$$

Suppose that the steady-state probability of Q is given by $\boldsymbol{y} = (\boldsymbol{y}_0, \boldsymbol{y}_1, \cdots)$. Then, we have (see section 8.6 of Volume 1):

$$\boldsymbol{y}_0 = \boldsymbol{y}_1[(-T)^{-1} \otimes \boldsymbol{S^0}], \quad \boldsymbol{y}_i = \boldsymbol{y}_1 R^{i-1}, \; i \geq 1,$$

where R satisfies: $R^2(I \otimes \boldsymbol{S^0\beta}) + R(T \oplus S) + \boldsymbol{T^0\alpha} \otimes I = 0$.

One can solve for the steady-state vector along the lines described for $MAP/PH/1$ by taking $D_0 = T$ and $D_1 = \boldsymbol{T^0\alpha}$ and use the expressions given in theorem 3.1 or use a different approach by first computing the vector $\boldsymbol{y}_1$ as:

$$\boldsymbol{y}_1[\boldsymbol{e\alpha} \otimes \boldsymbol{S^0\beta} + T \oplus S + R(I \otimes \boldsymbol{S^0\beta})] = \boldsymbol{0},$$

$$\boldsymbol{y}_1[((-T)^{-1}\boldsymbol{e} \otimes \boldsymbol{S^0}) + (I - R)^{-1}\boldsymbol{e}] = 1.$$

The computation of R is discussed in section 8.6 of volume 1. Here, we show how to exploit the structure of the coefficient matrices, especially if m and n are large and cause memory issues on the computing device one uses.

Let $\tilde{R} = R^2$ and partition them into blocks of matrices of dimension n as:

$$R = \begin{bmatrix} R_{11} & \cdots & R_{1m} \\ \vdots & \cdots & \vdots \\ R_{m1} & \cdots & R_{mm} \end{bmatrix}, \quad \tilde{R} = \begin{bmatrix} \tilde{R}_{11} & \cdots & \tilde{R}_{1m} \\ \vdots & \cdots & \vdots \\ \tilde{R}_{m1} & \cdots & \tilde{R}_{mm} \end{bmatrix}.$$

Equations for R are as follows:

$$\tilde{R}_{ij}\boldsymbol{S^0\beta} + R_{ij}S + \sum_{k=1}^{m} R_{ik}T_{kj} + T_i^0\alpha_j I = 0, 1 \leq i, j \leq m,$$

which are rewritten as ($\theta = \max_j |T_{jj}|$):

$$R_{ij} = \left[\tilde{R}_{ij}\boldsymbol{S^0\beta} + \theta R_{ij} + \sum_{k=1}^{m} R_{ik}T_{kj} + T_i^0\alpha_j I\right](\theta I - S)^{-1}, 1 \leq i, j \leq m.$$

REMARK 3.8.– Once the steady-state vector $\boldsymbol{y}$ is obtained, we can get the steady-state probability vectors at arrivals and at departure epochs.

EXAMPLE 3.4.– In this example, we look at two arrival processes, ERA: Erlang of order 2 and HEA: hyperexponential with parameters 10 and 1 with mixing probabilities, respectively, given by 0.9 and 0.1. For services we look at the two mentioned in example 3.3. Again, the arrival rate is fixed at 1. Similar to example 3.3, we look at two traffic intensities, namely, $\rho = 0.80$ and $\rho = 0.95$. In Figure 3.3, we display the plots of the probability distribution of the sojourn time in the system of a customer at an arrival epoch for $\rho = 0.80$ and $\rho = 0.95$.

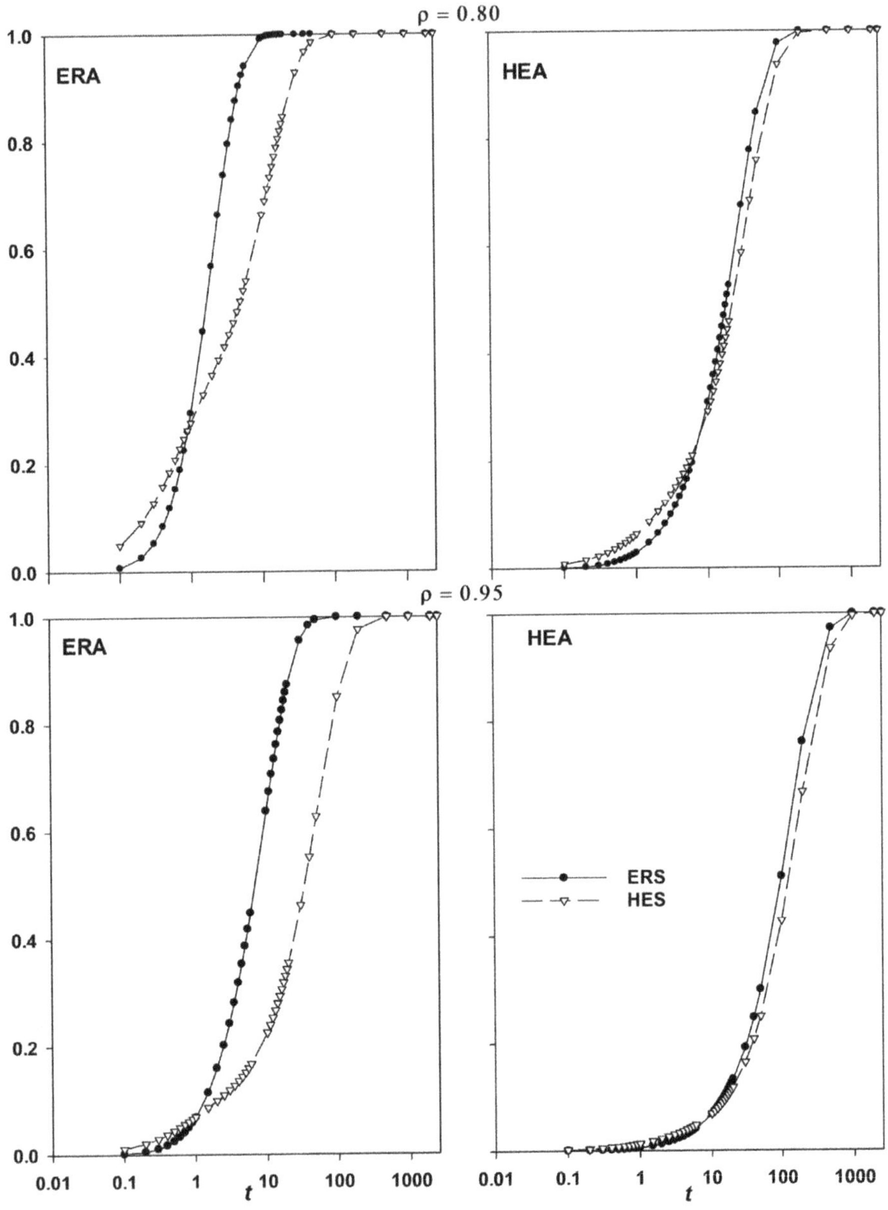

Figure 3.3. *Probability distribution of the waiting time in system when $\rho = 0.80$ and $\rho = 0.95$ for example 3.4*

In Table 3.2, we display some selected measures under various scenarios of arrival/service and traffic intensity.

ρ	ToS	ToA	μ'_{W_s}	P_{wait}	25th	50th	75th	95th
0.8	ERS	ERA	2.292	0.7262	0.868	1.700	3.088	6.305
		HEA	26.722	0.9742	7.852	18.594	36.953	79.59
	HES	ERA	9.520	0.7534	0.816	4.887	13.750	34.34
		HEA	34.482	0.9627	7.922	22.930	48.594	108.20
0.95	ERS	ERA	9.838	0.9298	3.078	6.930	13.516	28.79
		HEA	140.135	0.9941	40.508	97.217	194.141	419.14
	HES	ERA	50.764	0.9373	11.688	33.809	71.631	159.57
		HEA	181.546	0.9917	50.098	124.902	252.734	549.81

Table 3.2. *Selected measures for the waiting time in system under various scenarios for example 3.4*

Like in example 3.3, here also we see stochastic dominance in the waiting time in the system. It can be seen that $F(t; ERA, ERS) \leq F(t; HEA, ERS), \forall\, t \geq 0$, and $F(t; ERA, HES) \leq F(t; HEA, HES), \forall\, t \geq 0$. Furthermore, when comparing Erlang services to hyperexponential ones, we notice a cut-off point, say, $t^* > 0$, such that for $t \leq t^*, F(t; ERA, ERS) \leq F(t; ERA, HES)$, and $F(t; ERA, HES) < F(t; ERA, ERS)$, $t > t^*$. Likewise, there is a cut-off point, say, $\tilde{t}^* > 0$, such that for $t \leq \tilde{t}^*$, $F(t; HEA, ERS) \leq F(t; HEA, HES)$, and for $t > \tilde{t}^*$, we notice $F(t; HEA, HES) < F(t; HEA, ERS)$.

3.4. *PH/M/1* queue

Consider a single-server queueing model in which the customers arrive singly according to a CPH process with PH representation given by $(\boldsymbol{\alpha}, T)$ of order m. The rate, λ, of arrivals is then given by $\lambda = [\boldsymbol{\alpha}(-T)^{-1}\mathbf{e}]^{-1}$. The service times are exponential with parameter μ. Denote $N(t)$ and $J(t)$, respectively, to be the number of customers in the system and the phase of the arrival process at time t. The process $\{(N(t), J(t)) : t \geq 0\}$ completely describes the queueing model under study and this process is a CTMC on state space given by $\Omega = \{(i, j) : 1 \leq j \leq m,\ i \geq 0\}$.

Note that here the level $\boldsymbol{i} = \{(i, 1), \cdots, (i, m)\}$ consists of m states to correspond to the phase of the arrival process. The process is a QBD process and its generator is obtained by first noting the following:

1) When the process is in level **0** (i.e. the system is idle), in an infinitesimal time interval, an arrival may occur and so the system gets into level **1** whereupon the service will start and the arrival process will start in one of m phases to generate the next

arrival. This results in the entry $\boldsymbol{T}^{\mathbf{0}}\boldsymbol{\alpha}$ in the first (block) row of Q. The entry T in the first block corresponds to the transitions within level **0** without having an arrival.

2) When the system is level i, in an infinitesimal interval, either (1) an arrival occurs or (2) a service completion occurs; or (3) the system remains in level **0**. The rates of these three events are, respectively, given by $\boldsymbol{T}^{\mathbf{0}}\boldsymbol{\alpha}$, μI, and $T - \mu I$.

Thus, the generator Q is of the form:

$$Q = \begin{bmatrix} T & \boldsymbol{T}^{\mathbf{0}}\boldsymbol{\alpha} & & & \\ \mu I & T - \mu I & \boldsymbol{T}^{\mathbf{0}}\boldsymbol{\alpha} & & \\ & \mu I & T - \mu I & \boldsymbol{T}^{\mathbf{0}}\boldsymbol{\alpha} & \\ & & \mu I & T - \mu I & \boldsymbol{T}^{\mathbf{0}}\boldsymbol{\alpha} \\ & & & \ddots & \ddots \end{bmatrix}. \quad [3.19]$$

Assuming that the queue is stable (i.e. $\lambda < \mu$), we perform the steady-state analysis. Let $\boldsymbol{y}$, partitioned as $\boldsymbol{y} = (\boldsymbol{y}_0, \boldsymbol{y}_1, \cdots)$, be the steady state probability vector of Q. That is, $\boldsymbol{y}_i$ gives the steady-state probability that at an arbitrary time there are i customers in the system, and $\boldsymbol{y}$ satisfies:

$$\boldsymbol{y}Q = \mathbf{0} \textbf{ and } \boldsymbol{ye} = 1. \quad [3.20]$$

The equations in [3.20] can be rewritten as:

$$\begin{aligned} &\boldsymbol{y}_0 T + \mu \boldsymbol{y}_1 = \mathbf{0}, \\ &\boldsymbol{y}_{i-1}\boldsymbol{T}^{\mathbf{0}}\boldsymbol{\alpha} + \boldsymbol{y}_i(T - \mu I) + \mu \boldsymbol{y}_{i+1} = \mathbf{0},\ i \geq 1, \\ &\sum_{i=0}^{\infty} \boldsymbol{y}_i \boldsymbol{e} = 1. \end{aligned} \quad [3.21]$$

Suppose that ξ is the solution in $(0, 1)$ of the following equation (see Neuts (1995)):

$$\boldsymbol{\alpha}[\mu(1 - \xi)I - T]^{1}\boldsymbol{T}^{\mathbf{0}} = \xi. \quad [3.22]$$

Defining:

$$M = \boldsymbol{T}^{\mathbf{0}}\boldsymbol{\alpha}[\mu(1 - \xi)I - T]^{-1}, \quad [3.23]$$

it can readily be verified the following:

$$\begin{aligned} &\mu M^2 + M(T - \mu I) + \boldsymbol{T}^{\mathbf{0}}\boldsymbol{\alpha} = 0,\ \ M^i = \xi^{i-1}M,\ i \geq 1, \\ &M\boldsymbol{T}^{\mathbf{0}} = \xi \boldsymbol{T}^{\mathbf{0}},\ \ (I - M)^{-1} = I + \frac{1}{1 - \xi}M. \end{aligned} \quad [3.24]$$

Observe that:

$$\boldsymbol{y}_i = \boldsymbol{y}_{i-1}M,\ i \geq 1, \quad [3.25]$$

satisfies the second equation in [3.21]. From equation [3.25], the normalizing equation (see the third equation in [3.21]), and the fact that in steady-state the arrival process is in phase $j, 1 \leq j \leq m$, is given by the jth component of $\lambda\boldsymbol{\alpha}(-T)^{-1}$, we get:

$$\sum_{i=0}^{\infty} \boldsymbol{y}_i = \boldsymbol{y}_0(I - M)^{-1} = \lambda\boldsymbol{\alpha}(-T)^{-1}, \qquad [3.26]$$

which immediately implies that:

$$\boldsymbol{y}_0\boldsymbol{T}^{\mathbf{0}} = \lambda(1 - \xi). \qquad [3.27]$$

Thus, we can summarize (see Neuts (1995)) that the steady-state probability vector is explicitly in terms of the root ξ (see equation [3.22]) as follows (note that $(-T)^{-1}[\mu(1 - \xi)I - T]^{-1} = [\mu(1 - \xi)I - T]^{-1}(-T)^{-1}$ due to their commutativity):

$$\begin{aligned} \boldsymbol{y}_0 &= \lambda\mu(1-\xi)\boldsymbol{\alpha}(-T)^{-1}[\mu(1-\xi)I - T]^{-1}, \\ \boldsymbol{y}_i &= \lambda(1-\xi)\xi^{i-1}\boldsymbol{\alpha}[\mu(1-\xi)I - T]^{-1}, \ i \geq 1. \end{aligned} \qquad [3.28]$$

It is easy to verify that the steady-state probabilities, $\boldsymbol{x} = (x_0, x_1, \cdots)$, at arrival epochs are obtained as:

$$x_k = \frac{1}{\lambda}\boldsymbol{y}_k\boldsymbol{T}^{\mathbf{0}} = (1-\xi)\xi^k, \ \ k \geq 0, \qquad [3.29]$$

which, as it should be, agrees with the result established in section 7.3 of Volume 1.

The marginal probabilities at arbitrary time are given as:

$$\boldsymbol{y}_k\boldsymbol{e} = \begin{cases} 1-\rho, \ \ k = 0, \\ \rho(1-\xi)\xi^{k-1}, \ \ k \geq 1. \end{cases} \qquad [3.30]$$

REMARK 3.9.– Some additional points worthy of registering for the $PH/M/1$ model are as follows:

1) The rate matrix R is same as M as it can be seen from the matrix-quadratic equation along with the fact that M is the minimal nonnegative solution to the matrix-quadratic equation (see equation [3.24]). Thus, R is explicitly known in terms of one root, which is obtained as the solution to a nonlinear (scalar) equation given in equation [3.22].

2) The root, ξ, is also the spectral radius of R. Thus, the spectral radius can be obtained without the actual knowledge of R. This is due to the fact that one can use equation [3.22], which does not depend on R, to get ξ.

3) Recall that G is the unique solution to:

$$\boldsymbol{T^0}\boldsymbol{\alpha}G^2 + (T - \mu I)G + \mu I = 0,$$

which, on simple matrix manipulations, yields:

$$G = \mu(\mu I - T - \rho^{-1}\boldsymbol{T^0}\boldsymbol{\pi})^{-1} + \frac{1-\rho^{-1}}{\boldsymbol{\alpha}(\mu I - T)^{-1}\boldsymbol{T^0}}(\mu I - T)^{-1}\boldsymbol{T^0}\boldsymbol{g}, \quad [3.31]$$

where $\boldsymbol{g}$ is the left eigenvector of G as well as for $\mu(\mu I - T - \rho^{-1}\boldsymbol{T^0}\boldsymbol{\pi})^{-1}$ (corresponding to the maximal eigenvalue of 1 since G is stochastic) and $\boldsymbol{\pi} = \lambda\boldsymbol{\alpha}(-T)^{-1}$.

4) Recall (see equation [2.40]) that the waiting time distribution at an arrival epoch is exponential with parameter depending on the service rate and ξ. Specifically (verify), if Y_q and Y_s denote the waiting time in the queue and in the system of an arriving customer, respectively, then we have:

$$P(Y_q \le t) = 1 - \xi e^{-\mu(1-\xi)t}, \ t \ge 0,$$

$$P(Y_s \le t) = 1 - e^{-\mu(1-\xi)t}, \ t \ge 0,$$

5) The virtual waiting time , $\tilde{W}_q$ (i.e. the waiting time distribution at an arbitrary epoch), in the queue, is obtained as (see Neuts (1981)):

$$P(\tilde{W}_q \le t) = 1 - \boldsymbol{y}_0 R(I - R)^{-1} e^{-\mu(I-R)t}\boldsymbol{e}, \ t \ge 0. \quad [3.32]$$

6) The Laplace transform (LT), $\tilde{w}^*(s)$, of the virtual waiting time is given by:

$$\tilde{w}^*(s) = 1 - \rho + \mu\boldsymbol{y}_0 R[sI + \mu(I - R)]^{-1}\boldsymbol{e}, \ Re(s) \ge 0.$$

EXAMPLE 3.5.– Suppose that the inter-arrival times are modeled using hyperexponential distribution with rates 1.7, 0.85 and 0.17 with mixing probabilities, respectively, given by 0.7, 0.25 and 0.05. Suppose that the service rate is $\mu = \frac{100}{95}$. Noting that $\rho = 0.95$ and:

$$\boldsymbol{\alpha} = (0.70, 0.25, 0.05), \quad T = \begin{pmatrix} -1.7 & 0 & 0 \\ 0 & -0.85 & 0 \\ 0 & 0 & -0.17 \end{pmatrix},$$

verify the following in that order:

$$\boldsymbol{\pi} = \lambda\boldsymbol{\alpha}(-T)^{-1} = (0.4118, 0.2941, 0.2941),$$

$$\mu(\mu I - T - \rho^{-1}\boldsymbol{T^0}\boldsymbol{\pi})^{-1} = \begin{pmatrix} 0.5682 & 0.1920 & 0.2988 \\ 0.1344 & 0.6921 & 0.2161 \\ 0.0418 & 0.0432 & 0.9282 \end{pmatrix},$$

the invariant vector $\boldsymbol{g}$ is:

$$\boldsymbol{g} = (0.1225, 0.1750, 0.7024),$$

and hence:

$$G = \begin{pmatrix} 0.5600 & 0.1803 & 0.2519 \\ 0.1316 & 0.6881 & 0.1999 \\ 0.0415 & 0.0427 & 0.9261 \end{pmatrix}.$$

Finally, verify that $\xi = 0.977297$ and that the rate matrix R is:

$$R = \boldsymbol{T^0}\boldsymbol{\alpha}[\mu(1-\xi)I - T]^{-1} = \begin{pmatrix} 0.6903 & 0.4863 & 0.4384 \\ 0.3451 & 0.2432 & 0.2192 \\ 0.0690 & 0.0486 & 0.0438 \end{pmatrix}.$$

3.5. *M/PH/1* queue

Consider a single-server queueing model in which the customers arrive singly according to a Poisson process with rate λ. The service times are of phase type with representation $(\boldsymbol{\beta}, S)$ of order n. Denote $N(t)$ and $J(t)$, respectively, to be the number of customers in the system and the phase of the service at time t. The process $\{(N(t), J(t)) : t \geq 0\}$ completely describes the queueing model under study and this process is a CTMC on state space given by $\Omega = \{\{0\} \bigcup \{(i,j) : 1 \leq j \leq n, i \geq 1\}\}$.

For use in the following, we denote $\boldsymbol{i} = \{(i,1), \cdots, (i,n)\}$ to be the level corresponding to having i customers in the system with the phase of the service process in j, $1 \leq j \leq n$. The process is a QBD process and its generator is obtained by first noting the following:

1) When the process is in state 0 (i.e. the system is idle), in an infinitesimal time interval, an arrival may occur and so the system gets into state $(1, j)$ whereupon the service will start in phase j with probability β_j. This results in the entry $\lambda\boldsymbol{\beta}$ in the first (block) row of Q corresponding to the transition from state 0 to level $\mathbf{1}$. Otherwise the system stays in state 0.

2) When the system is in level $\boldsymbol{i}$, in an infinitesimal interval, either (1) an arrival occurs (and of course the phase of the service process cannot change as the probability of having two or more events in an infinitesimal interval is zero) or (2) a service completion occurs and the server will become idle when $i = 1$; or (3) a service completion occurs and a new service will begin when $i > 1$. The rates of these three events are, respectively, given by λI, $\lambda \boldsymbol{S^0}$ and $\lambda \boldsymbol{S^0}\boldsymbol{\beta}$. Note that neither an arrival nor a service completion occurring, which corresponds to the transitions within level $\boldsymbol{i}$, are governed by the $(S - \lambda I)$ when away from state 0.

3) The diagonal elements are such that the row sums of Q are zero. Of course, one can view the elements $S - \lambda I$ governing transitions corresponding to no arrival or no service completions and only changes in the service phase are allowed.

Thus, the generator Q is of the form:

$$Q = \begin{bmatrix} -\lambda & \lambda\boldsymbol{\beta} & & & \\ \boldsymbol{S}^{\mathbf{0}} & S-\lambda I & \lambda I & & \\ & \boldsymbol{S}^{\mathbf{0}}\boldsymbol{\beta} & S-\lambda I & \lambda I & \\ & & \boldsymbol{S}^{\mathbf{0}}\boldsymbol{\beta} & S-\lambda I & \lambda I \\ & & & \ddots & \ddots \end{bmatrix}. \qquad [3.33]$$

Assuming that the queue is stable (i.e. $\lambda > \mu$), we perform the steady-state analysis. Let $\boldsymbol{y}$, partitioned as $\boldsymbol{y} = (y_0, \boldsymbol{y}_1, \boldsymbol{y}_2, \cdots)$, be the steady state probability vector of Q. That is:

$$\boldsymbol{y}Q = \mathbf{0} \text{ and } \boldsymbol{y}\boldsymbol{e} = 1. \qquad [3.34]$$

The equations in [3.34] can be rewritten as:

$$\lambda y_0 = \boldsymbol{y}_1\boldsymbol{S}^{\mathbf{0}}, \qquad [3.35]$$

$$\boldsymbol{y}_1(\lambda I - S) = \lambda y_0\boldsymbol{\beta} + \boldsymbol{y}_2\boldsymbol{S}^{\mathbf{0}}\boldsymbol{\beta}, \qquad [3.36]$$

$$\boldsymbol{y}_i(\lambda I - S) = \lambda \boldsymbol{y}_{i-1} + \boldsymbol{y}_{i+1}\boldsymbol{S}^{\mathbf{0}}\boldsymbol{\beta},\ i \geq 2, \qquad [3.37]$$

$$y_0 + \sum_{i=1}^{\infty} \boldsymbol{y}_i\boldsymbol{e} = 1. \qquad [3.38]$$

It is easy to verify from equations [3.35] through [3.37] the following:

$$\boldsymbol{y}_{i+1}\boldsymbol{S}^{\mathbf{0}} = \lambda\boldsymbol{y}_i\boldsymbol{e},\ i \geq 1. \qquad [3.39]$$

Substituting equation [3.39] into equations [3.36] and [3.37], and using the fact that $(\lambda I - \lambda\boldsymbol{e}\boldsymbol{\beta} - S)^{-1}$ exists, we get:

$$\boldsymbol{y}_1 = \lambda y_0\boldsymbol{\beta}(\lambda I - \lambda\boldsymbol{e}\boldsymbol{\beta} - S)^{-1},\ \boldsymbol{y}_{i+1} = \lambda\boldsymbol{y}_i(\lambda I - \lambda\boldsymbol{e}\boldsymbol{\beta} - S)^{-1},\ i \geq 1, \qquad [3.40]$$

where y_0 is obtained as follows. Suppose that $M = \lambda(\lambda I - \lambda\boldsymbol{e}\boldsymbol{\beta} - S)^{-1}$. Now adding equations [3.35], [3.36] and [3.37] over i, we get:

$$\sum_{i=1}^{\infty} \boldsymbol{y}_i = \sum_{i=1}^{\infty} \boldsymbol{y}_i\boldsymbol{S}^{\mathbf{0}}\boldsymbol{\beta}(-S)^{-1}, \qquad [3.41]$$

which implies:

$$\sum_{i=1}^{\infty} \boldsymbol{y}_i\boldsymbol{e} = \frac{1}{\mu}\sum_{i=1}^{\infty} \boldsymbol{y}_i\boldsymbol{S}^{\mathbf{0}}. \qquad [3.42]$$

Using the fact that $M\boldsymbol{S}^{\mathbf{0}} = \lambda \boldsymbol{e}$ and $\sum_{i=1}^{\infty} \boldsymbol{y}_i = y_0 \boldsymbol{\beta}(I - M)^{-1} M$, we get:

$$\sum_{i=1}^{\infty} \boldsymbol{y}_i \boldsymbol{S}^{\mathbf{0}} = \lambda y_0 \boldsymbol{\beta}(I - M)^{-1} \boldsymbol{e}. \qquad [3.43]$$

The normalizing equation [3.38] can be rewritten as:

$$y_0 + y_0 \boldsymbol{\beta}(I - M)^{-1} M \boldsymbol{e} = y_0 \boldsymbol{\beta}(I - M)^{-1} \boldsymbol{e} = 1. \qquad [3.44]$$

From equations [3.42] through [3.44], we get an explicit (but intuitively obvious) expression for $y_0 = 1 - \rho$ and thus the steady-state probability vector is explicitly obtained as:

$$y_0 = 1 - \rho, \;\; \boldsymbol{y}_i = (1 - \rho)\boldsymbol{\beta}[\lambda(\lambda I - \lambda \boldsymbol{e}\boldsymbol{\beta} - S)^{-1}]^i, \; i \geq 1. \qquad [3.45]$$

Note from above that the rate matrix, R, is given by $R = \lambda(\lambda I - \lambda \boldsymbol{e}\boldsymbol{\beta} - S)^{-1}$. This is what we got when analyzing this model using the embedded approach (see section 2.3).

3.6. *M/M/1* queue

The simplest queue in continuous-time is the one involving the Poisson arrivals and the exponential services. Suppose we assume that the arrival rate is λ and the service rate is μ. We also assume that the capacity of the system is infinite so that no arriving customer will be lost. Denoting $X(t)$ to be the number of customers in the system at time t, we can study the stochastic process $\{X(t) : t \geq 0\}$ as a CTMC on the state space $\Omega = \{0, 1, 2, \cdots\}$. Noting that in an infinitesimal interval the system can increase by 1 through an arrival and (if the server is busy) the system can go down 1 through a service completion, the generator Q of this CTMC is given by:

$$Q = \begin{pmatrix} -\lambda & \lambda & & & \\ \mu & -(\lambda + \mu) & \lambda & & \\ & \mu & -(\lambda + \mu) & \lambda & \\ & & \ddots & \ddots & \ddots \end{pmatrix}. \qquad [3.46]$$

The irreducible generator Q is a QBD process and we can apply the results of section 8.5 of Volume 1, and get the steady-state probability vector, $\boldsymbol{x}$, partitioned as $\boldsymbol{x} = (x_0, x_1, \cdots)$, of Q. Assume that the system is stable. That is, $\lambda < \mu$ or $\rho = \frac{\lambda}{\mu} < 1$). Then, we have:

$$x_n = (1 - \rho)\rho^n, \; n \geq 0. \qquad [3.47]$$

The probability mass function of the number of customers, say, N_q, in the queue is given by:

$$P(N_q = n) = \begin{cases} x_0 + x_1 = 1 - \rho^2, \; n = 0, \\ x_{n+1} = (1-\rho)\rho^{n+1}, \; n \geq 1. \end{cases} \quad [3.48]$$

The mean numbers, $E(N_s)$ and $E(N_q)$, are calculated to be:

$$E(N_q) = \frac{\rho^2}{1-\rho} \; and \; E(N_s) = E(N_q) + \rho = \frac{\rho}{1-\rho} \quad [3.49]$$

The variances, $V(N_s)$ and $V(N_q)$, are computed to be:

$$V(N_s) = \frac{\rho}{(1-\rho)^2} \; and \; V(N_q) = \frac{\rho^2(1+\rho-\rho^2)}{(1-\rho)^2}. \quad [3.50]$$

Suppose that W_q denotes the time that an arriving customer spends in the queue. Then, we have (apply equation [3.22] with $\xi = \rho$ and also note that for Poisson arrivals the arbitrary time probabilities are identical to those at arrival epochs):

$$P(W_q \leq t) = 1 - \rho e^{-(\mu-\lambda)t}, \; t \geq 0. \quad [3.51]$$

If W_s denotes the time that an arriving customer spends in the system, then we have (easy to derive from that of the time spent in the queue):

$$P(W_s \leq t) = 1 - e^{-(\mu-\lambda)t}, \; t \geq 0. \quad [3.52]$$

Note that the means and variances of W_q and W_s of the waiting times are:

$$V(W_s) = \frac{1}{(\mu-\lambda)^2} \; and \; V(W_q) = \frac{(2\lambda\mu - \lambda^2)}{\mu^2(\mu-\lambda)^2}. \quad [3.53]$$

3.7. Exercises

Exercise 3.1. Referring to example 3.1, verify the results displayed for various matrices by writing a code.

Exercise 3.2. Referring to example 3.2, verify the results displayed for various matrices by writing a code.

Exercise 3.3. Referring to example 3.3, verify the results displayed for various matrices by writing a code.

Exercise 3.4. Prove the expressions given in equation [3.16].

Exercise 3.5. Prove the expressions given in equation [3.24].

Exercise 3.6. For the $MAP/PH/1$ queueing model, show that the steady-state vector at arrivals is given by $\boldsymbol{z}_i = \frac{1}{\lambda}\boldsymbol{y}_0(\boldsymbol{\beta} \otimes D_1)\hat{R}^{i-1}, \ i \geq 1$, where $\hat{R} = -(A_1 + RA_2)^{-1}A_0$.

Exercise 3.7. For the $MAP/PH/1$ queueing model, prove that the equation: $\boldsymbol{y}_0 + \sum_{i=1}^{\infty} \boldsymbol{y}_i(\boldsymbol{e} \otimes I) = \boldsymbol{\delta}$ holds good. Here, $\boldsymbol{\delta}$ is the invariant vector of $D_0 + D_1$.

Exercise 3.8. For the $MAP/PH/1$ queueing model, prove the expression given in equation [3.14].

Exercise 3.9. For the $MAP/PH/1$ queueing model, apply Ozawa's method to prove the expression given in equation [3.16].

Exercise 3.10. For the same queueing model, show that $D_0 + (\boldsymbol{\beta} \otimes I)\, R(\boldsymbol{S^0} \otimes I)$ is an infinitesimal generator.

Exercise 3.11. Prove the expressions given in equation [3.30].

Exercise 3.12. For the $PH/M/1$ queue, show that:

$$G = \mu(\mu I - T - \rho^{-1}\boldsymbol{T^0}\boldsymbol{\pi})^{-1} + \frac{1-\rho^{-1}}{\boldsymbol{\alpha}(\mu I - T)^{-1}\boldsymbol{T^0}}(\mu I - T)^{-1}\boldsymbol{T^0}\boldsymbol{g},$$

and that $\boldsymbol{g}$ is the left eigenvector of G as well as for $\mu(\mu I - T - \rho^{-1}\boldsymbol{T^0}\boldsymbol{\pi})^{-1}$.

Exercise 3.13. For the $PH/M/1$ queueing model, show that the CDF of the waiting time in the queue (Y_q) and the waiting time in the system (Y_s) are given by:

$$P(Y_q \leq t) = 1 - \xi e^{-\mu(1-\xi)t}, \ t \geq 0,$$

$$P(Y_s \leq t) = 1 - e^{-\mu(1-\xi)t}, \ t \geq 0,$$

Exercise 3.14. For the $PH/M/1$ queueing model, show that the CDF of the virtual waiting time, $\tilde{W}_q$, and its LT are given by:

$$P(\tilde{W}_q \leq t) = 1 - \boldsymbol{y}_0 R(I - R)^{-1} e^{-\mu(I-R)t}\boldsymbol{e}, \ t \geq 0,$$

$$\tilde{w}^*(s) = 1 - \rho + \mu\boldsymbol{y}_0 R[sI + \mu(I - R)]^{-1}\boldsymbol{e}, \ Re(s) \geq 0.$$

Exercise 3.15. Referring to example 3.5, verify the results displayed there.

Exercise 3.16. For the $M/M/1$ queue, prove the expressions given in equations [3.47] through [3.53].

Exercise 3.17. Suppose that the generator of $M/G/1$-type queue is of the form:

$$Q = \begin{pmatrix} C_0 & C_1 & C_2 & C_3 & \cdots \\ F & H_1 & H_2 & H_3 & \cdots \\ 0 & H_0 & H_1 & H_2 & \cdots \\ 0 & 0 & H_0 & H_1 & \cdots \\ \vdots & \vdots & \vdots & \vdots & \ddots \end{pmatrix},$$

where C_0 is a square matrix of order m, $C_i,\ i \geq 1$, are of dimension $m \times n$, F is of dimension $n \times m$ and $H_i,\ i \geq 0$, are square matrices of dimension n. Assume that the generator is irreducible. Let $H = \sum_{n=0}^{\infty} H_n$. [Note that $C_0\boldsymbol{e} + \sum_{n=1}^{\infty} C_n\boldsymbol{e} = \boldsymbol{0}$ and $\sum_{n=0}^{\infty} H_n\boldsymbol{e} = \boldsymbol{0} = F\boldsymbol{e} + \sum_{n=1}^{\infty} H_n\boldsymbol{e} = \boldsymbol{0}$.] Taking $G(z,s), \Phi(z,s)$ and $K(z,s)$, respectively, to be the joint transform matrices of the first passage times from level $\boldsymbol{i+1}$ to level $\boldsymbol{i}$ (away from the boundary), from level $\mathbf{1}$ to level $\mathbf{0}$ and from level $\mathbf{0}$ to $\mathbf{0}$, prove the following results. For $Re(s) \geq 0,\ |z| \leq 1$, we have:

$$G(z,s) = z(sI - H_1)^{-1}H_0 + (sI - H_1)^{-1}\sum_{n=2}^{\infty} H_n G^n(z,s),$$

$$\Phi(z,s) = z(sI - H_1)^{-1}F + (sI - H_1)^{-1}\sum_{n=2}^{\infty} H_n G^{n-1}(z,s)\Phi(z,s),$$

$$K(z,s) = (sI - C_0)^{-1}\sum_{n=1}^{\infty} C_n G^{n-1}\Phi(z,s).$$

Exercise 3.18. Referring to exercise 3.17, prove the following results related to vector means. In the following, we denote $\boldsymbol{h}$ to be a column vector of dimension n given by:

$$\boldsymbol{h} = \sum_{r=2}^{\infty}(r-1)H_r\boldsymbol{e}.$$

$$\tilde{\boldsymbol{\mu}}_G = \frac{\partial}{\partial z}G(z,s)\Big]_{s=0,z=1}\boldsymbol{e} = (I - G + \boldsymbol{e}\,\boldsymbol{g})[(H_0\boldsymbol{e} - \boldsymbol{h})\boldsymbol{g} - H]^{-1}H_0\boldsymbol{e},$$

$$\hat{\boldsymbol{\mu}}_G = -\frac{\partial}{\partial s}G(z,s)\Big]_{s=0,z=1}\boldsymbol{e} = (I - G + \boldsymbol{e}\,\boldsymbol{g})[(H_0\boldsymbol{e} - \boldsymbol{h})\boldsymbol{g} - H]^{-1}\boldsymbol{e}.$$

Exercise 3.19. Referring to exercise 3.17, prove the following results:

$$\boldsymbol{g}\tilde{\boldsymbol{\mu}}_G = \frac{1}{1-\rho} \text{ and } \boldsymbol{g}\hat{\boldsymbol{\mu}}_G = \frac{1}{(1-\rho)\boldsymbol{\pi}H_0\boldsymbol{e}},$$

Exercise 3.20. For the $BMAP/PH/1$ queue, assuming that the service time PH distribution follows an irreducible representation given by $(\boldsymbol{\beta}, S)$ of order n, prove the following relationship between the G matrices obtained using embedded approach and arbitrary approach:

$$G^{(e)} = (\boldsymbol{\beta} \otimes I)G^{(a)}(\mathbf{e} \otimes I).$$

Exercise 3.21. For the $BMAP/PH/1$ queue, assuming that the service time PH distribution follows an irreducible representation given by $(\boldsymbol{\beta}, S)$ of order n, show that:

$$K = \sum_{k=1}^{\infty}(\boldsymbol{\beta} \otimes (-D_0)^{-1}D_k)[G^{(a)}]^{k-1}L,$$

where L is as given in equation [3.3].

Exercise 3.22. Referring to example 3.1, obtain $G^{(a)}, \hat{K}$ and K matrices for other services considered. You may want to refer to example 1.9 for more details.

Exercise 3.23. Referring to example 3.2, obtain $G^{(a)}, \hat{K}$ and K matrices for other services considered. You may want to refer to example 1.10 for more details.

Exercise 3.24. Consider the $PH/PH/1$ queueing model. Analyze this model using the tools of the (a) $M/G/1$-type queues; (b) $GI/M/1$-type queues; and (c) QBD-type queues.

Exercise 3.25. For the $PH/PH/1$ queueing model, obtain expressions for the steady-state probability vectors at arrival epochs in terms of the steady-state probability at an arbitrary epoch.

Exercise 3.26. For the $PH/PH/1$ queueing model, obtain expressions for the steady-state probability vectors at departure epochs in terms of the steady-state probability at an arbitrary epoch.

Exercise 3.27. For the $PH/M/1$ queueing model, prove the expressions given in equation [3.30] for the marginal probabilities at an arbitrary time.

Exercise 3.28. For the $PH/M/1$ queueing model, if Y_q and Y_s, respectively, denote the waiting time in the queue and in the system of an arriving customer, show that (see section 3.4 for the definition of the notation):

$$P(Y_q \leq t) = 1 - \xi e^{-\mu(1-\xi)t}, \ t \geq 0,$$

$$P(Y_s \leq t) = 1 - e^{-\mu(1-\xi)t}, \ t \geq 0,$$

Exercise 3.29. For the $PH/M/1$ queueing model, using the arbitrary time approach (see the generator in equation [3.19]), and assuming that the queue is stable (i.e. $\lambda < \mu$), answer the following questions:

a) Verify that the steady-state probability vector of the generator given in equation [3.19] can be obtained differently as:

$$\boldsymbol{y}_i \lambda \boldsymbol{\alpha}(-T)^{-1}(I - R)R^i, \; i \geq 0.$$

b) Show that $\boldsymbol{y}_0 \boldsymbol{e} = 1 - \rho$.

c) Show that $\mu R\boldsymbol{e} = \boldsymbol{T^0}$.

d) Show that the virtual waiting time, $\tilde{W}_q$, in the queue, is obtained as:

$$P(\tilde{W}_q \leq t) = 1 - \boldsymbol{y}_0 R(I - R)^{-1} e^{-\mu(I-R)t}\boldsymbol{e}, \;\; t \geq 0.$$

e) Show that the LT, $\tilde{w}^*(s)$, of the virtual waiting time is given by:

$$\tilde{w}^*(s) = 1 - \rho + \mu \boldsymbol{y}_0 R[sI + \mu(I - R)]^{-1}\boldsymbol{e}, \; Re(s) \geq 0.$$

f) Show that the matrix M as given in equation [3.23] is the minimal non-negative solution to the matrix-quadratic equation given in [3.24]. Thus, $M = R$, where R is the rate matrix.

Exercise 3.30. Suppose that the customers arrive in batches whose PMF is modeled using a geometric distribution with parameter 2/3. That is $a_k = (2/3)(1/3)^{k-1}$, $k \geq 1$. The inter-arrival times are $i.i.d$ with a common distribution that has the PDF given by a hyperexponential with parameters 1.4, 0.7, 0.35 and 0.175 with mixing probabilities, respectively, given by 0.5, 0.3, 0.15 and 0.05. Let the service times be hyperexponential with parameters 2.375 and 0.2375, with mixing probabilities given by 0.9 and 0.1, respectively. Assuming that there is a single server offering services on an FCFS basis, answer the following questions:

a) Using the arbitrary time approach, study the queueing model and compute G and its invariant vector $\boldsymbol{g}$ for the model.

b) Using the answer in (a) compute $G^{(e)}$, the corresponding G matrix if the model is studied using the embedded (at departure epochs) approach.

c) Compute the matrix L containing the conditional probabilities of eventually reaching the empty system by visiting the state $(0, j)$ starting from state $(1, i)$ (see equation [3.3]).

d) Compute the matrix K (of the embedded approach) containing the conditional probabilities of eventual return to an empty system starting from an empty system.

Exercise 3.31. Suppose that the customers arrive in batches whose PMF is modeled using a DPH with representation given by:

$$\alpha = (0.7397, 0.2603), \quad T = \begin{pmatrix} 0.1000 & 0.2000 \\ 0.1512 & 0.2341 \end{pmatrix},$$

That is, $a_k = \alpha T^{k-1} \boldsymbol{T}^{\mathbf{0}}, \ k \geq 1$. The inter-arrival times are $i.i.d$ with a common distribution that has the PDF given by a hyperexponential with parameters 1.4, 0.7, 0.35 and 0.175 with mixing probabilities, respectively, given by 0.5, 0.3, 0.15 and 0.05. Let the service times be hyperexponential with parameters 2.375 and 0.2375, with mixing probabilities given by 0.9 and 0.1, respectively. Assuming that there is a single server offering services on an FCFS basis, answer the following questions:

a) Using the arbitrary time approach study the queueing model, compute G and its invariant vector $\boldsymbol{g}$ for the model.

b) Using the answer in (a) compute $G^{(e)}$, the corresponding G matrix if the model is studied using the embedded (at departure epochs) approach.

c) Compute the matrix, L, containing the conditional probabilities of eventually reaching the empty system by visiting the state $(0, j)$ starting from state $(1, i)$ (see equation [3.3]).

d) Compute the matrix K (of the embedded approach) containing the conditional probabilities of eventual return to an empty system starting from an empty system.

Exercise 3.32. Suppose that the customers arrive to a single-server system according to a MAP with representation (D_0, D_1) given by:

$$D_0 = \begin{pmatrix} -1.19 & 1.19 & 0 & 0 \\ 0 & -1.19 & 0 & 0 \\ 0 & 0 & -5.95 & 0 \\ 0 & 0 & 0 & -0.595 \end{pmatrix},$$

$$D_1 = \begin{pmatrix} 0 & 0 & 0 & 0 \\ 0.595 & 0 & 0.5355 & 0.0595 \\ 2.975 & 0 & 2.6775 & 0.2975 \\ 0.2975 & 0 & 0.26775 & 0.02975 \end{pmatrix}.$$

The service times are Erlang of order 2 with parameter $2/0.95$ in each of two states so that the mean service time is 0.95. Analyze this queueing model by setting it up as a QBD process in continuous-time and answer the following questions:

a) Find the mean and the standard deviation of the inter-arrival times.

b) Find the rate matrix R.

c) Find the stochastic matrix G.

d) Find the mean waiting time in the system of an arriving customer.

e) Find the probability that an arriving customer has to wait to receive the service.

f) Find the 25th, 50th, 75th and 95th percentiles of the waiting time in the system of an arriving customer.

g) Compute the probability distribution function of the sojourn time (W_s) at 0.1, 0.5, 1, 2, 5, 10, 20, 50 and 100.

Exercise 3.33. Suppose that the customers arrive to a single-server system according to a MAP with representation (D_0, D_1) given by:

$$D_0 = \begin{pmatrix} -1.51726 & 0 & 0.61929 & 0 \\ 0 & -0.98776 & 0 & 0.61929 \\ 0 & 0 & -1.51726 & 0 \\ 0 & 0 & 0 & -0.98776 \end{pmatrix},$$

$$D_1 = \begin{pmatrix} 0.80817 & 0.0890 & 0 & 0 \\ 0.33163 & 0.03684 & 0 & 0 \\ 1.36553 & 0.15173 & 0 & 0 \\ 0.88899 & 0.09877 & 0 & 0 \end{pmatrix}.$$

The service times are Erlang of order 2 with parameter $2/0.95$ in each of two states so that the mean service time is 0.95. Analyze this queueing model by setting it up as a QBD process in continuous-time and answer the following questions:

a) Find the mean and the standard deviation of the inter-arrival times.

b) Find the rate matrix R.

c) Find the stochastic matrix G.

d) Find the mean waiting time in the system of an arriving customer.

e) Find the probability that an arriving customer has to wait to receive the service.

f) Find the 25th, 50th, 75th and 95th percentiles of the waiting time in the system of an arriving customer.

g) Compute the probability distribution function of the sojourn time (W_s) at 0.1, 0.5, 1, 2, 5, 10, 20, 50 and 100.

Exercise 3.34. Suppose that the customers arrive to a single-server system according to a MAP with representation (D_0, D_1) given by:

$$D_0 = \begin{pmatrix} -1.19 & 1.19 & 0 & 0 \\ 0 & -1.19 & 0 & 0 \\ 0 & 0 & -5.95 & 0 \\ 0 & 0 & 0 & -0.595 \end{pmatrix},$$

$$D_1 = \begin{pmatrix} 0 & 0 & 0 & 0 \\ 0.595 & 0 & 0.5355 & 0.0595 \\ 2.975 & 0 & 2.6775 & 0.2975 \\ 0.2975 & 0 & 0.26775 & 0.02975 \end{pmatrix}.$$

The service times are hyperexponential with parameters 2 and 0.2 with mixing probabilities, respectively, given by 0.9 and 0.1, so that the mean service time is 0.95. Analyze this queueing model by setting it up as a QBD process in continuous-time and answer the following questions:

a) Find the mean and the standard deviation of the inter-arrival times.

b) Find the rate matrix R.

c) Find the stochastic matrix G.

d) Find the mean waiting time in the system of an arriving customer.

e) Find the probability that an arriving customer has to wait to receive the service.

f) Find the 25th, 50th, 75th and 95th percentiles of the waiting time in the system of an arriving customer.

g) Compute the probability distribution function of the sojourn time (W_s) at 0.1, 0.5, 1, 2, 5, 10, 20, 50 and 100.

Exercise 3.35. Suppose that the customers arrive to a single-server system according to a MAP with representation (D_0, D_1) given by:

$$D_0 = \begin{pmatrix} -1.51726 & 0 & 0.61929 & 0 \\ 0 & -0.98776 & 0 & 0.61929 \\ 0 & 0 & -1.51726 & 0 \\ 0 & 0 & 0 & -0.98776 \end{pmatrix},$$

$$D_1 = \begin{pmatrix} 0.80817 & 0.0898 & 0 & 0 \\ 0.33163 & 0.03685 & 0 & 0 \\ 1.36553 & 0.15173 & 0 & 0 \\ 0.88899 & 0.09878 & 0 & 0 \end{pmatrix}.$$

The service times are hyperexponential with parameters 2 and 0.2 with mixing probabilities, respectively, given by 0.9 and 0.1, so that the mean service time is 0.95.

Analyze this queueing model by setting it up as a QBD process in continuous-time and answer the following questions:

a) Find the mean and the standard deviation of the inter-arrival times.

b) Find the rate matrix R.

c) Find the stochastic matrix G.

d) Find the mean waiting time in the system of an arriving customer.

e) Find the probability that an arriving customer has to wait to receive the service.

f) Find the 25th, 50th, 75th and 95th percentiles of the waiting time in the system of an arriving customer.

g) Compute the probability distribution function of the sojourn time (W_s) at 0.1, 0.5, 1, 2, 5, 10, 20, 50 and 100.

Exercise 3.36. Suppose that the customers arrive to a single-server system according to a MAP with representation (D_0, D_1) given by:

$$D_0 = \begin{pmatrix} -1.75 & 1.75 & 0 & 0 \\ 0 & -1.75 & 1.75 & 0 \\ 0 & 0 & -1.75 & 0 \\ 0 & 0 & 0 & -3.5 \end{pmatrix}, D_1 = \begin{pmatrix} 0 & 0 & 0 & 0 \\ 0 & 0 & 0 & 0 \\ 0.0175 & 0 & 0 & 1.7325 \\ 3.465 & 0 & 0 & 0.035 \end{pmatrix}.$$

The service times are Erlang of order 2 with parameter $2/0.95$ in each of two states so that the mean service time is 0.95. Analyze this queueing model by setting it up as a QBD process in continuous-time and answer the following questions:

a) Find the mean and the standard deviation of the inter-arrival times.

b) Find the rate matrix R.

c) Find the stochastic matrix G.

d) Find the mean waiting time in the system of an arriving customer.

e) Find the probability that an arriving customer has to wait to receive the service.

f) Find the 25th, 50th, 75th and 95th percentiles of the waiting time in the system of an arriving customer.

g) Compute the probability distribution function of the sojourn time (W_s) at 0.1, 0.5, 1, 2, 5, 10, 20, 50 and 100.

Exercise 3.37. Suppose that the customers arrive to a single-server system according to a MAP with representation (D_0, D_1) given by:

$$D_0 = \begin{pmatrix} -1.75 & 1.75 & 0 & 0 \\ 0 & -1.75 & 1.75 & 0 \\ 0 & 0 & -1.75 & 0 \\ 0 & 0 & 0 & -3.5 \end{pmatrix}, D_1 = \begin{pmatrix} 0 & 0 & 0 & 0 \\ 0 & 0 & 0 & 0 \\ 1.7325 & 0 & 0 & 0.0175 \\ 0.035 & 0 & 0 & 3.465 \end{pmatrix}.$$

The service times are Erlang of order 2 with parameter $2/0.95$ in each of two states so that the mean service time is 0.95. Analyze this queueing model by setting it up as a QBD process in continuous-time and answer the following questions:

a) Find the mean and the standard deviation of the inter-arrival times.

b) Find the rate matrix R.

c) Find the stochastic matrix G.

d) Find the mean waiting time in the system of an arriving customer.

e) Find the probability that an arriving customer has to wait to receive the service.

f) Find the 25th, 50th, 75th and 95th percentiles of the waiting time in the system of an arriving customer.

g) Compute the probability distribution function of the sojourn time (W_s) at 0.1, 0.5, 1, 2, 5, 10, 20, 50 and 100.

Exercise 3.38. Suppose that the customers arrive to a single-server system according to a MAP with representation (D_0, D_1) given by:

$$D_0 = \begin{pmatrix} -1.75 & 1.75 & 0 & 0 \\ 0 & -1.75 & 1.75 & 0 \\ 0 & 0 & -1.75 & 0 \\ 0 & 0 & 0 & -3.5 \end{pmatrix}, D_1 = \begin{pmatrix} 0 & 0 & 0 & 0 \\ 0 & 0 & 0 & 0 \\ 0.0175 & 0 & 0 & 1.7325 \\ 3.465 & 0 & 0 & 0.035 \end{pmatrix}.$$

The service times are hyperexponential with parameters 2 and 0.2 with mixing probabilities, respectively, given by 0.9 and 0.1, so that the mean service time is 0.95. Analyze this queueing model by setting it up as a QBD process in continuous-time and answer the following questions:

a) Find the mean and the standard deviation of the inter-arrival times.

b) Find the rate matrix R.

c) Find the stochastic matrix G.

d) Find the mean waiting time in the system of an arriving customer.

e) Find the probability that an arriving customer has to wait to receive the service.

f) Find the 25th, 50th, 75th and 95th percentiles of the waiting time in the system of an arriving customer.

g) Compute the probability distribution function of the sojourn time (W_s) at 0.1, 0.5, 1, 2, 5, 10, 20, 50 and 100.

Exercise 3.39. Suppose that the customers arrive to a single-server system according to a MAP with representation (D_0, D_1) given by:

$$D_0 = \begin{pmatrix} -1.75 & 1.75 & 0 & 0 \\ 0 & -1.75 & 1.75 & 0 \\ 0 & 0 & -1.75 & 0 \\ 0 & 0 & 0 & -3.5 \end{pmatrix}, D_1 = \begin{pmatrix} 0 & 0 & 0 & 0 \\ 0 & 0 & 0 & 0 \\ 1.7325 & 0 & 0 & 0.0175 \\ 0.035 & 0 & 0 & 3.465 \end{pmatrix}.$$

The service times are hyperexponential with parameters 2 and 0.2 with mixing probabilities, respectively, given by 0.9 and 0.1, so that the mean service time is 0.95. Analyze this queueing model by setting it up as a QBD process in continuous-time and answer the following questions:

a) Find the mean and the standard deviation of the inter-arrival times.

b) Find the rate matrix R.

c) Find the stochastic matrix G.

d) Find the mean waiting time in the system of an arriving customer.

e) Find the probability that an arriving customer has to wait to receive the service.

f) Find the 25th, 50th, 75th and 95th percentiles of the waiting time in the system of an arriving customer.

g) Compute the probability distribution function of the sojourn time (W_s) at 0.1, 0.5, 1, 2, 5, 10, 20, 50 and 100.

Exercise 3.40. For the $BMAP/PH/1$ queueing model, prove the results mentioned in item 11.

Exercise 3.41. For the $BMAP/PH/1$ queueing model, prove the results mentioned in item 12.

Exercise 3.42. For the $BMAP/PH/1$ queueing model, prove the results mentioned in item 13.

Exercise 3.43. Suppose that the customers arrive singly such that the inter-arrival times are of phase type obtained as the convolution of an Erlang of order 2 with parameter 4 and a hyperexponential with parameters 3.8 and 0.38 with mixing probabilities, respectively, given by 0.9 and 0.1. Suppose that the service times are exponential with parameter μ such that the traffic intensity is $\rho = 0.99$. Analyze this model and obtain the rate matrix R.

4

Busy Period in Queues

The analysis of the busy period (BP) in queues is more complicated (see Bhat (2015); Brill (2017); Gross and Harris (1998); Kleinrock (1975)) and very involved in terms of the tools needed as compared to the analysis of other items like queue lengths and waiting times. This is not due to the choice but rather to the inherent nature of the study of the BP. Even for the classical $M/M/1$ queue, the probability distribution of the BP is known in terms of the modified Bessel functions. In queueing literature, several approaches such as complex analysis, combinatorics, lattice path and matrix-analytic methods (MAM) have been applied to study some selected queueing models by several authors. The study of BP got revived since the introduction of MAM by Neuts (1981, 1989) in the context of $M/G/1$ and $GI/M/1$ paradigms, which were presented in Volume 1. A summary of such approaches as well as some interesting results on the BP in the context of $GI/G/c$ and $MAP/G/c$ queues via simulation can be seen in Chakravarthy (2019a).

Recall that the BP is defined as the duration of the time interval that begins with an arrival of a customer to an empty system and ends with the system becoming empty again at the departure of a customer in the case of a single-server system. In the case of multi-server systems, there are two ways to define them. One is partial and the other is full. In this book, we define a *partial busy period* to be the duration of the time interval that begins with an arrival of a customer to an empty system and ends with the system becoming empty again at the departure of a customer. A *full busy period* is one in which a BP starts from all servers becoming busy and ends when at least one server becomes free.

This chapter is organized as follows. We look at the $M/G/1$-type for both scalar and matrix versions in sections 4.1 and 4.2. In sections 4.3 and 4.4, respectively, we look at the $GI/M/1$-type for both scalar and matrix versions. The BP of the continuous-time version of the $BMAP/PH/1$-type is considered in section 4.5. The

BP in the case of quasi-birth-and-death (QBD) queues, which produces additional simplifications due to the tri-diagonal structure, is discussed in section 4.6. Finally, in section 4.7 we look at the case when the fundamental period does not correspond to a BP.

4.1. Busy period in *M/G/1*-type queues (scalar case)

Suppose we look at the $M/G/1$ queue by denoting $F(.)$ to be the (cumulative) distribution function (CDF) of the inter-arrival times and $H(.)$ to be that of the services. Assume that the service times have a finite mean, μ'_H, and a finite variance, σ_H^2.

Let Y denote the BP of the queueing system under study and let $\Psi(.)$ and $\vartheta(.)$, respectively, denote the probability distribution and the density function of Y. We denote N_Y to be the number of customers served during the BP, Y. Let μ'_Y and σ_Y, respectively, denote the mean and the standard deviation of the BP; let μ'_N and σ_N, respectively, denote the mean and standard deviation of the number of customers served during the BP. Let the traffic intensity be denoted by $\rho = \lambda\mu'_H$.

Let $f^*(s)$, $h^*(s)$, and $\vartheta^*(s)$, respectively, denote the LST of $F(.)$, $H(.)$, and $\Psi(.)$. That is:

$$
\begin{aligned}
f^*(s) &= \int_0^\infty e^{-st} dF(t), \quad h^*(s) = \int_0^\infty e^{-st} dH(t), \\
\vartheta^*(s) &= \int_0^\infty e^{-st} d\Psi(t), \quad Re(s) \geq 0.
\end{aligned}
\tag{4.1}
$$

The probability generating function, $\tilde{N}(z) = E[z^{N_Y}]$, is:

$$
\tilde{N}(z) = \sum_{k=1}^{\infty} z^k P(N_Y = k), \quad |z| \leq 1. \tag{4.2}
$$

For this model, observing that the BP does not depend on the type of service discipline such as first come, first served (FCFS), last come, first served (LCFS) and service in random order (SIRO), Takacs (1962) introduced a novel idea to obtain an expression for $\Psi(.)$ and $\vartheta^*(s)$. Very briefly, we outline Takacs's novel idea.

The BP starts with a new arrival and supposes that the time to serve this customer is x_1 units. Suppose that M denotes the number of customers who arrive during x_1 time units. That is, M is the number of Poisson arrivals during the first service time in a BP. Denote $T_1, T_2, \cdots, T_M$ to be the M sub-BPs generated by the first BP. Noting that the

BP ends when the last sub-BP concludes, and that $Y = x_1 + \sum_{i=1}^{M} T_i$, Takacs established (see Kleinrock (1975)) the following relationship between $\vartheta^*(s)$ and $h^*(s)$:

$$\vartheta^*(s) = \int_0^\infty E[e^{-sY}|x_1 = t]dH(t)$$

$$\Rightarrow \vartheta^*(s) = \int_0^\infty \sum_{k=0}^{\infty} E[e^{-sY}|x_1 = t, M = k]P(M = k)dH(t) \qquad [4.3]$$

$$\Rightarrow \vartheta^*(s) = \int_0^\infty e^{-[s+\lambda(1-\vartheta^*(s))]t}dH(t) = h^*[s + \lambda(1 - \vartheta^*(s))].$$

Similarly, Takacs obtained the following relationship between $\tilde{N}(z)$ and $h^*(s)$. Suppose that K denotes the number of customers arriving during the first service time in the BP. Then, we have:

$$\tilde{N}(z) = \sum_{k=0}^{\infty} E[z^N|K = k]P(K = k)$$

$$\Rightarrow \tilde{N}(z) = \sum_{k=0}^{\infty} E[z^{1+T_1+\cdots+T_k}]P(K = k) \qquad [4.4]$$

$$\Rightarrow \tilde{N}(z) = z\sum_{k=0}^{\infty} A_k N^k(z) = zA(\tilde{N}(z)) = zh^*[\lambda(1 - \tilde{N}(z)],$$

where A_k is as given in equation [1.80] corresponding to the scalar case of $M/G/1$-type queue and $A(z)$ is the probability generating function (PGF) of $\{A_n\}$.

It is easy to verify, from the Laplace-Stieltjes transform (LST) and the PGF, that the means and the standard deviations associated with the BP are obtained as follows:

$$\mu'_Y = \frac{\mu'_H}{1-\rho}, \text{ and } \sigma_Y^2 = \frac{\sigma_H^2 + \rho\mu_H'^2}{(1-\rho)^3},$$

$$\mu'_{N_Y} = \frac{1}{1-\rho}, \text{ and } \sigma_{N_Y}^2 = \frac{\rho(1-\rho) + \lambda^2(\sigma_H^2 + \mu_H'^2)}{(1-\rho)^3}. \qquad [4.5]$$

REMARK 4.1.– The busy cycle is defined as the BP plus the idle period. Noting that the idle period is exponentially distributed with mean $\frac{1}{\lambda}$, we get the mean busy cycle, μ'_{BC}, as $\mu'_{BC} = \frac{1}{\lambda} + \mu'_H = \frac{1}{\lambda(1-\rho)}$.

We now focus on the joint probability function of Y and N_Y. Suppose we define:

$$g_k(y) = P(Y \le y, N_Y = k), y \ge 0, k \ge 1. \qquad [4.6]$$

That is, $g_k(y)$ is the joint probability that the BP will end by time y and that during that time k customers are served. Its joint transform matrix, $g(z, s)$, is:

$$g(z,s) = \sum_{k=1}^{\infty} z^k \int_0^{\infty} e^{-st} dg_k(t), \ |z| \leq 1, Re(s) \geq 0. \tag{4.7}$$

The (scalar) joint transform matrix, $g(z, s)$, satisfies (see equation [1.29] for the matrix analog; we will revisit this later on!):

$$g(z,s) = z \sum_{r=0}^{\infty} a_r^*(s) g^r(z,s), |z| \leq 1, Re(s) \geq 0, \tag{4.8}$$

where $a_k^*(s) = \int_0^{\infty} e^{-st} dA_k(t),\ k \geq 0$, and $A_k(t)$ are the probabilities of seeing k Poisson arrivals during a service time that takes no longer than t units of time. The transform $g(z, s)$ can be rewritten as:

$$g(z,s) = zh^*[s + \lambda(1 - g(z,s))], \tag{4.9}$$

which is the classical Takacs equation for the joint transform of the duration of and the number of customers served during a BP for $M/G/1$ queue.

Similar to the matrix case, the solution to the joint transform is constructed through a sequence of transform functions:

$$\begin{aligned} g_0(z,s) &= 0, \\ g_{n+1}(z,s) &= z \sum_{r=0}^{\infty} a_r^*(s) g_n^r(z,s),\ n \geq 0. \end{aligned} \tag{4.10}$$

$g_n(z, s)$ is a transform of the form:

$$g_n(z,s) = \sum_{k=1}^{\infty} z^k \int_0^{\infty} e^{-st} dg_k(n,t), \tag{4.11}$$

where $g_k(n,t),\ k \geq 1,\ t \geq 0$, are probability mass functions (PMFs). From the functions:

$$\begin{aligned} g(z,0) &= z \sum_{r=0}^{\infty} a_r g^r(z,0), \\ g(1,s) &= \sum_{r=0}^{\infty} a_r(s) g^r(1,s),\ n \geq 0, \end{aligned} \tag{4.12}$$

through differentiation and letting $z \uparrow 1$ and $s \downarrow 0$, we obtain the means (see earlier equations on these):

$$\mu'_{N_Y} = \frac{1}{1-\rho} \text{ and } \mu'_Y = \frac{1}{1-\rho} \int_0^{\infty} t\, dF(t) = \frac{\mu'_H}{1-\rho}, \tag{4.13}$$

which agree with the one given in equation [4.5].

REMARK 4.2.– The role of the matrix analog of the joint transform matrix, $g(z,s)$, was seen in the theory Markov renewal process (MRP) of the $M/G/1$-type queues in section 1.1.

REMARK 4.3.– In the simplest of all queueing models, namely, $M/M/1$, noting that:

$$F(t) = 1 - e^{-\lambda t}, \quad H(t) = 1 - e^{-\mu t},\ t \geq 0,$$

$$f^*(s) = \frac{\lambda}{\lambda + s}, \quad h^*(s) = \frac{\mu}{\mu + s},\ Re(s)\ \geq 0. \qquad [4.14]$$

it is easy to verify the following (see Takacs (1962)):

1) The LST of the BP is given by:

$$\vartheta^*(s) = \frac{\mu}{\mu + s + \lambda(1 - \phi^*(s))}$$
$$\Rightarrow \vartheta^*(s) = \tfrac{1}{2\lambda}\left[(\lambda + \mu + s) - \sqrt{[(\lambda + \mu + s)^2 - 4\lambda\mu]}\right]. \qquad [4.15]$$

2) The PGF of N_Y is given by:

$$\tilde{N}(z) = \frac{z}{1 + \rho(1 - \tilde{N}(z))}. \qquad [4.16]$$

3) The PGF can be obtained explicitly as (see Kleinrock (1975); Takacs (1962)):

$$P(N_Y = n) = \frac{1}{n}\binom{2n-2}{n-1}\rho^{n-1}(1+\rho)^{1-2n},\ n \geq 1. \qquad [4.17]$$

Leguesdron et al. (1993) obtained explicit expressions for the probability distribution function of Y and N_Y. Defining $\theta = \lambda + \mu$, $a = \frac{\mu}{\theta}$ and $b = 1 - a$, the probability distribution function of Y is obtained using Bessel function as:

$$P(Y \leq t) = \sum_{n=1}^{\infty} e^{-\theta t}\frac{(\theta t)^k}{k!}\sum_{k=0}^{\lfloor\frac{n-1}{2}\rfloor}\binom{2k}{k}\frac{a^{k+1}b^k}{k+1},\ t \geq 0, \qquad [4.18]$$

and:

$$P(N_Y = n) = \binom{2n-2}{n-1}\frac{a^n b^{n-1}}{n},\ n \geq 1. \qquad [4.19]$$

Other approaches to the study of BP for $M/M/1$ queue can be seen in Bhat (2015); Bailey (1954); Kherani (2006); Kinateder and Lee (2000); Prabhu (1960).

4.2. Busy period in *M/G/1*-type queues (matrix case)

First, we look at the $M/G/1$-type in matrix form as given in equation [1.5]. In this case, the fundamental period is also the BP. Thus, the joint transform matrix to study the BP and the number of customers served during a BP is as given in equation [1.29]. The discussion pertaining to the transform matrix $G(z,s)$ as well as G is applicable here. Recall from equation [1.28] that the transform matrix $G(z,s)$ of the joint (matrix) probability function $G(n,t)$ is such that the (i,j)th element of $G(n,t)$ gives the joint probability that the BP ends no later than t units of time by visiting the state $(0,j)$ given that the BP starts in state $(1,i)$. This assumes that the BP starts in $(1,i)$, which is the case when the arrivals occur singly. However, if the arrivals occur in batches such as the one seen in a batch Markovian arrival process (BMAP) process, then we need to look at the matrix transform $G^k(z,s)$ of the (matrix) probability function $G^k(n,t)$ such that the (i,j)th element of $G^k(n,t)$ gives the joint probability that the BP ends no later than t units of time by visiting the state $(0,j)$ given that the BP starts in state (k,i), $k \geq 1$. A pictorial description of this is given in Figure 4.1. In the following, we discuss the mean BP as well as the mean number of customers served during a BP for $BMAP/G/1$ queue.

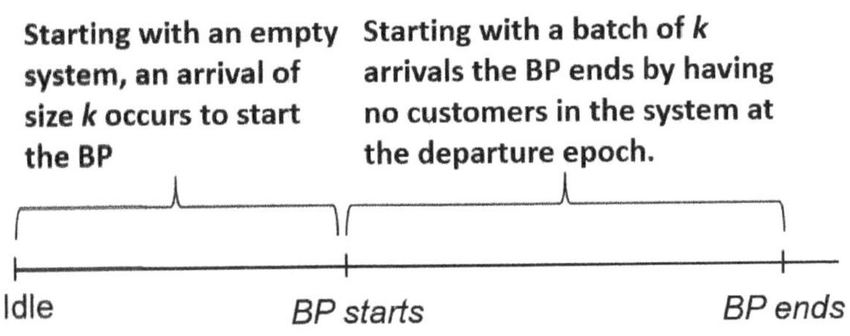

Figure 4.1. *Pictorial description of BP*

The mean BP, μ'_Y, and the mean number of customers served, μ'_{N_Y}, during a BP are given in the following result.

RESULT 4.1.– In the case of $BMAP/G/1$ queue, μ'_Y and μ'_{N_Y} are given by:

$$\mu'_{N_Y} = \frac{1}{1-\rho}\boldsymbol{g}(-D_0)^{-1}\sum_{k=1}^{\infty} kD_k\boldsymbol{e} + \boldsymbol{g}[(-D_0)^{-1}D - K][I - A + (\boldsymbol{e} - \boldsymbol{\beta}^*)\boldsymbol{g}]^{-1}\boldsymbol{e},$$

$$\mu'_Y = \frac{1}{\mu}\mu'_{N_Y},$$

where K and $\boldsymbol{g}$ are, respectively, given in equations [1.42] and [1.43].

PROOF.– First, note that the conditional probability (conditioned on the fact that at an arbitrary time the system is empty) that the BP starts with k customers is $\frac{1}{1-\rho}\boldsymbol{y}_0(-D_0)^{-1}D_k = \boldsymbol{g}(-D_0)^{-1}D_k$. Using this fact, the joint transform, say, $\xi^*(z,s)$, of the duration of the BP and the number served during the BP is then given by:

$$\xi^*(z,s) = \boldsymbol{g}\sum_{k=1}^{\infty}(-D_0)^{-1}D_kG^k(z,s)\boldsymbol{e},\ Re(s) \geq 0, |z| \leq 1. \qquad [4.20]$$

To get μ'_{N_Y}, we differentiate $\xi^*(z,s)$ as follows:

$$\begin{aligned}\mu'_{N_Y} &= \frac{\partial\,\xi^*(z,s)}{\partial z}\Big]_{z=1,s=0}\\ &= \boldsymbol{g}\sum_{k=1}^{\infty}(-D_0)^{-1}D_k\sum_{r=0}^{k-1}G^r(z,s)\frac{\partial}{\partial z}G(z,s)\Big]_{z=1,s=0}G^{k-1-r}(z,s)\boldsymbol{e} \qquad [4.21]\\ &= \boldsymbol{g}\sum_{k=1}^{\infty}(-D_0)^{-1}D_k\sum_{r=0}^{k-1}G^r\tilde{\boldsymbol{\mu}}_{\boldsymbol{G}}.\end{aligned}$$

Now substituting the expression for $\tilde{\boldsymbol{\mu}}_{\boldsymbol{G}}$ (see equation [1.44]) and using the following facts:

$$\boldsymbol{g}[I-A+(\boldsymbol{e}-\boldsymbol{\beta}^*)\boldsymbol{g}]^{-1}\boldsymbol{e} = \frac{1}{1-\rho},\ \sum_{r=0}^{k-1}G^r\boldsymbol{e}\,\boldsymbol{g} = r\boldsymbol{e}\,\boldsymbol{g},\ \sum_{r=0}^{k-1}G^r(I-G) = (I-G^r),$$

$$\sum_{k=1}^{\infty}(-D_0)^{-1}D_k\sum_{r=0}^{k-1}G^r(I-G) = [(-D_0)^{-1}D-K][I-A+(\boldsymbol{e}-\boldsymbol{\beta}^*)\boldsymbol{g}]^{-1}\boldsymbol{e},$$

we get the stated result for μ'_{N_Y}. □

EXAMPLE 4.1.– Referring to example 1.1, find $\tilde{\boldsymbol{\mu}}_{\boldsymbol{G}}$, μ'_{N_Y} and μ'_Y.

Having obtained the G, K, and the invariant vector $\boldsymbol{g}$ of G, simple matrix calculations yield:

$$\tilde{\boldsymbol{\mu}}_{\boldsymbol{G}} = (2.5926, 6.1265, 13.5223),\ \ \mu'_{N_Y} = 14.4757,\ \ \mu'_Y = 13.0281.$$

EXAMPLE 4.2.– Referring to example 1.2, find $\tilde{\boldsymbol{\mu}}_{\boldsymbol{G}}$, μ'_{N_Y} and μ'_Y.

Having obtained the G, K, and the invariant vector $\boldsymbol{g}$ of G, simple matrix calculations yield:

$$\tilde{\boldsymbol{\mu}}_{\boldsymbol{G}} = 10,\ \ \mu'_{N_Y} = 25,\ \ \mu'_Y = 22.5.$$

EXAMPLE 4.3.– Referring to example 1.3, find $\tilde{\boldsymbol{\mu}}_G$, μ'_{N_Y} and μ'_Y.

Having obtained the G, K, and the invariant vector $\boldsymbol{g}$ of G, simple matrix calculations yield:

$$\tilde{\boldsymbol{\mu}}_G = (36.9777, 21.4357, 5.5483), \quad \mu'_{N_Y} = 70.2830, \quad \mu'_Y = 63.2547.$$

EXAMPLE 4.4.– Referring to example 1.4, find $\tilde{\boldsymbol{\mu}}_G$, μ'_{N_Y} and μ'_Y.

Having obtained the G, K, and the invariant vector $\boldsymbol{g}$ of G, simple matrix calculations yield for the negatively correlated, NC, arrival process:

$$\tilde{\boldsymbol{\mu}}_G = (3.2408, 12.3278, 11.9333), \quad \mu'_{N_Y} = 24.3263, \quad \mu'_Y = 21.8936,$$

and for the positively correlated, PC, case, we have:

$$\tilde{\boldsymbol{\mu}}_G = (3.1550, 11.9214, 750.0519), \quad \mu'_{N_Y} = 29.7454, \quad \mu'_Y = 26.7708.$$

EXAMPLE 4.5.– Referring to example 1.5, find $\tilde{\boldsymbol{\mu}}_G$, μ'_{N_Y} and μ'_Y under various scenarios. In Table 4.1, we display these measures.

BMAP	$\tilde{\boldsymbol{\mu}}_G$	μ'_{N_Y}	μ'_Y
ER	(2.8364, 6.4177, 13.2922)	13.004	11.7036
EX	10	22.5000	20.2500
HE	(36.1640, 21.2203, 5.5499)	62.4876	56.2388
NC	(3.4469, 12.2885, 11.8171)	21.9551	19.7596
PC	(3.3491, 11.8696, 723.3187)	26.8058	24.1252

Table 4.1. *$\tilde{\boldsymbol{\mu}}_G$, μ'_{N_Y} and μ'_Y for example 4.5*

EXAMPLE 4.6.– Referring to example 1.6, find $\tilde{\boldsymbol{\mu}}_G$, μ'_{N_Y} and μ'_Y under various scenarios. In Table 4.2, we display these measures.

BMAP	$\tilde{\boldsymbol{\mu}}_G$	μ'_{N_Y}	μ'_Y
ER	(2.4947, 6.0779, 13.9672)	16.1359	14.5223
EX	10	26.7347	24.0612
HE	(34.5504, 19.9105, 5.1804)	68.9042	62.0138
NC	(3.1476, 12.5108, 12.3506)	26.1406	23.5266
PC	(3.005, 11.7851, 721.8041)	32.4300	29.1870

Table 4.2. *$\tilde{\boldsymbol{\mu}}_G$, μ'_{N_Y} and μ'_Y for $BMAP/D/1$ model with DPH batches*

EXAMPLE 4.7.– Referring to example 1.8, find $\tilde{\boldsymbol{\mu}}_G$, μ'_{N_Y} and μ'_Y under various scenarios. In Table 4.3, we display these measures.

ToS	$Batch$	$BMAP$	$\tilde{\boldsymbol{\mu}}_{\boldsymbol{G}}$	μ'_{N_Y}	μ'_Y
ARS	UB	ER	(3.1341, 6.4275, 13.533)	14.9264	13.4338
		EX	10	25	22.5
		HE	(34.7507, 20.5981, 5.492)	67.2386	60.5147
		NC	(3.6104, 12.284, 11.962)	24.3268	21.8941
		PC	(3.4822, 11.7356, 687.547)	30.1052	27.0946
	PB	ER	(3.381, 6.6966, 13.332)	13.4562	12.1106
		EX	10	22.5	20.25
		HE	(33.8614, 20.329, 5.491)	59.5688	53.6119
		NC	(3.8228, 12.2445, 11.863)	21.9509	19.7558
		PC	(3.6769, 11.6708, 660.112)	27.1581	24.4423
UFS	UB	ER	(2.8474, 6.3372, 13.515)	14.712	13.2408
		EX	10	25	22.5
		HE	(36.0847, 21.124, 5.525)	69.0898	62.1808
		NC	(3.4058, 12.3072, 11.921)	24.2936	21.8642
		PC	(3.3, 11.8373, 720.803)	29.9111	26.9200
	PB	ER	(3.1102, 6.6237, 13.302)	13.2528	11.9275
		EX	10	22.5	20.25
		HE	(35.2257, 20.885, 5.524)	61.3249	55.1924
		NC	(3.6198, 12.2668, 11.81)	21.9165	19.7248
		PC	(3.4984, 11.7768, 692.881)	26.9738	24.2764

Table 4.3. *$\tilde{\boldsymbol{\mu}}_{\boldsymbol{G}}$, μ'_{N_Y} and μ'_Y for $BMAP/G/1$ under various scenarios*

EXAMPLE 4.8.– Referring to example 1.9, find $\tilde{\boldsymbol{\mu}}_{\boldsymbol{G}}$, μ'_{N_Y} and μ'_Y under various scenarios. In Table 4.4, we display these measures.

BMAP	$\tilde{\boldsymbol{\mu}}_{\boldsymbol{G}}$	μ'_{N_Y}	μ'_Y
ER	(3.588, 6.95, 13.301)	13.6876	12.3189
EX	10	22.5	20.25
HE	(33.477, 20.239, 5.481)	59.1029	53.1927
NC	(3.941, 12.223, 11.805)	21.8666	19.6800
PC	(3.772, 11.605, 639.172)	27.2863	24.5577

Table 4.4. *$\tilde{\boldsymbol{\mu}}_{\boldsymbol{G}}$, μ'_{N_Y} and μ'_Y for $BMAP/ER/1$ with Poisson batches*

EXAMPLE 4.9.– Referring to example 1.10, find $\tilde{\boldsymbol{\mu}}_{\boldsymbol{G}}$, μ'_{N_Y} and μ'_Y under various scenarios. In Table 4.5, we display these measures.

BMAP	$\tilde{\mu}_G$	μ'_{N_Y}	μ'_Y
ER	(5.914, 8.137, 12.999)	15.9939	14.3946
EX	10	22.5	20.25
HE	(21.866, 15.121, 5.483)	42.4143	38.1728
NC	(5.941, 11.802, 11.867)	22.0471	19.8424
PC	(5.446, 10.42, 329.934)	29.2422	26.3180

Table 4.5. *$\tilde{\mu}_G$, μ'_{N_Y} and μ'_Y for $BMAP/HE/1$ with Poisson batches*

EXAMPLE 4.10.– Referring to example 1.11, find $\tilde{\mu}_G$ and μ'_{N_Y} under various scenarios mentioned in that example. Here, we use the notational convention that $\tilde{\mu}_G{}^{(a,b,c)}$ denotes the $\tilde{\mu}_G$ vector when the arrival is a, the service process is b and the batch size distribution is c. Thus, in abbreviated form MM, MS, CT, ER, HE, PB, GB, UB and PB, denote MMPP, MSPP, CTS, ERS, HES, POI, GEO, UNF and DPH, respectively.

$$\tilde{\mu}_G{}^{(MM,CT,PB)} = (1.8812, 4.3013, 8.2118, 12.6939),\ \mu'_{N_Y} = 13.3757.$$

$$\tilde{\mu}_G{}^{(MM,CT,GB)} = (1.8925, 4.2546, 7.9883, 12.2256),\ \mu'_{N_Y} = 13.1584.$$

$$\tilde{\mu}_G{}^{(MM,CT,UB)} = (1.8775, 4.32, 8.3007, 12.8797),\ \mu'_{N_Y} = 13.4598.$$

$$\tilde{\mu}_G{}^{(MM,CT,DB)} = (1.8953, 4.2443, 7.9385, 12.1210),\ \mu'_{N_Y} = 13.1087.$$

$$\tilde{\mu}_G{}^{(MM,ER,PB)} = (1.9540, 4.3620, 8.0090, 11.9430),\ \mu'_{N_Y} = 13.2019.$$

$$\tilde{\mu}_G{}^{(MM,ER,GB)} = (1.9630, 4.3190, 7.8200, 11.5600),\ \mu'_{N_Y} = 13.0068.$$

$$\tilde{\mu}_G{}^{(MM,ER,UB)} = (1.9510, 4.3790, 8.0830, 12.0940),\ \mu'_{N_Y} = 13.2769.$$

$$\tilde{\mu}_G{}^{(MM,ER,DB)} = (1.9650, 4.3090, 7.7770, 11.4740),\ \mu'_{N_Y} = 12.9620.$$

$$\tilde{\mu}_G{}^{(MM,HE,PB)} = (2.4460, 4.4790, 6.9060, 9.0410),\ \mu'_{N_Y} = 12.2484.$$

$$\tilde{\mu}_G{}^{(MM,HE,GB)} = (2.4490, 4.4630, 6.8450, 8.9150),\ \mu'_{N_Y} = 12.1532.$$

$$\tilde{\mu}_G{}^{(MM,HE,UB)} = (2.4460, 4.4850, 6.9290, 9.0890),\ \mu'_{N_Y} = 12.2836.$$

$$\tilde{\mu}_G{}^{(MM,HE,DB)} = (2.4500, 4.4590, 6.8300, 8.8850),\ \mu'_{N_Y} = 12.1306.$$

$$\tilde{\mu}_G{}^{(MS,CT,PB)} = (10.8212, 7.7332, 6.1842, 4.4810),\ \mu'_{N_Y} = 12.8665.$$

$$\tilde{\mu}_G{}^{(MS,CT,GB)} = (10.7195, 7.6882, 6.1441, 4.4408),\ \mu'_{N_Y} = 12.6418.$$

$$\tilde{\mu}_G{}^{(MS,CT,UB)} = (10.8612, 7.7514, 6.2001, 4.4967),\ \mu'_{N_Y} = 12.9566.$$

$$\tilde{\mu}_G{}^{(MS,CT,DB)} = (10.6969, 7.6784, 6.1354, 4.4318),\ \mu'_{N_Y} = 12.5927.$$

$$\tilde{\mu}_G{}^{(MS,ER,PB)} = (10.2570, 7.4910, 6.0820, 4.4730),\ \mu'_{N_Y} = 12.6173.$$

$$\tilde{\mu}_G{}^{(MS,ER,GB)} = (10.1730, 7.4540, 6.0480, 4.4380),\ \mu'_{N_Y} = 12.4148.$$

$$\tilde{\mu}_G^{(MS,ER,UB)} = (10.2910, 7.5060, 6.0960, 4.4870), \; \mu'_{N_Y} = 12.6988.$$

$$\tilde{\mu}_G^{(MS,ER,DB)} = (10.1540, 7.4450, 6.0410, 4.4300), \; \mu'_{N_Y} = 12.3704.$$

$$\tilde{\mu}_G^{(MS,HE,PB)} = (8.4470, 6.6550, 5.6910, 4.5030), \; \mu'_{N_Y} = 11.7208.$$

$$\tilde{\mu}_G^{(MS,HE,GB)} = (8.4100, 6.6410, 5.6770, 4.4870), \; \mu'_{N_Y} = 11.6080.$$

$$\tilde{\mu}_G^{(MS,HE,UB)} = (8.4610, 6.6610, 5.6970, 4.5090), \; \mu'_{N_Y} = 11.7653.$$

$$\tilde{\mu}_G^{(MS,HE,DB)} = (8.4020, 6.6380, 5.6740, 4.4830), \; \mu'_{N_Y} = 11.5826.$$

4.3. Busy period in *GI/M/1*-type queues (scalar case)

Here, we look at the general independent and identically distributed inter-arrival times and exponential services. That is, the service times have:

$$H(t) = 1 - e^{-\mu t}, \; t \geq 0, \text{ and } h^*(s) = \frac{\mu}{\mu + s}, \; Re(s) \geq 0. \qquad [4.22]$$

Verify from section 2.1 (by scaling it down to the scalar case) that the quantity $r(t), \; t \geq 0$, defined as $r(t) = \sum_{k=1}^{\infty} i p_{i,i+1}^{(k)}(t)$ gives the expected number of visits to state $i+1$ during $\min\{t, \tau\}$ units of time before the first return to state i given that the MRP started in state i. Here, τ is defined as the epoch at which the first return to state i occurs starting from state i. Then, it is easy to see (see Takacs (1962); Ramaswami (1982)) that $\vartheta^*(s)$ satisfies:

$$\vartheta^*(s) = \frac{\mu(1 - r^*(s))}{s + \mu(1 - r^*(s))}, \; Re(s) \geq 0, \qquad [4.23]$$

where $r^*(s), \; Re(s) \geq 0$, is obtained as the solution to:

$$r^*(s) = \sum_{n=0}^{\infty} [r^*(s)]^n \int_0^{\infty} e^{-(s+\mu)t} \frac{(\mu t)^n}{n!} dF(t) = f^*[s + \mu(1 - r^*(s))]. \qquad [4.24]$$

REMARK 4.4.– Now looking at $M/M/1$ queue via $GI/M/1$-type approach, we first note that $F(t) = 1 - e^{-\lambda t}, \; t \geq 0$, and $f^*(s) = \frac{\lambda}{s+\lambda}$. For this special case, we have:

$$r^*(s) = \frac{\lambda}{\lambda + s + \mu(1 - r^*(s))}, \qquad [4.25]$$

from which the simplified expression for $\vartheta^*(s) = \frac{\mu}{\lambda} r^*(s)$ is obtained as:

$$\vartheta^*(s) = \frac{1}{2\lambda}\left[(\lambda + \mu + s) - \sqrt{[(\lambda + \mu + s)^2 - 4\lambda\mu]}\right]. \qquad [4.26]$$

The above (as it should be) is the same as the one we got earlier (see equation [4.15]).

EXAMPLE 4.11.– In this example, we look at three types of arrivals: (1) CTA: constant with parameter 1; (2) UTA: uniform $(0.1, 1.9)$; and (3) ABA: inter-arrival times have a discrete PMF having masses at 3.0, 0.5 and 1.75, with respective probabilities given by 0.1, 0.7 and 0.2. All these arrivals have a mean of 1 and the standard deviations are, respectively, given by 0, 0.51962 and 0.82916. The service times are assumed to follow an exponential distribution with parameter μ. We look at the impact of the traffic intensity ρ and the spectral radius, η, of the R matrix associated with the model. The plots of the mean BP as a function of ρ for these three arrivals are displayed in Figure 4.2, and in Figure 4.3 the mean BP as a function of η is plotted.

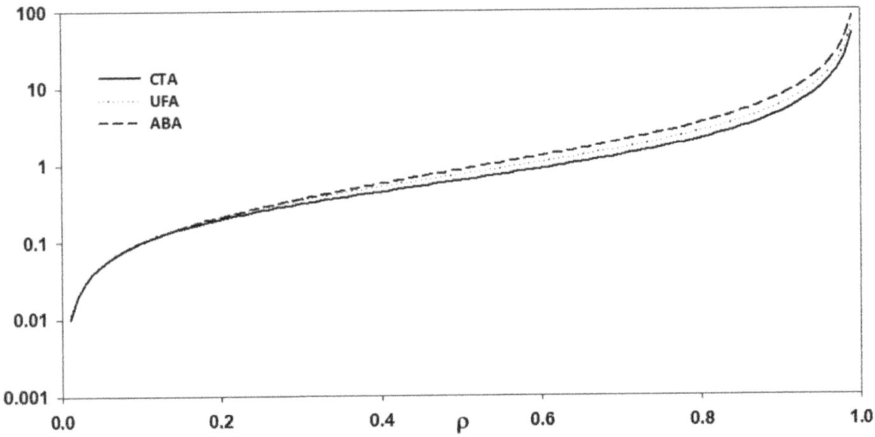

Figure 4.2. *Mean busy period versus ρ for example 4.11*

4.4. Busy period in *GI/PH/1*-type queues

As mentioned earlier, BP analysis tends to be very complicated, and the first study, to the best of our knowledge, to deal with non-exponential services was carried out by Conolly (1960). To study the BP in the context of $GI/E_k/1$ queue, Conolly utilized the difference equation approach that he earlier applied (see Conolly (1959)) to study $GI/M/1$ queue.

It took more than 30 years from the first study on $GI/E_k/1$ queue to rekindle an interest in the BP analysis. Using probabilistic arguments, popularized by Neuts in the context of MAM, Ramaswami (1982) presented methods to study a large class of queues of the type $GI/PH/1$ (recall that such queues have matrix-geometric steady-state probability vectors). We briefly outline the major steps here, and for full details we refer to Ramaswami (1982).

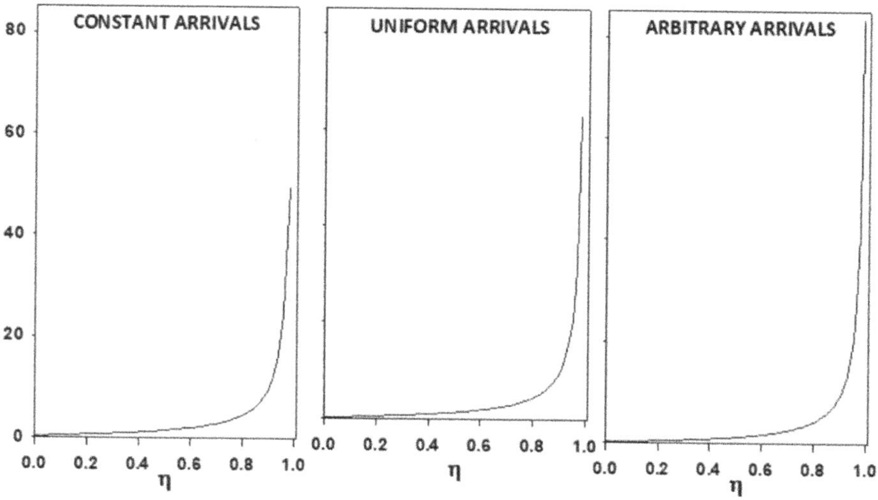

Figure 4.3. *Mean busy period versus η for example 4.11*

We look at the $GI/PH/1$ queue in which the services are of phase type with an irreducible representation $(\boldsymbol{\beta}, S)$ of order n. Let the average service rate be denoted by μ so that $\mu = [\boldsymbol{\beta}(-S)^{-1}\boldsymbol{e}]^{-1}$. We assume that $\lambda < \mu$ so as to have a stable queue. Recall from Chapter 4 of Volume 1 the following:

$$h(t) = \boldsymbol{\beta}e^{St}\boldsymbol{S^0},\ H(t) = 1 - \boldsymbol{\beta}e^{St}\mathbf{e}, t \geq 0,\ h^*(s) = \boldsymbol{\beta}(sI - S)^{-1}\boldsymbol{S^0}, Re(s) \geq 0.$$

The notation set for the counting process as well as for the transition probability matrix (TPM) of the MRP for the $GI/PH/1$ queue in section 2.1 will be used.

Referring to the TPM, $\tilde{P}(t)$, of the MRP governing the $GI/PH/1$ model (see equation [2.3]), suppose we define $r_{j,l}^{(k)}(t),\ 1 \leq j,l \leq m, k \geq 0$, as the expected number of visits during the time interval $(0, \min(t,\tau)]$ (here τ is the epoch of the first return to level $\boldsymbol{i}$) to the state $(i+k, l)$ starting in state (i,j) before the first return to level $\boldsymbol{i}$. That is, we define:

$$r_{j,l}^{(k)}(t) = \sum_{n=0}^{\infty} \boldsymbol{i}p_{(i,j),(i+k,l)}^{(n)}(t),\ 1 \leq j,l \leq m, k \geq 0, \qquad [4.27]$$

and their associated matrices to be $R^{(k)}(t)$. That is, $R^{(k)}(t) = \{(r_{j,l}^{(k)}(t))\},\ 1 \leq j,l \leq m, k \geq 0$. We adopt the normal convention that $R^{(0)}(t) = I$, an identity matrix of dimension m.

Recall that ${}_{\boldsymbol{i}}p^{(n)}_{(i,j),(i+k,l)}(t)$ gives the conditional probability that the MRP, $P(.)$, starting at time 0 in state (i, j) makes n transitions during $(0, t]$ by avoiding the level $\boldsymbol{i}$ and that the nth transition results in MRP being in state $(i + k, l)$. The structure of $P(.)$ makes the taboo probabilities to be independent of i and depends only on the submatrix obtained from Q by deleting all rows and all columns with indices $(l', j'), l' \leq i, 1 \leq j' \leq m$. Also, note that these submatrices are identical for all $i \geq 0$.

For use in the following, we define $R^*(s)$ to denote the LST of $R(t)$ and we take $R^{(1)}(t) = R(t)$. Let $A_n^*(s)$ denote the LST of $A_n(t)$ (see equation [2.1] for details on $A_n(t)$). Then, it is known (see Ramaswami (1982)) that $R^*(s)$ is obtained as the minimal non-negative solution to the nonlinear matrix equation:

$$R^*(s) = \sum_{n=0}^{\infty}(R^*(s))^n A_n^*(s), \; Re(s) \geq 0, \qquad [4.28]$$

and that the rate matrix R (see section 2.1) is given to be $R = R^*(0) = R(\infty)$.

In order to derive the expressions for the LST of the BP distribution as well as the mean BP and the mean number served during the BP, we need to establish a number of auxiliary matrices (see Ramaswami (1982)). Let $c_{i,j}^{(n)}(t)$ denote the (conditional) probability that the BP finishes by time t with exactly $n + 1$ service completions and that the next BP starts in phase j, conditioned on the fact that at time 0, the arriving customer sees n in the system with the BP starting with the service phase in i. Define $C_n(t)$ to be the matrix whose (i, j)th element is given by $c_{i,j}^{(n)}(t)$ and let its matrix LST be denoted by $C_n^*(s)$. The matrices $C_n(t)$ and $C_n^*(s)$ are given by (see Ramaswami (1982)):

$$C_n(t) = \int_0^t P(n, u)\boldsymbol{e}\boldsymbol{\beta}[1 - F(u)]du, \; n \geq 0, \qquad [4.29]$$

$$\begin{aligned} C_0^*(s) &= [I - A_0^*(s)](sI - S)^{-1}\boldsymbol{S^0}\boldsymbol{\beta}, \\ C_n^*(s) &= [C_{n-1}^*(s) - A_n^*(s)](sI - S)^{-1}\boldsymbol{S^0}\boldsymbol{\beta}, \; n \geq 1. \end{aligned} \qquad [4.30]$$

For $1 \leq i \leq m$, let Y_i denote the duration of the BP that started in phase i and let $\vartheta_i^*(s)$ denote its LST. Let $\boldsymbol{\vartheta}^*(\mathrm{s})$ denote the vector LST whose ith component is given by $\vartheta_i^*(s)$. Ramaswami (1982) derived an explicit expression for this vector LST as follows:

$$\begin{aligned} \boldsymbol{\vartheta}^*(\mathrm{s}) &= \sum_{n=0}^{\infty}[R^*(s)]^n C_n^*(s)\boldsymbol{e} \\ &= [I - R^*(s)][I - h^*(s)R^*(s)]^{-1}(sI - S)^{-1}\boldsymbol{S^0}, \; Re(s) \geq 0. \end{aligned} \qquad [4.31]$$

Suppose that:

$$\boldsymbol{\mu}'_{\vartheta} = -\frac{d}{ds}\boldsymbol{\vartheta}^*(s)\Big]_{s=0}. \qquad [4.32]$$

Then, verify that (see Ramaswami (1982)):

$$\boldsymbol{\mu}'_{\vartheta} = (-S)^{-1}\boldsymbol{e} + \frac{1}{\mu}(I - R)^{-1}R, \qquad [4.33]$$

and hence we have:

$$\mu'_{N_Y} = \boldsymbol{\beta\mu}'_{\vartheta} = \boldsymbol{\beta}(I - R)^{-1}\boldsymbol{e}, \;\; \mu'_Y = \frac{1}{\mu}\mu'_{N_Y}, \qquad [4.34]$$

REMARK 4.5.– It is worth noting that the expression for μ'_{N_Y} was pointed out using a probabilistic argument in remark 2.10.

REMARK 4.6.– While the BP expressions for the $GI/M/1$ were obtained using scalar versions earlier, one can easily verify those expressions by particularizing from the $GI/PH/1$ case. Very briefly, we see for the $GI/M/1$ queue that the expression given in equation [4.31] can be simplified due to the scalar nature of the underlying functions as follows. First, obtain $R^*(s)$ as the solution to:

$$R^*(s) = \sum_{n=0}^{\infty}(R^*(s))^n \int_0^\infty e^{-(s+\mu)t}\frac{(\mu t)^n}{n!}dF(t)$$

$$\Rightarrow R^*(s) = f^*[s + \mu(1 - R^*(s))], \; Re(s) \geq 0.$$

Once $R^*(s)$ is obtained, then the LST of the BP is computed as:

$$\vartheta^*(s) = \frac{\mu(1 - R^*(s))}{s + \mu(1 - R^*(s))}, \; Re(s) \geq 0,$$

which agrees with equation [4.23] obtained for the scalar case.

REMARK 4.7.– The BP expressions for $M/M/1$ are now obtained by particularizing from $GI/PH/1$ queue. First, verify that, for this model, the following are true. The Laplace transform (LT) of the inter-arrival times is given by $f^*(s) = \dfrac{\lambda}{s+\lambda}$. The scalar quantity, $R^*(s)$, is obtained as the solution to:

$$R^*(s) = \frac{\lambda}{\lambda + s + \mu(1 - R^*(s))}$$

$$\Rightarrow \; R^*(s) = \tfrac{1}{2\mu}\left[(\lambda + \mu + s) - \sqrt{[(\lambda + \mu + s)^2 - 4\lambda\mu]}\right],$$

from which it can be seen that $\vartheta^*(s) = \frac{\mu}{\lambda}R^*(s)$ is obtained as:

$$\vartheta^*(s) = \frac{1}{2\lambda}\left[(\lambda+\mu+s) - \sqrt{[(\lambda+\mu+s)^2 - 4\lambda\mu]}\right],$$

which again agrees with [4.23] obtained for the scalar case.

EXAMPLE 4.12.– Referring to example 2.1, compute the vector $\boldsymbol{\mu}'_{\vartheta}$, the mean BP and the mean number served during a BP.

Using the R matrix and the matrix S (which corresponds to the Erlang of order 3), from that example, verify that:

$$\boldsymbol{\mu}'_{\vartheta} = (6.5872,\ 2.8730,\ 0.7986),\ \ \mu'_Y = 6.5872,\ \ \mu'_{N_Y} = \frac{6.5872}{0.95} = 6.9339.$$

EXAMPLE 4.13.– Referring to example 2.2, compute the vector $\boldsymbol{\mu}'_{\vartheta}$, the mean BP and the mean number served during a BP.

Using the R matrix and the matrix S (which corresponds to the Erlang of order 3), from that example, verify that:

$$\boldsymbol{\mu}'_{\vartheta} = (10.9740,\ 5.6175,\ 1.7823),\ \ \mu'_Y = 10.9740,\ \ \mu'_{N_Y} = \frac{10.9740}{0.95} = 11.5516.$$

EXAMPLE 4.14.– Referring to example 2.3, compute the vector $\boldsymbol{\mu}'_{\vartheta}$, the mean BP and the mean number served during a BP.

Using the R matrix and the matrix S (which corresponds to the Erlang of order 3), from that example, verify that:

$$\boldsymbol{\mu}'_{\vartheta} = (10.0812,\ 5.7197,\ 2.2525),\ \ \mu'_Y = 10.0812,\ \ \mu'_{N_Y} = \frac{10.0812}{0.95} = 10.6118.$$

EXAMPLE 4.15.– Referring to example 2.4, compute the vector $\boldsymbol{\mu}'_{\vartheta}$, the mean BP and the mean number served during a BP.

Using the R matrix and the matrix S (which corresponds to the Erlang of order 3), from that example, verify that:

$$\boldsymbol{\mu}'_{\vartheta} = (11.1203,\ 6.1171,\ 2.2185),\ \ \mu'_Y = 11.1203,\ \ \mu'_{N_Y} = \frac{11.1203}{0.95} = 11.7056.$$

EXAMPLE 4.16.– Referring to example 2.6, discuss the effect of the traffic intensity as well as the spectral radius, ρ, on the mean BP, μ'_Y.

Toward this end, the plots of μ'_Y versus ρ and μ'_Y versus η are displayed in Figures 4.4 and 4.5, respectively.

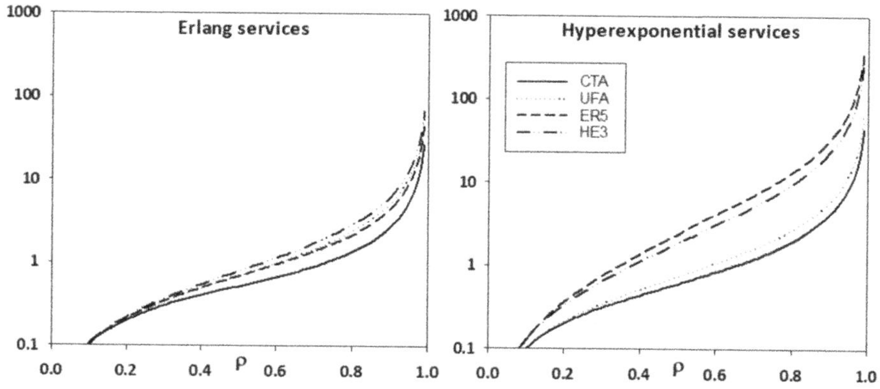

Figure 4.4. *Mean busy period versus ρ under various scenarios for example 4.16*

EXAMPLE 4.17.– Referring to the data given in example 2.7, discuss the behavior of the mean BP. In Table 4.6, we display the three measures and μ'_Y under various scenarios. It is obvious that the mean number in the system for Erlang arrivals is smaller than the hyperexponential ones. We notice a similar pattern for the mean BP. This, as pointed out earlier, indicates the upsurges in the hyperexponential case leading to a higher mean BP for the hyperexponential case.

$Service$	$ER\,5\ arrivals$			$HE\,5\ arrivals$		
PH	L_s	η	μ'_Y	L_s	η	μ'_Y
CNV	14.6434	0.9410	10.1960	39.6009	0.9758	36.4909
MXT	14.9578	0.9379	12.8923	40.1402	0.9758	32.4157
MIN	10.0798	0.9092	11.2809	35.0279	0.9722	36.0113
MAX	19.6261	0.9558	10.5043	44.6887	0.9788	35.0931

Table 4.6. *L_s, η and μ'_Y for example 4.17*

EXAMPLE 4.18.– Referring to example 2.9, discuss the effect of the traffic intensity, ρ, as well as the spectral radius, η, on the mean BP, μ'_Y. Toward this end, the plots of μ'_Y versus ρ and μ'_Y versus η are displayed in Figures 4.6 and 4.7, respectively. From these figures it is clear that the patterns are similar for both Erlang and hyperexponential services. The dependence of the mean BP on the type of arrivals as well as on the traffic intensity is also seen for both types of arrivals. The upsurges in the hyperexponential services are seen clearly for all but constant arrivals. In the case of constant arrivals, the upsurges are not significant.

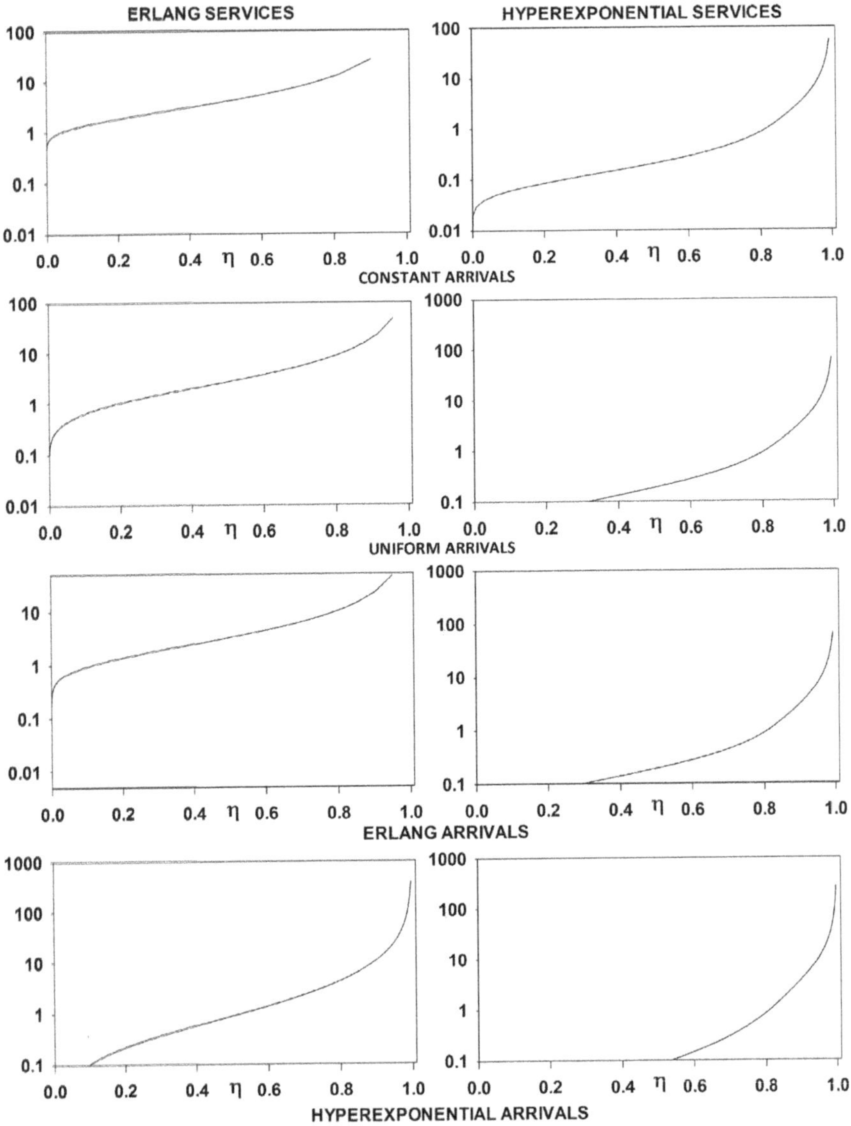

Figure 4.5. *Mean busy period versus η under various scenarios for example 4.16*

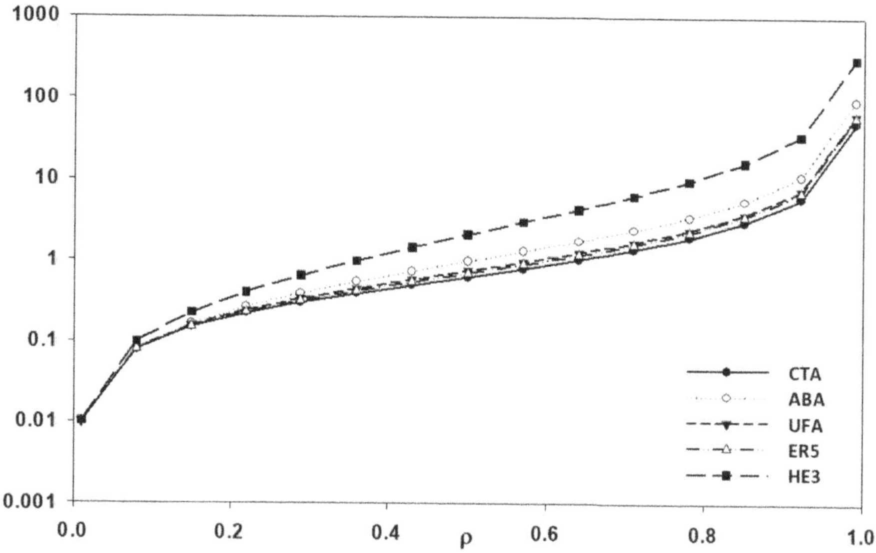

Figure 4.6. *Mean busy period versus ρ under various scenarios for example 4.18*

EXAMPLE 4.19.– Referring to example 2.10, compute the vector $\boldsymbol{\mu}'_{\vartheta}$, the mean BP and the mean number served during a BP.

Using the R matrix and the matrix S (which corresponds to the Erlang of order 3), from that example, verify that:

$$\boldsymbol{\mu}'_{\vartheta} = \left(\frac{1}{2}, \frac{1}{3}, \frac{1}{6}\right), \quad \mu'_Y = \frac{1}{2}, \quad \mu'_{N_Y} = 6.$$

EXAMPLE 4.20.– Revisit example 2.11 and compare the mean number in system (μ_s), the spectral radius (η) and the mean BP (μ'_Y). Recall that the models considered in this example are of $M/PH/1$-type using the $GI/PH/1$ tools, namely, looking at embedded epochs at arrivals, and hence we know (see equation [4.5]) that the mean BP depends only on ρ and can serve as another internal accuracy check in numerical computations. In Table 4.7, selected measures for the waiting time in queue under various scenarios for the M/PH/1 model are displayed.

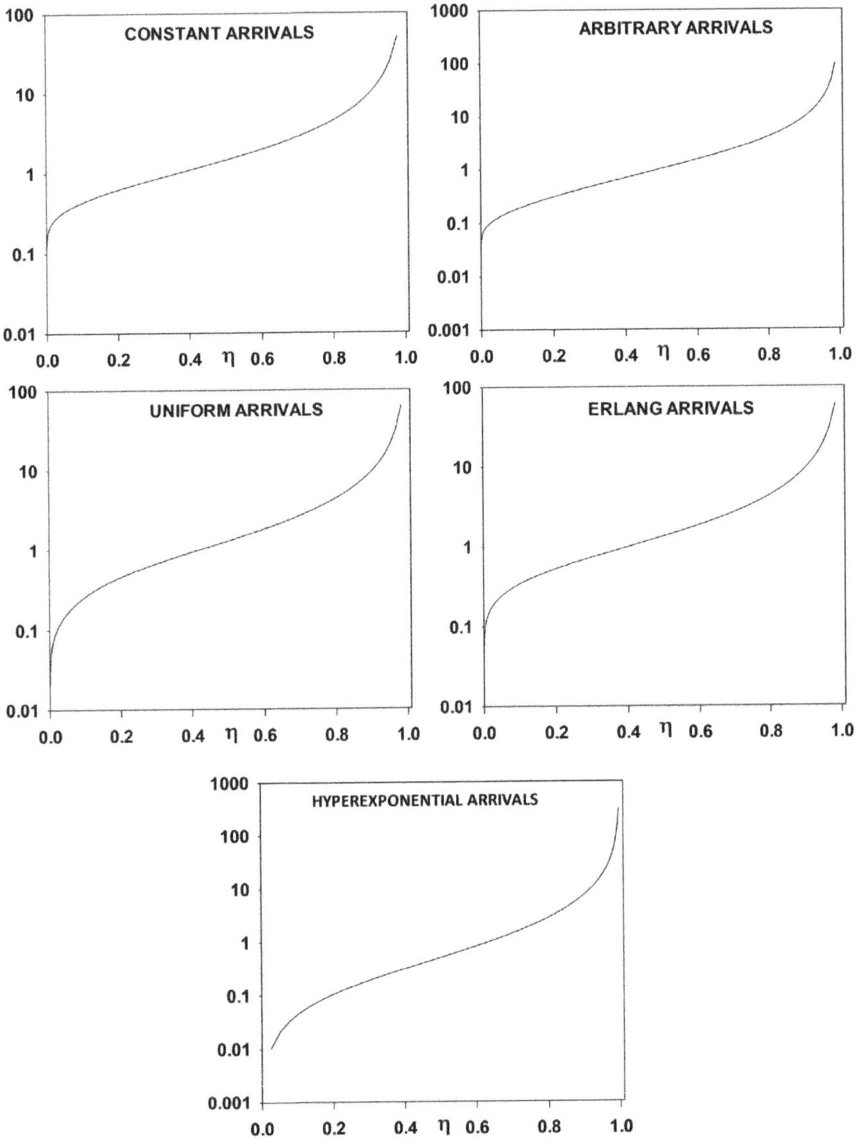

Figure 4.7. *Mean busy period versus η under various scenarios for example 4.18*

4.5. *BMAP/PH/1* queue

We will revisit $BMAP/PH/1$ queue studied in continuous-time (see section 3.1) with regard to the first passage times. Using the same notation as in section 3.1, the generator of the process governing the queueing model is given by (see equation [3.1]):

$$Q = \begin{pmatrix} C_0 & C_1 & C_2 & C_3 & C_4 & \cdots \\ F & H_1 & H_2 & H_3 & H_4 & \cdots \\ & H_0 & H_1 & H_2 & H_3 & \cdots \\ & & H_0 & H_1 & H_2 & \cdots \\ & & & \ddots & \ddots & \ddots \end{pmatrix}, \quad [4.35]$$

where:

$$\begin{aligned} &C_0 = D_0,\ H_0 = \boldsymbol{S^0}\boldsymbol{\beta} \otimes I, C_k = \boldsymbol{\beta} \otimes D_k,\ k \geq 1, \\ &F = \boldsymbol{S^0} \otimes I,\ H_1 = S \oplus D_0,\ \ H_k = I \otimes D_{k-1},\ k \geq 2. \end{aligned} \quad [4.36]$$

TOS	$\rho = 0.50$			$\rho = 0.80$			$\rho = 0.95$			$\rho = 0.99$		
	μ'_{W_s}	η	μ'_Y	μ'_{W_s}	η	μ'_Y	μ'_{W_s}	η	μ'_Y	μ'_{W_s}	η	μ'_Y
$ER\,1$	1	0.5	1	4	0.8	4	19	0.95	19	98.99	0.99	99
$ER\,2$	0.875	0.4101	1	3.2	0.7453	4	14.488	0.9341	19	74.49	0.9867	99
$ER\,5$	0.8	0.3401	1	2.72	0.6955	4	11.78	0.9185	19	59.79	0.9834	99
$HE\,1$	2.072	0.8036	1	10.86	0.9332	4	57.698	0.9839	19	308.9	0.9968	99
$HE\,2$	2.362	0.8334	1	12.71	0.9438	4	68.152	0.9865	19	365.56	0.9973	99
$HE\,3$	3.295	0.9271	1	18.69	0.9703	4	101.852	0.9915	19	548.6	0.9982	99
MXT	1.107	0.5458	1	4.68	0.8282	4	22.848	0.9584	19	119.87	0.9917	99
MIN	0.971	0.4851	1	3.81	0.7897	4	17.937	0.947	19	93.22	0.9894	99
MAX	1.234	0.6969	1	5.5	0.8752	4	27.461	0.9668	19	144.91	0.9932	99

Table 4.7. *Selected measures for the waiting time in queue under various scenarios for the $M/PH/1$ model*

For use in the follwing, we define:

$$Q^* = \sum_{k=0}^{\infty} D_k,\ \boldsymbol{d} = \sum_{k=1}^{\infty} kD_k\boldsymbol{e},\ H = \sum_{n=0}^{\infty} H_n,\ \boldsymbol{h} = \sum_{k=2}^{\infty}(k-1)H_k\boldsymbol{e}, \quad [4.37]$$

and let $\boldsymbol{y} = (\boldsymbol{y}_0, \boldsymbol{y}_1, \cdots)$ denote the steady-state probability vector of Q.

Suppose that $\boldsymbol{\pi}$ is the steady-state probability vector of $H = (S + \boldsymbol{S^0}\boldsymbol{\beta}) \oplus Q^*$. Let $G(z,s)$, $\Phi(z,s)$ and $K(z,s)$, respectively, denote the joint transform matrices of the first passage times from level $i+1$ to level i (away from the boundary), from level $\mathbf{1}$ to level $\mathbf{0}$ and from level $\mathbf{0}$ to level $\mathbf{0}$.

We state a number of results for the $BMAP/PH/1$ queue in continuous-time, whose proofs (similar to what we have seen in earlier sections for the $M/G/1$-type queues in continuous-time), are left as exercises, to keep the continuity of our discussion on the BP. We assume that the queue is stable and the generator given in equation [4.35] is irreducible.

1) The vector $\boldsymbol{\pi}$ is given explicitly in terms of the input parameters as follows:

$$\boldsymbol{\pi} = \mu\boldsymbol{\beta}(-S)^{-1} \otimes \boldsymbol{\delta}. \qquad [4.38]$$

2) For $Re(s) \geq 0,\ |z| \leq 1$, we have:

$$G(z,s) = z(sI - H_1)^{-1}H_0 + (sI - H_1)^{-1}\sum_{n=2}^{\infty} H_n G^n(z,s),$$

$$\Phi(z,s) = z(sI - H_1)^{-1}F + (sI - H_1)^{-1}\sum_{n=2}^{\infty} H_n G^{n-1}(z,s)\Phi(z,s), \qquad [4.39]$$

$$K(z,s) = (sI - C_0)^{-1}\sum_{n=1}^{\infty} C_n G^{n-1}\Phi(z,s).$$

3) The matrix $G = G(1,0)$ is obtained as the minimal non-negative solution to the first equation below, and with the knowledge of G the other two equations yield K and L explicitly:

$$\sum_{n=0}^{\infty} H_n G^n = 0,$$

$$\Phi = \Phi(1,0) = \left[I - (-H_1)^{-1}\sum_{n=2}^{\infty} H_n G^{n-1}\right]^{-1}(-H_1)^{-1}F, \qquad [4.40]$$

$$K = K(1,0) = \left[\sum_{n=1}^{\infty}(\boldsymbol{\beta}\otimes(-D_0)^{-1}D_n)G^{n-1}\right]\Phi.$$

4) The vectors, $\tilde{\boldsymbol{\mu}}_G = \frac{\partial}{\partial z}G(z,s)\mathbf{e}\Big]_{s=0,z=1}$ and $\hat{\boldsymbol{\mu}}_G = -\frac{\partial}{\partial s}G(z,s)\mathbf{e}\Big]_{s=0,z=1}$, are given by:

$$\tilde{\boldsymbol{\mu}}_G = (I - G + \mathbf{e}\,\mathbf{g})[(H_0\mathbf{e} - \mathbf{h})\mathbf{g} - H]^{-1}(\mathbf{S^0}\otimes\mathbf{e}),$$

$$\hat{\boldsymbol{\mu}}_G = (I - G + \mathbf{e}\,\mathbf{g})[(H_0\mathbf{e} - \mathbf{h})\mathbf{g} - H]^{-1}\mathbf{e}, \qquad [4.41]$$

where $\mathbf{g}$ is the steady-state probability vector of G.

5) The following results will serve as internal accuracy checks in the numerical computation of G and its moment vectors:

$$\boldsymbol{g}\tilde{\boldsymbol{\mu}}_G = \frac{1}{1-\rho} \text{ and } \boldsymbol{g}\hat{\boldsymbol{\mu}}_G = \frac{1}{(1-\rho)\boldsymbol{\pi} H_0 \boldsymbol{e}}.$$

6) The mean recurrence time, μ'_{RT}, of level **0** is given by:

$$\mu'_{RT} = \boldsymbol{y}^* \left[\boldsymbol{e} + \frac{1}{\mu(1-\rho)}\boldsymbol{d} + \sum_{r=1}^{\infty}(\boldsymbol{\beta} \otimes D_r)[I - G^r]\left([(H_0\boldsymbol{e} - \boldsymbol{h}) - (\boldsymbol{e} \otimes \boldsymbol{d})]\boldsymbol{g} - H\right)^{-1}\boldsymbol{e}\right],$$

where $\boldsymbol{y}^*$ is given by:

$$\boldsymbol{y}^* = [\boldsymbol{y}_1(\boldsymbol{S}^{\mathbf{0}} \otimes \boldsymbol{e})]^{-1}\boldsymbol{y}_1(\boldsymbol{S}^{\mathbf{0}} \otimes (-D_0)^{-1}).$$

Note that the quantity $[\boldsymbol{y}_1(\boldsymbol{S}^{\mathbf{0}} \otimes \boldsymbol{e})]^{-1}\boldsymbol{y}_1(\boldsymbol{S}^{\mathbf{0}} \otimes I)$ gives the conditional probability that a departure leaves the system empty with the arrival process in one of m phases.

REMARK 4.8.– Note that the mean recurrence time μ'_{RT} should be equal to the inverse of the probability that a departure will leave the system empty. This fact will also serve as an accuracy check in the numerical computation.

EXAMPLE 4.21.– In this example, we revisit examples 1.9 and 3.1, and calculate the vectors $\tilde{\boldsymbol{\mu}}_G$ and $\hat{\boldsymbol{\mu}}_G$ along with the measures: μ'_Y, μ'_{N_Y}, μ'_K, and μ'_I, the mean BP, the mean number served during a BP, the mean recurrence time and the mean idle time. These are computed using the formulas of this section. Thus, in this example, we look at five BMAPs with the batch size modeled using Poisson distribution with parameter 1.25, and the service times are assumed to be Erlang of order 3. The arrival process will be normalized to have a rate of 1 (i.e $\lambda = 1$) and the service time will have a mean of 0.9 so that the traffic intensity is 0.9. We use the superscript in the measures to identify the nature of the arrival process for which the vectors are displayed. First, we display the vectors as follows:

$$\boldsymbol{g}^{(ER)} = (0.1279, 0.3241, 0.5479, 0, 0, 0, 0, 0, 0),$$
$$\tilde{\boldsymbol{\mu}}_G^{(ER)} = (3.588,\ 6.95,\ 13.301,\ 2.243,\ 4.41,\ 9.643,\ 1.376,\ 2.317,\ 5.608),$$
$$\hat{\boldsymbol{\mu}}_G^{(ER)} = (0.444,\ 1.111,\ 1.299,\ 0,\ 0.52,\ 1.369,\ 13.688,\ 12.319,\ 13.688).$$

$$\boldsymbol{g}^{(EX)} = (1,\ 0,\ 0),\ \ \tilde{\boldsymbol{\mu}}_G^{(EX)} = (10,\ 7,\ 4),\ \ \hat{\boldsymbol{\mu}}_G^{(EX)} = (9,\ 6,\ 3).$$

$$\boldsymbol{g}^{(HE)} = (0.1095,\ 0.0985,\ 0.7920,\ 0,\ 0,\ 0,\ 0,\ 0,\ 0),$$
$$\tilde{\boldsymbol{\mu}}_G^{(HE)} = (33.478,\ 20.24,\ 5.481,\ 23.442,\ 13.81,\ 3.88,\ 12.647,\ 7.385,\ 2.384),$$
$$\hat{\boldsymbol{\mu}}_G^{(HE)} = (30.13,\ 18.216,\ 4.933,\ 20.797,\ 12.129,\ 3.192,\ 10.783,\ 6.047,\ 1.545).$$

$$\boldsymbol{g}^{(NC)} = (0.2640,\ 0.6488,\ 0.0872,\ 0,\ 0,\ 0,\ 0,\ 0,\ 0),$$
$$\tilde{\boldsymbol{\mu}}_G^{(NC)} = (3.941,\ 12.223,\ 11.805,\ 2.56,\ 8.605,\ 8.502,\ 1.553,\ 4.869,\ 4.942),$$
$$\hat{\boldsymbol{\mu}}_G^{(NC)} = (3.547,\ 11.001,\ 10.624,\ 2.004,\ 7.444,\ 7.352,\ 0.797,\ 3.782,\ 3.848).$$

$\boldsymbol{g}^{(PC)} = (0.3420,\ 0.6562,\ 0.0017,\ 0,\ 0,\ 0,\ 0,\ 0,\ 0),$
$\tilde{\boldsymbol{\mu}}_G^{(PC)} = (3.772,\ 11.605,\ 639.174,\ 2.467,\ 8.158,\ 469.074,\ 1.518,\ 4.627,\ 259.464),$
$\hat{\boldsymbol{\mu}}_G^{(PC)} = (3.395,\ 10.445,\ 575.257,\ 1.92,\ 7.042,\ 421.867,\ 0.766,\ 3.564,\ 232.917).$

In Table 4.8, selected measures for the busy period for the $BMAP/ER/1$ model are displayed.

BMAP	μ_I'	μ_Y'	μ_{N_Y}'	μ_K'
ER	1.3688	12.3188	13.6876	13.6876
EX	2.25	20.25	22.5	22.5
HE	5.9103	53.193	59.1033	59.1033
NC	2.1867	19.6800	21.8666	21.8666
PC	2.7286	24.5577	27.2863	27.2863

Table 4.8. *Selected measures for the busy period for the $BMAP/ER/1$ model*

EXAMPLE 4.22.– In this example, we revisit examples 1.10 and 3.2, and calculate the vectors $\tilde{\boldsymbol{\mu}}_G$ and $\hat{\boldsymbol{\mu}}_G$ along with the measures: μ_Y', μ_{N_Y}' and μ_{RT}', the mean BP, the mean number served during a BP and the mean recurrence times. These are computed using the formulas of this section. Thus, in this example, we look at five BMAPs with the batch size modeled using Poisson distribution with parameter 1.25, and the service times are assumed to be hyperexponential with parameters 10 and 1 with mixing probabilities, respectively, given by 0.9 and 0.1. The arrival process will be normalized to have a rate of 1 (i.e $\lambda = 1$) and the service time will have a mean of 0.9 so that the traffic intensity is 0.9. We use the superscript in the measures to identify the nature of the arrival process for which the vectors are displayed. First, we display the vectors as follows:

$$\boldsymbol{g}^{(ER)} = (0.1688,\ 0.3092,\ 0.422,\ 0.0188,\ 0.0344,\ 0.0469),$$

$$\tilde{\boldsymbol{\mu}}_G^{(ER)} = (2.1266,\ 3.9105,\ 8.5187,\ 39.9978,\ 46.1725,\ 53.3248),$$

$$\hat{\boldsymbol{\mu}}_G^{(ER)} = (1.4876,\ 3.0931,\ 7.2405,\ 39.8349,\ 45.3921,\ 51.8292).$$

$$\boldsymbol{g}^{(EX)} = (0.9,\ 0.1),\ \ \tilde{\boldsymbol{\mu}}_G^{(EX)} = (5.7368,\ 48.3684),\ \ \hat{\boldsymbol{\mu}}_G^{(EX)} = (4.7368,\ 47.3684).$$

$$\boldsymbol{g}^{(HE)} = (0.1803,\ 0.1152,\ 0.6044,\ 0.02,\ 0.0128,\ 0.0672),$$

$$\tilde{\boldsymbol{\mu}}_G^{(HE)} = (14.1407,\ 8.5174,\ 2.6995,\ 91.3961,\ 74.5576,\ 30.5389),$$

$$\hat{\boldsymbol{\mu}}_G^{(HE)} = (12.3003,\ 7.2394,\ 2.0032,\ 86.0934,\ 70.9387,\ 31.3218).$$

$$g^{(NC)} = (0.278,\ 0.5092,\ 0.1128,\ 0.0309,\ 0.0566,\ 0.0125),$$

$$\tilde{\mu}_G^{(NC)} = (2.2942,\ 7.2124,\ 7.5598,\ 38.7621,\ 53.1117,\ 50.6285),$$

$$\hat{\mu}_G^{(NC)} = (1.6385,\ 6.0648,\ 6.3775,\ 38.7227,\ 51.6374,\ 49.4025).$$

$$g^{(PC)} = (0.3503,\ 0.5455,\ 0.0043,\ 0.0389,\ 0.0606,\ 0.0005),$$

$$\tilde{\mu}_G^{(PC)} = (2.0671,\ 6.122,\ 257.6084,\ 35.8576,\ 49.1036,\ 980.8926),$$

$$\hat{\mu}_G^{(PC)} = (1.4341,\ 5.0835,\ 231.4212,\ 36.1087,\ 48.03,\ 886.6402).$$

In Table 4.9, selected measures for the busy period for the $BMAP/HE/1$ model are displayed.

BMAP	μ'_I	μ'_Y	μ'_{N_Y}	μ'_K
ER	1.5994	14.3945	15.9939	15.9939
EX	2.25	20.25	22.5	22.5
HE	4.2415	38.1731	42.4145	42.4146
NC	2.2047	19.8424	22.0471	22.0471
PC	2.9242	26.3180	29.2422	29.2422

Table 4.9. *Selected measures for the busy period for the $BMAP/HE/1$ model*

4.6. Busy period in QBD-type queues

Generally speaking, it is fairly easy to set up a queueing model (subject to certain conditions enabling Markov property to hold good) as a QBD process. Due to the special structure of the infinitesimal generator the study of BP and the waiting distribution becomes much easier.

In earlier sections (see sections 7.6 and 8.6 of Volume 1 and others in the context of queueing models studied via the arbitrary time approach), QBD models were discussed both in the general context as well as in selected queueing models. Here, we revisit QBD-type queues from the BP point of view. Suppose that the infinitesimal generator of the underlying QBD process is given by:

$$Q = \begin{bmatrix} C_0 & C_1 & & & \\ F & H_1 & H_2 & & \\ & H_0 & H_1 & H_2 & \\ & & H_0 & H_1 & H_2 \\ & & & \ddots & \ddots & \ddots \end{bmatrix}, \qquad [4.42]$$

where C_0 is a square matrix of order m, C_1 is of dimension $m \times p$, F is of dimension $p \times m$ and H_0, H_1 and H_2 are square matrices of dimension p. We assume that the generator is irreducible, which is the case in most applications. For use in the following, let $\boldsymbol{y} = (\boldsymbol{y}_0, \boldsymbol{y}_1, \cdots)$ denote the steady-state probability vector of Q.

First, we look at the first passage time from level $\boldsymbol{i+r}$ to level $\boldsymbol{i}$ away from the boundary. Then, we look at the first passage time from level $\mathbf{1}$ to level $\mathbf{0}$, and finally the first passage time (or recurrence time) from level $\mathbf{0}$ to level $\mathbf{0}$.

First passage time from level $\boldsymbol{i+r}$ *to level* $\boldsymbol{i}$ *away from the boundary*: Suppose that the (matrix) joint probabilities, $\{G^r(n,t) : r \geq 1, n \geq 1, t \geq 0\}$, are such that the (j,k)th element of $G^r(n,t)$ gives the conditional probability that starting from state $(i+r, j)$, it takes exactly n left transitions to reach the state (i,k) and the total duration does not exceed t units of time.

It is easy to verify the following results:

$$\begin{aligned} G(1,t) &= \int_0^\infty e^{H_1 u} H_0 dt, \ t \geq 0, \\ G(k,t) &= \int_0^\infty e^{H_1 u} H_2 G^2(k, t-u) du, \ k \geq 2, \ t \geq 0, \end{aligned} \qquad [4.43]$$

from which using the Leibnitz's rule and taking the LST, we get:

$$\begin{aligned} G(1,s) &= (sI - H_1)^{-1} H_0, \ Re(s) \geq 0, \\ G(k,s) &= (sI - H_1)^{-1} H_2 G^2(k,s) du, \ k \geq 2, \ Re(s) \geq 0. \end{aligned} \qquad [4.44]$$

The joint transform, $G(z,s)$ of $G(n,t)$, for $Re(s) \geq 0, \ |z| \leq 1$, is then given by (see Neuts (1981)):

$$G(z,s) = z(sI - H_1)^{-1} H_0 + (sI - H_1)^{-1} H_2 G^2(z,s). \qquad [4.45]$$

The above equation can be intuitively explained as follows: the process can go from level $\boldsymbol{i+1}$ to level $\boldsymbol{i}$ without visiting any level to the right, and this case is governed by the first term; or it can go one level to the right and then go from there to level $\boldsymbol{i}$, which is governed by the second term.

Using equation [4.45] we see:

$$G = (-H_1)^{-1} H_0 + (-H_1)^{-1} H_2 G^2, \qquad [4.46]$$

whose solution has been discussed in section 7.6 of Volume 1.

The mean vectors, $\tilde{\mu}_G$ and $\hat{\mu}_G$ of $G(z,s)$ (see section 1.1) are, for the current case, given by:

$$\begin{aligned}\tilde{\mu}_G &= \frac{\partial}{\partial z}G(z,s)\Big]_{s=0,z=1} e = (I - G + e\,g)[(H_0 e - H_2 e)g - H]^{-1}H_0 e,\\ \hat{\mu}_G &= -\frac{\partial}{\partial s}G(z,s)\Big]_{s=0,z=1} e = (I - G + e\,g)[(H_0 e - H_2 e)g - H]^{-1}e,\end{aligned} \quad [4.47]$$

where $H = H_0 + H_1 + H_2$ and g is the left eigenvector of G corresponding to 1.

Denoting by π the invariant probability vector of H, the following results play a key role:

$$g\tilde{\mu}_G = \frac{1}{1-\rho} \text{ and } g\hat{\mu}_G = \frac{1}{(1-\rho)\pi H_0 e}. \quad [4.48]$$

First passage time from level **1** *to level* **0**: Suppose we denote by $L(n,t) = \{l_{i,j}(n,t)\}$ the matrix of dimension $n \times m$ whose (i,j)th element gives the conditional probability that the QBD process with generator (see equation [4.42]) starting in state $(1,i)$ at time 0 will reach the state $(0,j)$ by time t and it occurs by making exactly n left transitions. Using the law of total probability, it is easy to verify that the (matrix) joint transform, $L(z,s)$ of $\{L(n,t) : n \geq 1, t \geq 0\}$, defined as:

$$L(z,s) = \sum_{n=1}^{\infty}\int_{0}^{\infty} z^n e^{-st} dL(n,t), \ |z| \leq 1, \quad [4.49]$$

satisfies the following (intuitively obvious) equation, for $Re(s) \geq 0$, $|z| \leq 1$,

$$L(z,s) = z(sI - H_1)^{-1}F + (sI - H_1)^{-1}H_2 G(z,s)L(z,s). \quad [4.50]$$

From equation [4.50], it is easy to verify that the matrix L, whose (i,j)th element gives the conditional probability of eventually hitting the state $(0,k)$ starting from state $(1,j)$, is given by:

$$L = [I - (-H_1)^{-1}H_2 G]^{-1}(-H_1)^{-1}F. \quad [4.51]$$

Note that under the assumption of the QBD being irreducible, the matrix L is such that the row sums add up to 1.

The mean vectors, $\tilde{\mu}_L$ and $\hat{\mu}_L$, of $L(z,s)$, which, respectively, correspond to the conditional means of the number of left transitions and the times spent before hitting level **0** starting from level **1**, are given by:

$$\tilde{\mu}_L = \frac{\partial}{\partial z}L(z,s)\Big]_{s=0,z=1} e = \tilde{\mu}_G \text{ and } \hat{\mu}_L = -\frac{\partial}{\partial s}G(z,s)\Big]_{s=0,z=1} e = \hat{\mu}_G. \quad [4.52]$$

Note that the above result is intuitively obvious on noting that $F\boldsymbol{e} = H_0\boldsymbol{e}$.

First passage time from level **0** *to level* **0**: Suppose we denote by $K(n,t) = \{k_{i,j}(n,t)\}$ the matrix of dimension m whose (i,j)th element gives the conditional probability that the QBD process with generator (see equation [4.42]) starting in state $(0,i)$ at time 0 will reach the state $(0,j)$ by time t and it occurs by making exactly n left transitions. Using the law of total probability, it is easy to verify that the (matrix) joint transform, $K(z,s)$ of $\{K(n,t) : n \geq 1, t \geq 0\}$, defined as:

$$K(z,s) = \sum_{n=1}^{\infty} \int_0^{\infty} z^n e^{-st} dK(n,t), Re(s) \geq 0, \qquad [4.53]$$

satisfies the following (intuitively obvious) equation:

$$K(z,s) = (sI - C_0)^{-1} C_1 L(z,s), \; Re(s) \geq 0, \; |z| \leq 1. \qquad [4.54]$$

From equation [4.54], we get:

$$K = K(1,0) = (-C_0)^{-1} C_1 L. \qquad [4.55]$$

The mean vector, $\hat{\boldsymbol{\mu}}_K$, of $K(z,s)$, is such that the jth component gives the conditional mean recurrence time by returning to level **0** starting from state $(0,j)$ (see Neuts (1981)).

$$\hat{\boldsymbol{\mu}}_K = -\frac{\partial}{\partial s} K(z,s)\Big]_{s=0,z=1} \boldsymbol{e} = (-C_0)^{-1}\boldsymbol{e} + (-C_0)^{-1} C_1 \hat{\boldsymbol{\mu}}_G. \qquad [4.56]$$

$$\tilde{\boldsymbol{\mu}}_K = -\frac{\partial}{\partial z} K(z,s)\Big]_{s=0,z=1} \boldsymbol{e} = (-C_0)^{-1} C_1 \tilde{\boldsymbol{\mu}}_G. \qquad [4.57]$$

Equation [4.56] has an interesting probabilistic interpretation. The first term on the right-hand side gives the mean time before leaving the level **0** starting from any one of the states in $(0,j)$, $1 \leq j \leq m$. The second quantity is such that the (i,j)th term of $(-C_0)^{-1}C_1$ gives the probability of reaching state $(1,j)$ from state $(0,i)$, and hence the vector $(-C_0)^{-1}C_1\hat{\boldsymbol{\mu}}_G$ gives the mean time spent before returning to level **0** starting from state $(1,j)$.

The mean number served, μ'_{N_Y}, during a BP and the mean BP, μ'_Y, are given by:

$$\mu'_{N_Y} = \frac{1}{1-\rho} + c\boldsymbol{y}_0(-C_0)^{-1}C_1[I-G][(H_0\boldsymbol{e} - H_2\boldsymbol{e})\boldsymbol{g} - H]^{-1}H_0\boldsymbol{e},$$
$$[4.58]$$
$$\mu'_Y = \frac{1}{(1-\rho)\boldsymbol{\pi}H_0\boldsymbol{e}} + c\boldsymbol{y}_0(-C_0)^{-1}C_1[I-G][(H_0\boldsymbol{e} - H_2\boldsymbol{e})\boldsymbol{g} - H]^{-1}\boldsymbol{e},$$

where $c = [\boldsymbol{y}_0\boldsymbol{e}]^{-1}$.

The mean recurrence time, μ'_K, is given by:

$$\mu'_K = \boldsymbol{y}^*(-C_0)^{-1}\boldsymbol{e} + \boldsymbol{y}^*(-C_0)^{-1}C_1\hat{\boldsymbol{\mu}}_G, \quad [4.59]$$

where $\boldsymbol{y}^* = [\boldsymbol{y}_1 F\boldsymbol{e}]^{-1}\boldsymbol{y}_1 F$ stands for the conditional probability of starting the idle period in one of m states given that the QBD process hits level **0** from level **1**.

It is worth recalling at this point that when dealing with $BMAP/G/1$-type queues one of the key components is the G matrix (see section 1.1). When dealing with $GI/PH/1$-type queues (see section 2.1) it is the R matrix that plays an important role. As mentioned before, QBD processes can be viewed as the intersection of the two paradigms. In the following, we discuss both R and G matrices in the context of the same example.

EXAMPLE 4.23.– Suppose that the customers arrive singly with inter-arrival times modeled using an Erlang of order 3. Let the arrival rate be $\lambda = 1$. The service times are assumed to be Erlang order 3. The parameter of this Erlang is chosen such that the mean service time is taken to be 0.95. Verify that (after normalization) the standard deviation of the inter-arrival times and the service times are 0.5774 and 0.5485, respectively. This queueing model can be set up as a QBD process. This was seen in section 3.2. The matrices G and R are displayed below:

$$R = \begin{pmatrix} 0 & 0 & 0 & 0 & 0 & 0 & 0 & 0 & 0 \\ 0 & 0 & 0 & 0 & 0 & 0 & 0 & 0 & 0 \\ 0.7206 & 0.5872 & 0.5174 & 0.3695 & 0.4811 & 0.4997 & 0.1895 & 0.3391 & 0.4214 \\ 0 & 0 & 0 & 0 & 0 & 0 & 0 & 0 & 0 \\ 0 & 0 & 0 & 0 & 0 & 0 & 0 & 0 & 0 \\ 0.1514 & 0.2130 & 0.2330 & 0.5648 & 0.3844 & 0.3068 & 0.2897 & 0.3382 & 0.3221 \\ 0 & 0 & 0 & 0 & 0 & 0 & 0 & 0 & 0 \\ 0 & 0 & 0 & 0 & 0 & 0 & 0 & 0 & 0 \\ 0.0632 & 0.0823 & 0.0855 & 0.0324 & 0.0580 & 0.0721 & 0.5038 & 0.2752 & 0.1710 \end{pmatrix}$$

$$G = \begin{pmatrix} 0.1995 & 0.3569 & 0.4436 & 0 & 0 & 0 & 0 & 0 & 0 \\ 0.0885 & 0.3578 & 0.5537 & 0 & 0 & 0 & 0 & 0 & 0 \\ 0.1104 & 0.2861 & 0.6034 & 0 & 0 & 0 & 0 & 0 & 0 \\ 0.3049 & 0.3560 & 0.3391 & 0 & 0 & 0 & 0 & 0 & 0 \\ 0.0676 & 0.4259 & 0.5065 & 0 & 0 & 0 & 0 & 0 & 0 \\ 0.1010 & 0.2484 & 0.6506 & 0 & 0 & 0 & 0 & 0 & 0 \\ 0.5303 & 0.2897 & 0.1800 & 0 & 0 & 0 & 0 & 0 & 0 \\ 0.0359 & 0.5946 & 0.3695 & 0 & 0 & 0 & 0 & 0 & 0 \\ 0.0737 & 0.1678 & 0.7585 & 0 & 0 & 0 & 0 & 0 & 0 \end{pmatrix}$$

The key vectors are as follows:

$$\boldsymbol{g} = (0.1136,\ 0.3169,\ 0.5695,\ 0,\ 0,\ 0,\ 0,\ 0,\ 0),$$

$\tilde{\mu}_G = (11.7056,\ 16.8986,\ 23.38,\ 6.7723,\ 10.7412,\ 16.6696,\ 3.0019,$
$5.1092,\ 9.4348),$

$\hat{\mu}_G = (11.1204,\ 16.0537,\ 22.211,\ 6.1171,\ 9.8875,\ 15.5194,\ 2.2185,$
$4.2204,\ 8.3297),$

$\tilde{\mu}_K = (11.7056,\ 11.7056,\ 11.7056),\ \hat{\mu}_K = (2.1204,\ 11.7870,\ 11.4537).$

The means, μ'_Y, μ'_{N_Y}, μ'_K and μ'_I (recall that these are the mean BP, the mean number served during a BP, the mean recurrence time and the mean idle time, respectively), are calculated to be:

$$\mu'_Y = 11.1204,\ \mu'_{N_Y} = 11.7056,\ \mu'_K = 11.7056,\ \mu'_I = 0.5853,\ \eta = 0.3286.$$

EXAMPLE 4.24.– This example is similar to example 4.23 except now the inter-arrival times are modeled using a hyperexponential distribution with parameters 10, 5 and 1 with mixing probabilities given, respectively, by 0.6, 0.3 and 0.1. The parameters of the arrival process will be normalized so that the arrival rate is 1.0. Verify that (after normalization) the standard deviation of the inter-arrival times and the service times are 1.9688 and 0.5485, respectively. This queueing model can be set up as a QBD process. This was seen in section 3.2. The matrices G and R are displayed below:

$$R = \begin{pmatrix} 0.3772 & 0.3319 & 0.4383 & 0.2223 & 0.2461 & 0.4098 & 0.1310 & 0.1825 & 0.3831 \\ 0.1886 & 0.1659 & 0.2192 & 0.1111 & 0.1231 & 0.2049 & 0.0655 & 0.0913 & 0.1916 \\ 0.0377 & 0.0332 & 0.0438 & 0.0222 & 0.0246 & 0.0410 & 0.0131 & 0.0183 & 0.0383 \\ 0.1236 & 0.1520 & 0.2850 & 0.3192 & 0.2677 & 0.3315 & 0.1881 & 0.1986 & 0.3100 \\ 0.0618 & 0.0760 & 0.1425 & 0.1596 & 0.1339 & 0.1658 & 0.0941 & 0.0993 & 0.1550 \\ 0.0124 & 0.0152 & 0.0285 & 0.0319 & 0.0268 & 0.0332 & 0.0188 & 0.0199 & 0.0310 \\ 0.0896 & 0.0993 & 0.1653 & 0.0528 & 0.0736 & 0.1545 & 0.2775 & 0.2096 & 0.2096 \\ 0.0448 & 0.0496 & 0.0826 & 0.0264 & 0.0368 & 0.0772 & 0.1388 & 0.1048 & 0.1048 \\ 0.0090 & 0.0099 & 0.0165 & 0.0053 & 0.0074 & 0.0154 & 0.0278 & 0.0210 & 0.0210 \end{pmatrix}$$

$$G = \begin{pmatrix} 0.2808 & 0.1642 & 0.5550 & 0 & 0 & 0 & 0 & 0 & 0 \\ 0.0589 & 0.5321 & 0.4090 & 0 & 0 & 0 & 0 & 0 & 0 \\ 0.0190 & 0.0390 & 0.9420 & 0 & 0 & 0 & 0 & 0 & 0 \\ 0.4216 & 0.1450 & 0.4334 & 0 & 0 & 0 & 0 & 0 & 0 \\ 0.0520 & 0.6506 & 0.2973 & 0 & 0 & 0 & 0 & 0 & 0 \\ 0.0148 & 0.0283 & 0.9568 & 0 & 0 & 0 & 0 & 0 & 0 \\ 0.6449 & 0.0974 & 0.2577 & 0 & 0 & 0 & 0 & 0 & 0 \\ 0.0350 & 0.8029 & 0.1621 & 0 & 0 & 0 & 0 & 0 & 0 \\ 0.0088 & 0.0154 & 0.9757 & 0 & 0 & 0 & 0 & 0 & 0 \end{pmatrix}$$

The key vectors are as follows:

$$\boldsymbol{g} = (0.0303,\ 0.0844,\ 0.8853,\ 0,\ 0,\ 0,\ 0,\ 0,\ 0),$$

$$\tilde{\mu}_G = (66.901,\ 49.211,\ 15.612,\ 50.442,\ 34.819,\ 10.393,\ 29.300,\ 18.805,\ 5.489),$$

$$\hat{\mu}_G = (63.556,\ 46.750,\ 14.832,\ 47.604,\ 32.761,\ 9.557,\ 27.201,\ 17.232,\ 4.581),$$

$$\tilde{\mu}_K = (56.4653,\ 56.4653,\ 56.4653),\quad \hat{\mu}_K = (54.0966,\ 54.5511,\ 58.1875).$$

The means, μ'_Y, μ'_{N_Y}, μ'_K and μ'_I, are calculated to be:

$$\mu'_Y = 53.642,\ \mu'_{N_Y} = 56.4653,\ \ \mu'_K = 56.4653,\ \mu'_I = 2.8233,\ \ \eta = 0.9754.$$

EXAMPLE 4.25.– This example is similar to example 4.23 except now the services are modeled using a hyperexponential with parameters 10, 5 and 1, with mixing probabilities, respectively, given by 0.6, 0.3 and 0.1. The parameters of this service process will be normalized to make the service mean 0.95. Verify that (after normalization) the standard deviation of the inter-arrival times and the service times are 0.5774 and 1.8703, respectively. This queueing model can be set up as a QBD process. This was seen in section 3.2. The matrices G and R are displayed below:

$$R = \begin{pmatrix} 0 & 0 & 0 & 0 & 0 & 0 & 0 & 0 & 0 \\ 0 & 0 & 0 & 0 & 0 & 0 & 0 & 0 & 0 \\ 0.6116 & 0.3787 & 0.2388 & 0.0302 & 0.0432 & 0.0472 & 0.0129 & 0.0212 & 0.0266 \\ 0 & 0 & 0 & 0 & 0 & 0 & 0 & 0 & 0 \\ 0 & 0 & 0 & 0 & 0 & 0 & 0 & 0 & 0 \\ 0.0873 & 0.1250 & 0.1365 & 0.7773 & 0.6093 & 0.4817 & 0.0239 & 0.0430 & 0.0580 \\ 0 & 0 & 0 & 0 & 0 & 0 & 0 & 0 & 0 \\ 0 & 0 & 0 & 0 & 0 & 0 & 0 & 0 & 0 \\ 0.1589 & 0.2604 & 0.3260 & 0.1016 & 0.1824 & 0.2461 & 0.9719 & 0.9491 & 0.9301 \end{pmatrix}$$

$$G = \begin{pmatrix} 0.2908 & 0.1864 & 0.1227 & 0.1454 & 0.0932 & 0.0614 & 0.0485 & 0.0311 & 0.0205 \\ 0.0522 & 0.3303 & 0.2175 & 0.0261 & 0.1651 & 0.1087 & 0.0087 & 0.0550 & 0.0362 \\ 0.0925 & 0.1221 & 0.3854 & 0.0463 & 0.0611 & 0.1927 & 0.0154 & 0.0204 & 0.0642 \\ 0.2215 & 0.2010 & 0.1775 & 0.1108 & 0.1005 & 0.0887 & 0.0369 & 0.0335 & 0.0296 \\ 0.0755 & 0.2786 & 0.2460 & 0.0377 & 0.1393 & 0.1230 & 0.0126 & 0.0464 & 0.0410 \\ 0.1046 & 0.1545 & 0.3409 & 0.0523 & 0.0773 & 0.1704 & 0.0174 & 0.0258 & 0.0568 \\ 0.1421 & 0.2068 & 0.2510 & 0.0711 & 0.1034 & 0.1255 & 0.0237 & 0.0345 & 0.0418 \\ 0.1068 & 0.2228 & 0.2704 & 0.0534 & 0.1114 & 0.1352 & 0.0178 & 0.0371 & 0.0451 \\ 0.1150 & 0.1937 & 0.2913 & 0.0575 & 0.0968 & 0.1457 & 0.0192 & 0.0323 & 0.0486 \end{pmatrix}$$

The key vectors are as follows:

$$\boldsymbol{g} = (0.1175,\ 0.2063,\ 0.2762,\ 0.0587,\ 0.1031,\ 0.1381,\ 0.0196,\ 0.0344,\ 0.046),$$

$$\tilde{\boldsymbol{\mu}}_G = (4.838,\ 7.800,\ 13.049,\ 12.096,\ 16.379,\ 22.314,\ 79.487,$$
$$85.545,\ 92.0725),$$
$$\hat{\boldsymbol{\mu}}_G = (4.077,\ 6.892,\ 11.878,\ 11.405,\ 15.473,\ 21.112,\ 78.881,$$
$$84.636,\ 90.836),$$
$$\tilde{\boldsymbol{\mu}}_K = (14.4799,\ 14.4799,\ 14.4799),,\ \ \hat{\boldsymbol{\mu}}_K = (14.7559,\ 14.4226,\ 14.0893).$$

The means, μ'_Y, μ'_{N_Y}, μ'_K and μ'_I, are calculated to be:

$$\mu'_Y = 13.7559,\ \mu'_{N_Y} = 14.4799,\ \ \mu'_K = 14.4799,\ \mu'_I = 0.7240,\ \ \eta = 0.9764.$$

EXAMPLE 4.26.– This example is similar to example 4.25 except now the inter-arrival times are modeled using a hyperexponential with parameters 10, 5 and 1, with mixing probabilities, respectively, given by 0.6, 0.3 and 0.1. The parameters of this service process will be normalized to make the service mean 0.95. Verify that (after normalization) the standard deviation of the inter-arrival times and the service times are 1.9688 and 1.8703, respectively. This queueing model can be set up as a QBD process. This was seen in section 3.2. The matrices G and R are displayed below:

$$R = \begin{pmatrix} 0.3672 & 0.2744 & 0.1898 & 0.0504 & 0.0614 & 0.0948 & 0.0232 & 0.0347 & 0.0964 \\ 0.1836 & 0.1372 & 0.0949 & 0.0252 & 0.0307 & 0.0474 & 0.0116 & 0.0173 & 0.0482 \\ 0.0367 & 0.0274 & 0.0190 & 0.0050 & 0.0061 & 0.0095 & 0.0023 & 0.0035 & 0.0096 \\ 0.0876 & 0.1067 & 0.1648 & 0.4520 & 0.3730 & 0.3113 & 0.0271 & 0.0456 & 0.1543 \\ 0.0438 & 0.0534 & 0.0824 & 0.2260 & 0.1865 & 0.1557 & 0.0136 & 0.0228 & 0.0771 \\ 0.0088 & 0.0107 & 0.0165 & 0.0452 & 0.0373 & 0.0311 & 0.0027 & 0.0046 & 0.0154 \\ 0.0801 & 0.1199 & 0.3331 & 0.0539 & 0.0907 & 0.3065 & 0.5677 & 0.5469 & 0.7990 \\ 0.0401 & 0.0599 & 0.1666 & 0.0269 & 0.0453 & 0.1533 & 0.2838 & 0.2735 & 0.3995 \\ 0.0080 & 0.0120 & 0.0333 & 0.0054 & 0.0091 & 0.0307 & 0.0568 & 0.0547 & 0.0799 \end{pmatrix}$$

$$G = \begin{pmatrix} 0.3869 & 0.0892 & 0.1238 & 0.1935 & 0.0446 & 0.0619 & 0.0645 & 0.0149 & 0.0206 \\ 0.0524 & 0.4658 & 0.0819 & 0.0262 & 0.2329 & 0.0409 & 0.0087 & 0.0776 & 0.0136 \\ 0.0141 & 0.0159 & 0.5700 & 0.0071 & 0.0079 & 0.2850 & 0.0024 & 0.0026 & 0.0950 \\ 0.2961 & 0.1130 & 0.1908 & 0.1481 & 0.0565 & 0.0954 & 0.0494 & 0.0188 & 0.0318 \\ 0.0664 & 0.3917 & 0.1419 & 0.0332 & 0.1959 & 0.0710 & 0.0111 & 0.0653 & 0.0237 \\ 0.0217 & 0.0275 & 0.5507 & 0.0109 & 0.0138 & 0.2754 & 0.0036 & 0.0046 & 0.0918 \\ 0.1317 & 0.1160 & 0.3523 & 0.0659 & 0.0580 & 0.1762 & 0.0220 & 0.0193 & 0.0587 \\ 0.0681 & 0.2102 & 0.3217 & 0.0340 & 0.1051 & 0.1609 & 0.0113 & 0.0350 & 0.0536 \\ 0.0402 & 0.0624 & 0.4974 & 0.0201 & 0.0312 & 0.2487 & 0.0067 & 0.0104 & 0.0829 \end{pmatrix}$$

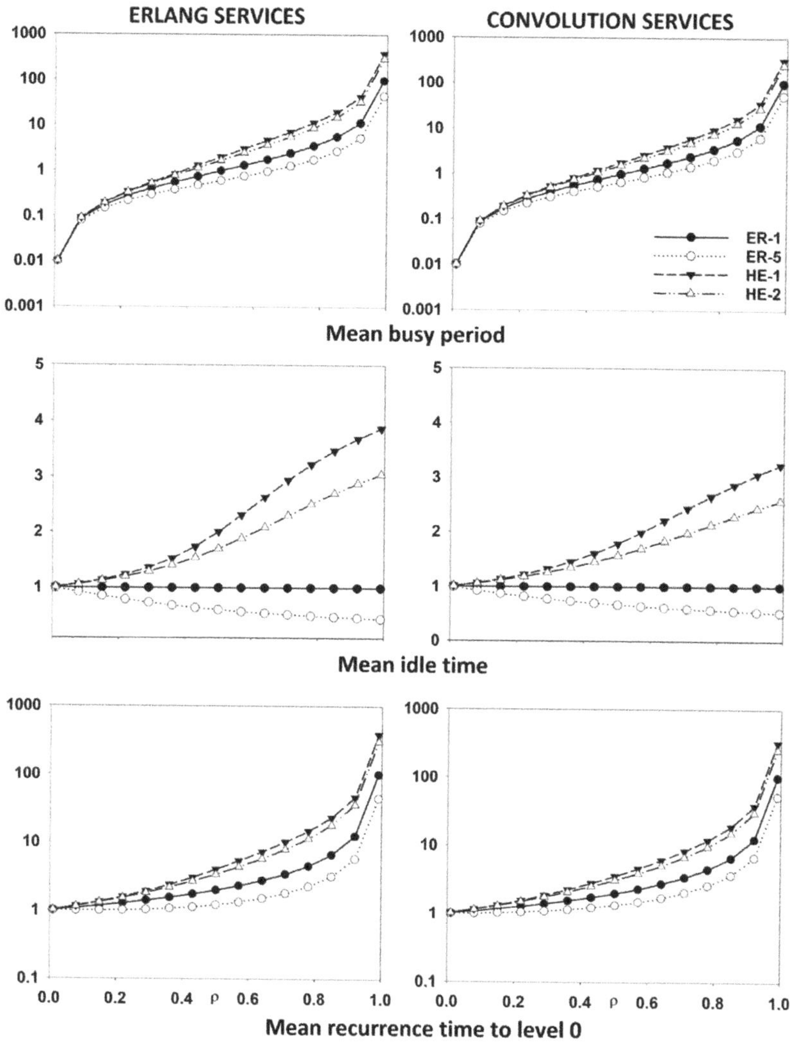

Figure 4.8. *Plots of μ'_Y, μ'_K and μ'_I under various scenarios for example 4.27*

The key vectors are as follows:

$$\boldsymbol{g} = (0.0523,\ 0.089,\ 0.4587,\ 0.0261,\ 0.0445,\ 0.2294,\ 0.0087,\ 0.0148,\ 0.0765),$$

$$\tilde{\boldsymbol{\mu}}_G = (27.876,\ 18.766,\ 5.786,\ 46.010,\ 34.469,\ 11.969,\ 137.167,\ 125.326,\ 74.321),$$

$\hat{\mu}_G = (25.964,\ 17.309,\ 4.979,\ 43.623,\ 32.659,\ 11.284,\ 133.677,\ 122.428,\ 73.973),$

$\tilde{\mu}_K = (38.2363,\ 38.2363,\ 38.2363), \quad \hat{\mu}_K = (36.779,\ 37.2335,\ 40.8699).$

The means, μ'_Y, μ'_{N_Y}, μ'_K and μ'_I, are calculated to be:

$$\mu'_Y = 36.3244,\ \mu'_{N_Y} = 38.2363,\ \mu'_K = 38.2363,\ \mu'_I = 1.9118,\ \eta = 0.9869.$$

REMARK 4.9.– Looking at examples 4.23 through 4.26, it is clear the role of the variability in the inter-arrival times on the measures such as the mean BP, the mean number served, the mean recurrence time and the mean idle time.

EXAMPLE 4.27.– In this example, we look at four different arrival processes: (1) ER 1: exponential; (2) ER 5: Erlang of order 5; (3) HE 1: hyperexponential with parameters 10 and 1 with mixing probabilities 0.9 and 0.1, respectively; and (4) HE 2: hyperexponential with parameters 100, 10 and 1 with mixing probabilities, respectively, given by 0.6, 0.3 and 0.1. These processes will be normalized to have an arrival rate of 1. For services we look at two PH distributions. These are (1) ERS: Erlang of order 5 and (2) HES: hyperexponential with parameters 10 and 1 with mixing probabilities 0.9 and 0.1, respectively. Again, these service distributions will be normalized to have a mean of 0.95. Note that these arrival and services are qualitatively different. In Figure 4.8, we display the plots of μ'_Y, μ'_K and μ'_I.

EXAMPLE 4.28.– In this example, we look the arrival process that has a 1-lag negative correlation for the inter-arrival times. Specifically, the MAP has the representation matrices D_0 and D_1 of order 3 given by:

$$D_0 = \begin{pmatrix} -1.25 & 1.25 & 0 \\ 0 & -1.25 & 1.25 \\ 0 & 0 & -2.5 \end{pmatrix}, \quad D_1 = \begin{pmatrix} 0 & 0 & 0 \\ 0.01250 & 0 & 1.2375 \\ 2.4750 & 0 & 0.0250 \end{pmatrix}.$$

The service times are assumed to be Erlang of order 3 and the parameter will be normalized to have a mean of 0.95. Verify that (after normalization) the standard deviation of the inter-arrival times and the service times are 1.0392 and 0.5485, respectively; the 1-lag correlation coefficient is given by -0.3267. This queueing model can be set up as a QBD process. This was seen in section 3.2. The matrices G and R are displayed below:

$$R = \begin{pmatrix} 0 & 0 & 0 & 0 & 0 & 0 & 0 & 0 & 0 \\ 0.2106 & 0.2691 & 0.2837 & 0.1508 & 0.2355 & 0.1583 & 0.1081 & 0.1994 & 0.0884 \\ 0.7009 & 0.3782 & 0.1086 & 0.5021 & 0.4133 & 0.0606 & 0.3597 & 0.3981 & 0.0338 \\ 0 & 0 & 0 & 0 & 0 & 0 & 0 & 0 & 0 \\ 0.186 & 0.2052 & 0.0502 & 0.1361 & 0.1856 & 0.2467 & 0.0975 & 0.1606 & 0.1377 \\ 0.0870 & 0.1305 & 0.0701 & 0.6238 & 0.2704 & 0.0435 & 0.4469 & 0.3205 & 0.0243 \\ 0 & 0 & 0 & 0 & 0 & 0 & 0 & 0 & 0 \\ 0.1293 & 0.1174 & 0.0294 & 0.0926 & 0.1104 & 0.0164 & 0.0692 & 0.0987 & 0.2279 \\ 0.0369 & 0.0523 & 0.0320 & 0.0265 & 0.0450 & 0.0179 & 0.5804 & 0.1968 & 0.0144 \end{pmatrix}$$

$$G = \begin{pmatrix} 0.4562 & 0.5029 & 0.0409 & 0 & 0 & 0 & 0 & 0 & 0 \\ 0.1745 & 0.7507 & 0.0749 & 0 & 0 & 0 & 0 & 0 & 0 \\ 0.2712 & 0.5037 & 0.2251 & 0 & 0 & 0 & 0 & 0 & 0 \\ 0.5678 & 0.4048 & 0.0275 & 0 & 0 & 0 & 0 & 0 & 0 \\ 0.1355 & 0.7993 & 0.0653 & 0 & 0 & 0 & 0 & 0 & 0 \\ 0.2431 & 0.4057 & 0.3512 & 0 & 0 & 0 & 0 & 0 & 0 \\ 0.7389 & 0.2486 & 0.0125 & 0 & 0 & 0 & 0 & 0 & 0 \\ 0.0792 & 0.8767 & 0.0441 & 0 & 0 & 0 & 0 & 0 & 0 \\ 0.1691 & 0.2494 & 0.5814 & 0 & 0 & 0 & 0 & 0 & 0 \end{pmatrix}$$

The key vectors are as follows:

$$\boldsymbol{g} = (0.253,\ 0.669,\ 0.078,\ 0,\ 0,\ 0,\ 0,\ 0,\ 0),$$

$$\tilde{\boldsymbol{\mu}}_G = (11.458,\ 23.082,\ 21.333,\ 6.857,\ 16.087,\ 15.039,\ 3.204,\ 8.772,\ 8.416),$$

$$\hat{\boldsymbol{\mu}}_G = (10.885,\ 21.928,\ 20.266,\ 6.198,\ 14.966,\ 13.971,\ 2.41,\ 7.7,\ 7.362),$$

$$\tilde{\boldsymbol{\mu}}_K = (21.234,\ 21.234,\ 11.557),\ \ \hat{\boldsymbol{\mu}}_K = (21.772,\ 20.972,\ 11.379).$$

The means, μ'_Y, μ'_{N_Y}, μ'_K and μ'_I, are calculated to be:

$$\mu'_Y = 18.4339,\ \mu'_{N_Y} = 19.4041,\ \mu'_K = 19.4041,\ \mu'_I = 0.9702,\ \ \eta = 0.5269.$$

EXAMPLE 4.29.– In this example, we look at the arrival process that has a 1-lag positive correlation for the inter-arrival times. Specifically, the MAP has the representation matrices D_0 and D_1 of order 3 given by:

$$D_0 = \begin{pmatrix} -1.25 & 1.25 & 0 \\ 0 & -1.25 & 1.25 \\ 0 & 0 & -2.5 \end{pmatrix},\quad D_1 = \begin{pmatrix} 0 & 0 & 0 \\ 1.2375 & 0 & 0.01250 \\ 0.0250 & 0 & 2.4750 \end{pmatrix}.$$

The service times are assumed to be Erlang of order 3 and the parameter will be normalized to have a mean of 0.95. Verify that (after normalization) the standard deviation of the inter-arrival times and the service times are 1.0392 and 0.5485, respectively; the 1-lag correlation coefficient is given by 0.3267. This queueing model can be set up as a QBD process. This was seen in section 3.2. The matrices G and R are displayed below:

$$R = \begin{pmatrix} 0 & 0 & 0 & 0 & 0 & 0 & 0 & 0 & 0 \\ 0.3813 & 0.1713 & 0.0039 & 0.2731 & 0.2002 & 0.0022 & 0.1957 & 0.1989 & 0.0012 \\ 0.3911 & 0.5555 & 0.6163 & 0.2802 & 0.4774 & 0.3440 & 0.2007 & 0.3990 & 0.1920 \\ 0 & 0 & 0 & 0 & 0 & 0 & 0 & 0 & 0 \\ 0.0640 & 0.0535 & 0.0016 & 0.3266 & 0.1309 & 0.0031 & 0.2340 & 0.1601 & 0.0017 \\ 0.3108 & 0.4456 & 0.2005 & 0.2283 & 0.3840 & 0.5493 & 0.1636 & 0.3215 & 0.3066 \\ 0 & 0 & 0 & 0 & 0 & 0 & 0 & 0 & 0 \\ 0.0276 & 0.0209 & 0.0011 & 0.0198 & 0.0206 & 0.0006 & 0.2949 & 0.0984 & 0.0025 \\ 0.1913 & 0.2736 & 0.1690 & 0.1371 & 0.2349 & 0.0943 & 0.1039 & 0.1977 & 0.4901 \end{pmatrix}$$

$$G = \begin{pmatrix} 0.4968 & 0.5025 & 0.0007 & 0 & 0 & 0 & 0 & 0 & 0 \\ 0.2488 & 0.7500 & 0.0012 & 0 & 0 & 0 & 0 & 0 & 0 \\ 0.2511 & 0.5040 & 0.2449 & 0 & 0 & 0 & 0 & 0 & 0 \\ 0.5950 & 0.4046 & 0.0004 & 0 & 0 & 0 & 0 & 0 & 0 \\ 0.2002 & 0.7988 & 0.0010 & 0 & 0 & 0 & 0 & 0 & 0 \\ 0.2027 & 0.4061 & 0.3912 & 0 & 0 & 0 & 0 & 0 & 0 \\ 0.7513 & 0.2485 & 0.0002 & 0 & 0 & 0 & 0 & 0 & 0 \\ 0.1230 & 0.8764 & 0.0007 & 0 & 0 & 0 & 0 & 0 & 0 \\ 0.1249 & 0.2498 & 0.6253 & 0 & 0 & 0 & 0 & 0 & 0 \end{pmatrix}$$

The key vectors are as follows:

$$\boldsymbol{g} = (0.331,\ 0.668,\ 0.001,\ 0,\ 0,\ 0,\ 0,\ 0,\ 0),$$

$$\tilde{\boldsymbol{\mu}}_G = (11.176,\ 22.536,\ 918.012,\ 6.679,\ 15.649,\ 737.722,\ 3.129,\ 8.506,\ 452.963),$$

$$\hat{\boldsymbol{\mu}}_G = (10.617,\ 21.41,\ 872.112,\ 6.028,\ 14.55,\ 700.519,\ 2.339,\ 7.447,\ 429.682),$$

$$\tilde{\boldsymbol{\mu}}_K = (20.244,\ 20.244,\ 908.944),\quad \hat{\boldsymbol{\mu}}_K = (20.832,\ 20.032,\ 863.897).$$

The means, μ'_Y, μ'_{N_Y}, μ'_K and μ'_I, are calculated to be:

$$\mu'_Y = 22.6685,\ \mu'_{N_Y} = 23.8616,\ \mu'_K = 23.8616,\ \mu'_I = 1.1931,\quad \eta = 0.7634.$$

EXAMPLE 4.30.– This example is similar to example 4.28 except for the service time, which is assumed to be hyperexponential having parameters 10, 5 and 1, with mixing probabilities, respectively, given by 0.6, 0.3 and 0.1. The parameters will be normalized to have a mean of 0.95. Verify that (after normalization) the standard deviation of the inter-arrival times and the service times are 1.0392 and 1.8703, respectively; recall that the 1-lag correlation coefficient is given by -0.3267. This queueing model can be set up as a QBD process. This was seen in section 3.2. The matrices G and R are displayed below:

$$R = \begin{pmatrix} 0 & 0 & 0 & 0 & 0 & 0 & 0 & 0 & 0 \\ 0.0978 & 0.0831 & 0.2730 & 0.0699 & 0.0724 & 0.0106 & 0.0378 & 0.0515 & 0.0047 \\ 0.7268 & 0.2808 & 0.0312 & 0.0242 & 0.0318 & 0.0171 & 0.0131 & 0.0215 & 0.0077 \\ 0 & 0 & 0 & 0 & 0 & 0 & 0 & 0 & 0 \\ 0.1106 & 0.1188 & 0.0251 & 0.0871 & 0.1045 & 0.3548 & 0.0444 & 0.0695 & 0.0074 \\ 0.0702 & 0.0906 & 0.0462 & 1.0799 & 0.6094 & 0.0373 & 0.0282 & 0.0502 & 0.0136 \\ 0 & 0 & 0 & 0 & 0 & 0 & 0 & 0 & 0 \\ 0.1202 & 0.1807 & 0.0432 & 0.0890 & 0.1488 & 0.0284 & 0.0567 & 0.1034 & 0.4657 \\ 0.1758 & 0.2779 & 0.0856 & 0.1302 & 0.2277 & 0.0563 & 1.7410 & 1.5557 & 0.0343 \end{pmatrix}$$

$$G = \begin{pmatrix} 0.4140 & 0.1664 & 0.0196 & 0.2070 & 0.0832 & 0.0098 & 0.0690 & 0.0277 & 0.0033 \\ 0.0695 & 0.4746 & 0.0560 & 0.0347 & 0.2373 & 0.0280 & 0.0116 & 0.0791 & 0.0093 \\ 0.1491 & 0.1381 & 0.3128 & 0.0746 & 0.0690 & 0.1564 & 0.0249 & 0.0230 & 0.0521 \\ 0.3424 & 0.2227 & 0.0349 & 0.1712 & 0.1114 & 0.0174 & 0.0571 & 0.0371 & 0.0058 \\ 0.1037 & 0.4291 & 0.0672 & 0.0519 & 0.2145 & 0.0336 & 0.0173 & 0.0715 & 0.0112 \\ 0.1746 & 0.1976 & 0.2278 & 0.0873 & 0.0988 & 0.1139 & 0.0291 & 0.0329 & 0.0380 \\ 0.2311 & 0.3043 & 0.0646 & 0.1155 & 0.1521 & 0.0323 & 0.0385 & 0.0507 & 0.0108 \\ 0.1628 & 0.3607 & 0.0766 & 0.0814 & 0.1803 & 0.0383 & 0.0271 & 0.0601 & 0.0128 \\ 0.1890 & 0.2950 & 0.1160 & 0.0945 & 0.1475 & 0.0580 & 0.0315 & 0.0492 & 0.0193 \end{pmatrix}$$

The key vectors are as follows:

$$\boldsymbol{g} = (0.188,\ 0.335,\ 0.078,\ 0.094,\ 0.167,\ 0.039,\ 0.031,\ 0.056,\ 0.013),$$

$$\tilde{\boldsymbol{\mu}}_G = (4.804,\ 11.851,\ 11.757,\ 11.628,\ 21.473,\ 20.522,\ 77.944,\ 92.199,\ 89.275),$$

$$\hat{\boldsymbol{\mu}}_G = (4.046,\ 10.74,\ 10.651,\ 10.96,\ 20.313,\ 19.409,\ 77.415,\ 90.957,\ 88.179),$$

$$\tilde{\boldsymbol{\mu}}_K = (22.058,\ 22.058,\ 14.245),\ \ \hat{\boldsymbol{\mu}}_K = (22.555,\ 21.755,\ 13.933).$$

The means, μ'_Y, μ'_{N_Y}, μ'_K and μ'_I, are calculated to be:

$$\mu'_Y = 18.5952,\ \mu'_{N_Y} = 19.5738,\ \ \mu'_K = 19.5738,\ \mu'_I = 0.9787,\ \ \eta = 0.9783.$$

EXAMPLE 4.31.– This example is similar to example 4.29 except for the service time, which is assumed to be hyperexponential having 10, 5 and 1, with mixing probabilities, respectively, given by 0.6, 0.3 and 0.1. The parameters will be normalized to have a mean of 0.95. Verify that (after normalization) the standard deviation of the inter-arrival times and the service times are 1.0392 and 1.8703, respectively; recall that the 1-lag correlation coefficient is given by 0.3267. This queueing model can be set up as a QBD process. This was seen in section 3.2. The matrices G and R are displayed below:

$$R = \begin{pmatrix} 0 & 0 & 0 & 0 & 0 & 0 & 0 & 0 & 0 \\ 0.3722 & 0.1425 & 0.0037 & 0.0187 & 0.0185 & 0.0008 & 0.0101 & 0.0133 & 0.0003 \\ 0.1012 & 0.1371 & 0.6801 & 0.0697 & 0.1114 & 0.1094 & 0.0378 & 0.0726 & 0.0488 \\ 0 & 0 & 0 & 0 & 0 & 0 & 0 & 0 & 0 \\ 0.0504 & 0.0499 & 0.0015 & 0.5513 & 0.3101 & 0.0044 & 0.0202 & 0.0300 & 0.0004 \\ 0.1435 & 0.2066 & 0.1677 & 0.1166 & 0.1763 & 0.7870 & 0.0575 & 0.1112 & 0.0493 \\ 0 & 0 & 0 & 0 & 0 & 0 & 0 & 0 & 0 \\ 0.1145 & 0.1501 & 0.0019 & 0.0848 & 0.1255 & 0.0013 & 0.8812 & 0.7876 & 0.0051 \\ 0.2471 & 0.3603 & 0.0802 & 0.1829 & 0.2976 & 0.0528 & 0.1160 & 0.2076 & 0.9296 \end{pmatrix}$$

$$G = \begin{pmatrix} 0.4288 & 0.1707 & 0.0005 & 0.2144 & 0.0854 & 0.0002 & 0.0715 & 0.0285 & 0.0001 \\ 0.1116 & 0.4870 & 0.0013 & 0.0558 & 0.2435 & 0.0007 & 0.0186 & 0.0812 & 0.0002 \\ 0.0742 & 0.1106 & 0.4153 & 0.0371 & 0.0553 & 0.2076 & 0.0124 & 0.0184 & 0.0692 \\ 0.3673 & 0.2318 & 0.0010 & 0.1836 & 0.1159 & 0.0005 & 0.0612 & 0.0386 & 0.0002 \\ 0.1517 & 0.4465 & 0.0018 & 0.0758 & 0.2232 & 0.0009 & 0.0253 & 0.0744 & 0.0003 \\ 0.1128 & 0.1694 & 0.3179 & 0.0564 & 0.0847 & 0.1589 & 0.0188 & 0.0282 & 0.053 \\ 0.2732 & 0.3245 & 0.0023 & 0.1366 & 0.1622 & 0.0012 & 0.0455 & 0.0541 & 0.0004 \\ 0.2127 & 0.3846 & 0.0028 & 0.1063 & 0.1923 & 0.0014 & 0.0354 & 0.0641 & 0.0005 \\ 0.1938 & 0.2942 & 0.1120 & 0.0969 & 0.1471 & 0.0560 & 0.0323 & 0.0490 & 0.0187 \end{pmatrix}$$

The key vectors are as follows:

$$\boldsymbol{g} = (0.236,\ 0.36,\ 0.003,\ 0.118,\ 0.18,\ 0.002,\ 0.039,\ 0.06,\ 0.001),$$

$$\tilde{\boldsymbol{\mu}}_G = (4.16,\ 10.015,\ 376.166,\ 10.176,\ 18.676,\ 578.051,\ 74.064,\ 87.6,\ 1055.491),$$

$$\hat{\boldsymbol{\mu}}_G = (3.434,\ 8.996,\ 356.84,\ 9.581,\ 17.656,\ 549.062,\ 73.729,\ 86.588,\ 1006.085),$$

$$\tilde{\boldsymbol{\mu}}_K = (17.872,\ 17.872,\ 499.747),\quad \hat{\boldsymbol{\mu}}_K = (18.579,\ 17.779,\ 475.16).$$

The means, μ'_Y, μ'_{N_Y}, μ'_K and μ'_I, are calculated to be:

$$\mu'_Y = 24.8678,\ \mu'_{N_Y} = 26.1766,\ \mu'_K = 26.1766,\ \mu'_I = 1.3088,\ \eta = 0.9976.$$

4.7. When the fundamental period is not the busy period

In the last section, the fundamental period happened to be the BP. However, there are models (with complicated boundaries) in which the fundamental period is not the BP. This case (see Neuts (1989)) can be handled as follows.

Suppose that we look at the case where the fundamental period, which is studied through the sequence of the matrix joint transforms $\{G^k(z, s)\}$, does not correspond to the BP. This occurs in models where the boundary states need to be grouped or when models in continuous-time, like $BMAP/PH/1$ (see section 3.1), need to be studied as $M/G/1$-type queue. Specifically, suppose we look at the $M/G/1$-type queue such that the MRP on the state space $\Omega = \{(0, j) : 1 \leq j \leq m\} \bigcup \{(i, j) : 1 \leq j \leq n : i \geq 1\}$ with the TPM is given by:

$$P(t) = \begin{bmatrix} B_0(t) & B_1(t) & B_2(t) & B_3(t) & B_4(t) & \cdots \\ E(t) & A_1(t) & A_2(t) & A_3(t) & A_4(t) & \cdots \\ & A_0(t) & A_1(t) & A_2(t) & A_3(t) & \cdots \\ & & A_0(t) & A_1(t) & A_2(t) & \cdots \\ & & & A_0(t) & A_1(t) & \cdots \\ & & & & A_0(t) & \cdots \\ & & & & & \ddots \end{bmatrix}, \qquad [4.60]$$

where $B_0(t)$ is a square matrix of dimension m, $B_i(t),\ i \geq 1$, are of dimension $m \times p$, E is of dimension $p \times m$ and $A_i,\ i \geq 0$, are square matrices of dimension p. The fundamental period that was discussed earlier will not correspond to the BP anymore since the structure is different for level **1** as compared to level $\boldsymbol{i}$, for $i \geq 1$. However, the BP is studied by looking at the first passage time from level **1** to level **0**. This first passage time has been thoroughly studied by Neuts (1989) and we summarize some of the key results here and for full details we refer the reader to Neuts (1989).

Let $\Phi(n, t) = \{\phi_{i,j}(n, t)\}$, of dimension $p \times m$, be such that $\phi_{i,j}(n, t),\ 1 \leq i \leq n,\ 1 \leq j \leq m$, denotes the conditional probability the MRP with TPM given in equation [4.60] will reach level **0** by time t and in n transitions by visiting the state $(0, j)$ given that the process started in state $(1, i)$.

Define $E^*(s)$ to be the LST of $E(t)$ and $A_n^*(s)$ to be that of $A_n(t)$. If $\Phi(z, s)$, $|z| \leq 1$, $Re(s) \geq 0$, is the joint transform of $\Phi(n, t)$, then we have the following result.

RESULT 4.2.– The joint transform, $\Phi(z, s)$, satisfies:

$$\Phi(z, s) = zE^*(s) + z\sum_{r=1}^{\infty} A_r^*(s)G^{r-1}(z, s)\Phi(z, s),\ |z| \leq 1, Re(s) \geq 0, \qquad [4.61]$$

and, furthermore, the matrix $\Phi = \Phi(1, 0)$ is obtained explicitly as:

$$\Phi = (I - \tilde{A}_1)^{-1}E, \qquad [4.62]$$

where $\tilde{A}_1 = \sum_{r=1}^{\infty} A_r G^{r-1}$ (see also equation [1.68]).

REMARK 4.10.– If one is interested to see (especially when dealing with batch arrivals) the first passage times from state (k, i) to level **0**, then it is easy to verify (see Neuts (1989)) that the transform matrix, say, $\Phi^{[k]}(z, s)$, is given by:

$$\Phi^{[k]}(z,s) = G^{k-1}(z,s)\Phi(z,s),\ |z| \leq 1, Re(s) \geq 0,\ k \geq 1. \qquad [4.63]$$

REMARK 4.11.– Suppose in equation [4.60], instead of having a simple set of boundary states, we have more complex boundary states. For example, in equation [4.60], there are $M = m_1 + m_2 + \cdots + m_k$ boundary states based on k sets of states. In this case, we can partition the state space in such a way that the TPM is of the form given in equation [4.60] but the dimensions of the boundary states will be accordingly adjusted. To be more specific, let us write the original TPM for a model with more complex boundary states as:

$$P(t) = \begin{bmatrix} \tilde{B}_0(t) & \tilde{B}_1(t) & \tilde{B}_2(t) & \tilde{B}_3(t) & \tilde{B}_4(t) & \cdots \\ \tilde{E}(t) & A_1(t) & A_2(t) & A_3(t) & A_4(t) & \cdots \\ & A_0(t) & A_1(t) & A_2(t) & A_3(t) & \cdots \\ & & A_0(t) & A_1(t) & A_2(t) & \cdots \\ & & & A_0(t) & A_1(t) & \cdots \\ & & & & A_0(t) & \cdots \\ & & & & & \ddots \end{bmatrix}, \qquad [4.64]$$

where $\tilde{B}_0(t)$ is a square matrix of dimension M, $\tilde{B}_i(t),\ i \geq 1$, are of dimension $M \times p$, $\tilde{E}$ is of dimension $p \times M$ and $A_i,\ i \geq 0$, are square matrices of dimension p. We can then apply the tools used earlier but with a careful book-keeping. We refer the reader to Neuts (1989) for more details on this.

4.8. *GI/G/1* queues

The study of single-server queues in which the inter-arrival times are independent and identically distributed and the service times are generally distributed becomes complex as compared to the models seen so far. This is not due to the choice but rather due to inherent difficulties. Unlike the approaches such as embedded points or arbitrary points based on the type of models, the study of $GI/G/1$ queues requires more book-keeping. Realizing this difficulty, several researchers resort to approximating or obtaining bounds for key system performance measures. However, study of such queues can be carried out either through (1) using PH distributions as the class of PH distributions is dense in the class of all distributions on the non-negative real line and/or through (2) simulation. The simulation approach will be discussed in Chapter 7. However, here we briefly, for the sake of completeness, outline some key concepts in such books as Bhat (2015), Bladt and Nielsen (2017), Gross and Harris (1998) and Kleinrock (1975).

Suppose that the inter-arrival times are independent and identically distributed with CDF, say, $F(.)$. The service times are independent of the arrivals and assume that their CDF is given by $H(.)$. Assume that arrival rate is λ and the service rate is μ. Let $\rho = \frac{\lambda}{\mu}$. For the rest of this section, we assume that $\rho < 1$ so that the queue is always stable.

Let $\tau_i, i \geq 1$, denote the inter-arrival times between the $(i-1)$st and the ith customers, so that their common CDF is $F(.)$. Suppose that s_i denotes the service time of the ith customer. Let $w_q^{(i)}$ denote the waiting time in the queue of the ith customer. From the definition of τ_i, s_i and $w_i^{(q)}$, it is easy to verify (see Kleinrock (1975)) that:

$$w_{i+1}^{(q)} = \begin{cases} w_i^{(q)} + s_i - \tau_{i+1}, \text{ if } w_i^{(q)} + s_i - \tau_{i+1} \geq 0, \\ 0, \text{ elsewhere.} \end{cases} \quad [4.65]$$

Note that the quantity, $w_i^{(q)} + s_i - \tau_{i+1}$, gives the sum of the unfinished work by the system $(w_i^{(q)})$ and the service time of the ith arrival (s_i) less the time duration (τ_{i+1}) of the next arrival.

If we denote $u_i = s_i - \tau_{i+1}$, then we can write:

$$w_{i+1}^{(q)} = \max\{0, w_i^{(q)} + u_i\} = (w_i^{(q)} + u_i)^+, \; i \geq 0. \quad [4.66]$$

Note that due to the fact that τ_i and s_i are independent, $\{u_i\}$ is a sequence of $i.i.d$ random variables. Let $U_i = \sum_{k=0}^{i-1} u_k, \; i \geq 1$ with $U_0 = 0$. Verify that:

$$\begin{aligned} w_1^{(q)} &= \max\{0, w_0^{(q)} + u_0\}, \\ w_2^{(q)} &= \max\{0, w_1^{(q)} + u_1\} = \max\{0, u_1, u_1 + u_0 + w_0^{(q)}\}, \end{aligned} \quad [4.67]$$

$$\vdots$$

$$w_i = \max\{0, w_{i-1}^{(q)} + u_{i-1}\} = \max\left\{0, u_{i-1}, \sum_{k=i-2}^{i-1} u_k, \cdots, \sum_{k=1}^{i-1} u_k, \right. \quad [4.68]$$

$$\left. w_0^{(q)} + \sum_{k=0}^{i-1} u_k \right\}, \; i \geq 1.$$

If we define the sequence $\{\hat{w}_i\}$ to be:

$$\hat{w}_i = \max\{U_0, U_1, \cdots, U_{i-1}, U_i + w_0\}, \; i \geq 1, \quad [4.69]$$

then it is easy to check that the sequences $\{w_i^{(q)}\}$ and $\{\hat{w}_i\}$ have identical probability distributions.

When $w_0^{(q)} = 0$, and if $\hat{w}_i \to \tilde{w}$ as $i \to \infty$, we have $\tilde{w} = \sup_i U_i$.

While observing the $GI/G/1$ queue at an arbitrary time t is not sufficient to study the model (as we need to keep track of the elapsed times of the arrival and the service process, if the system is non-empty), the system does indeed possess an imbedded Markov chain in $\{w_i\}$. This is due to the fact that $w_{i+1}^{(q)}$ depends only on $w_i^{(q)}$ and u_i, and that u_i is independent of $w_k^{(q)}$, $k \leq i$. Thus, $\{w_i^{(q)}\}$ is a discrete-time Markov chain (DTMC).

The DTMC, $\{w_i^q\}$, is Ergodic if $\tilde{w}$ is finite with probability 1. Lindley (1952) proved that for $0 < E(|u_i|) < \infty$, the system is stable if and only if $E(u_i) < 0$. This condition holds good since $E(u_i) = \frac{1}{\lambda}(\rho - 1)$ and we assumed that $\rho < 1$.

Suppose that $\tilde{H}(t) = P(u_i \leq t)$. It is easy to verify that:

$$\tilde{H}(t) = P(u_i \leq t) = P(s_i - \tau_{i+1} \leq t) = \int_0^\infty P(s_i \leq t + r | \tau_{i+1} = r) dF(r),$$

which implies:

$$\tilde{H}(t) = \int_0^\infty H(t+r) dF(r). \qquad [4.70]$$

Let $W(t) = \lim_{i\to\infty} P(w_i^{(q)} \leq t)$, $t \geq 0$. That is, $W(t)$ is the stationary waiting time (in the queue) distribution. Thus, if $W_{i+1}(t) = P(w_{i+1}^{(q)} \leq t)$, then we have:

$$W_{i+1}(t) = P(w_{i+1}^{(q)} \leq t) = P(w_i^{(q)} + u_i \leq t) = \int_{0-}^\infty P(u_i \leq t - r | w_i^{(q)} = r) dW_i(r),$$

from which it is easy to see (noting that u_i and w_i are independent) that:

$$W(t) = \int_{0-}^\infty \tilde{H}(t-r) dW(r). \qquad [4.71]$$

The above equation is referred to as *Lindley's integral equation* and is written as:

$$W(t) = \begin{cases} \int_{0-}^\infty H(t-r) dW(r), \ t \geq 0, \\ 0, \ t < 0. \end{cases} \qquad [4.72]$$

or equivalently:

$$W(t) = \begin{cases} \int_{0-}^\infty W(t-r) dH(r), \ t \geq 0, \\ 0, \ t < 0. \end{cases} \qquad [4.73]$$

REMARK 4.12.– Obviously, looking at the expressions for the waiting time distributions for the $GI/G/1$ queues, it is clear that one has to rely on some approximations or simulation to analyze such queueing models. Talking about approximations there is a literature on this for single- and multiple-server systems. For example, in Kimura (1991), two-moment approximation formulas for the mean waiting time as well as for the probability that an arrival has to wait are derived for a $GI/G/c$ queue. These formulas are derived such that they yield exact results for $M/M/c$, $M/D/c$ and $D/M/c$ queues. In Whitt (1993), approximations for the mean waiting time based on the first two moments of the arrival and service times are provided. This paper also has extensive insight into such approximations as well as a literature survey on numerical methods to approximate such queueing models. However, since we discuss simulation in a later chapter, our focus will be on results based on simulated models. In any case, readers should know the approximations are available in the literature. Other key papers dealing with approximations include Bertsimas (1990) and Shore (1988).

4.9. Exercises

Exercise 4.1. Prove the expressions given in equation [4.5].

Exercise 4.2. Referring to example 4.1, verify the results reported there.

Exercise 4.3. Referring to example 4.2, verify the results reported there.

Exercise 4.4. Referring to example 4.3, verify the results reported there.

Exercise 4.5. Referring to example 4.4, verify the results reported there.

Exercise 4.6. Referring to example 4.5, verify the results reported there.

Exercise 4.7. Referring to example 4.6, verify the results reported there.

Exercise 4.8. Referring to example 4.7, verify the results reported there.

Exercise 4.9. Referring to example 4.8, verify the results reported there.

Exercise 4.10. Referring to example 4.10, verify the results reported there.

Exercise 4.11. Prove the expressions given in equations [4.25] and [4.26].

Exercise 4.12. Referring to example 4.11, bring out the key points based on the graphs displayed there.

Exercise 4.13. Prove the expressions given in equations [4.33] and [4.34].

Exercise 4.14. Referring to example 4.12, verify the results mentioned there.

Exercise 4.15. Referring to example 4.13, verify the results mentioned there.

Exercise 4.16. Referring to example 4.14, verify the results mentioned there.

Exercise 4.17. Referring to example 4.15, verify the results mentioned there.

Exercise 4.18. Referring to example 4.16, verify the results mentioned there.

Exercise 4.19. Referring to example 4.19, verify the results mentioned there.

Exercise 4.20. Referring to example 4.20, highlight some key points based on the results displayed in the table.

Exercise 4.21. Referring to example 4.21, verify the results mentioned there.

Exercise 4.22. Referring to example 4.22, verify the results mentioned there.

Exercise 4.23. Prove the expression given in equation [4.51].

Exercise 4.24. Prove the expression given in equation [4.52].

Exercise 4.25. Prove the expressions given in equation [4.43].

Exercise 4.26. Prove the expressions given in equations [4.44] and [4.45].

Exercise 4.27. Prove the expressions given in equations [4.58] and [4.59].

Exercise 4.28. Referring to example 4.23, verify the results mentioned there.

Exercise 4.29. Referring to example 4.24, verify the results mentioned there.

Exercise 4.30. Referring to example 4.25, verify the results mentioned there.

Exercise 4.31. Referring to example 4.26, verify the results mentioned there.

Exercise 4.32. Referring to example 4.27, highlight some key points based on the results displayed in the form of graphs.

Exercise 4.33. Referring to example 4.28, verify the results mentioned there.

Exercise 4.34. Referring to example 4.29, verify the results mentioned there.

Exercise 4.35. Referring to example 4.30, verify the results mentioned there.

Exercise 4.36. Referring to example 4.31, verify the results mentioned there.

Exercise 4.37. Prove the expression given in equation [4.62].

Note: For exercises needing computational work, your answers may differ due to the tolerance level chosen for the error components.

Exercise 4.38. Prove the expression given in equation [4.46].

Exercise 4.39. For the $BMAP/PH/1$ queueing model studied by looking at arbitrary time, prove the expression given in equation [4.38].

Exercise 4.40. For the $BMAP/PH/1$ queueing model studied by looking at arbitrary time, prove the expressions given in equation [4.39].

Exercise 4.41. For the $BMAP/PH/1$ queueing model studied by looking at arbitrary time, prove the expressions given in equation [4.40].

Exercise 4.42. For the $BMAP/PH/1$ queueing model studied by looking at arbitrary time, prove the expressions given in equation [4.41].

Exercise 4.43. For the $BMAP/PH/1$ queueing model studied by looking at arbitrary time, prove

$$\boldsymbol{g}\tilde{\boldsymbol{\mu}}_G = \frac{1}{1-\rho} \text{ and } \boldsymbol{g}\hat{\boldsymbol{\mu}}_G = \frac{1}{(1-\rho)\boldsymbol{\pi} H_0 \boldsymbol{e}},$$

Exercise 4.44. For the $BMAP/PH/1$ queueing model studied by looking at arbitrary time, taking $\boldsymbol{y}^* = [\boldsymbol{y}_1(\boldsymbol{S^0}\otimes\boldsymbol{e})]^{-1}\boldsymbol{y}_1(\boldsymbol{S^0}\otimes(-D_0)^{-1})$, prove that the mean recurrence time is given by:

$$\mu'_{RT} = \boldsymbol{y}^*\left[e + \frac{1}{\mu(1-\rho)}\boldsymbol{d} + \sum_{r=1}^{\infty}(\boldsymbol{\beta}\otimes D_r)[I - G^r]\left((H_0\boldsymbol{e} - \boldsymbol{h})\boldsymbol{g} - H\right)^{-1}\boldsymbol{e}\right].$$

Exercise 4.45. For the QBD-type queues in continuous times, taking $H = H_0 + H_1 + H_2$ and $\boldsymbol{g}$ is the left eigenvector of G corresponding to 1, prove that:

$$\tilde{\boldsymbol{\mu}}_G = \frac{\partial}{\partial z}G(z,s)\Big]_{s=0,z=1}\boldsymbol{e} = (I - G + \boldsymbol{e}\,\boldsymbol{g})[(H_0\boldsymbol{e} - H_2\boldsymbol{e})\boldsymbol{g} - H]^{-1}H_0\boldsymbol{e},$$

$$\hat{\boldsymbol{\mu}}_G = -\frac{\partial}{\partial s}G(z,s)\Big]_{s=0,z=1}\boldsymbol{e} = (I - G + \boldsymbol{e}\,\boldsymbol{g})[(H_0\boldsymbol{e} - H_2\boldsymbol{e})\boldsymbol{g} - H]^{-1}\boldsymbol{e}.$$

Exercise 4.46. For the QBD-type queues by looking at arbitrary times, prove that:

$$g\tilde{\mu}_G = \frac{1}{1-\rho} \text{ and } g\hat{\mu}_G = \frac{1}{(1-\rho)\pi H_0 e}.$$

Exercise 4.47. Suppose that $l_{i,j}(n,t)$ is the conditional probability that the QBD process with the generator given in equation [4.42], starting in state $(1,i)$ at time 0, will reach the state $(0,j)$ by time t and it occurs by making exactly n left transitions. If $L(n,t)$ denotes the matrix with (i,j)th element given by $l_{i,j}(n,t)$, prove that the matrix joint transform matrix $L(z,s)$ satisfies:

$$L(z,s) = z(sI - H_1)^{-1}F + (sI - H_1)^{-1}H_2G(z,s)L(z,s), \; Re(s) \geq 0, \; |z| \leq 1.$$

Exercise 4.48. Referring to the QBD process in continuous-time with the generator given in equation [4.42], suppose that the matrix L is such that its (i,j)th element gives the conditional probability of eventually hitting the state $(0,k)$ starting from state $(1,j)$. Prove that:

$$L = [I - (-H_1)^{-1}H_2G]^{-1}(-H_1)^{-1}F.$$

Exercise 4.49. Suppose we denote $K(n,t) = \{k_{i,j}(n,t)\}$ to be the matrix of dimension m whose (i,j)th element gives the conditional probability that the QBD process with the generator (see equation [4.42]) starting in state $(0,i)$ at time 0 will reach the state $(0,j)$ by time t and it occurs by making exactly n left transitions. Prove that the matrix joint transform, $K(z,s)$, satisfies:

$$K(z,s) = (sI - C_0)^{-1}C_1L(z,s), \; Re(s) \geq 0, \; |z| \leq 1.$$

and that $K = K(1,0)$ is given by:

$$K = K(1,0) = (-C_0)^{-1}C_1L.$$

Exercise 4.50. Prove the expression given in equation [4.50].

Exercise 4.51. For the QBD process with the generator given in equation [4.42], show that the vector (of the conditional probabilities that the process will begin an idle period) is given by:

$$[\boldsymbol{y}_1 F\boldsymbol{e}]^{-1}\boldsymbol{y}_1 F.$$

Exercise 4.52. For the QBD process with the generator given in equation [4.42], taking $\boldsymbol{y}^* = [\boldsymbol{y}_1 F\boldsymbol{e}]^{-1}\boldsymbol{y}_1 F$, prove that the mean recurrence time is:

$$\mu_K' = \boldsymbol{y}^*(-C_0)^{-1}\boldsymbol{e} + \boldsymbol{y}^*(-C_0)^{-1}C_1\hat{\mu}_G.$$

Exercise 4.53. For the QBD process with the generator given in equation [4.42], prove that the mean vectors of $K(z, s)$ are:

$$\hat{\mu}_K = -\frac{\partial}{\partial s}K(z, s)\Big]_{s=0,z=1} e = (-C_0)^{-1}e + (-C_0)^{-1}C_1\hat{\mu}_G,$$

$$\tilde{\mu}_K = -\frac{\partial}{\partial z}K(z, s)\Big]_{s=0,z=1} e = (-C_0)^{-1}C_1\tilde{\mu}_G.$$

Exercise 4.54. Referring to exercise 1.13, compute the (a) mean vector $\tilde{\mu}_G$ and (b) mean number of customers served during a busy period.

Exercise 4.55. Referring to exercise 1.14, compute the (a) mean vector $\tilde{\mu}_G$ and (b) mean number of customers served during a busy period.

Exercise 4.56. Referring to exercise 1.15, compute the (a) mean vector $\tilde{\mu}_G$ and (b) mean number of customers served during a busy period.

Exercise 4.57. Referring to exercise 1.16, compute the (a) mean vector $\tilde{\mu}_G$ and (b) mean number of customers served during a busy period.

Exercise 4.58. Referring to exercise 1.17, compute the (a) mean vector $\tilde{\mu}_G$ and (b) mean number of customers served during a busy period.

Exercise 4.59. Referring to exercise 1.18, compute the (a) mean vector $\tilde{\mu}_G$ and (b) mean number of customers served during a busy period.

Exercise 4.60. Referring to exercise 1.19, compute the (a) mean vector $\tilde{\mu}_G$ and (b) mean number of customers served during a busy period.

Exercise 4.61. Referring to exercise 1.20, compute the (a) mean vector $\tilde{\mu}_G$ and (b) mean number of customers served during a busy period.

Exercise 4.62. Referring to exercise 1.21, compute the (a) mean vector $\tilde{\mu}_G$ and (b) mean number of customers served during a busy period, under the four types of batch distributions.

Exercise 4.63. Referring to exercise 1.22, compute the (a) mean vector $\tilde{\mu}_G$ and (b) mean number of customers served during a busy period, under the four types of batch distributions.

Exercise 4.64. Referring to exercise 1.23, compute the (a) mean vector $\tilde{\mu}_G$ and (b) mean number of customers served during a busy period, under the four types of batch distributions.

Exercise 4.65. Referring to exercise 1.24, compute the (a) mean vector $\tilde{\mu}_G$ and (b) mean number of customers served during a busy period, under the four types of batch distributions.

Exercise 4.66. Referring to exercise 1.25, compute the (a) mean vector $\tilde{\mu}_G$ and (b) mean number of customers served during a busy period, under the four types of batch distributions.

Exercise 4.67. Referring to exercise 1.26, compute the (a) mean vector $\tilde{\mu}_G$ and (b) mean number of customers served during a busy period, under the four types of batch distributions.

Exercise 4.68. Referring to exercise 1.27, compute the (a) mean vector $\tilde{\mu}_G$ and (b) mean number of customers served during a busy period, under the four types of batch distributions.

Exercise 4.69. Referring to exercise 1.28, compute the (a) mean vector $\tilde{\mu}_G$ and (b) mean number of customers served during a busy period, under the four types of batch distributions.

Exercise 4.70. Referring to exercise 1.29, compute the (a) mean vector $\tilde{\mu}_G$ and (b) mean number of customers served during a busy period, under the four types of batch distributions.

Exercise 4.71. Referring to exercise 1.30, compute the (a) mean vector $\tilde{\mu}_G$ and (b) mean number of customers served during a busy period, under the four types of batch distributions.

Exercise 4.72. Referring to exercise 1.32, compute the (a) mean vector $\tilde{\mu}_G$ and (b) mean number of customers served during a busy period, under the four types of services.

Exercise 4.73. Referring to exercise 1.33, compute the (a) mean vector $\tilde{\mu}_G$ and (b) mean number of customers served during a busy period, under the four types of services.

Exercise 4.74. Referring to exercise 1.34, compute the (a) mean vector $\tilde{\mu}_G$ and (b) mean number of customers served during a busy period, under the four types of services.

Exercise 4.75. Referring to exercise 1.32, compute the (a) mean vector $\tilde{\mu}_G$ and (b) mean number of customers served during a busy period, under the four types of services.

Exercise 4.76. Prove the expression given in equation [4.61].

Exercise 4.77. Prove the expression given in equation [4.63].

Exercise 4.78. Referring to exercise 2.25, compute the mean busy period and the mean number served during a busy period.

Exercise 4.79. Referring to exercise 2.26, compute the mean busy period and the mean number served during a busy period.

Exercise 4.80. Referring to exercise 2.27, compute the mean busy period and the mean number served during a busy period.

Exercise 4.81. Referring to exercise 2.28, compute the mean busy period and the mean number served during a busy period.

Exercise 4.82. Referring to exercise 2.29, compute the mean busy period and the mean number served during a busy period.

Exercise 4.83. Referring to exercise 2.30, compute the mean busy period and the mean number served during a busy period.

Exercise 4.84. Referring to exercise 2.31, compute the mean busy period and the mean number served during a busy period.

Exercise 4.85. Referring to exercise 2.32, compute the mean busy period and the mean number served during a busy period.

Exercise 4.86. Referring to exercise 2.33, compute the mean busy period and the mean number served during a busy period.

Exercise 4.87. Referring to exercise 2.34, compute the mean busy period and the mean number served during a busy period.

Exercise 4.88. Referring to exercise 2.35, compute the mean busy period and the mean number served during a busy period.

Exercise 4.89. Referring to exercise 2.36, compute the mean busy period and the mean number served during a busy period.

Exercise 4.90. Referring to exercise 2.37, compute the mean busy period and the mean number served during a busy period.

Exercise 4.91. Referring to exercise 2.38, compute the mean busy period and the mean number served during a busy period.

Exercise 4.92. Referring to exercise 2.39, compute the mean busy period and the mean number served during a busy period.

Exercise 4.93. Referring to exercise 2.40, compute the mean busy period and the mean number served during a busy period.

Exercise 4.94. Referring to exercise 2.41, compute the mean busy period and the mean number served during a busy period.

Exercise 4.95. Referring to exercise 2.42, compute the mean busy period and the mean number served during a busy period.

Exercise 4.96. Referring to exercise 2.43, compute the mean busy period and the mean number served during a busy period.

Exercise 4.97. Referring to exercise 2.44, compute the mean busy period and the mean number served during a busy period.

Exercise 4.98. Referring to exercise 2.45, compute the mean busy period and the mean number served during a busy period.

Exercise 4.99. Referring to exercise 2.46, compute the mean busy period and the mean number served during a busy period.

Exercise 4.100. Referring to exercise 2.47, compute the mean busy period and the mean number served during a busy period.

Exercise 4.101. Referring to exercise 2.48, compute the mean busy period and the mean number served during a busy period.

Exercise 4.102. Referring to exercise 2.49, compute the mean busy period and the mean number served during a busy period.

Exercise 4.103. Referring to exercise 2.50, compute the mean busy period and the mean number served during a busy period.

Exercise 4.104. Referring to exercise 2.51, compute the mean busy period and the mean number served during a busy period.

Exercise 4.105. Referring to exercise 2.52, compute the mean busy period and the mean number served during a busy period.

Exercise 4.106. Referring to exercise 2.53, compute the mean busy period and the mean number served during a busy period.

Exercise 4.107. Referring to exercise 3.30, compute:

a) $\tilde{\mu}_G$;

b) $\hat{\mu}_G$;

c) the mean busy period;

d) the mean number served during a busy period;

e) the mean idle time;

f) the mean recurrence time to an empty system.

Exercise 4.108. Referring to exercise 3.31, compute:

a) $\tilde{\mu}_G$;

b) $\hat{\mu}_G$;

c) the mean busy period;

d) the mean number served during a busy period;

e) the mean idle time;

f) the mean recurrence time to an empty system.

Exercise 4.109. Referring to exercise 3.32, compute the mean busy period and the mean number served during a busy period.

Exercise 4.110. Referring to exercise 3.33, compute the mean busy period and the mean number served during a busy period.

Exercise 4.111. Referring to exercise 3.34, compute the mean busy period and the mean number served during a busy period.

Exercise 4.112. Referring to exercise 3.35, compute the mean busy period and the mean number served during a busy period.

Exercise 4.113. Referring to exercise 3.36, compute the mean busy period and the mean number served during a busy period.

Exercise 4.114. Referring to exercise 3.37, compute the mean busy period and the mean number served during a busy period.

Exercise 4.115. Referring to exercise 3.38, compute the mean busy period and the mean number served during a busy period.

Exercise 4.116. Referring to exercise 3.39, compute the mean busy period and the mean number served during a busy period.

5

Multi-Server Queues

Multi-server queueing systems are very common and useful in practice. While $BMAP/PH/c$ and $GI/PH/c$ queues can be studied using the techniques seen in the first three chapters of this volume, the restriction arises through the explosion of the state space. One needs to develop efficient algorithmic procedures to obtain the system performance measures. In this chapter, we discuss a few multi-server queueing models using the tools seen in those chapters along with algorithmic procedures by exploiting the coefficient matrices, and a few complex models will be analyzed through simulation in Chapter 7.

This chapter is organized as follows. In section 5.1, we discuss the complexity involved in the study of $BMAP/PH/c$ queues. In the context of $MAP/PH/3$ queues with heterogeneous servers, we present in section 5.2 the steady-state analysis in great detail along with a number of illustrative examples. The discussion of $BMAP/M/c$ queues is presented in section 5.3, and in section 5.4 a brief remark is made on $GI/PH/c$ queues as detailed illustrative examples on this along with others will be presented in the chapter dealing with simulation.

5.1. *BMAP/PH/c* queue

Suppose we look at a c-server system in which arrivals occur according to a batch Markovian arrival process (BMAP) with representation matrices $\{D_k\}$ of dimension m. First, we look at the heterogeneous server case and thus the service times are of continuous phase type (CPH) with representation $(\boldsymbol{\beta}(r), S(r))$ of dimension n_r for server r, $1 \leq r \leq c$. Defining the following random variables at time t: $N(t)$ to be the number of customers in the queue; $J_r(t)$ to be the phase of the server r; and $I(t)$

to be the phase of the BMAP, we can study $BMAP/PH/c$ as an $M/G/1$-type queue with the state space given by:

$$\Omega = \Omega_B \bigcup \{(i, j_1, j_2, \cdots, j_c, k) : 1 \leq j_r \leq n_r, 1 \leq r \leq c, 1 \leq k \leq m, i \geq 0\},$$

where Ω_B, the state space for the boundary states, is given by:

$$\Omega_B = \{(j_1, \cdots, j_c, k) : 0 \leq j_r \leq n_r, 1 \leq r \leq c, 1 \leq k \leq m, \ j_r = 0, \text{ for some } r\}.$$

Note that we used the notation $j_r = 0$ to indicate that the server r is idle. Also, the state corresponding to all servers being busy is included in $\Omega - \Omega_B$, the set containing non-boundary states, and hence the condition $j_r = 0$, for some r, mentioned in Ω_B guarantees that there is no overlap between the boundary and the non-boundary states.

It is clear that the state space grows exponentially as the number of servers and/or the dimension of the service phases increase. For example, the cardinality of Ω_B is $m[(n_1 + 1)(n_2 + 1) \cdots (n_c + 1) - n_1 n_2 \cdots n_c]$ and each one of the levels in the non-boundary states has $n_1 n_2 \cdots n_c$ states. Thus, when $c = 10, m = n_1 = \cdots = n_{10} = 2$, the number of boundary states is $2(3^{10} - 2^{10}) = 116,050$.

In the case of identical server system, one can define the state space differently to reduce the cardinality. There are two methods. The first method (see Dudin et al. (2020)) involves numbering the busy servers as follows. The server who has been busy the longest period of time will be numbered 1, then the next longest one will get 2, etc. That is, the server who initiated the service first and is still busy will get the lowest number and the server who initiated a service recently will get the highest number, and these numbers will be between 1 and c. Upon completion of a service the numbers will be reassigned, if needed, in such a way that only the servers who initiated a service after the server who just completed the service will have their numbers reduced by 1 and all others' numbers will not be affected since they started their services prior to the one who just finished. Under this scheme, it is shown that the number of boundary states is given by $m\, n^c$, where $n = n_1 = \cdots = n_c$.

In the second method, which is applicable only when the servers are identical, instead of keeping track of the phases of each of the c servers, we keep track of the number of servers who are busy in each of the n phases. In this method, the number of states to be considered decreases significantly. This idea was first implemented algorithmically by Mayhugh and McCormick (1968) in the context of $M/E_k/r$ queue. This approach was further explored in the context of $GI/PH/c$ queue by Ramaswami and Lucantoni (1985). While in Ramaswami (1985) the basic groundwork needed to handle the construction of the generator for the Markov process governing the parallel Markov chains (which correspond to the multiple servers offering services that are of phase type) is established, in Ramaswami and Lucantoni (1985) the focus is to apply that technique to analyze $GI/PH/c$ queue. It should be pointed out that approaching the analysis with reduced state space in the multi-server case requires more work in the construction of the entries of the transition probability matrix (TPM) of the queue (Ramaswami and Lucantoni 1985).

Using the approach suggested in Mayhugh and McCormick (1968), Ramaswami and Lucantoni (1985) and Ramaswami (1985), we illustrate the reduction in the state space. Toward this end, suppose we look at a c-server homogeneous queueing system with BMAP arrivals.

Suppose that $N(t)$ denotes the number of customers in the system; $J_r(t)$ denotes the number of severs that are in phase r; and $I(t)$ denotes the phase of the BMAP, at time t. The state space for this setup is given by:

$$\tilde{\Omega} = \tilde{\Omega}_B \bigcup \tilde{\Omega}_{NB},$$

where, for $0 \leq i < c$,

$$\Omega_B = \{(i, j_1, \cdots, j_n, k) : 0 \leq j_r \leq c, 1 \leq r \leq n, 1 \leq k \leq m, \sum_{r=1}^{n} j_r < c\},$$

and, for $i \geq c$,

$$\tilde{\Omega}_{NB} = \{(i, j_1, j_2, \cdots, j_n, k) : 0 \leq j_r \leq c,\ 1 \leq r \leq n,\ \sum_{r=1}^{n} j_r = c,\ 1 \leq k \leq m\}.$$

Suppose we denote the level $\boldsymbol{i}$ to be the set of states corresponding to $\boldsymbol{i} = \{(i, j_1, \cdots, j_n, k)$: $0 \leq j_r \leq c,\ 1 \leq r \leq n,\ 1 \leq k \leq m\}$, for $i \geq 0$. [Note that $\sum_{r=1}^{n} j_r$ is less than c or equal to c according to the boundary or non-boundary states, respectively.]

A classical combinatorial problem that deals with finding the number of ways to put c balls in n boxes with no restriction on how many balls a box can contain (i.e. a box may contain all c balls or no balls but between them all c balls should be accounted for) plays a role here to determine the number of states in each of the levels. It is easy to verify that the dimension of $\boldsymbol{i},\ 0 \leq i \leq c-1$, is given by $\binom{i+n}{n-1}$ and the dimension of $\boldsymbol{i},\ i \geq c-1$, is given by $\binom{c+n-1}{n-1}$. For example, if $c = 7$ and $n = 4$, then using the standard approach in which each server's phase of service is kept track, then the non-boundary levels will have dimension $4^7 = 16{,}384$; however, if we keep track of the number of servers in each of the n phases, the dimension of the non-boundary levels will be $\binom{7+4-1}{4-1} = 120$, a significant reduction in the dimension.

Essentially, analyzing $BMAP/PH/c$ queue reduces to exploiting the structure of the coefficient matrices, and in the next section we illustrate this using $MAP/PH/3$ queue.

We conclude this section with an illustrative example involving $BMAP/PH/2$ model, a two-server heterogeneous system with BMAP arrivals with representation $\{D_k\}$ of dimension and phase type services with representation $(\boldsymbol{\beta}(r), S(r))$ of dimension n_r for server r, $r = 1, 2$. When both servers are idle an arriving customer (if the batch has only one customer) will pick server r with probability p_r with $0 \leq p_r \leq 1$, $r = 1, 2$ and $p_1 + p_2 = 1$. Define the state space of the system along the lines mentioned above, and let the set of states, namely levels, be defined as follows:

1) $\mathbf{0}$ denotes the system is empty with the arrival process in one of m states.

2) $\hat{\mathbf{1}}$ of dimension $m\,n_1$ denotes the set of states $\{(1, j_1, k) : 1 \leq j_1 \leq n_1,\ 1 \leq k \leq m$, corresponding to the system having one customer who is being served by server 1 and the arrival and the service processes in various phases.

3) $\tilde{\mathbf{1}}$ of dimension $m\,n_2$ denotes the set of states $\{(1, j_2, k) : 1 \leq j_2 \leq n_2,\ 1 \leq k \leq m$, corresponding to the system having one customer who is being served by server 2 and the arrival and the service processes in various phases.

4) $\boldsymbol{i}$ of dimension $m\,n_1\,n_2$ denotes the set of states $\{(i, j_1, j_2, k) : 1 \leq j_r \leq n_r,\ r = 1, 2,\ 1 \leq k \leq m$, corresponding to the system having i customers with both servers busy, and the arrival and the services in various phases.

The generator of the Markov process governing $BMAP/PH/2$ system using arbitrary time approach is given by:

$$Q = \begin{pmatrix} D_0 & B_1 & B_2 & B_3 & B_4 & B_5 & B_6 & \cdots \\ C_0 & C_1 & 0 & C_3 & C_4 & C_5 & C_6 & \cdots \\ E_0 & 0 & E_2 & E_3 & E_4 & E_5 & E_6 & \cdots \\ 0 & F_1 & F_2 & H_1 & H_2 & H_3 & H_4 & \cdots \\ 0 & 0 & 0 & H_0 & H_1 & H_2 & H_3 & \cdots \\ \vdots & \vdots & \vdots & \vdots & \vdots & \vdots & \vdots & \ddots \end{pmatrix}, \qquad [5.1]$$

where:

$$B_1 = p_1\boldsymbol{\beta}(1) \otimes D_1,\ B_2 = p_2\boldsymbol{\beta}(2) \otimes D_1,\ B_k = \boldsymbol{\beta}(1) \otimes \boldsymbol{\beta}(2) \otimes D_{k-1},\ k \geq 3,$$

$$C_0 = \boldsymbol{S}^{\mathbf{0}}(1) \otimes I_m,\ C_1 = S(1) \oplus D_0,\ C_k = I_{n_1} \otimes \boldsymbol{\beta}(2) \otimes D_{k-2},\ k \geq 3,$$

$$E_0 = \boldsymbol{S}^{\mathbf{0}}(2) \otimes I_m,\ E_2 = S(2) \oplus D_0,\ E_k = \boldsymbol{\beta}(1) \otimes I_{n_2} \otimes D_{k-2},\ k \geq 3,$$

$$F_1 = I_{n_1} \otimes \boldsymbol{S}^{\mathbf{0}}(2) \otimes I_m,\ F_2 = \boldsymbol{S}^{\mathbf{0}}(1) \otimes I_{m\,n_1},\ H_1 = S(1) \oplus S(2) \oplus D_0,$$

$$H_0 = \boldsymbol{S}^{\mathbf{0}}(1)\boldsymbol{\beta}(1) \otimes I_{m\,n_2} + I_{n_1} \otimes \boldsymbol{S}^{\mathbf{0}}(2)\boldsymbol{\beta}(2) \otimes I_m,\ H_k = I_{n_1 n_2} \otimes D_k,\ k \geq 1.$$

The queueing system $BMAP/PH/2$ can now be studied as an $M/G/1$-type (see section 3.1 but with many boundary states to account for the levels $\tilde{\mathbf{1}}$ and $\hat{\mathbf{1}}$), and one needs to exploit the structure of the coefficient matrices for efficient computation. The details will be presented in the section as part of a different model. The reader can carry out such procedures for this model.

Now one can carry out such construction for more general c-server heterogenous systems involving many folds of Kronecker products and Kronecker sums. Note that in the case of a c-server (heterogeneous or homogeneous) queueing system, $MAP/PH/c$, one can set up the generator governing the system as a quasi-birth-and-death (QBD) process. Again one needs to exploit the structure of the matrices as the state space grows exponentially with the orders of the phase type (PH) distributions.

5.2. *MAP/PH/3* queue

Consider a three-server queueing system in which the customers arrive according to a Markovian arrival process (MAP) with representation matrices (D_0, D_1) of order m. We assume that the customers are served on a first come, first served (FCFS) basis and the servers are heterogeneous with the service times of phase type with representation $(\boldsymbol{\beta}^{(r)}, S^{(r)})$ of dimension n_r for server r, $1 \le r \le 3$. An arriving customer finding at least one free server will get service from the lowest numbered server among the free ones. Recall that the average service rates, μ_r, for these three servers are given by $\mu_r = [\boldsymbol{\beta}^{(r)}(-S^{(r)})^{-1}\mathbf{e}]^{-1}$. Since we are dealing with the heterogeneous servers, the book-keeping is a little more than that of the homogeneous case. These are listed along with the notation as follows:

– $N(t)$ is the number of customers in the queue at time t.

– $J_r(t),\ 1 \le r \le 3$, to be the phase of the server r at time t.

– $I(t)$ to be the phase of the MAP at time t.

– The state space is given by:

$$\Omega = \Omega_B \bigcup \{(i, j_3, j_2, j_1, k) : 1 \le j_r \le n_r, 1 \le r \le 3, 1 \le k \le m, i \ge 0\},$$

where Ω_B, the state space for the boundary states, is given by:

$$\Omega_B = \{\tilde{0}\} \bigcup \{\tilde{1}\} \bigcup \{\tilde{2}\} \bigcup \{\tilde{3}\} \bigcup \{\tilde{12}\} \bigcup \{\tilde{13}\} \bigcup \{\tilde{23}\},$$

where the levels in Ω_B have the following meaning:

1) By level $\tilde{\mathbf{0}}$ we indicate that the system is idle with no customers present.

2) By level $\tilde{\mathbf{1}}$ we indicate that only server 1 is busy.

3) By level $\tilde{\mathbf{2}}$ we indicate that only server 2 is busy.

4) By level $\tilde{\mathbf{3}}$ we indicate that only server 3 is busy.

5) By level $\tilde{\mathbf{12}}$ we indicate that only servers 1 and 2 are busy.

6) By level $\tilde{\mathbf{13}}$ we indicate that only servers 1 and 3 are busy.

7) By level $\tilde{\mathbf{23}}$ we indicate that only servers 2 and 3 are busy.

– We denote by level $\boldsymbol{i}$, for $i \geq 0$, the set of states $\{(i, j_3, j_2, j_1, k) : 1 \leq j_r \leq n_r, 1 \leq r \leq 3, 1 \leq k \leq m\}$, corresponding to all three servers being busy with i customers waiting in the queue.

The generator Q of the $MAP/PH/3$ queue is given by:

$$Q = \begin{pmatrix} B & C & 0 & 0 & 0 & \cdots \\ E & A^{(1)} & A^{(0)} & 0 & 0 & \cdots \\ 0 & A^{(2)} & A^{(1)} & A^{(0)} & \cdots & \\ 0 & 0 & A^{(2)} & A^{(1)} & A^{(0)} & \cdots \\ \vdots & \vdots & \vdots & \vdots & \vdots & \ddots \end{pmatrix}, \qquad [5.2]$$

where:

$$B = \begin{pmatrix} D_0 & B^{(01)} & 0 & 0 & 0 & 0 & 0 \\ B^{(10)} & B^{(1)} & 0 & 0 & B^{(11)} & 0 & 0 \\ B^{(20)} & 0 & B^{(2)} & 0 & B^{(21)} & 0 & 0 \\ B^{(30)} & 0 & 0 & B^{(3)} & 0 & B^{(31)} & 0 \\ 0 & B^{(42)} & B^{(43)} & 0 & B^{(4)} & 0 & 0 \\ 0 & B^{5(2)} & 0 & B^{(53)} & 0 & B^{(5)} & 0 \\ 0 & 0 & B^{(62)} & B^{(63)} & 0 & 0 & B^{(6)} \end{pmatrix}, \quad C = \begin{pmatrix} 0 \\ 0 \\ 0 \\ 0 \\ B^{(41)} \\ B^{(51)} \\ B^{(61)} \end{pmatrix}, \qquad [5.3]$$

$$E = \begin{pmatrix} 0 & 0 & 0 & 0 & A^{(21)} & A^{(22)} & A^{(23)} \end{pmatrix}.$$

In the following, $\boldsymbol{S}^0(r)$ is such that $S^{(r)}\boldsymbol{e} + \boldsymbol{S}^0(r) = \mathbf{0},\ 1 \leq r \leq 3$. The matrices not yet defined in equation [5.3] are described below:

$$B^{(01)} = \boldsymbol{\beta}^{(1)} \otimes D_1, \ \ B^{(10)} = \boldsymbol{S}^0(1) \otimes I_m, \ \ B^{(1)} = S^{(1)} \oplus D_0, \qquad [5.4]$$

$$B^{(11)} = \boldsymbol{\beta}^{(2)} \otimes I_{n_1} \otimes D_1, \ B^{(20)} = \boldsymbol{S}^0(2) \otimes I_m, \ B^{(2)} = S^{(2)} \oplus D_0, \qquad [5.5]$$

$$B^{(21)} = I_{n_2} \otimes \boldsymbol{\beta}^{(1)} \otimes D_1, \ B^{(30)} = \boldsymbol{S}^0(3) \otimes I_m, \ B^{(3)} = S^{(3)} \oplus D_0, \qquad [5.6]$$

$$B^{(31)} = I_{n_3} \otimes \boldsymbol{\beta}^{(1)} \otimes D_1, \ \ B^{(42)} = \boldsymbol{S}^0(2) \otimes I_{n_1} \otimes I_m, \qquad [5.7]$$

$$B^{(43)} = I_{n_2} \otimes \boldsymbol{S}^0(1) \otimes I_m, \ \ B^{(4)} = S^{(2)} \oplus S^{(1)} \oplus D_0, \qquad [5.8]$$

$$B^{(41)} = \boldsymbol{\beta}^{(3)} \otimes I_{n_2} \otimes I_{n_1} \otimes D_1, \ \ B^{(52)} = \boldsymbol{S}^0(3) \otimes I_{n_1} \otimes I_m, \qquad [5.9]$$

$$B^{(53)} = I_{n_3} \otimes \boldsymbol{S}^0(1) \otimes I_m, \ \ B^{(5)} = S^{(3)} \oplus S^{(1)} \oplus D_0, \qquad [5.10]$$

$$B^{(51)} = I_{n_3} \otimes \boldsymbol{\beta}^{(2)} \otimes I_{n_1} \otimes I_m, \ \ B^{(62)} = \boldsymbol{S}^0(3) \otimes I_{n_2} \otimes I_m, \qquad [5.11]$$

$$B^{(63)} = I_{n_3} \otimes \boldsymbol{S}^0(2) \otimes I_m, \ \ B^{(6)} = S^{(3)} \oplus S^{(2)} \oplus D_0, \qquad [5.12]$$

$$B^{(61)} = I_{n_3} \otimes \boldsymbol{S}^0(2) \otimes I_{n_1} \otimes I_m, \ A^{(21)} = \boldsymbol{S}^0(3) \otimes I_{n_2} \otimes I_{n_1} \otimes I_m, \ [5.13]$$

$$A^{(22)} = I_{n_3} \otimes \boldsymbol{S^0}(2) \otimes I_{n_1} \otimes I_m, \; A^{(23)} = I_{n_3} \otimes I_{n_2} \otimes \boldsymbol{S^0}(1) \otimes I_m, \quad [5.14]$$
$$A^{(0)} = I_{n_3} \otimes I_{n_2} \otimes I_{n_1} \otimes D_1, \; A^{(1)} = S^{(3)} \oplus S^{(2)} \oplus S^{(1)} \oplus D_1, \quad [5.15]$$
$$A^{(2)} = \boldsymbol{S^0}(3)\boldsymbol{\beta^{(3)}} \oplus \boldsymbol{S^0}(2)\boldsymbol{\beta^{(2)}} \oplus \boldsymbol{S^0}(1)\boldsymbol{\beta^{(1)}} \oplus I - I_{mn_1n_2n_2}. \quad [5.16]$$

Letting $\boldsymbol{y} = (\boldsymbol{y}^*, \boldsymbol{y}_0, \boldsymbol{y}_1, \cdots)$ denote the steady-state probability vector of Q, we can solve for $\boldsymbol{y}$ as follows. Noting that $\boldsymbol{y}^*$ is of dimension $m(1 + n_1 + n_2 + n_3 + n_1n_2 + n_1n_3 + n_2n_3)$ and $\boldsymbol{y}_i$, $i \geq 0$, is of dimension $mn_1n_2n_3$, we get:

$$\begin{aligned} &\boldsymbol{y}^*B + \boldsymbol{y}_0E = \boldsymbol{0}, \\ &\boldsymbol{y}^*C + \boldsymbol{y}_0[A^{(1)} + RA^{(2)}] = \boldsymbol{0}, \\ &\boldsymbol{y}^*\,\boldsymbol{e} + \boldsymbol{y}_0(I - R)^{-1}\boldsymbol{e} = 1. \end{aligned} \quad [5.17]$$

The matrix R is the minimal non-negative solution to the matrix-quadratic equation:

$$R^2A^{(2)} + RA^{(1)} + A^{(0)} = 0 \quad [5.18]$$

When solving for R and then $\boldsymbol{y}$, depending on the type of software, such as Fortran, and the memory requirements of the computer, one can either implement one of several numerical methods such as (block) Gauss-Seidel to arrive at solutions or exploit the structure of the coefficient matrices and then implement an appropriate algorithm for arriving at solutions. This is very important as the state space grows exponentially as m, n_1, n_2 and n_3 increase.

We briefly outline the ways to exploit the structure of the coefficient matrices. This will involve dealing with manipulation of matrices of dimension m. Toward this end, we denote the block matrices with appropriate suffixes and the context will indicate whether we are dealing with row or column or matrix blocks. That is, block matrices having only one suffix will be either row or column blocks and the ones with two suffixes will be a matrix block. Also, for convenience of displaying these matrices, the suffixes which need expanded description will be defined with supplementary notation.

Using the definition of Kronecker products and sums (recall these from section 1.3.6 of Volume 1), first verify the following. All the identity matrices are of dimension m and hence we suppress the index m.

1) $B^{(01)} = \{B_r^{(01)}\}$ with $B_r^{(01)} = \beta_r^{(1)} D_1, \; 1 \leq r \leq n_1.$

2) $B^{(l0)} = \{B_r^{(l0)}\}$ with $B_r^{(l0)} = S_r^0(l)I, \; 1 \leq r \leq n_l, \; l = 1, 2, 3.$

3) $B^{(l)} = \{B_{i,j}^{(l)}\}, \; l = 1, 2, 3$ with

$$B_{i,j}^{(l)} = \begin{cases} S_{i,j}^{(l)}I + D_0, & 1 \leq i, j \leq n_l, \; i = j, \\ S_{i,j}^{(l)}I, & 1 \leq i, j \leq n_l, \; i \neq j, \end{cases}$$

4) $B^{(11)} = \{B^{(11)}_{i,j}\}$, with

$$B^{(11)}_{i,i+(j-1)n_1} = \begin{cases} \beta^{(2)}_j D_1, & 1 \le i \le n_1 \ 1 \le j \le n_2, \\ 0, & \text{otherwise.} \end{cases}$$

5) $B^{(l1)} = \{B^{(l1)}_{i,j}\}$, with

$$B^{(l1)}_{i,j+(i-1)n_1} = \begin{cases} \beta^{(1)}_j D_1, & 1 \le i \le n_l, \ 1 \le j \le n_1, \ l = 2, 3, \\ 0, & \text{otherwise.} \end{cases}$$

6) $B^{(4)} = \{B^{(4)}_{i,j}\}$, with, for $1 \le i, j \le n_2$, $1 \le r_1, r_2 \le n_1$,

$$B^{(4)}_{(i-1)n_1+r_1,(j-1)n_1+r_2} = \begin{cases} (S^{(2)}_{i,j} + S^{(1)}_{r_1,r_2})I + D_0, & i = j, \ r_1 = r_2, \\ S^{(1)}_{r_1,r_2} I, & i = j, r_1 \ne r_2, \\ S^{(2)}_{i,j} I, & i \ne j, \ r_1 = r_2 \\ 0, & \text{otherwise.} \end{cases}$$

7) $B^{(41)} = \{B^{(41)}_{i,j}\}$, with

$$B^{(41)}_{i,i+(j-1)n_1n_2} = \begin{cases} \beta^{(3)}_j D_1, & 1 \le i \le n_1 n_2, \ 1 \le j \le n_3, \\ 0, & \text{otherwise.} \end{cases}$$

8) $B^{(42)} = \{B^{(42)}_{i,j}\}$, with

$$B^{(42)}_{(i-1)n_1+j,j} = \begin{cases} S^0_i(2)I, & 1 \le i \le n_2, \ 1 \le j \le n_1, \\ 0, & \text{otherwise.} \end{cases}$$

9) $B^{(43)} = \{B^{(43)}_{i,j}\}$, with

$$B^{(43)}_{(i-1)n_1+j,i} = \begin{cases} S^0_j(1)I, & 1 \le i \le n_2, \ 1 \le j \le n_1, \\ 0, & \text{otherwise.} \end{cases}$$

10) $B^{(5)} = \{B^{(5)}_{i,j}\}$, with, for $1 \le i, j \le n_3,\ 1 \le r_1, r_2 \le n_1$,

$$B^{(5)}_{(i-1)n_1+r_1,(j-1)n_1+r_2} = \begin{cases} (S^{(3)}_{i,j} + S^{(1)}_{r_1,r_2})I + D_0, \ i = j, \ r_1 = r_2, \\ S^{(1)}_{r_1,r_2} I, \ i = j, r_1 \neq r_2, \\ S^{(3)}_{i,j} I, \ \ i \neq j, \ r_1 = r_2 \\ 0, \ \ \text{otherwise.} \end{cases}$$

11) $B^{(51)} = \{B^{(51)}_{i,j}\}$, with, for $1 \le i \le n_3, \ \ 1 \le j \le n_2$,

$$B^{(51)}_{(i-1)n_1+r,(i-1)n_1 n_2+(j-1)n_1+r} = \begin{cases} \beta^{(2)}_j D_1, \ \ 1 \le r \le n_1, \\ 0, \ \ \text{otherwise.} \end{cases}$$

12) $B^{(52)} = \{B^{(52)}_{i,j}\}$, with

$$B^{(52)}_{(i-1)n_1+j,j} = \begin{cases} S^0_i(3)I, \ 1 \le i \le n_3, \ \ 1 \le j \le n_1, \\ 0, \ \ \text{otherwise.} \end{cases}$$

13) $B^{(53)} = \{B^{(53)}_{i,j}\}$, with

$$B^{(53)}_{(i-1)n_1+j,i} = \begin{cases} S^0_j(1)I, \ 1 \le i \le n_3, \ \ 1 \le j \le n_1, \\ 0, \ \ \text{otherwise.} \end{cases}$$

14) $B^{(6)} = \{B^{(6)}_{i,j}\}$, with, for $1 \le i, j \le n_3, \ 1 \le r_1, r_2 \le n_2$,

$$B^{(6)}_{(i-1)n_2+r_1,(j-1)n_2+r_2} = \begin{cases} (S^{(3)}_{i,j} + S^{(2)}_{r_1,r_2})I + D_0, \ i = j, \ r_1 = r_2, \\ S^{(2)}_{r_1,r_2} I, \ i = j, \ r_1 \neq r_2, \\ S^{(3)}_{i,j} I, \ \ i \neq j, \ r_1 = r_2 \\ 0, \ \ \text{otherwise.} \end{cases}$$

15) $B^{(61)} = \{B^{(61)}_{i,j}\}$, with

$$B^{(61)}_{(i-1)n_2+j,(i-1)n_1 n_2+(j-1)n_1+r} = \begin{cases} \beta^{(1)}_r D_1, 1 \le i \le n_3, 1 \le j \le n_2, \ 1 \le r \le n_1, \\ 0, \ \ \text{otherwise.} \end{cases}$$

16) $B^{(62)} = \{B^{(62)}_{i,j}\}$, with

$$B^{(62)}_{(i-1)n_2+j,j} = \begin{cases} S^0_i(3)I, \ 1 \le i \le n_3, \ \ 1 \le j \le n_2, \\ \qquad 0, \ \text{otherwise.} \end{cases}$$

17) $B^{(63)} = \{B^{(63)}_{i,j}\}$, with

$$B^{(63)}_{(i-1)n_2+j,i} = \begin{cases} S^0_j(2)I, \ 1 \le i \le n_3, \ \ 1 \le j \le n_2, \\ \qquad 0, \ \text{otherwise.} \end{cases}$$

18) $A^{(0)} = \{A^{(0)}_{i,j}\}$, with

$$A^{(0)}_{i,j} = \begin{cases} D_1, \ 1 \le i,j \le n_1n_2n_3, \ \ j = i, \\ \qquad 0, \ \text{otherwise.} \end{cases}$$

19) $A^{(21)} = \{A^{(21)}_{i,j}\}$, with, for $1 \le i \le n_3, \ \ 1 \le j \le n_1n_2$,

$$A^{(21)}_{(i-1)n_1n_2+j,j} = \begin{cases} S^0_i(3)I, \ 1 \le i \le n_3, \\ \ 0, \ \text{otherwise.} \end{cases}$$

20) $A^{(22)} = \{A^{(22)}_{i,j}\}$, with, for $1 \le i \le n_3, \ \ 1 \le j \le n_2$,

$$A^{(22)}_{(i-1)n_1n_2+(j-1)n_1+r,(i-1)n_1n_2+r} = \begin{cases} S^0_j(2)I, \ 1 \le r \le n_1, \\ \ 0, \ \text{otherwise.} \end{cases}$$

21) $A^{(23)} = \{A^{(23)}_{i,j}\}$, with, for $1 \le i \le n_3, \ \ 1 \le j \le n_2$,

$$A^{(23)}_{(i-1)n_1n_2+(j-1)n_1+r,(i-1)n_1n_2+j} = \begin{cases} S^0_r(1)I, \ 1 \le r \le n_1, \\ \ 0, \ \text{otherwise.} \end{cases}$$

22) Define $\tilde{A}^{(21)} = \{\tilde{A}^{(21)}_{i,j}\} = \boldsymbol{\beta}^{(3)} \otimes I_{mn_1n_2}$, $\tilde{A}^{(22)} = \{\tilde{A}^{(22)}_{i,j}\} = I_{n_3} \otimes \boldsymbol{\beta}^{(2)} \otimes I_{mn_1}$, and $\tilde{A}^{(23)} = \{\tilde{A}^{(23)}_{i,j}\} = I_{n_2n_3} \otimes \boldsymbol{\beta}^{(1)} \otimes I_m$. Then, we have:

$$\tilde{A}^{(21)}_{i,(j-1)n_1n_2+i} = \begin{cases} \beta^{(3)}_r I, \ 1 \le i \le n_1n_2, \ 1 \le j \le n_3, \\ \qquad 0, \ \text{otherwise.} \end{cases}$$

$$\tilde{A}^{(22)}_{(i-1)n_1+r,(i-1)n_1n_2+(j-1)n_1+r} = \begin{cases} \beta_j^{(2)} I, \ 1 \le i \le n_3, \ 1 \le j \le n_2, \ 1 \le r \le n_1, \\ 0, \ \text{otherwise.} \end{cases}$$

$$\tilde{A}^{(23)}_{(i-1)n_2+j,(i-1)n_1n_2+(j-1)n_1+r} = \begin{cases} \beta_r^{(1)} I, \ 1 \le i \le n_3, \ 1 \le j \le n_2, \ 1 \le r \le n_1, \\ 0, \ \text{otherwise.} \end{cases}$$

23) $A^{(2)} = \{A^{(2)}_{i,j}\}$ can be written in terms of the matrices defined earlier as follows: $A^{(2)}_{i,j} = \sum_{k=1}^{n_1n_2} A^{(21)}_{i,k} \tilde{A}^{(21)}_{k,j} + \sum_{k=1}^{n_1n_3} A^{(22)}_{i,k} \tilde{A}^{(22)}_{k,j} + \sum_{k=1}^{n_2n_3} A^{(23)}_{i,k} \tilde{A}^{(23)}_{k,j}$.

24) $A^{(1)} = \{A^{(1)}_{i,j}\}$, with, for $1 \le i,j \le n_3$, $1 \le r_1, r_2 \le n_1n_2$,

$$A^{(1)}_{(i-1)n_1n_2+r_1,(j-1)n_1n_2+r_2} = \begin{cases} (S^{(3)}_{i,j} I + B^{(4)}_{r_1,r_2}), \ i = j, \ r_1 = r_2, \\ S^{(3)}_{i,j} I, \ \ i \ne j, 1 \le r_1, r_2 \le n_1n_2. \\ 0, \ \text{otherwise.} \end{cases}$$

Suppose we partition the vector $\boldsymbol{y} = (\boldsymbol{y}^*, \boldsymbol{y}_0, \boldsymbol{y}_1, \cdots)$ such that the component vectors are further partitioned as follows:

$$\boldsymbol{y}^* = (\boldsymbol{u}_0, \boldsymbol{u}^{(1)}, \boldsymbol{u}^{(2)}, \cdots, \boldsymbol{u}^{(6)}),$$

$$\boldsymbol{y}_i = (\boldsymbol{y}_1^{(i)}, \boldsymbol{y}_2^{(i)}, \cdots, \boldsymbol{y}^{(i)}_{n_1 \, n_2 \, n_3}),$$

where $\boldsymbol{u}_0$ and $\boldsymbol{y}_j^{(i)}$, $1 \le j \le n_1 \, n_2 \, n_3$, $i \ge 0$, are of dimension m; $\boldsymbol{u}^{(r)}$, $1 \le r \le 6$, are of dimension $mn_1, mn_2, mn_3, mn_1 \, n_2, mn_1 \, mn_3$ and $mn_2 \, n_3$, respectively.

In terms of smaller dimensions, the steady-state probability vector $\boldsymbol{y}$ is solved using the following equations:

$$\boldsymbol{u}_0 = \left[\sum_{k=1}^{n_1} \boldsymbol{u}_k^{(1)} B_k^{(10)} + \sum_{k=1}^{n_2} \boldsymbol{u}_k^{(2)} B_k^{(20)} + \sum_{k=1}^{n_3} \boldsymbol{u}_k^{(3)} B_k^{(30)} \right] (-D_0)^{-1}.$$

For $1 \le r \le n_1$,

$$\boldsymbol{u}_r^{(1)} = \left[\boldsymbol{u}_0 B_r^{(01)} + \sum_{k \ne r, k=1}^{n_1} \boldsymbol{u}_k^{(1)} B^{(1)}_{k,r} + \sum_{k=1}^{n_1n_2} \boldsymbol{u}_k^{(4)} B^{(42)}_{k,r} + \sum_{k=1}^{n_1n_3} \boldsymbol{u}_k^{(5)} B_k^{(52)} \right] \left(-B^{(1)}_{r,r}\right)^{-1}.$$

For $1 \le r \le n_2$,

$$\boldsymbol{u}_r^{(2)} = \left[\sum_{k \ne r, k=1}^{n_2} \boldsymbol{u}_k^{(2)} B^{(2)}_{k,r} + \sum_{k=1}^{n_1n_2} \boldsymbol{u}_k^{(4)} B^{(43)}_{k,r} + \sum_{k=1}^{n_2n_3} \boldsymbol{u}_k^{(6)} B_k^{(62)} \right] \left(- B^{(2)}_{r,r}\right)^{-1}.$$

For $1 \le r \le n_3$,

$$\boldsymbol{u}_r^{(3)} = \left[\sum_{k \neq r, k=1}^{n_3} \boldsymbol{u}_k^{(3)} B_{k,r}^{(3)} + \sum_{k=1}^{n_1 n_3} \boldsymbol{u}_k^{(5)} B_{k,r}^{(53)} + \sum_{k=1}^{n_2 n_3} \boldsymbol{u}_k^{(6)} B_k^{(63)} \right] \left(- B_{r,r}^{(3)} \right)^{-1}.$$

$$\boldsymbol{u}_r^{(4)} = \left[\sum_{k=1}^{n_1} \boldsymbol{u}_k^{(1)} B_{k,r}^{(11)} + \sum_{k=1}^{n_2} \boldsymbol{u}_k^{(2)} B_{k,r}^{(21)} + \sum_{k \neq r, k=1}^{n_1 n_2} \boldsymbol{u}_k^{(4)} B_{k,r}^{(4)} \right.$$

$$\left. + \sum_{k=1}^{n_1 n_2 n_3} \boldsymbol{y}_k^{(0)} A_{k,r}^{(21)} \right] \left(- B_{r,r}^{(4)} \right)^{-1}, \ 1 \le r \le n_1 n_3.$$

For $1 \le r \le n_1 n_3$,

$$\boldsymbol{u}_r^{(5)} = \left[\sum_{k=1}^{n_3} \boldsymbol{u}_k^{(3)} B_{k,r}^{(31)} + \sum_{k \neq r, k=1}^{n_1 n_3} \boldsymbol{u}_k^{(5)} B_{k,r}^{(5)} + \sum_{k=1}^{n_1 n_2 n_3} \boldsymbol{y}_k^{(0)} A_{k,r}^{(22)} \right] \left(- B_{r,r}^{(5)} \right)^{-1}.$$

For $1 \le r \le n_2 n_3$,

$$\boldsymbol{u}_r^{(6)} = \left[\sum_{k \neq r, k=1}^{n_2 n_3} \boldsymbol{u}_k^{(6)} B_{k,r}^{(6)} + \sum_{k=1}^{n_1 n_2 n_3} \boldsymbol{y}_k^{(0)} A_{k,r}^{(23)} \right] \left(- B_{r,r}^{(6)} \right)^{-1}.$$

For $1 \le r \le n_1 n_2 n_3$,

$$\boldsymbol{y}_r^{(0)} = \left[\sum_{k=1}^{n_1 n_2} \boldsymbol{u}_k^{(4)} B_{k,r}^{(41)} + \sum_{k=1}^{n_1 n_3} \boldsymbol{u}_k^{(5)} B_{k,r}^{(51)} + \sum_{k=1}^{n_1 n_2} \boldsymbol{u}_k^{(6)} B_{k,r}^{(61)} \right.$$

$$\left. + \sum_{k \neq r, k=1}^{n_1 n_2 n_3} \boldsymbol{y}_k^{(0)} [A_{k,r}^{(1)} + \sum_{k_1=1}^{n_1 n_2 n_3} R_{k,k1} A_{k,r}^{(2)} \right] \left(- \left[A_{r,r}^{(1)} + \sum_{k_1=1}^{n_1 n_2 n_3} R_{k,k_1} A_{k,r}^{(2)} \right] \right)^{-1}.$$

The above equations along with the normalizing condition can be solved using one of the well-known methods including (block) Gauss-Seidel. It is important to have internal accuracy checks to guarantee that the computations are properly carried out. Before we display a few accuracy checks, we need to define a few quantities. Denote by $\boldsymbol{\eta}$ and $\boldsymbol{\delta}^{(i)}$, $1 \le i \le 3$, the steady-state probability vector of the MAP and the service processes, respectively. Also, let $\boldsymbol{\pi}$ denote the steady-state probability vector of $A = A^{(0)} + A^{(1)} + A^{(2)}$. That is:

$$\begin{aligned} &\boldsymbol{\eta}(D_0 + D_1) = \mathbf{0}, \ \boldsymbol{\eta}\, \boldsymbol{e} = 1, \\ &\boldsymbol{\delta}^{(i)}(S(i) + \boldsymbol{S^0}(\boldsymbol{i})\, \boldsymbol{\beta}^{(i)}) = \mathbf{0}, \ \boldsymbol{\delta}^{(i)}\, \boldsymbol{e} = 1, \ 1 \le i \le 3, \\ &\boldsymbol{\pi} A = \mathbf{0}, \ \boldsymbol{\pi}\, \boldsymbol{e} = 1. \end{aligned} \quad [5.19]$$

The following are a few results that serve as internal accuracy checks in numerical computations:

$$\begin{aligned} &\boldsymbol{\delta}^{(i)} = \mu_i \boldsymbol{\beta}^{(i)} \left(- S^{(i)} \right)^{-1}, \ 1 \le i \le 3, \\ &\boldsymbol{\pi} = \boldsymbol{\delta}^{(1)} \otimes \boldsymbol{\delta}^{(2)} \otimes \boldsymbol{\delta}^{(3)} \otimes \boldsymbol{\eta}. \end{aligned} \quad [5.20]$$

$$\boldsymbol{y}^*(\boldsymbol{e} \otimes I_m) + \boldsymbol{y}^{(0)}(\boldsymbol{e} \otimes I_m) = \boldsymbol{\eta}. \qquad [5.21]$$

Suppose that $\boldsymbol{\xi}^{(i)}$, $1 \leq i \leq 3$, denotes the vector of dimension n_i giving the service phase of server i when busy. If b_i, $1 \leq i \leq 3$, denotes the probability that server i is busy, then we have the following intuitively clear result:

$$\frac{1}{b_i}\boldsymbol{\xi}^{(i)} = \mu_i \boldsymbol{\beta}^{(i)}\big(-S^{(i)}\big)^{-1}, \ 1 \leq i \leq 3. \qquad [5.22]$$

The following intuitively obvious result says that the rate of customers leaving the system should be equal to the arrival rate in steady-state:

$$\sum_{i=1}^{3} \boldsymbol{\xi}^{(i)} \boldsymbol{S}^{\mathbf{0}}(i) = \lambda. \qquad [5.23]$$

The System Performance Measures: The qualitative analysis of any queueing model requires one or more system performance measures. While one can develop a number of such measures, here we give a few and leave the rest to interested readers.

1) *Probability that the system is idle.* The probability, P_{idle}, that the system is idle (i.e. there are no customers in the system) is given by $P_{idle} = \boldsymbol{u}_0\boldsymbol{e}$.

2) *Probability that only server r is busy.* The probability, $P_{busy}^{(r)}$, that only server r is busy is obtained as $P_{busy}^{(r)} = \boldsymbol{u}_r\boldsymbol{e}$, $r = 1, 2, 3$.

3) *Probability that exactly two specified servers are busy.* The probability, $P_{busy}^{(1,2)}$, that only servers 1 and 2 are busy is obtained as $P_{busy}^{(1,2)} = \boldsymbol{u}_4\boldsymbol{e}$; the probability, $P_{busy}^{(1,3)}$, that only servers 1 and 3 are busy is obtained as $P_{busy}^{(1,3)} = \boldsymbol{u}_5\boldsymbol{e}$; the probability, $P_{busy}^{(2,3)}$, that only servers 2 and 3 are busy is obtained as $P_{busy}^{(2,3)} = \boldsymbol{u}_6\boldsymbol{e}$.

4) *Probability that all servers are busy.* The probability, P_{busy}, that all servers are busy is calculated as $P_{busy} = \boldsymbol{y}_0(I - R)^{-1}\boldsymbol{e}$.

5) *Probability that server r is busy.* The probabilities, $\tilde{P}_{busy}^{(r)}$, $1 \leq r \leq 3$, that server r is busy irrespective of the status of the other servers are given by $\tilde{P}_{busy}^{(1)} = P_{busy}^{(1)} + P_{busy}^{(1,2)} + P_{busy}^{(1,3)} + P_{busy}$, $\tilde{P}_{busy}^{(2)} = P_{busy}^{(2)} + P_{busy}^{(1,2)} + P_{busy}^{(2,3)} + P_{busy}$, and $\tilde{P}_{busy}^{(3)} = P_{busy}^{(3)} + P_{busy}^{(1,3)} + P_{busy}^{(2,3)} + P_{busy}$.

6) *Mean number of busy servers.* The mean number, μ_{busy}, of busy servers is given by $P_{busy}^{(1)} + P_{busy}^{(2)} + P_{busy}^{(3)} + 2(P_{busy}^{(1,2)} + P_{busy}^{(1,3)} + P_{busy}^{(2,3)}) + 3P_{busy}^{(3)}$.

7) *Mean queue length.* The mean, L_q, queue length is given by $L_q = \boldsymbol{y}_0R(I - R)^{-2}\boldsymbol{e}$.

8) *Mean number in system.* The mean, L_s, number of customers in the system is given by $L_s = L_q + \mu_{busy}$.

9) *P(an arrival has to wait).* The probability, P_{wait}, that an arriving customer has to wait to receive service is given by $P_{wait} = 1 - \frac{1}{\lambda}\boldsymbol{y}^*(e \otimes D_1 \boldsymbol{e}) = \boldsymbol{y}_0(I - R)^{-1}\boldsymbol{e}$.

10) *Mean busy period.* The mean busy period, μ'_{BP}, which in the case of a multi-server system, is defined here as the full busy period that starts with all servers becoming busy until one of the servers becomes idle.

REMARK 5.1.– When solving this model numerically, it is very important to use a variety of internal accuracy checks. In this model, there are many such internal accuracy checks available. In addition to the above, one can incorporate the fact that $RA^{(2)}\boldsymbol{e} = A^{(0)}\boldsymbol{e}$, and the average number of busy servers should be equal to the ratio of the arrival rate to the average service rate. That is, $\mu_{busy} = \frac{3\lambda}{\mu_1+\mu_2+\mu_3}$.

REMARK 5.2.– The grouping of the boundary states for this model is done in such a way (see the generator given in equation [5.2]) that applying the busy period analysis of section 4.6 to the multi-server system here will yield the full busy period. Recall that the full busy period is the one that starts with all three servers becoming busy until one of the three servers becomes idle.

NOTE.– The scatter plots seen in this section for the illustrative examples do not have the values displayed for the axes. These are not needed due to making comparisons of various systems among themselves. This is similar to a probability plot.

EXAMPLE 5.1.– In this example, we look at nine arrival processes and for services we consider a combination of five scenarios, labeled as A through E. The nine arrival processes are as follows:

ERA: Inter-arrival times of batches follow an Erlang of order 5 with parameter 5.

EXA: Inter-arrival times of batches are exponential with parameter 1.

HEA: This is for hyperexponential inter-arrival times for the batches with parameters 2.18, 1.09, 0.545, 0.2725 and 0.13625 with their respective mixing probabilities given by 0.5, 0.3, 0.15, 0.04 and 0.01.

NC$_r$: Here, we look at three negatively correlated arrival processes by taking $r = 2, 3,$ and 4. The MAP representation matrices D_0 and D_1 are given by:

$$D_0 = \begin{pmatrix} T_r & 0 \\ 0 & -\lambda_2 \end{pmatrix}, \quad D_1 = \begin{pmatrix} p_1 \boldsymbol{T}_r^0 \boldsymbol{\alpha}_r & q_1 \boldsymbol{T}^0 \\ q_2 \lambda_2 \boldsymbol{\alpha}_r & p_2 \lambda_2 \end{pmatrix},$$

where (α_r, T_r) gives the PH representation for Erlang of order r with rate (in each of r phases given by $0.25 + 0.5\,r,\ r = 2, 3, 4$). The other parameters are: $\lambda_2 = 0.5 + r$, $p_1 = p_2 = 0.01$.

PC$_r$: Here, we look at three positively correlated arrival processes by setting $r = 2, 3$, and 4. The representation matrices are similar to the ones displayed for NC_r except that the probabilities are taken to be $p_1 = p_2 = 0.99$. All other parameters remain the same.

Note that the above nine arrival processes all have a mean arrival rate of 1. The standard deviations of the arrival processes are 0.44721, 1.0, 1.69710, 1.03923, 1.02020, 1.01227, 1.03923, 1.02020 and 1.01227, respectively. The 1-lag correlation coefficients are, respectively, given by 0, 0, 0, -0.32667, -0.48039, -0.57855, 0.32667, 0.48039 and 0.57855.

In order to compare a homogeneous system, namely, when all servers are identical with the same service time distribution, against a heterogeneous one, we consider the following scenarios for the services (the one labeled A is a homogeneous one and the other six are heterogeneous):

A: All three service times are assumed to be exponential with $\mu_1 = \mu_2 = \mu_3 = 0.35$.

B: Server 1 offers services that are assumed to be Erlang of order 3 with rate 1.05 in each phase, while the other two offer services that are assumed to be exponential with parameter 0.35.

C: Server 2 offers services that are assumed to be Erlang of order 3 with rate 1.05 in each phase, while the other two offer services that are assumed to be exponential with parameter 0.35.

D: Server 3 offers services that are assumed to be Erlang of order 3 with rate 1.05 in each phase, while the other two offer services that are assumed to be exponential with parameter 0.35.

E: All three servers offer services that are assumed to be Erlang of order 3 with rate 1.05 in each phase.

All of the above service processes have the same rate of 0.35 for each server, so the total service rate (when all servers are busy) is 1.05, which is greater than the arrival rate of 1, guaranteeing the stability of the system under study. Also, note that the service processes described in A and E are for homogeneous servers. While in A the service times of all are exponential, the ones in E all have Erlang services. We compare two sets of systems. In the first one, we compare A versus the ones given in B, C and D. In the second one, the comparison will use E as the base against B, C and D. The idea behind this is to compare homogeneous systems against their respective heterogeneous ones. It is important to observe that there are various ways to look at homogeneous versus heterogeneous systems and astute readers will realize this key fact and develop on their own such comparisons to get a full picture of the systems' behavior, especially in practice.

In Figures 5.1 through 5.3, we display the scatter plots of selected measures.

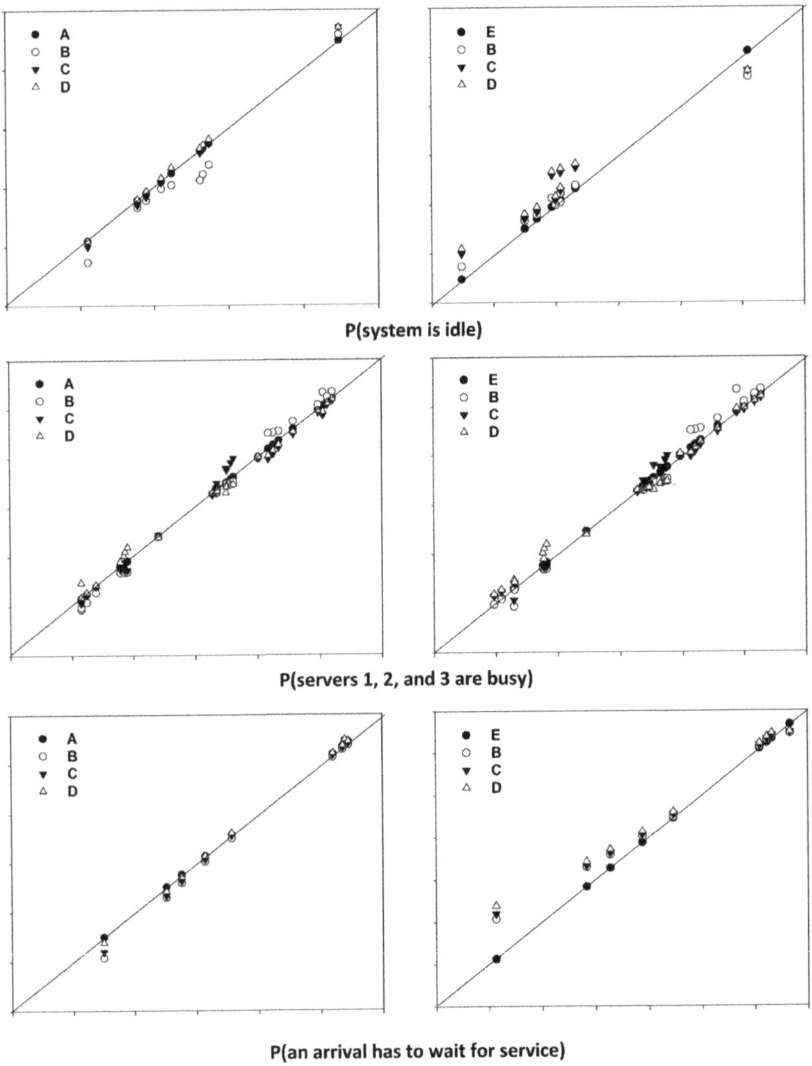

Figure 5.1. *Scatter plots of P_{idle}, $\tilde{P}^{(r)}_{busy}$ and P_{wait} under various scenarios for example 5.1*

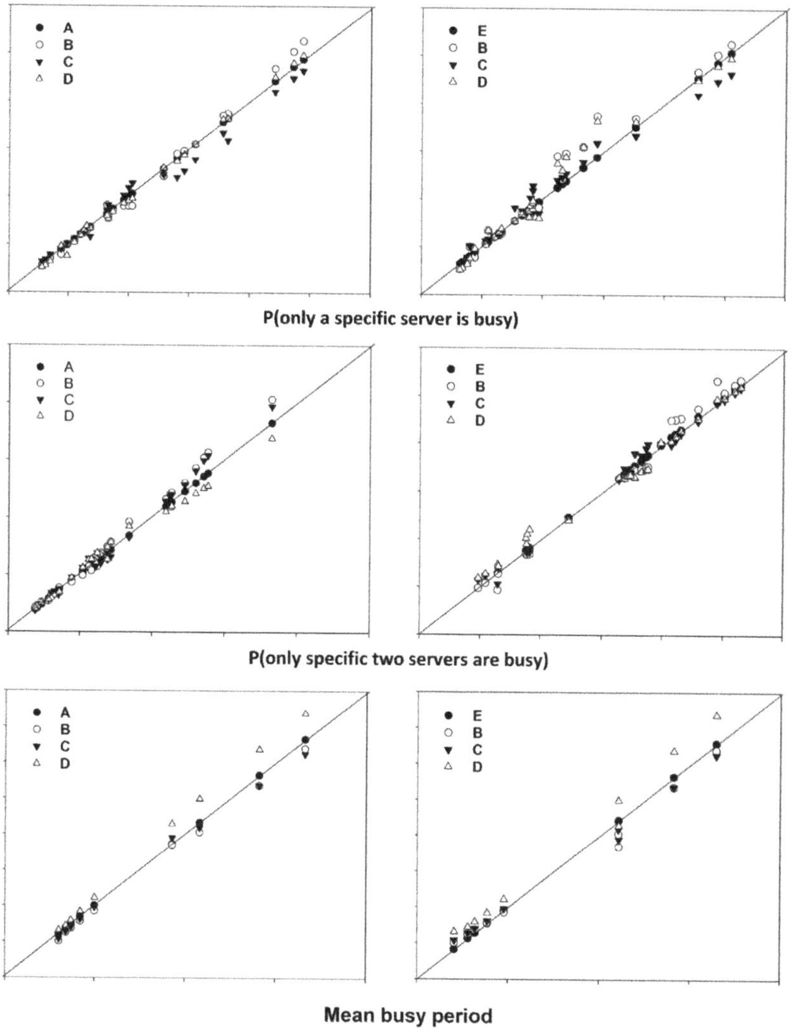

Figure 5.2. *Scatter plots of* $P^{(r)}_{busy}$, $P^{(i_1,i_2)}_{busy}$, $P^{(r)}_{busy}$ *and* μ'_{BP} *under various scenarios for example 5.1*

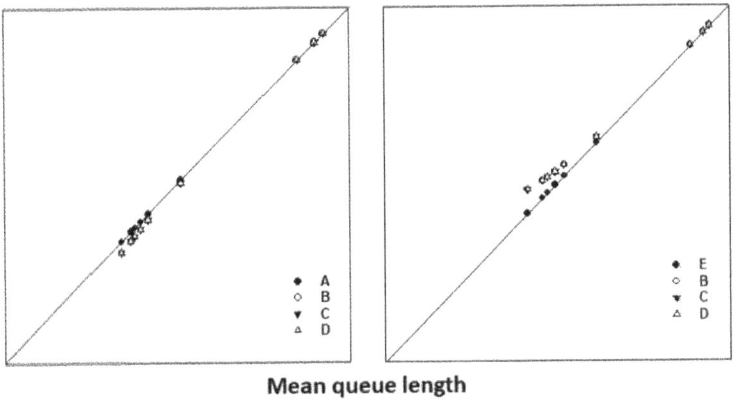

Figure 5.3. *Scatter plots of L_q under various scenarios for example 5.1*

Some key observations based on these three figures are as follows:

1) When the system idle probability is less than 0.015 or so, we notice that system A dominates that of B, C and D. However, when this idle probability is more than 0.02, A is dominated by the other three. With regard to the comparison of system E against B, C and D, we notice the behavior to be the exact opposite of A versus B, C and D.

2) The phenomenon observed for the system idle probability appears to hold good for the probability that an arrival has to wait for service.

3) It appears that the mean queue length is higher for the homogeneous system A compared with other three heterogeneous systems as well as the homogeneous system with Erlang services. However, this measure is smallest for the homogeneous system E when compared with all other systems.

4) It is interesting to see that the mean busy period for system D has the highest value compared to all other systems. Note that in system D server 3 has Erlang services.

EXAMPLE 5.2.– This example is similar to example 5.1 except here we look at hyperexponential services instead of Erlang services. The parameters of the hyperexponential services are 2.87, 0.287 and 0.0287, with mixing probabilities 0.7, 0.25 and 0.05, respectively. All other parameters remain the same. We use the same notation. Thus, the definitions of A through E for the type systems considered here need to replace Erlang with hyperexponential as described here.

In Figures 5.4 through 5.6, we display the scatter plots of selected measures.

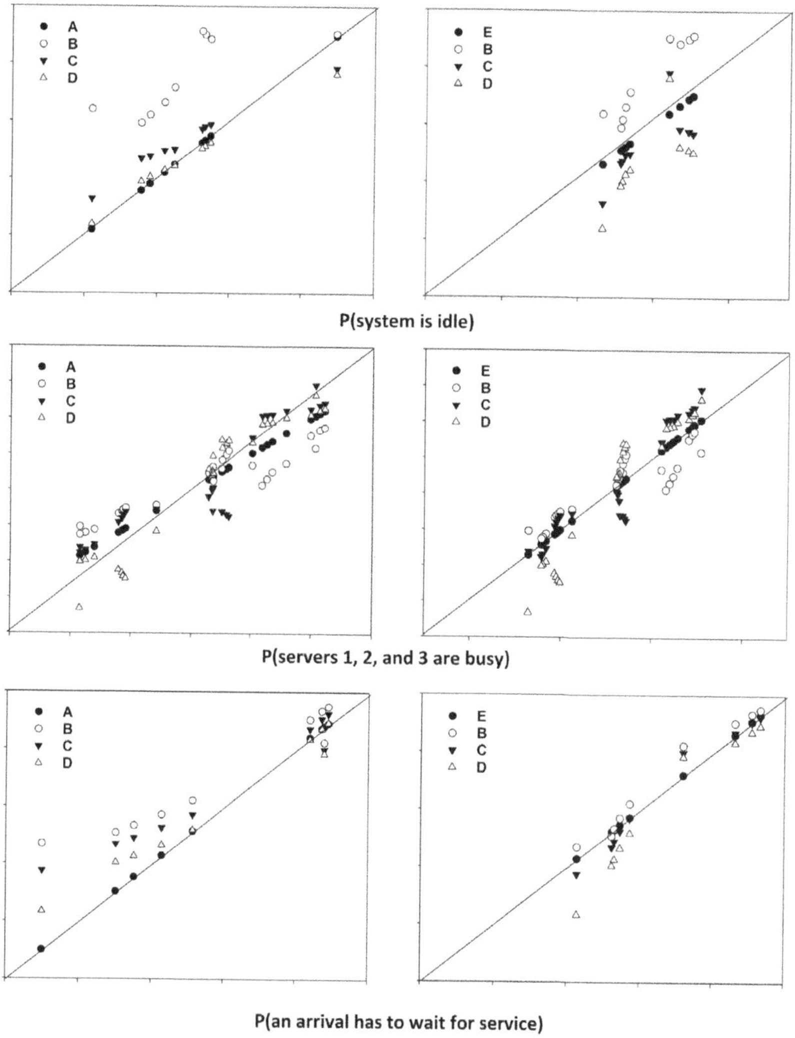

Figure 5.4. *Scatter plots of P_{idle}, $\tilde{P}^{(r)}_{busy}$ and P_{wait} under various scenarios for example 5.6*

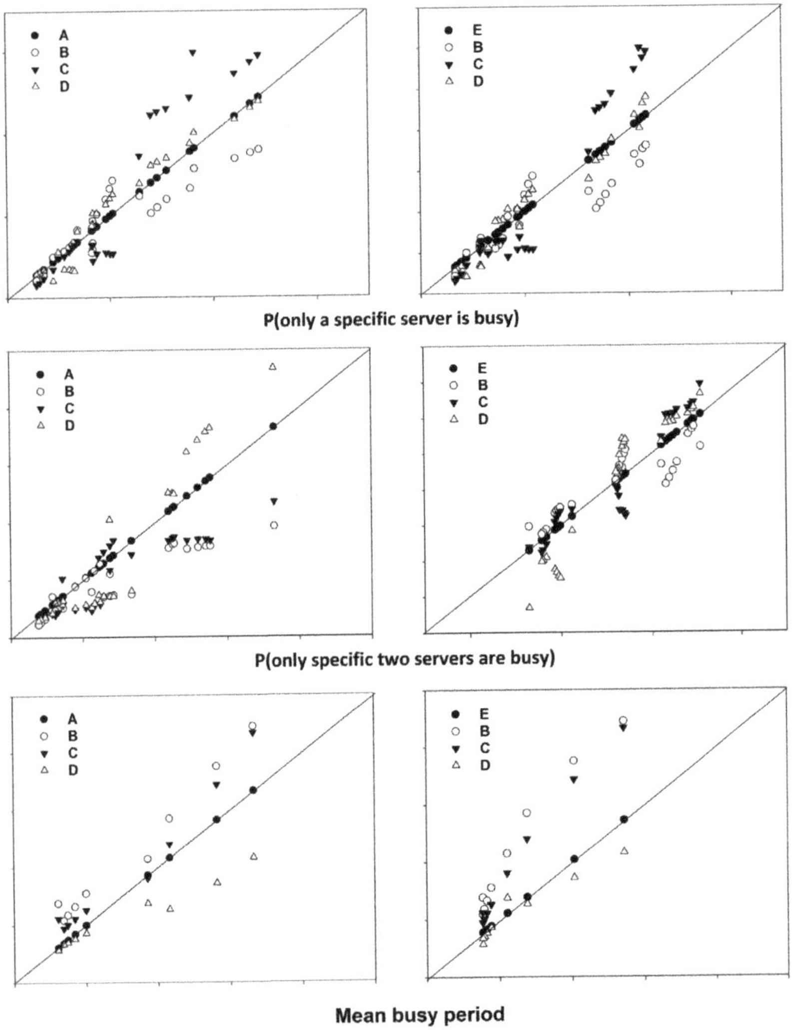

Figure 5.5. *Scatter plots of $P_{busy}^{(r)}$, $P_{busy}^{(i_1,i_2)}$, $P_{busy}^{(r)}$ and μ'_{BP} under various scenarios for example 5.6*

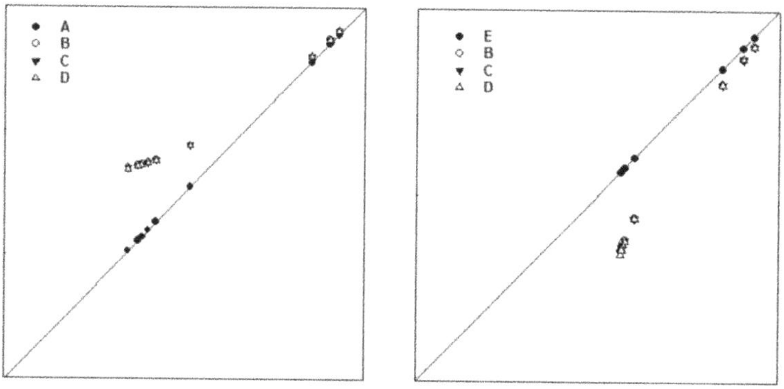

Figure 5.6. *Scatter plots of L_q under various scenarios for example*

The graphs reveal some interesting observations and are left as an exercise.

EXAMPLE 5.3.– Here, our focus is to see the impact of the various system measures when the servers are serving at different rates. Toward this end, we consider seven scenarios for the service rates. In this example, we assume that all servers offer exponential services. We use the same nine different arrival processes as mentioned in example 5.1.

A: Here, we have $\mu_1 = \mu_2 = \mu_3 = 0.35$.

B: Here, we have $\mu_1 = 0.6, \mu_2 = 0.3$ and $\mu_3 = 0.15$.

C: Here, we have $\mu_1 = 0.6, \mu_2 = 0.15$ and $\mu_3 = 0.3$.

D: Here, we have $\mu_1 = 0.3, \mu_2 = 0.6$ and $\mu_3 = 0.15$.

E: Here, we have $\mu_1 = 0.3, \mu_2 = 0.15$ and $\mu_3 = 0.6$.

F: Here, we have $\mu_1 = 0.15, \mu_2 = 0.6$ and $\mu_3 = 0.3$.

G: Here, we have $\mu_1 = 0.15, \mu_2 = 0.3$ and $\mu_3 = 0.6$

We now compare the system performance measures by keeping A as the base one (since here all service rates are the same) against the other six schemes. Note that the service rates for all service schemes, except A, are such that the fastest server is twice as fast as the next fastest and four times as fast as the slowest. For a better display of the scatter plots, we compare two service schemes at a time against that of A.

In Figures 5.7 through 5.9, we display the scatter plots of selected measures.

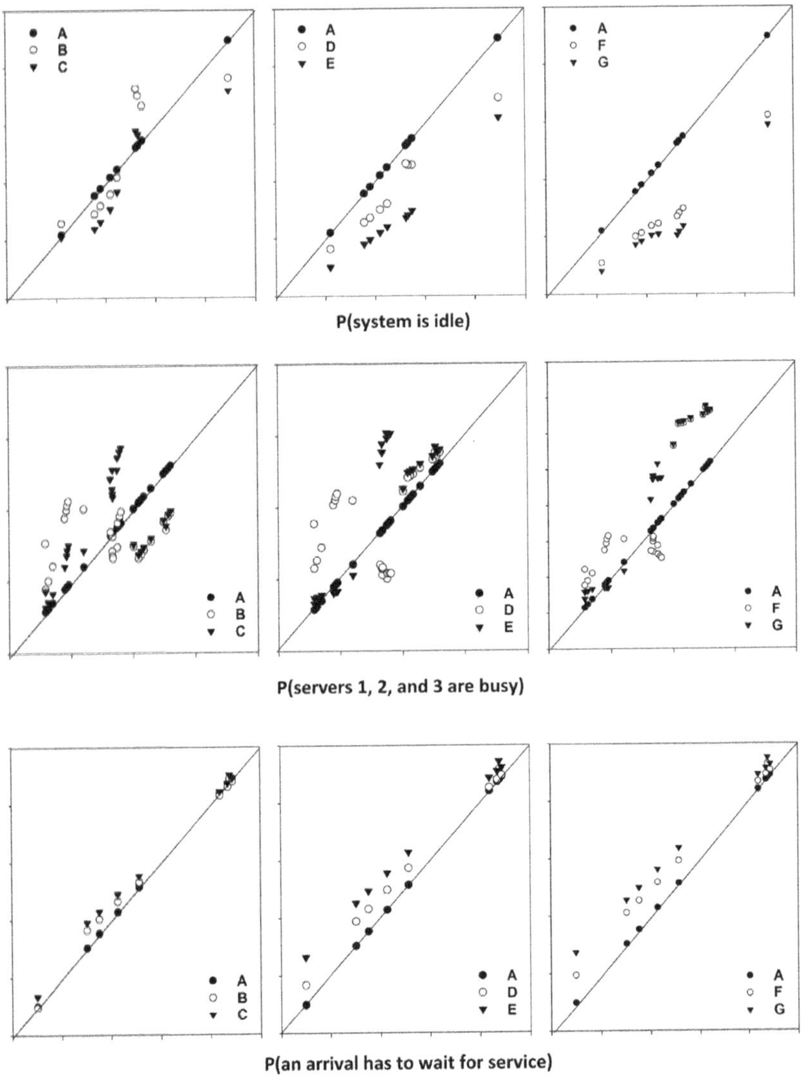

Figure 5.7. *Scatter plots of P_{idle}, $\tilde{P}^{(r)}_{busy}$ and P_{wait} under various scenarios for example 5.3*

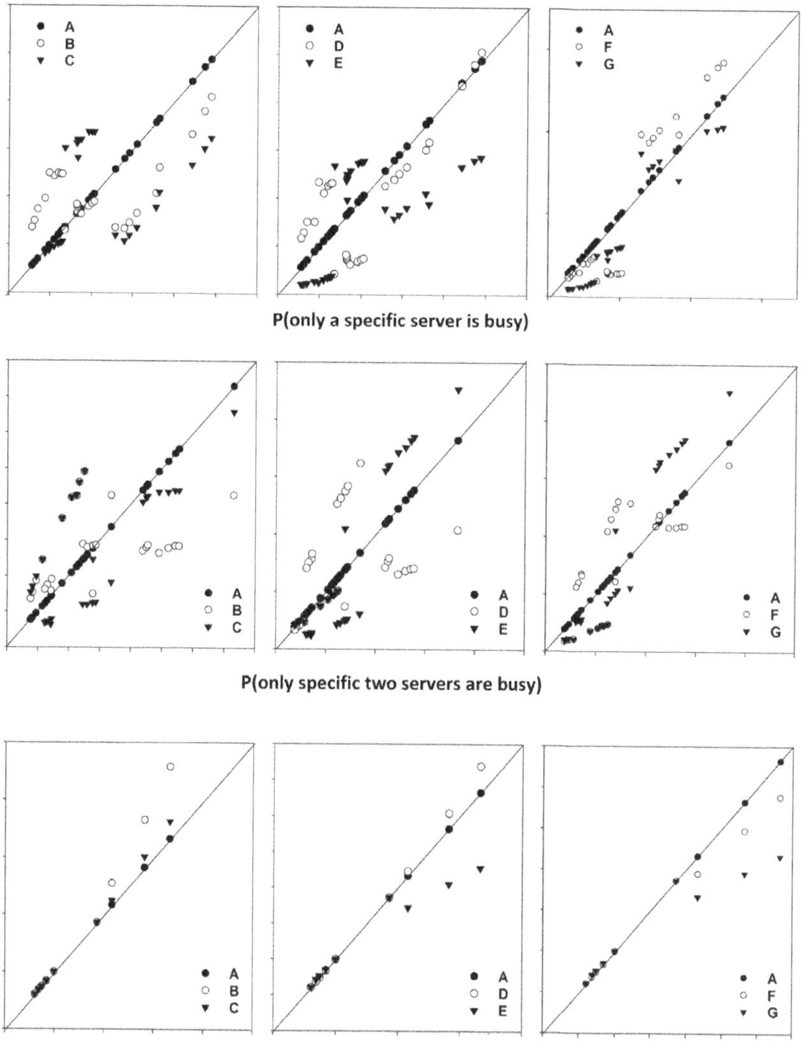

Figure 5.8. *Scatter plots of* $P_{busy}^{(r)}$, $P_{busy}^{(i_1,i_2)}$, $P_{busy}^{(r)}$ *and* μ'_{BP} *under various scenarios for example 5.3*

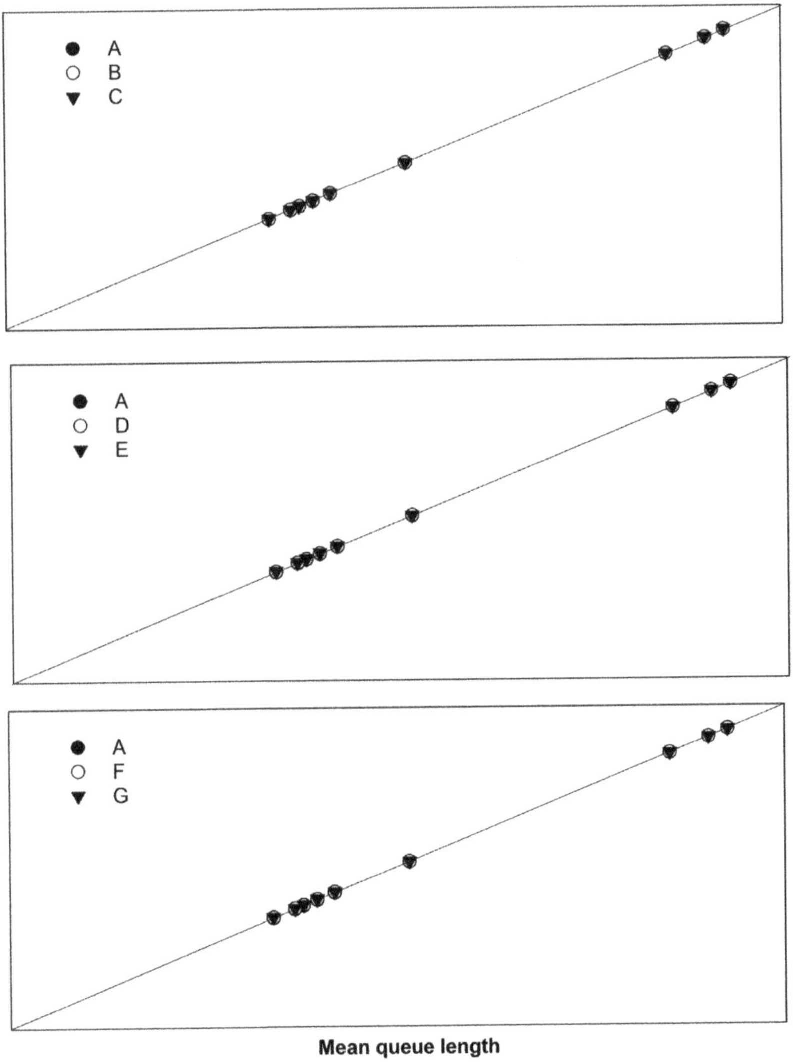

Figure 5.9. *Scatter plots of L_q under various scenarios for example 5.3*

The graphs reveal some interesting observations and shed light on the impact of the variability in the services.

EXAMPLE 5.4.– This example is similar to example 5.3 except we assume that server 1 offers Erlang services and the other two servers have exponential processing times.

All other parameters remain the same. Thus, we have nine different arrival processes and seven service schemes. In Figures 5.10 through 5.12, we display the scatter plots of selected measures. The graphs reveal some interesting observations and shed light on the impact of the variability in the services.

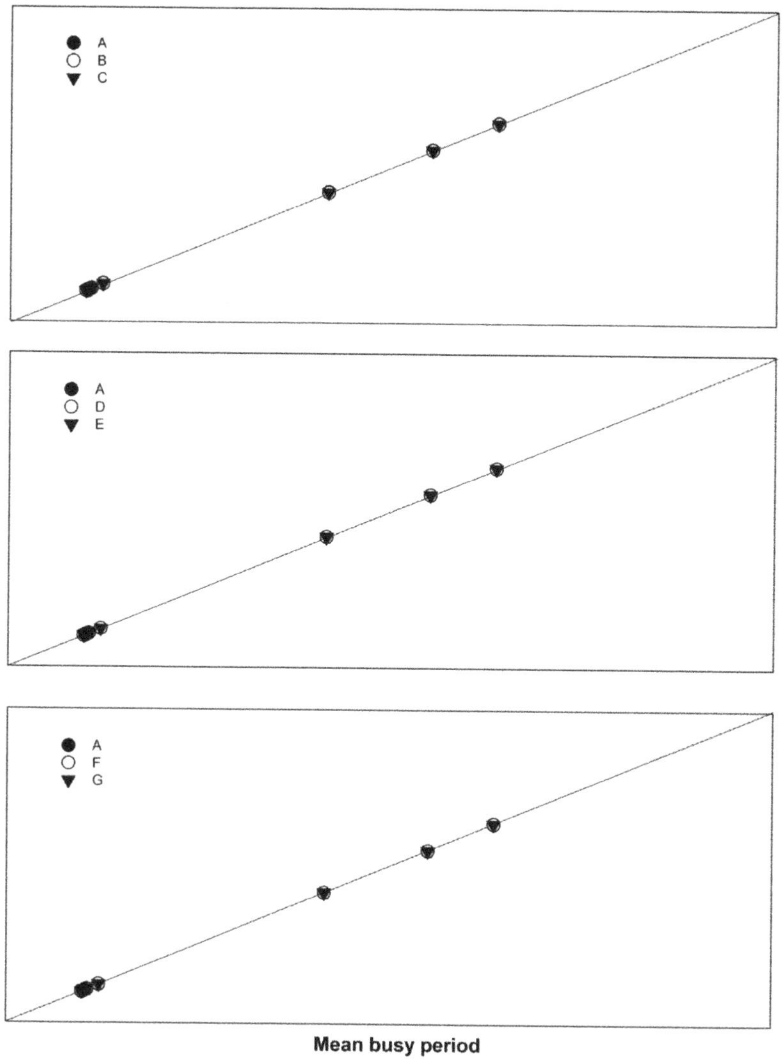

Figure 5.10. *Scatter plots of L_q under various scenarios for example 5.4*

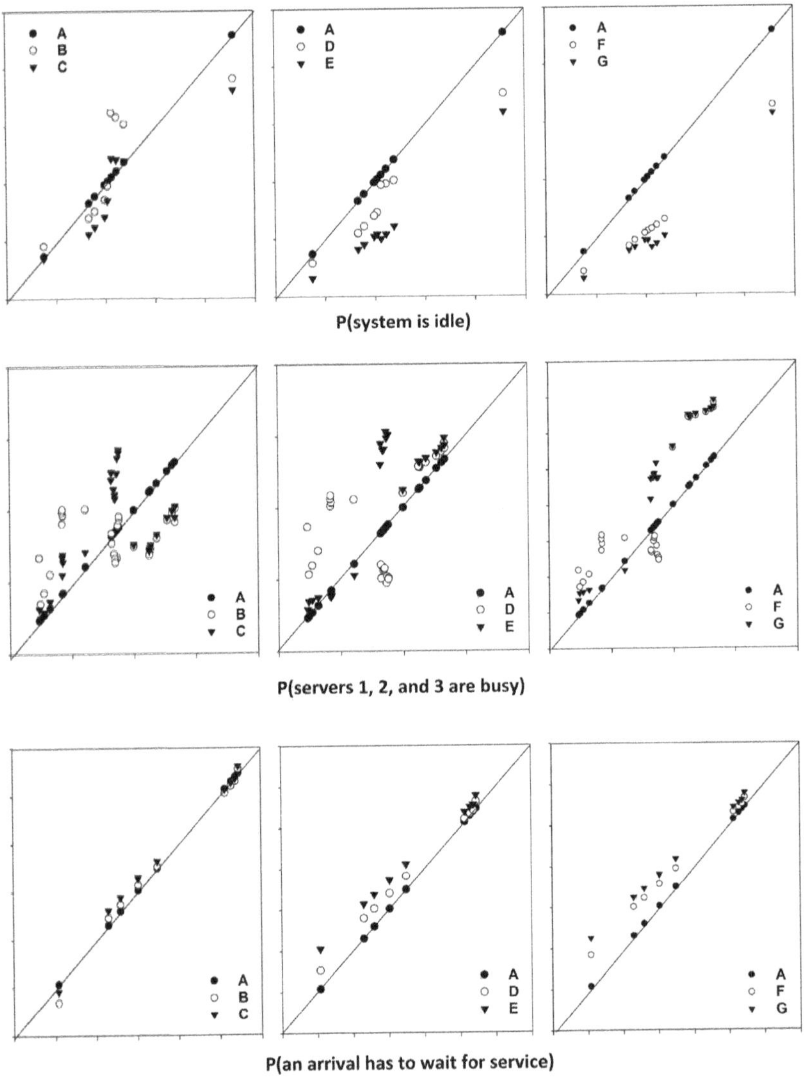

Figure 5.11. *Scatter plots of P_{idle}, $\tilde{P}^{(r)}_{busy}$ and P_{wait} under various scenarios for example 5.4*

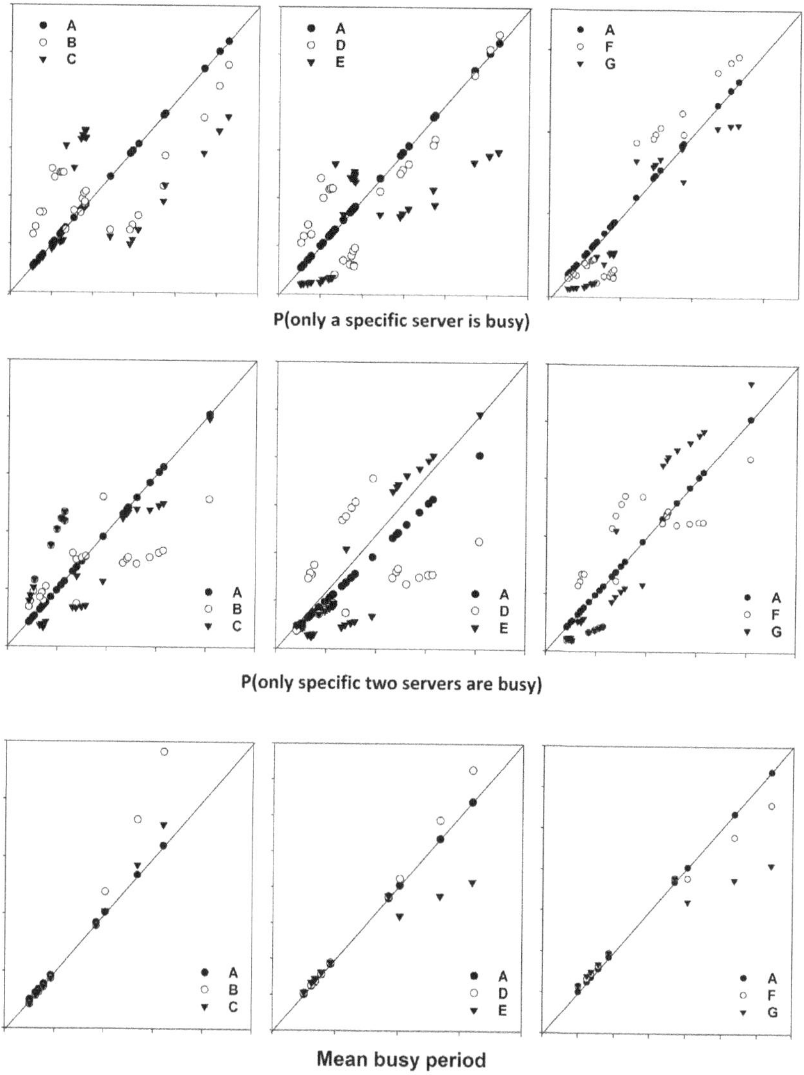

Figure 5.12. *Scatter plots of $P^{(r)}_{busy}$, $P^{(i_1,i_2)}_{busy}$, $P^{(r)}_{busy}$ and μ'_{BP} under various scenarios for example 5.4*

EXAMPLE 5.5.– This example is similar to example 5.3 except we assume that server 2 offers Erlang services and the other two servers have exponential processing times. All other parameters remain the same. Thus, we have nine different arrival processes and seven service schemes. In Figures 5.13 through 5.15, we display the scatter plots of selected measures. These graphs reveal some interesting observations and shed light on the impact of the variability in the services.

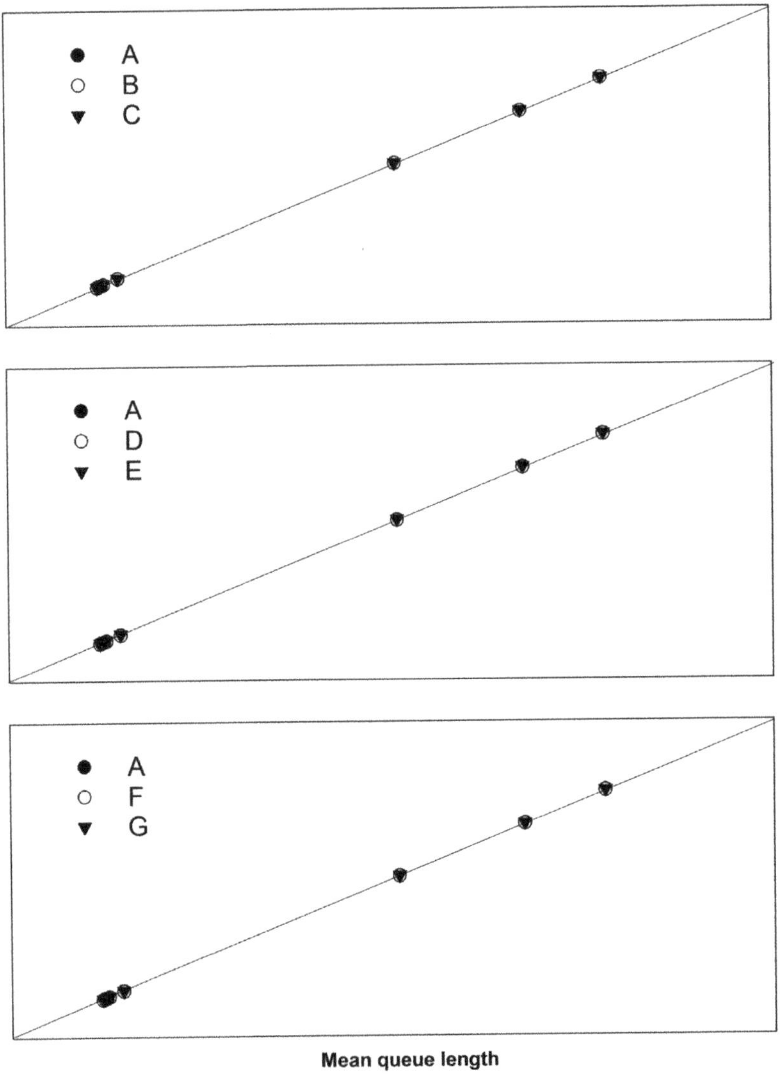

Figure 5.13. *Scatter plots of L_q under various scenarios for example 5.5*

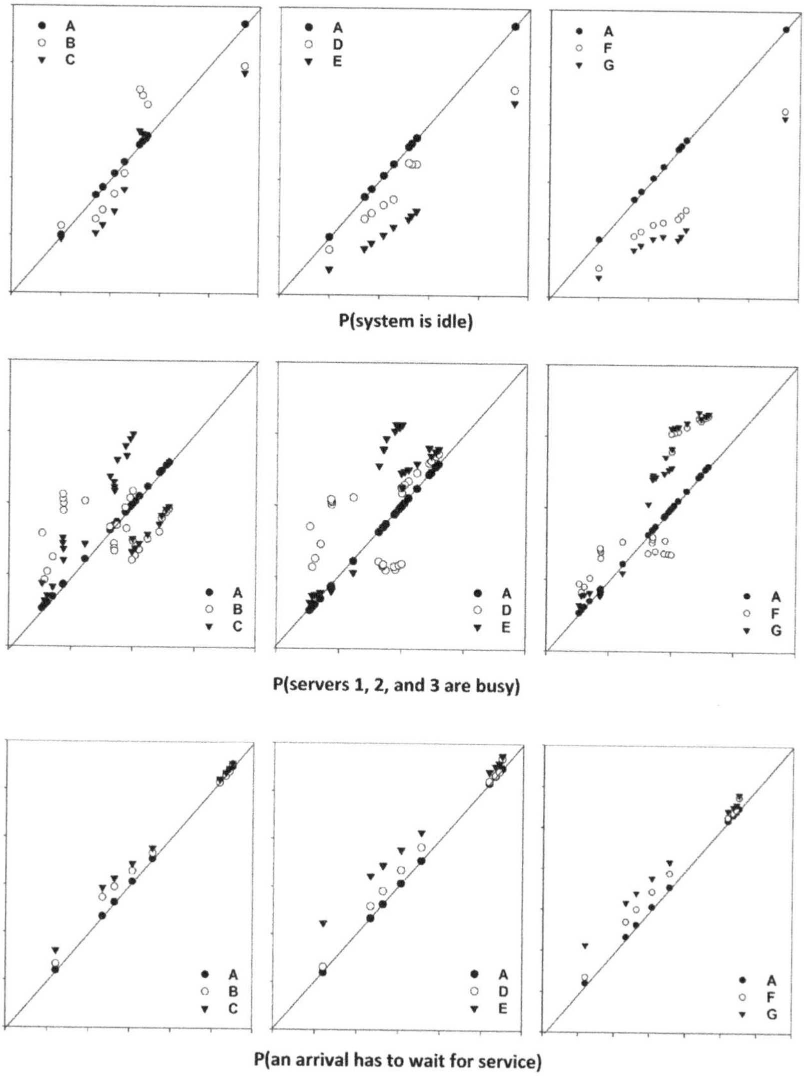

Figure 5.14. *Scatter plots of P_{idle}, $\tilde{P}^{(r)}_{busy}$ and P_{wait} under various scenarios for example 5.5*

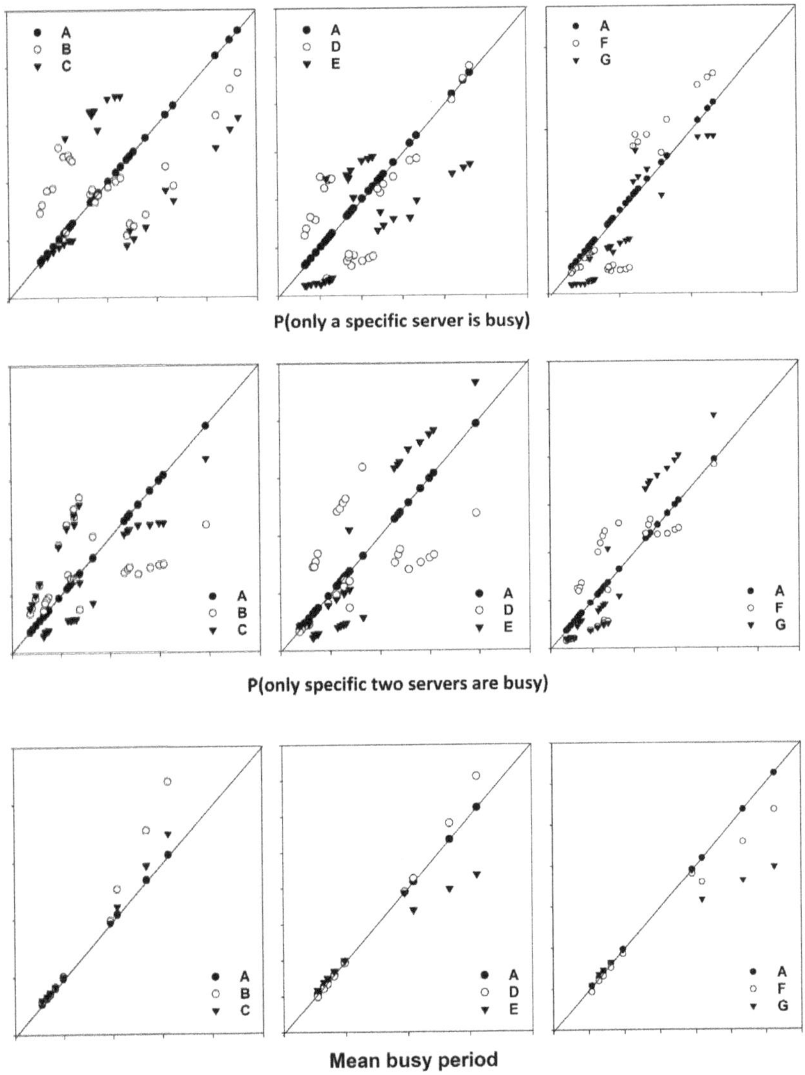

Figure 5.15. *Scatter plots of $P^{(r)}_{busy}$, $P^{(i_1,i_2)}_{busy}$, $P^{(r)}_{busy}$ and μ'_{BP} under various scenarios for example 5.5*

EXAMPLE 5.6.– This example is similar to example 5.3 except we assume that server 3 offers Erlang services and the other two servers have exponential processing times. All other parameters remain the same. Thus, we have nine different arrival processes and seven service schemes. In Figures 5.16 through 5.18, we display the scatter plots of selected measures. These graphs reveal some interesting observations and shed light on the impact of the variability in the services.

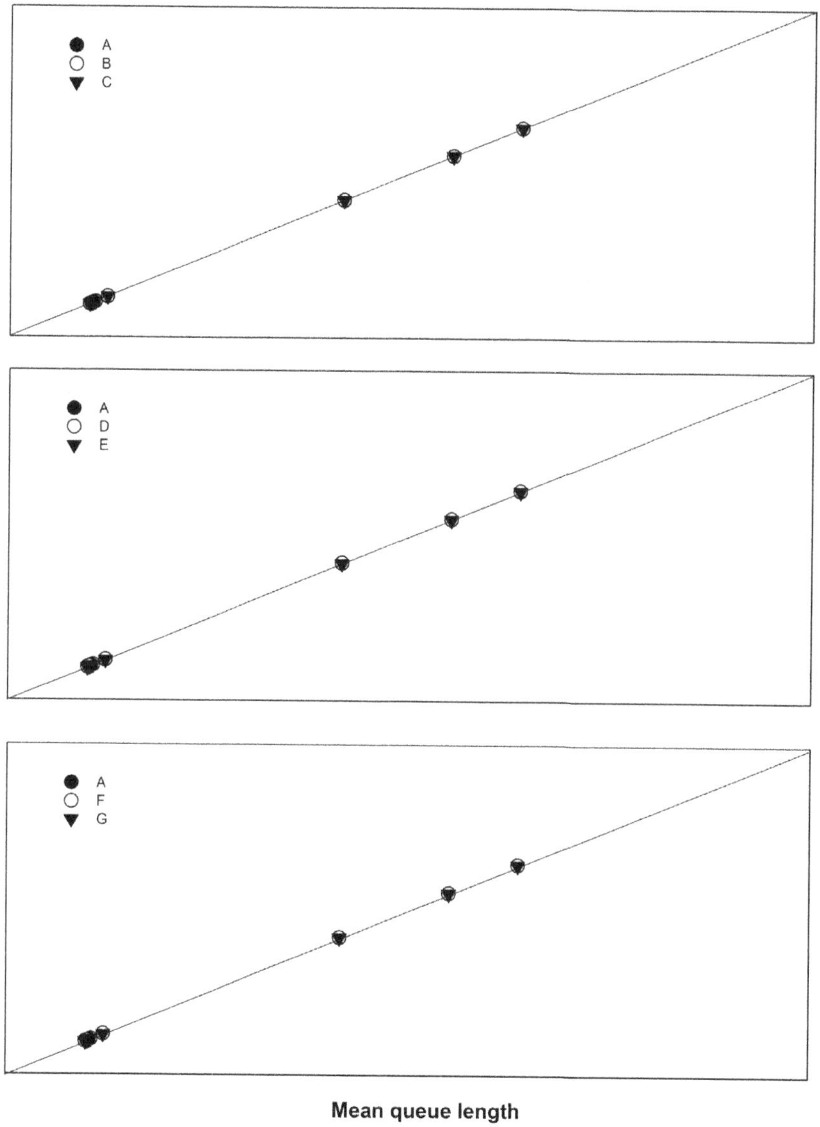

Figure 5.16. *Scatter plots of L_q under various scenarios for example 5.6*

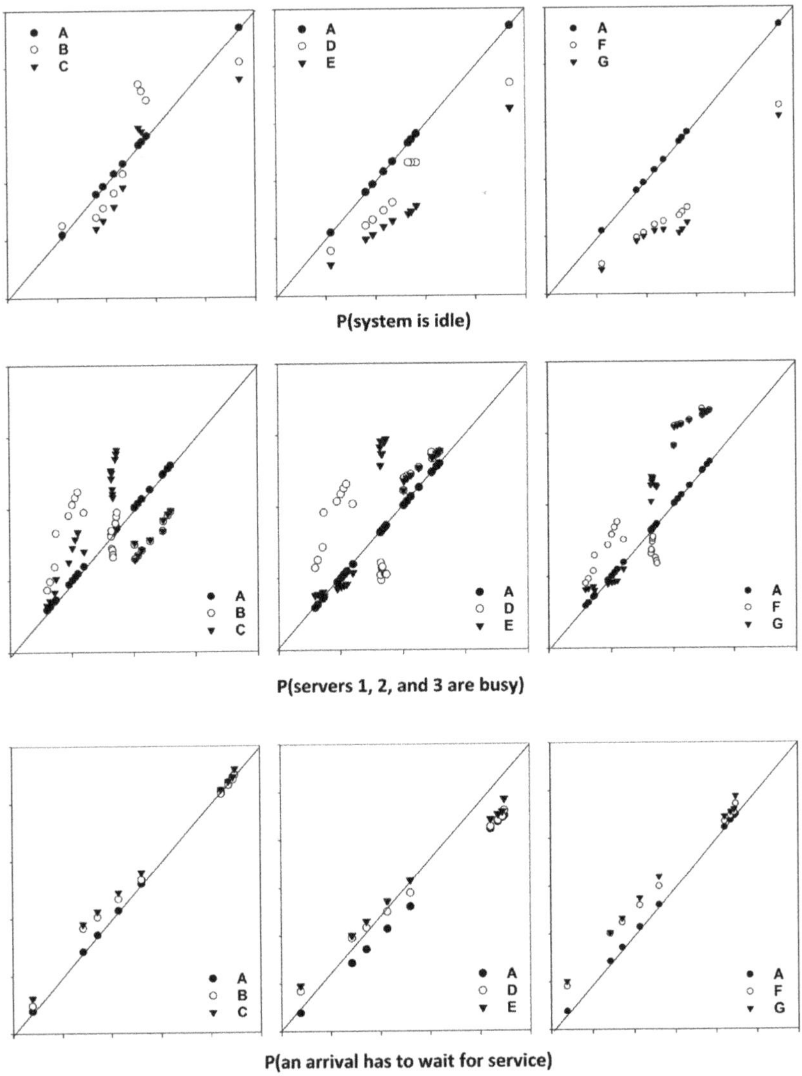

Figure 5.17. *Scatter plots of P_{idle}, $\tilde{P}^{(r)}_{busy}$ and P_{wait} under various scenarios for example 5.6*

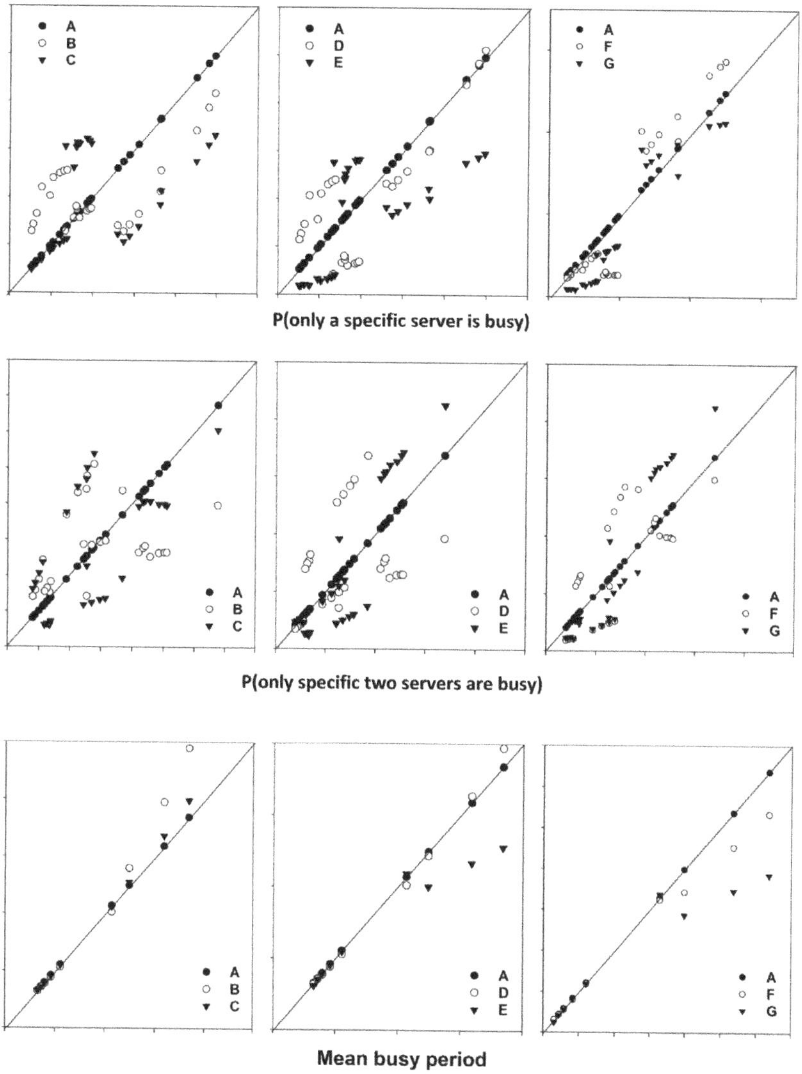

Figure 5.18. *Scatter plots of* $P^{(r)}_{busy}$, $P^{(i_1,i_2)}_{busy}$, $P^{(r)}_{busy}$ *and* μ'_{BP} *under various scenarios for example 5.6*

EXAMPLE 5.7.– This example is similar to example 5.3 except we assume that server 1 offers hyperexponential services and the other two servers have exponential processing times. The parameters of the hyperexponential services are 2.87, 0.287 and 0.0287, with mixing probabilities 0.7, 0.25 and 0.05, respectively. All other parameters remain the same. Thus, we have nine different arrival processes and seven service schemes. In Figures 5.19 through 5.21, we display the scatter plots of selected measures. These graphs reveal some interesting observations and shed light on the impact of the variability in the services.

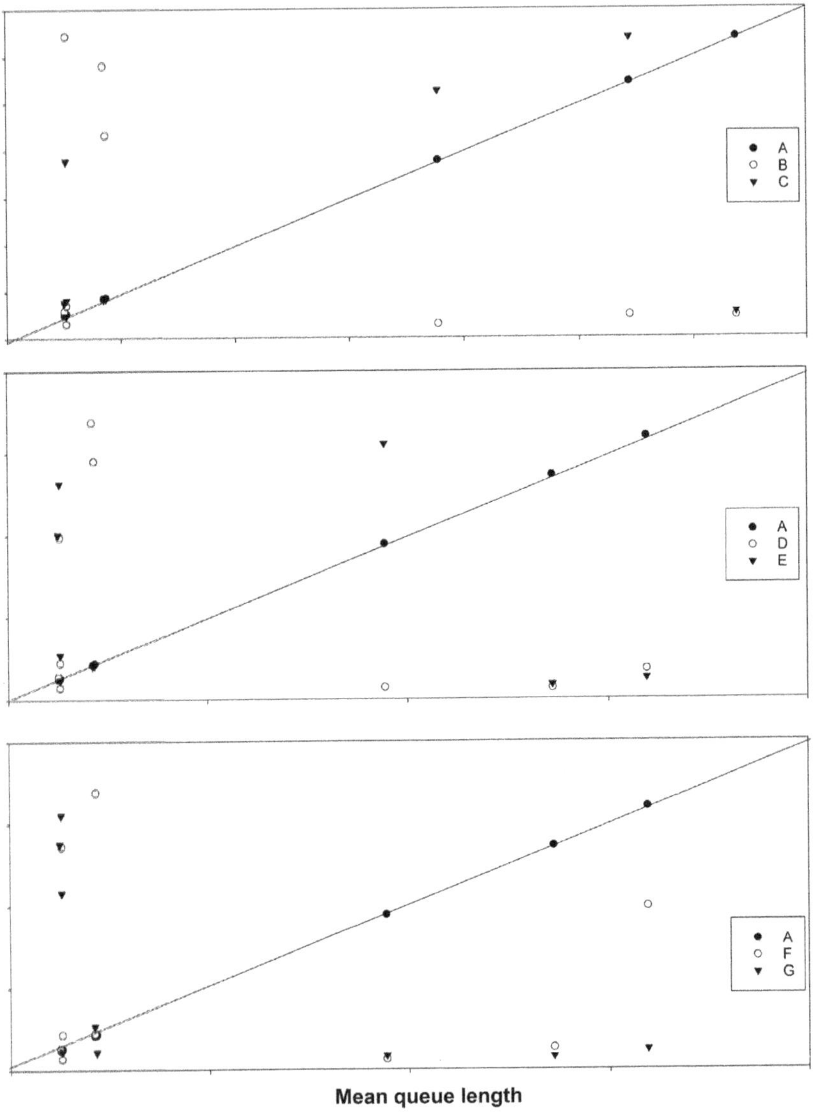

Figure 5.19. *Scatter plots of L_q under various scenarios for example 5.7*

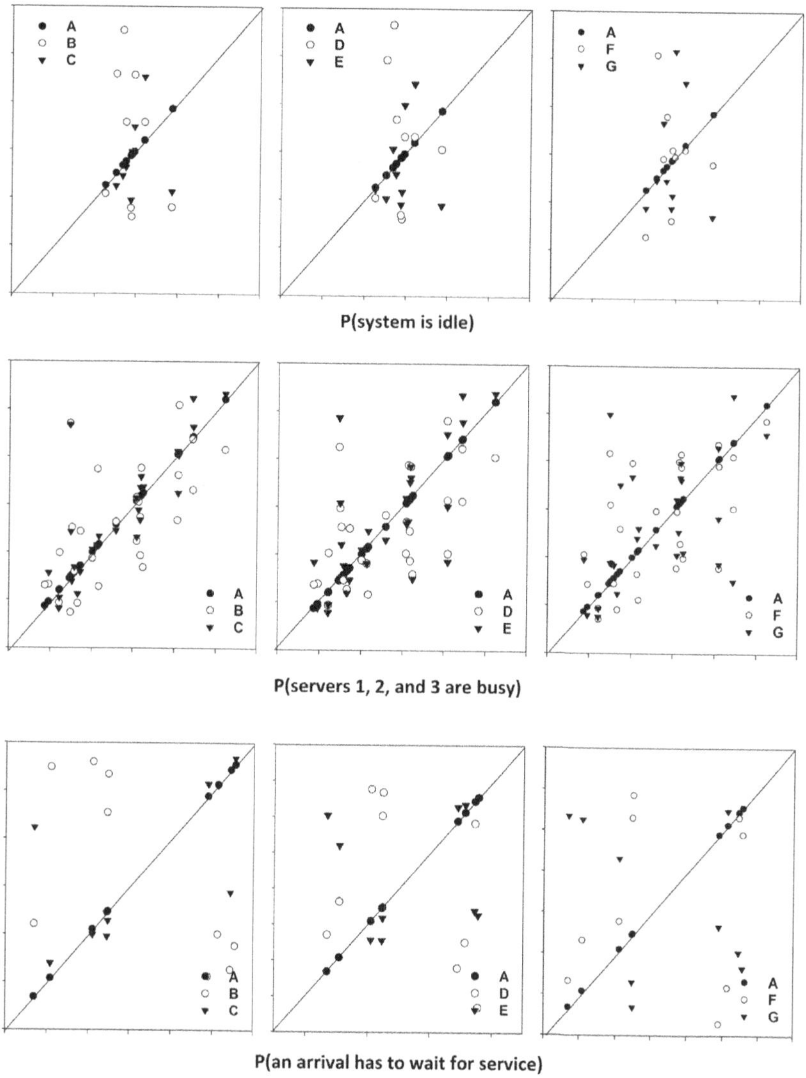

Figure 5.20. *Scatter plots of P_{idle}, $\tilde{P}_{busy}^{(r)}$ and P_{wait} under various scenarios for example 5.7*

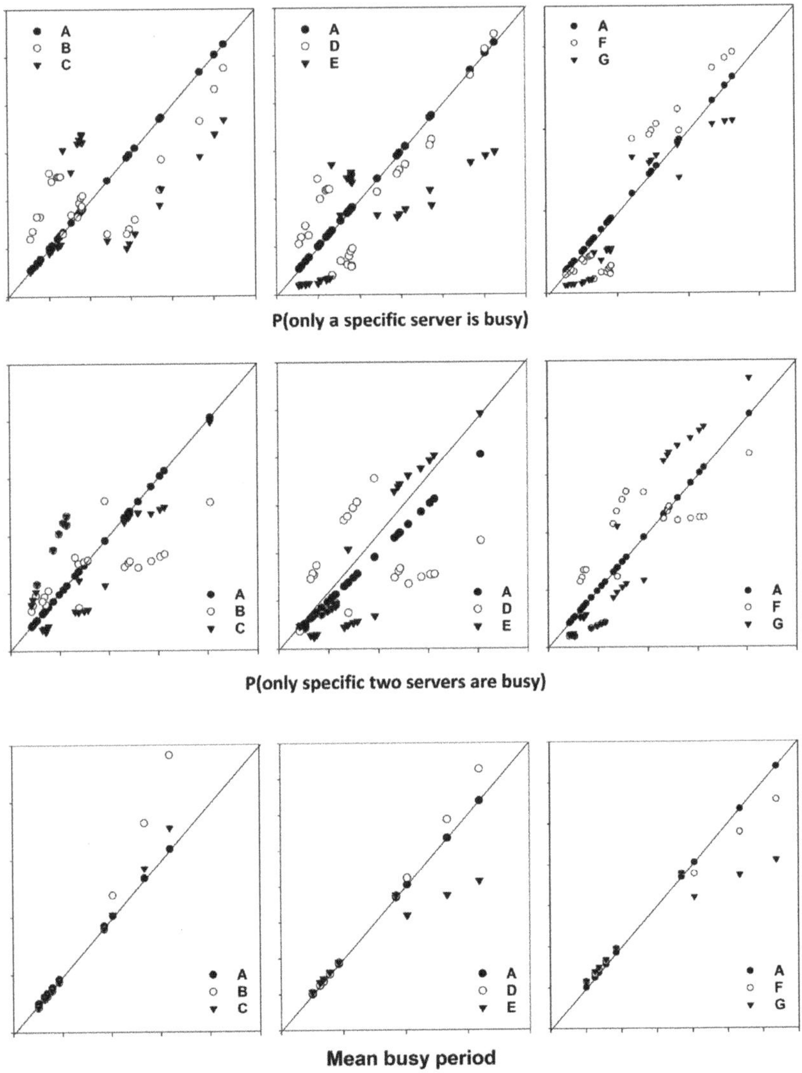

Figure 5.21. *Scatter plots of $P^{(r)}_{busy}$, $P^{(i_1,i_2)}_{busy}$, $P^{(r)}_{busy}$ and μ'_{BP} under various scenarios for example 5.7*

EXAMPLE 5.8.– This example is similar to example 5.7 except we assume that server 2 offers hyperexponential services and the other two servers have exponential processing times. All other parameters remain the same. Thus, we have nine different arrival processes and seven service schemes. In Figures 5.22 through 5.24, we display the scatter plots of selected measures. These graphs reveal some interesting observations and shed light on the impact of the variability in the services.

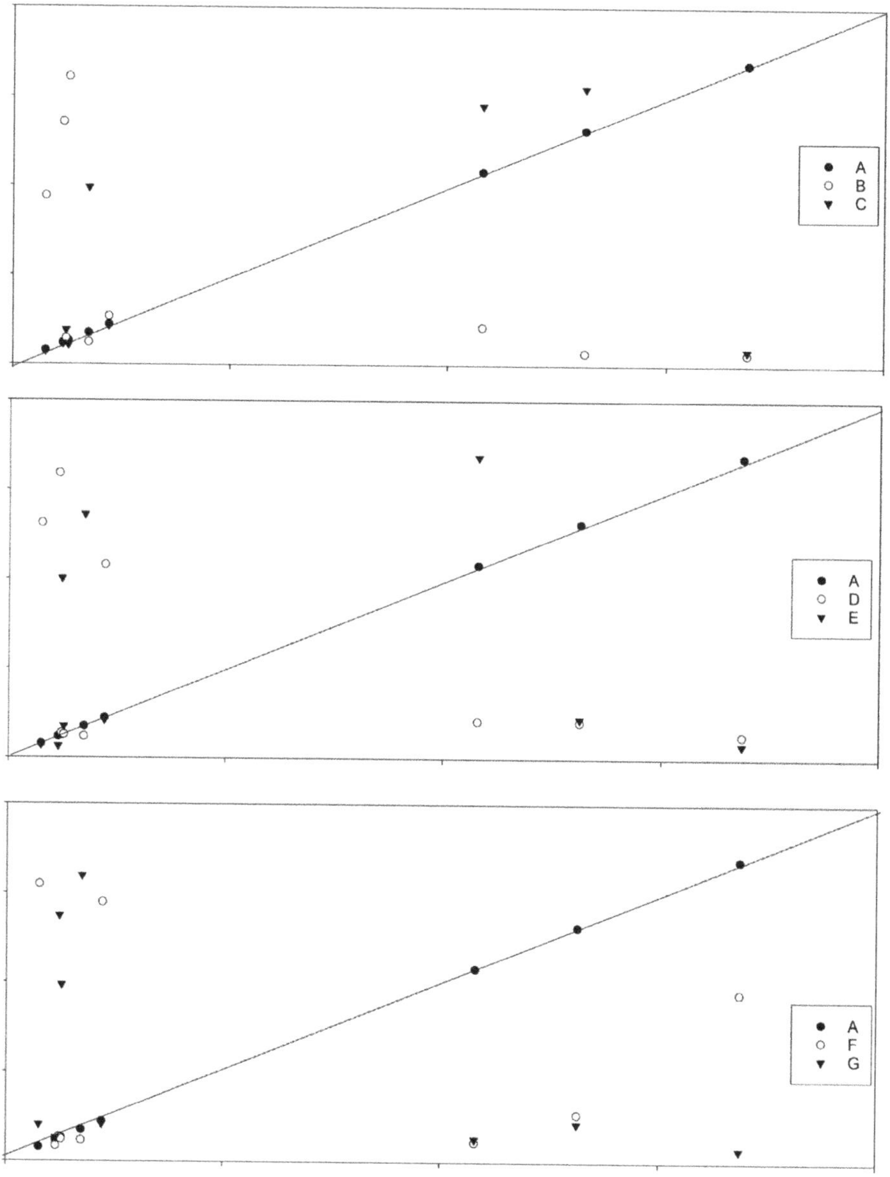

Figure 5.22. *Scatter plots of L_q under various scenarios for example 5.8*

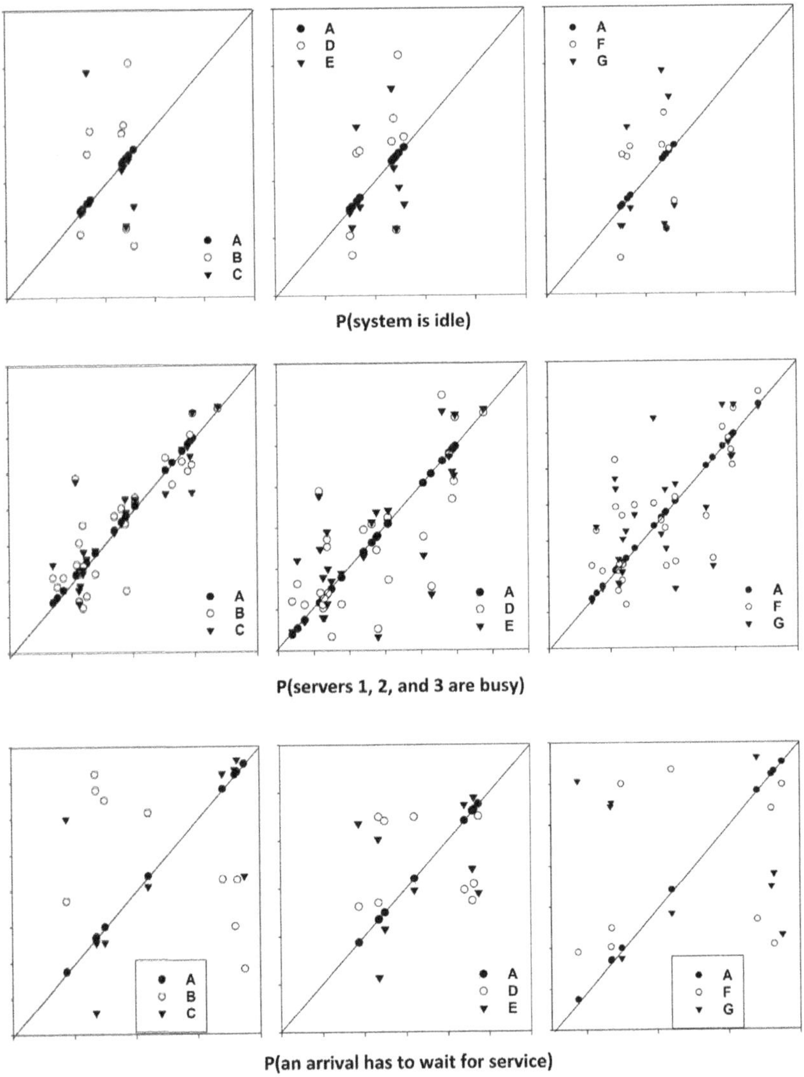

Figure 5.23. *Scatter plots of* P_{idle}, $\tilde{P}_{busy}^{(r)}$ *and* P_{wait} *under various scenarios for example 5.8*

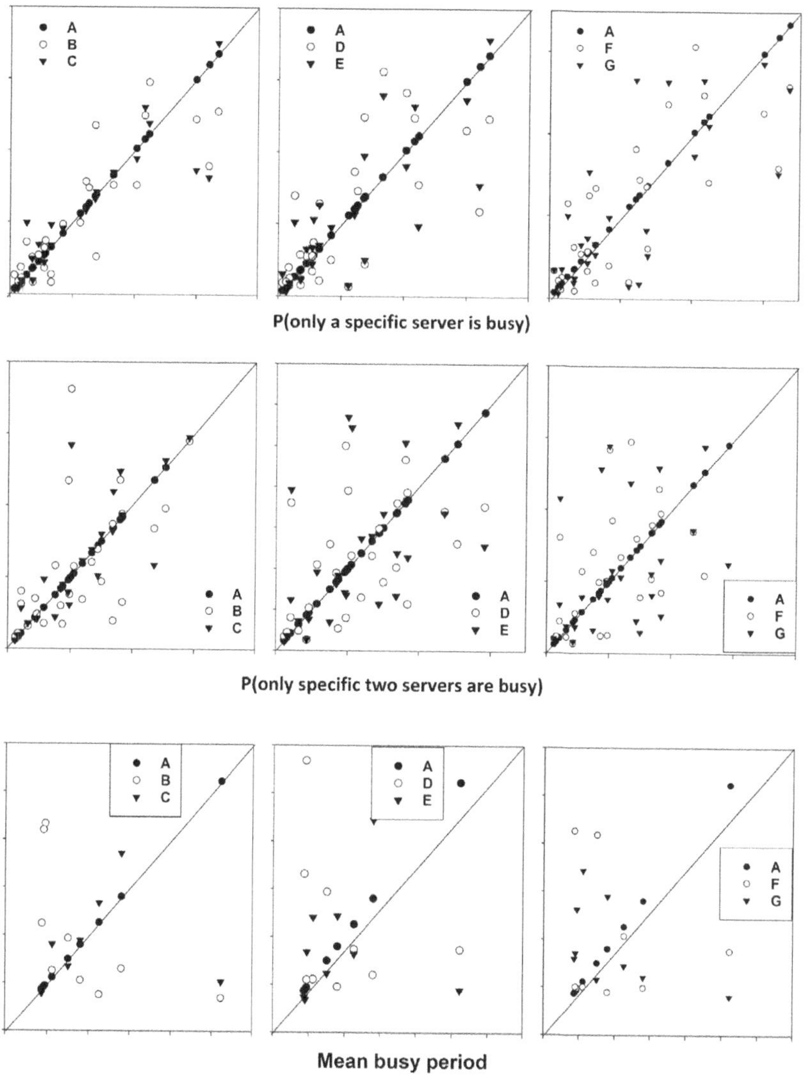

Figure 5.24. *Scatter plots of $P_{busy}^{(r)}$, $P_{busy}^{(i_1,i_2)}$, $P_{busy}^{(r)}$ and μ'_{BP} under various scenarios for example 5.8*

EXAMPLE 5.9.– This example is similar to example 5.7 except we assume that server 3 offers hyperexponential services and the other two servers have exponential processing times. All other parameters remain the same. Thus, we have nine different arrival processes and seven service schemes. In Figures 5.25 through 5.27, we display the scatter plots of selected measures. These graphs reveal some interesting observations and shed light on the impact of the variability in the services.

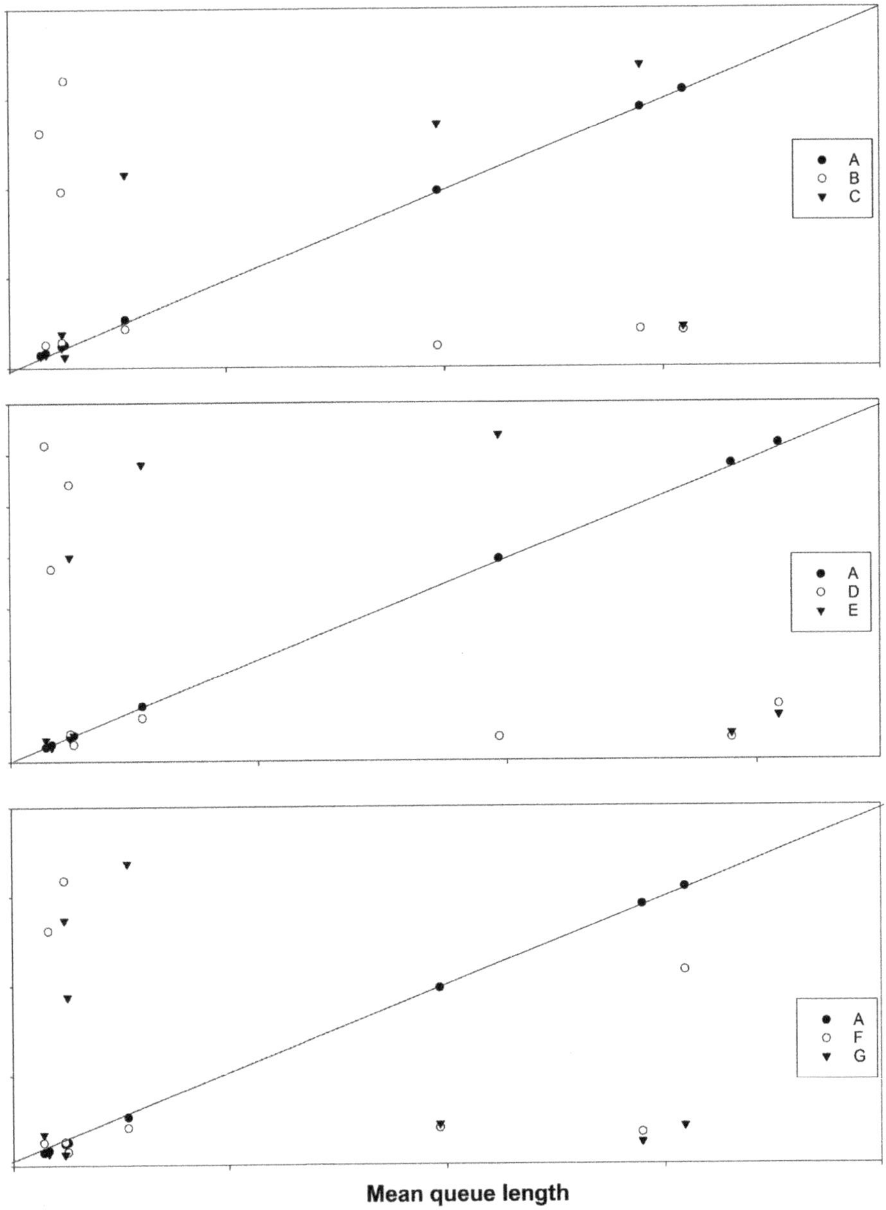

Figure 5.25. *Scatter plots of L_q under various scenarios for example 5.9*

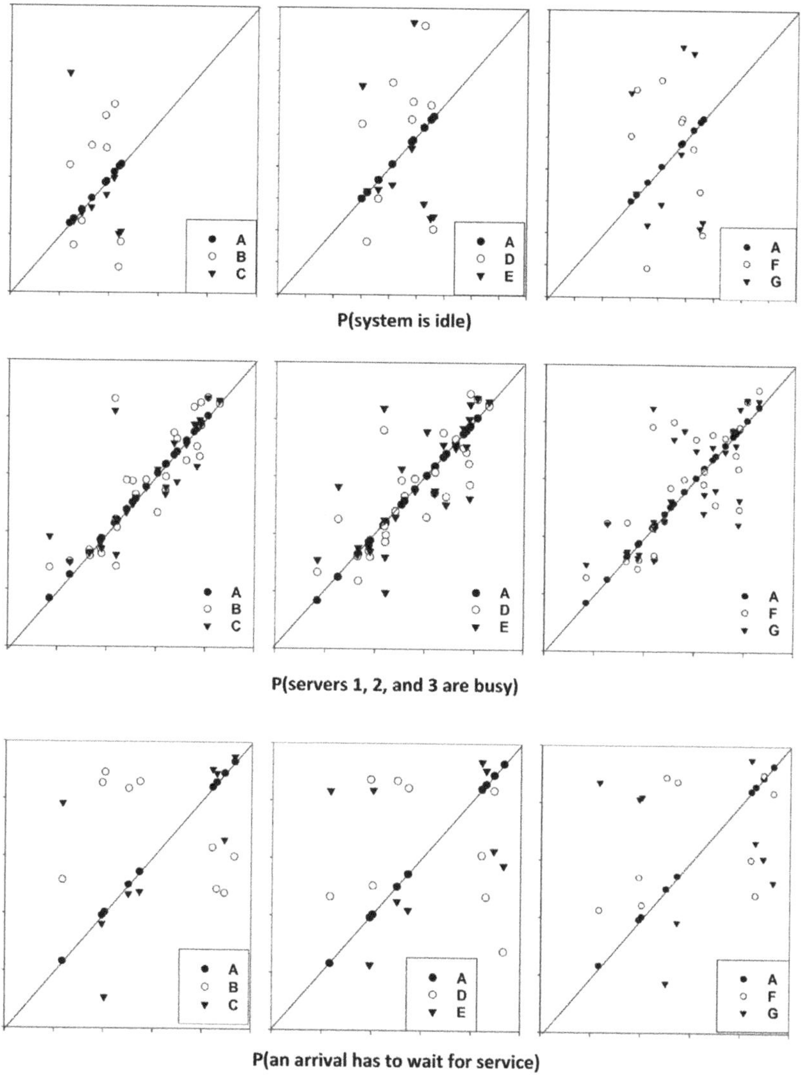

Figure 5.26. *Scatter plots of P_{idle}, $\tilde{P}_{busy}^{(r)}$ and P_{wait} under various scenarios for example 5.9*

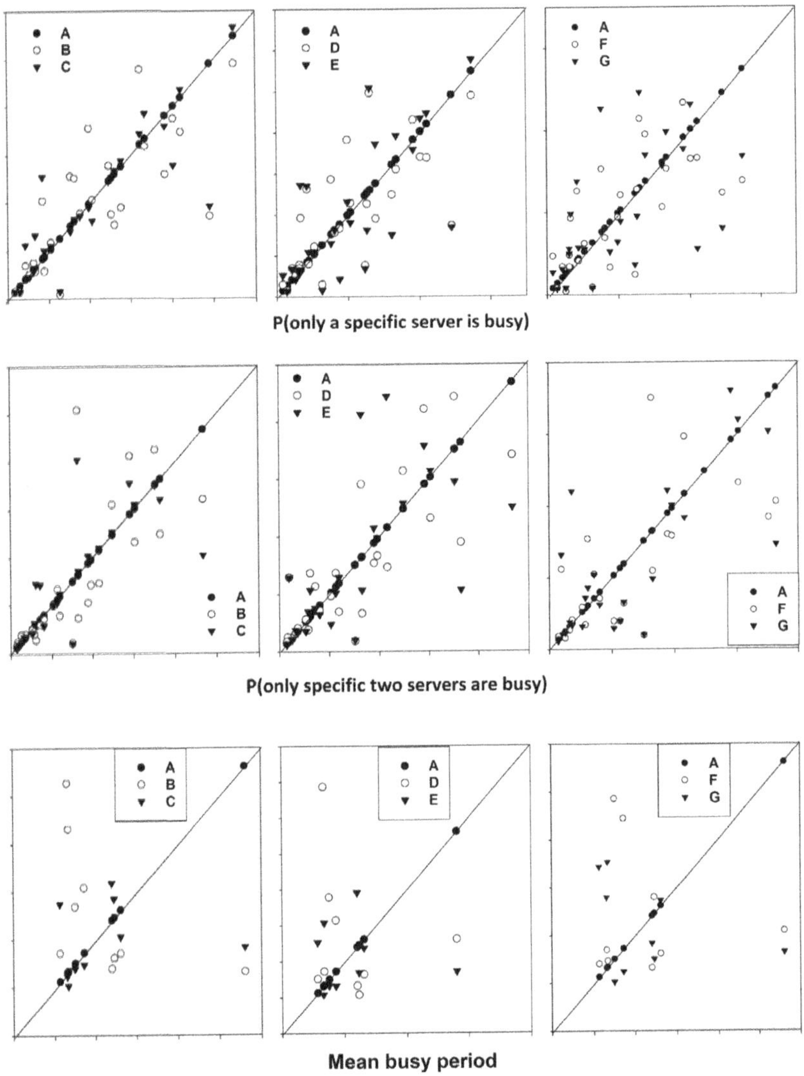

Figure 5.27. *Scatter plots of $P^{(r)}_{busy}$, $P^{(i_1,i_2)}_{busy}$, $P^{(r)}_{busy}$ and μ'_{BP} under various scenarios for example 5.9*

5.3. *BMAP/M/c* queue

Suppose we look at a c-server system in which the arrivals occur according to a BMAP with representation matrices $\{D_k\}$ of dimension m. We assume that all servers are homogeneous having a common service time distribution that is exponentially distributed with parameter μ. Define $N(t)$ to be the number of customers in the system and $J(t)$ to be the phase of the arrival process at time t. The model can be studied as an $M/G/1$-type queue in continuous-time with state space given by:

$$\Omega = \{(i,j) : 1 \leq j \leq m, i \geq 0\}.$$

Suppose we group the boundary levels as:

$$\tilde{\mathbf{0}} = \{(i,j) : 1 \leq j \leq m, 0 \leq i \leq c-1\},$$

and denote the non-boundary levels corresponding to the system having c or more customers as:

$$\boldsymbol{i} = \{(i,j) : 1 \leq j \leq m\},\ i \geq c.$$

The generator Q of the $BMAP/M/c$ queue is given by:

$$Q = \begin{pmatrix} C_0 & C_1 & C_2 & C_3 & \cdots \\ F & H_1 & H_2 & H_3 & \cdots \\ 0 & H_0 & H_1 & H_2 & \cdots \\ 0 & 0 & H_0 & H_1 & \cdots \\ \vdots & \vdots & \vdots & \vdots & \ddots \end{pmatrix}, \qquad [5.24]$$

where the matrices appear in the generator as follows:

$$C_0 = \begin{pmatrix} D_0 & D_1 & D_2 & D_3 & \cdots & D_{c-2} & D_{c-1} \\ \mu I & D_0 - \mu I & D_1 & D_2 & \cdots & D_{c-3} & D_{c-2} \\ 0 & 2\mu I & D_0 - 2\mu I & D_1 & \cdots & D_{c-4} & D_{c-3} \\ \vdots & \vdots & \vdots & \vdots & \cdots & \vdots & \vdots \\ 0 & 0 & 0 & 0 & \cdots & (c-1)\mu I & D_0 - (c-1)\mu I \end{pmatrix}, \qquad [5.25]$$

$$C_r = \begin{pmatrix} D_{c+r-1} \\ D_{c+r-2} \\ D_{c+r-3} \\ \vdots \\ D_r \end{pmatrix},\ r \geq 1,\ \ F = (0, 0, \cdots, 0, c\mu I), \qquad [5.26]$$

and:

$$H_0 = c\mu I,\ \ H_1 = D_0 - c\mu I,\ \ H_r = D_{r-1},\ r \geq 2. \qquad [5.27]$$

Note that C_0, of dimension cm, governs transitions within level $\tilde{\mathbf{0}}$ indicating that the system has at least one free server. The matrices C_r, $r \geq 1$, are of dimension $cm \times m$, F is of dimension $m \times cm$ and H_r, $r \geq 0$, are all square matrices of dimension m.

We can apply the results for the continuous-time $M/G/1$-type queue here. Some specific details are given as part of exercises at the end of this chapter.

REMARK 5.3.– If one is interested to look at the steady-state analysis done (without the busy period calculations) by restricting (or even truncating) the batch size to not exceed a certain value, say, N, then we can study the model as a QBD process. Under this restriction (that the maximum batch size is N) and after rewriting the state space as:

$$\Omega = \{\mathbf{0}\}\bigcup\{\mathbf{1},\cdots,\boldsymbol{N}\}\bigcup\{\boldsymbol{N}+\mathbf{1},\cdots,\mathbf{2}\boldsymbol{N}\}\bigcup\cdots,$$

where level $\boldsymbol{i} = \{(\boldsymbol{i},\mathbf{1}),\cdots,(\boldsymbol{i},\boldsymbol{N})\}$, for $i \geq 1$, with each of $(\boldsymbol{i},\boldsymbol{j})$, $1 \leq j \leq N$, is of dimension m keeping track of the phase of BMAP, the generator Q of the $BMAP/M/c$ queue is now of the form:

$$Q = \begin{pmatrix} D_0 & \tilde{B}_0 & 0 & 0 & 0 & \cdots \\ \tilde{B}_2 & \tilde{B}_1 & E_0 & 0 & 0 & \cdots \\ 0 & E_2 & E_1 & E_0 & 0 & \cdots \\ 0 & 0 & E_2 & E_1 & E_0 & \cdots \\ \vdots & \vdots & \vdots & \vdots & \ddots & \end{pmatrix}. \qquad [5.28]$$

The dimensions of the matrices appearing in Q depend on whether $N > c$ or $c \geq N$. We display the entries when $N > c$ and leave the other case as an exercise.

When $N > c$, the dimensions of $\tilde{B}_0, \tilde{B}_1$ and $\tilde{B}_2$ are $m \times N\,m$, $N\,m \times m$ and $N\,m \times N\,m$, respectively. The matrices E_0, E_1 and E_2 are square matrices of dimension $N\,m$. These matrices are given by:

$$\tilde{B}_0 = (D_1, D_2, \cdots, D_{N-1}, D_N),$$

$$\tilde{B}_2 = \begin{pmatrix} \mu I \\ 0 \\ \vdots \\ 0 \end{pmatrix}, \quad \tilde{B}_1 = \begin{pmatrix} D_0 - \mu I & D_1 & D_2 & D_3 & \cdots & D_{c-2} & D_{N-1} \\ 2\mu I & D_0 - 2\mu I & D_1 & D_2 & \cdots & D_{N-2} & D_{N-2} \\ \vdots & \vdots & \vdots & \vdots & \cdots & \vdots & \vdots \\ 0 & 0 & 0 & 0 & \cdots & c\mu I & D_0 - c\mu I \end{pmatrix}.$$

$$E_0 = \begin{pmatrix} D_N & 0 & 0 & 0 \\ D_{N-1} & D_N & 0 & 0 \\ \vdots & \vdots & \cdots & \vdots \\ D_1 & D_2 & \cdots & D_N \end{pmatrix}, \quad E_1 = \begin{pmatrix} D_0 - c\mu I & D_1 & D_2 & \cdots & D_{N-1} \\ c\mu I & D_0 - c\mu I & D_1 & \cdots & D_{N-2} \\ 0 & c\mu I & D_0 - c\mu I & \cdots & D_{N-3} \\ \vdots & \vdots & \vdots & \cdots & \vdots \\ 0 & 0 & 0 & \cdots & D_0 - c\mu I \end{pmatrix},$$

$$E_2 = \begin{pmatrix} 0 & 0 & \cdots & 0 & c\mu I \\ 0 & 0 & \cdots & 0 & 0 \\ \vdots & \vdots & \cdots & \vdots & \vdots \\ 0 & 0 & \cdots & 0 & 0 \end{pmatrix}.$$

Now, one can apply the continuous-time QBD method to carry out the steady-state analysis. Suppose that $\boldsymbol{y} = (\boldsymbol{y}_0, \boldsymbol{y}_1, \cdots)$ denotes the steady-state probability vector such that $\boldsymbol{y}_i$, of dimension m, gives the steady-state probability that the system has i, $i \geq 0$, customers in the system with the arrival process in one of m phases.

Defining $\boldsymbol{u} = (\boldsymbol{y}_0, \boldsymbol{u}_1, \boldsymbol{u}_2, \cdots)$ with $\boldsymbol{u}_i = (\boldsymbol{y}_{(i-1)N+1}, \boldsymbol{y}_{(i-1)N+2}, \cdots, \boldsymbol{y}_{(i-1)N+N})$, for $i \geq 1$, we first solve for $\boldsymbol{u}$ as follows:

$$\begin{aligned} &\boldsymbol{y}_0 D_0 + \boldsymbol{u}_1 \tilde{B}_2 = \mathbf{0}, \\ &\boldsymbol{y}_0 \tilde{B}_0 + \boldsymbol{u}_1 (\tilde{B}_1 + R E_2) = \mathbf{0}, \\ &\boldsymbol{u}_i = \boldsymbol{u}_1 R^i, \ i \geq 1, \\ &\boldsymbol{y}_0 + \boldsymbol{u}_1 (I - R)^{-1} \boldsymbol{e} = 1, \end{aligned} \qquad [5.29]$$

where R is obtained by solving the matrix-quadratic equation:

$$R^2 E_2 + R E_1 + E_0 = 0.$$

The equations given in [5.29] and the matrix R can be efficiently solved either directly (when N, c and m are of reasonable size) or by exploiting the structure of the coefficient matrices.

The performance measures needed to study the model qualitatively can be defined. Here, we list a few.

The System Performance Measures: While there are many measures available to qualitatively study and compare several models, here we give a few and leave the rest to the interested readers.

1) *Probability that the system is idle.* The probability, P_{idle}, that the system is idle (i.e. there are no customers in the system) is given by:

$$P_{idle} = \boldsymbol{y}_0 \boldsymbol{e}.$$

2) *Mean number in system.* The mean, L_s, number of customers in the system is given by:

$$L_s = N \boldsymbol{u}_1 R (I - R)^{-2} \boldsymbol{e} + \sum_{j=1}^{N} j \boldsymbol{a}_j \boldsymbol{e}, \text{ where } \boldsymbol{a} = \boldsymbol{u}_1 (I - R)^{-1}.$$

3) *Mean queue length.* The mean, μ_q, queue length is given by $\mu_q = L_s - \frac{\lambda}{\mu}$.

4) *P(an arriving group has to wait).* The probability, P_{wait}, that an arriving group of customers has to wait to receive service is given by:

$$P_{wait} = \frac{1}{\lambda_g}\sum_{i=0}^{\infty} \boldsymbol{y}_{c+i} D_1 \boldsymbol{e} = \frac{1}{\lambda_g}\big[\boldsymbol{a} - (\boldsymbol{y}_0, \cdots, \boldsymbol{y}_{c-1}, \boldsymbol{0})\big](\boldsymbol{e} \otimes D_1 \boldsymbol{e}),$$

where λ_g is the rate of arrivals of the groups in BMAP process and is given by $\lambda_g = \boldsymbol{\eta}(-D_0)\boldsymbol{e}$, and $\boldsymbol{\eta}$ is the invariant vector of the irreducible generator of the underlying generator of the continuous-time Markov chain (CTMC) governing the BMAP process.

5) *Mean number of busy servers.* The mean, μ_{BS}, number of busy servers is given by:

$$\mu_{BS} = \sum_{j=1}^{c} j\boldsymbol{y}_j \boldsymbol{e} + c\Big[\boldsymbol{u}_1 R(I - R)^{-1}\boldsymbol{e} + \sum_{j=c+1}^{N} \boldsymbol{y}_j \boldsymbol{e}\Big].$$

6) *Probability queue is empty.* The probability, PQ_0, that the queue size is zero, which is same as the probability that at most c customers are in the system at an arbitrary time, is given by:

$$PQ_0 = \sum_{j=0}^{c} \boldsymbol{y}_j \boldsymbol{e}$$

7) *Probability that all servers are busy.* The probability, P_{busy}, that all servers are busy is obtained as:

$$P_{busy} = \sum_{j=c}^{\infty} \boldsymbol{y}_j \boldsymbol{e}.$$

REMARK 5.4.– Recall that in any computational procedures it is always important to incorporate accuracy checks. In this model, we have many such accuracy checks. These were mentioned earlier (e.g., when computing R in the context of QBD process we know that $RA_2\boldsymbol{e} = A_0\boldsymbol{e}$) and here one can use the following, which can be verified.

In steady-state, the vector of the phase of the underling CTMC governing the BMAP process can be computed directly. This vector should be equal to the one obtained from the steady-state vector of Q. That is, $\sum_{r=0}^{\infty} \boldsymbol{y}_r = \boldsymbol{\eta}$. Also, the mean number of busy servers (see item 5 in the list of performance measures given above) should be equal to $\frac{\lambda}{\mu}$.

EXAMPLE 5.10.– Suppose that the customers arrive to a three-server system according to a BMAP with representation matrices, $\{D_k\}$, given by:

$$D_0 = -\frac{1}{2.25}, \ D_k = \frac{1}{2.25} e^{-1.25} \frac{(1.25)^{k-1}}{(k-1)!}, \ k \geq 1.$$

That is, we assume that a doubly Poisson process is used to model the arrival process, which is to say that we have a Poisson process with parameter $\frac{1}{2.25}$ to model the arrivals of groups along with the assumption that the group size is a Poisson random variable such that the rate of arrivals of customers is 1. Recall that using a Poisson random variable with parameter 1.25 results in the average number of customers in a batch being 2.25 (the "1" needs to be added to the 1.25 as a Poisson random variable has a mass at the origin and hence an arrival of a batch with size 1 in the batch is given by $e^{-1.25}$).

Assume that the (homogeneous) servers offer services that are exponentially distributed with parameter μ. Let $\rho = \frac{1}{3\mu}$.

Suppose that $\rho = 0.95$. Using $\epsilon = 10^{-12}$ in the computation of R and $\epsilon_1 = 10^{-8}$ for the truncation of the Poisson probabilities to find the maximum batch size, we obtain the following:

1) The maximum batch size is truncated at $N = 12$.

2) The probability mass function (PMF) of the batch size with the support on $\{1, 2, \cdots, 12\}$ is then given by (only positive probabilities up to five decimal places are displayed here):

$$(0.2865, 0.35813, 0.22383, 0.09326, 0.02914, 0.00729,$$

$$0.00152, 0.000270, 0.00004, 0.00001)$$

3) The matrix G has only one non-zero column, which occurs in the last column, and the entries are given by the unit vector $\boldsymbol{e}$.

4) The steady-state probability vectors y_0 (which is a scalar now since $m = 1$) and $\boldsymbol{u}_1$ are given by:

$$y_0 = 0.0221,$$

$$\boldsymbol{u}_1 = (0.028, 0.0277, 0.0235, 0.0237, 0.0231, 0.0225, \\ 0.0219, 0.0213, 0.0208, 0.0202, 0.0197, 0.0192),$$

5) The rate matrix R is given by:

$$R = \begin{pmatrix} 0 & 0 & 0 & 0 & 0 & 0 & 0 & 0 & 0 & 0 & 0 & 0 \\ 0 & 0 & 0 & 0 & 0 & 0 & 0 & 0 & 0 & 0 & 0 & 0 \\ 0 & 0 & 0 & 0 & 0 & 0 & 0 & 0 & 0 & 0 & 0 & 0 \\ 0 & 0 & 0 & 0 & 0 & 0 & 0 & 0 & 0 & 0 & 0 & 0 \\ 0 & 0 & 0 & 0 & 0 & 0 & 0 & 0 & 0 & 0 & 0 & 0 \\ 0.001 & 0 & 0 & 0 & 0 & 0 & 0 & 0 & 0 & 0 & 0 & 0 \\ 0.004 & 0.002 & 0.002 & 0.002 & 0.002 & 0.002 & 0.002 & 0.002 & 0.002 & 0.002 & 0.002 & 0.002 \\ 0.016 & 0.011 & 0.01 & 0.01 & 0.01 & 0.01 & 0.009 & 0.009 & 0.009 & 0.009 & 0.008 & 0.008 \\ 0.056 & 0.040 & 0.037 & 0.037 & 0.036 & 0.035 & 0.034 & 0.033 & 0.032 & 0.031 & 0.031 & 0.03 \\ 0.105 & 0.119 & 0.112 & 0.109 & 0.107 & 0.104 & 0.101 & 0.098 & 0.096 & 0.093 & 0.091 & 0.088 \\ 0.301 & 0.277 & 0.263 & 0.256 & 0.250 & 0.243 & 0.237 & 0.23 & 0.224 & 0.218 & 0.213 & 0.207 \\ 0.422 & 0.480 & 0.480 & 0.466 & 0.453 & 0.441 & 0.429 & 0.418 & 0.407 & 0.396 & 0.386 & 0.375 \end{pmatrix}.$$

6) The accuracy checks on R yielded an error to the tune of 10^{-16}.

7) Selected system performance measures are: $P_{idle} = 0.0221$, $L_s = 36.8956$, $\mu_{BS} = 2.85$, $P_{busy} = 0.9222$, $PQ_0 = 0.1013$ and $P_{wait} = 0.9222$.

EXAMPLE 5.11.– This example refers to example 5.10. Here, we compare different systems consisting of c servers with $c = 1, 2, 3, 5, 10$. In order to compare various systems properly, we set the traffic intensity of 0.95 to all of these systems. That is, we take $\mu = \dfrac{1}{0.95\, c}$. All other parameters remain the same as in example 5.10.

First, note that the matrices G and R for this example are identical to their respective ones obtained in example 5.10. This is due to the fact that we have taken $c\mu$ to be fixed by choosing μ accordingly. Thus, the matrices E_0, E_1 and E_2 do not change. Again, this is due to choice of the parameters chosen and not due to the structure. If, for example, the systems are compared by offering varying arrival rate (of groups) to compensate for varying c when μ is fixed, then G and R will differ. In Table 5.1, we display the selected measures to compare various c-server system. For the sake of convenience we also include the three-server system output from example 5.10.

c	P_{idle}	L_s	μ_{BS}	P_{busy}	PQ_0	L_q	P_{wait}
1	0.0500	36.1528	0.95	0.9500	0.0711	35.2028	0.9500
2	0.0352	36.4497	1.90	0.9352	0.0880	34.5497	0.9352
3	0.0221	36.8956	2.85	0.9222	0.1013	34.0456	0.9222
5	0.0077	38.0038	4.75	0.9011	0.1222	33.2538	0.9011
10	0.0004	41.3276	9.50	0.8627	0.1598	31.8276	0.8627

Table 5.1. *Selected measures for example 5.11 for a doubly Poisson process with exponential services by varying* c

It is worth pointing out that the measure P_{wait} is same as the probability that all servers are busy due to PASTA (Poisson arrivals see time averages) property. Also, note that while L_s increases as c increases, the measure L_q decreases as (keeping in mind the traffic load is the same as c is varied) c is increased. This is due to the fact that the number of servers busy increases with c.

EXAMPLE 5.12.– Suppose that the inter-arrival times of the batches of customers arriving to a c-server system are modeled as the maximum of Erlang of order 2 (with parameter 2) and a hyperexponential (with parameters 1.9 and 0.19 with mixing probabilities 0.9 and 0.1, respectively). Recall that (see example 4.5 in Volume 1) the inter-arrival times are modeled as a CPH with dimension 8. We assume the batch size to follow a Poisson distribution like in example 5.10. Like in example 5.11 we compare different systems consisting of c servers with $c = 1, 2, 3, 5, 10$. We take $\rho = 0.95$ and $\mu = \dfrac{1}{0.95\,c}$. All other parameters remain the same as in example 5.10.

In Table 5.2, we display the selected measures to compare various c-server systems. For the sake of convenience we also include the three-server system output from example 5.10.

c	P_{idle}	L_s	μ_{BS}	P_{busy}	PQ_0	L_q	P_{wait}
1	0.0500	53.3149	0.95	0.9500	0.0640	52.3649	0.9668
2	0.0403	53.6167	1.90	0.9403	0.0753	51.7167	0.9565
3	0.0312	54.0511	2.85	0.9315	0.0845	51.2011	0.9474
5	0.0183	55.1076	4.75	0.9164	0.0995	50.3576	0.9320
10	0.0044	58.2513	9.50	0.8874	0.1282	48.7513	0.9025

Table 5.2. *Selected measures for example 5.12*

EXAMPLE 5.13.– This example is similar to example 5.11 except we model the inter-arrival times using Erlang distribution of order k, where k is varied from 1 to 10. The purpose is to see how the system performance measures are affected when the coefficient of variation is reduced. We use the same parameter values of example 5.11 but here we vary $c = 1, 2, 3$, and $c = 5$. The standard deviations of these 10 Erlang distributions (of orders 1 through 10) are, respectively, given by 2.25, 1.591, 1.299, 1.125, 1.0062, 0.9186, 0.8504, 0.7955, 0.75 and 0.7115. In Figures 5.28 and 5.29, we display the plots of a few measures as functions of the order of Erlang under different scenarios. Recall that $\mu = \dfrac{1}{0.95\,c}$.

As is to be expected, we notice that as the variability in the services decreases (i.e. as the order of Erlang increases), the measures – the system idle probability, the probability that all servers are busy, the probability that an arriving group has to wait, and the mean number in the system – all show a non-increasing trend, whereas the probability that the queue is empty increases. This is the case for all c considered.

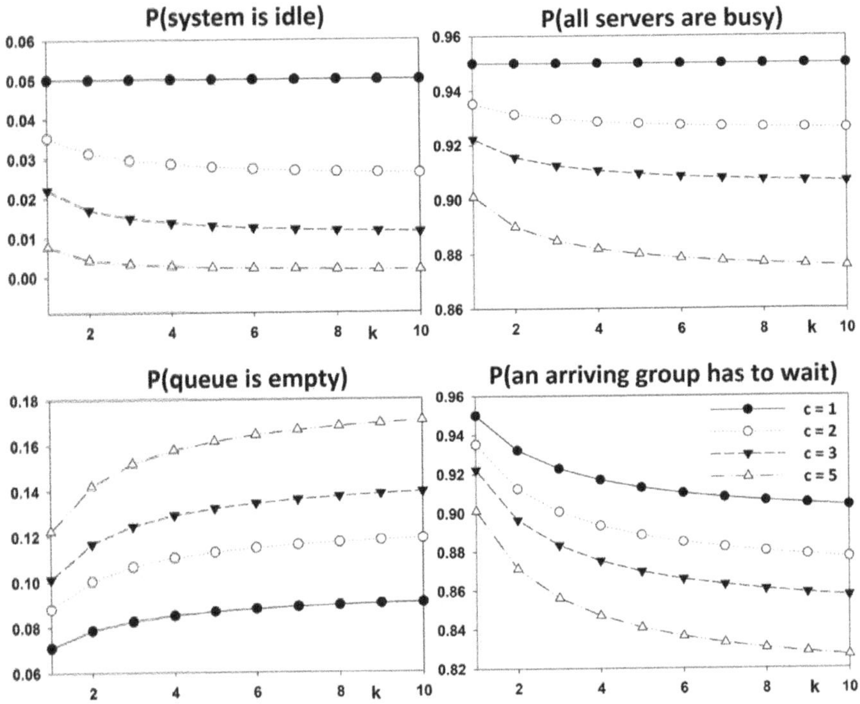

Figure 5.28. *Selected measures for example 5.13*

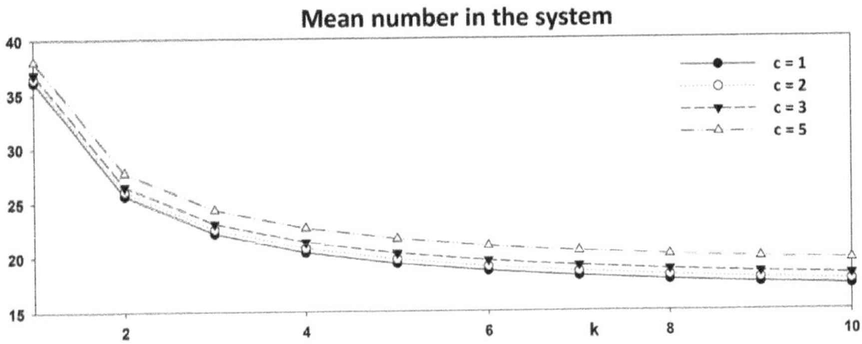

Figure 5.29. *Mean number in the system for example 5.13*

EXAMPLE 5.14.– This example is similar to example 5.13 except we model the inter-arrival times using hyperexponential distribution of order k, where k is varied from 1 to 10. Specifically, the 10 hyperexponentials are displayed below in the format $\boldsymbol{\alpha}$ and $\boldsymbol{a}$ with the former giving the mixing probabilities and the latter giving the parameters of the exponentials which are used to form the hyperexponential. Recall that $\mu = \frac{1}{0.95\,c}$.

HE_1: $\boldsymbol{\alpha} = 1$ and $\boldsymbol{a} = (1)$.

HE_2: $\boldsymbol{\alpha} = (0.9, 0.1)$ and $\boldsymbol{a} = (10, 5)$.

HE_3: $\boldsymbol{\alpha} = (0.7, 0.2, 0.1)$ and $\boldsymbol{a} = (10, 5, 2.5)$.

HE_4: $\boldsymbol{\alpha} = (0.6, 0.2, 0.15, 0.05)$ and $\boldsymbol{a} = (10, 5, 2.5, 1.25)$.

HE_5: $\boldsymbol{\alpha} = (0.5, 0.3, 0.15, 0.04, 0.01)$ and $\boldsymbol{a} = (20, 10, 5, 2.5, 1.25)$.

HE_6: $\boldsymbol{\alpha} = (0.5, 0.2, 0.15, 0.10, 0.04, 0.01)$ and $\boldsymbol{a} = (30, 15, 10, 7.5, 5, 1)$.

HE_7: $\boldsymbol{\alpha} = (0.35, 0.20, 0.17, 0.14, 0.09, 0.04, 0.01)$ and $\boldsymbol{a} = (40, 25, 15, 10, 5, 1, 0.5)$.

HE_8: $\boldsymbol{\alpha} = (0.30, 0.20, 0.17, 0.14, 0.09, 0.05, 0.04, 0.01)$ and $\boldsymbol{a} = (50, 40, 25, 15, 10, 5, 1, 0.5)$.

HE_9: $\boldsymbol{\alpha} = (0.34, 0.20, 0.15, 0.10, 0.09, 0.06, 0.03, 0.02, 0.01)$ and $\boldsymbol{a} = (50, 40, 25, 15, 10, 5, 1, 0.5, 0.1)$.

HE_{10}: $\boldsymbol{\alpha} = (0.30, 0.20, 0.15, 0.11, 0.09, 0.07, 0.04, 0.025, 0.01, 0.005)$ and $\boldsymbol{a} = (100, 50, 40, 25, 15, 10, 5, 1, 0.5, 0.1)$.

The parameters appearing in $\boldsymbol{a}$ in the above hyperexponential distributions will be normalized so as to achieve the arrival rate (of groups) to be $\frac{1}{2.25}$. The standard deviations of these 10 distributions, in that order, after normalization (to have the rate be $\frac{1}{2.25}$) are, respectively, given by 2.25, 2.4116, 2.9812, 3.5576, 3.8185, 4.827, 7.4518, 8.3951, 15.4254 and 18.1477.

The purpose is to see how the system performance measures are affected when the variability is increased. We use the same parameter values of example 5.11 but here we vary $c = 1, 2, 3$, and $c = 5$. In Figures 5.30 and 5.31, we display the plots of a few measures under different scenarios. The x-axis is labeled with k to denote the type of hyperexponential being considered. These graphs reveal some interesting observations and shed light on the impact of the variability in the inter-arrival times.

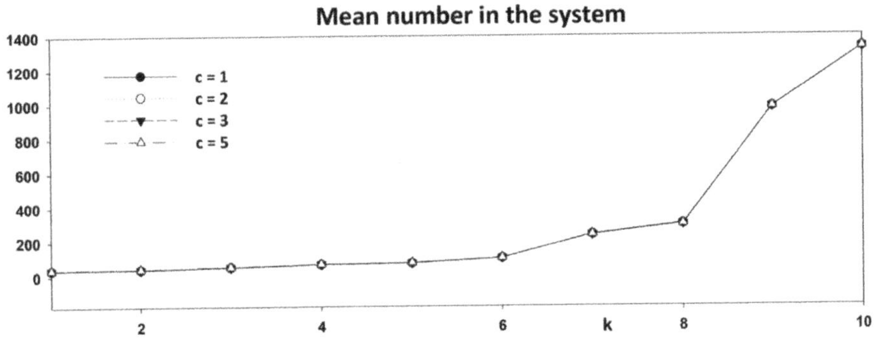

Figure 5.30. *Mean number in the system for example 5.14*

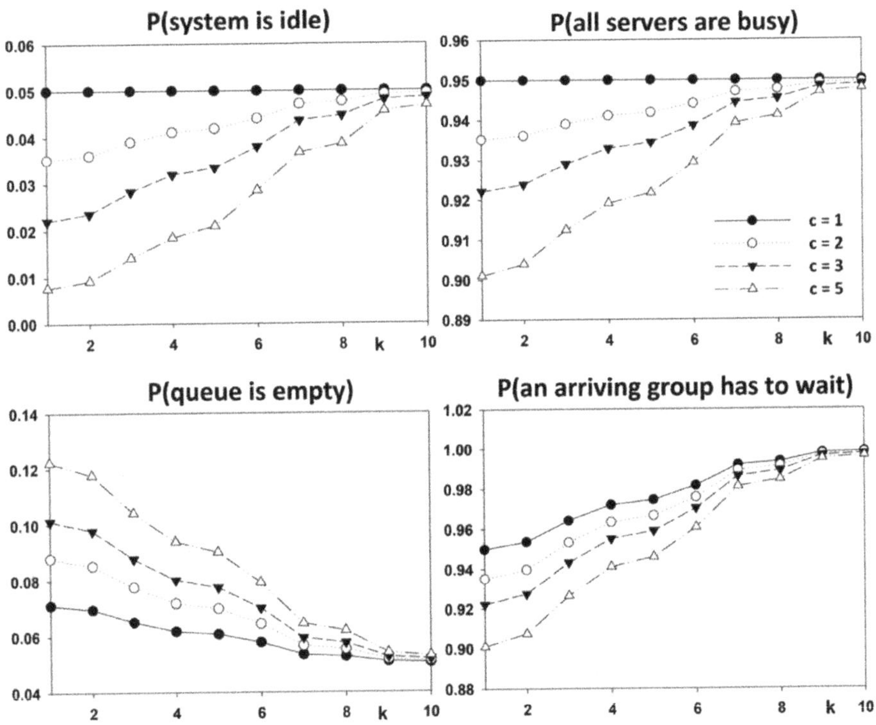

Figure 5.31. *Selected measures for example 5.14*

EXAMPLE 5.15.– This example is similar to example 5.13 except we model the inter-arrival times using a negatively correlated arrival process, labeled as NC_r, that was considered in example 5.1. We vary r from 2 through 9. All other parameter values remain the same as in example 5.10. However, these BMAPs will be normalized so that the (group) arrival rate will be $\frac{1}{2.25}$.

The standard deviations of these 8 distributions, in that order, after normalization (to have the rate be $\frac{1}{2.25}$) are, respectively, given by 2.3383, 2.2955, 2.2776, 2.2685, 2.2633, 2.26, 2.2578 and 2.2562. The 1-lag correlation coefficients are, respectively, given by -0.3267, -0.4804, -0.5786, -0.6454, -0.6935, -0.7296, -0.7577, and -0.7802.

The purpose is to see how the system performance measures are affected when the (negative) correlation coefficients are varied. We use the same parameter values of example 5.11 but here we vary $c = 1, 2, 3$, and $c = 5$. In Figures 5.32 and 5.33, we display the plots of a few measures as functions of the type, NC_r, of a negatively correlated process. Recall that $\mu = \dfrac{1}{0.95\,c}$. These graphs reveal some interesting observations and shed light on the impact of the variability as well as the 1-lag correlation in the inter-arrival times.

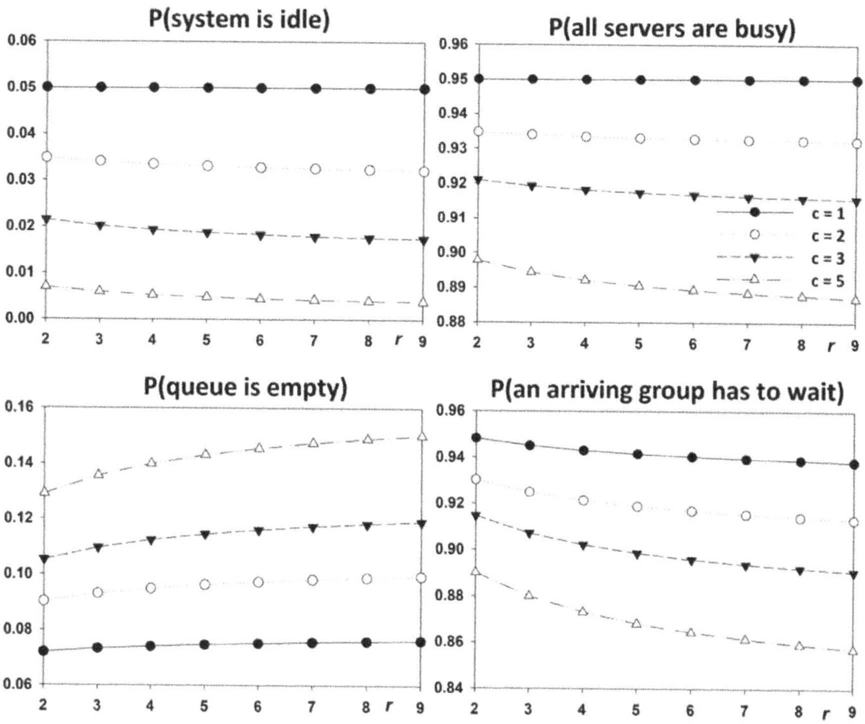

Figure 5.32. *Selected measures for example 5.15*

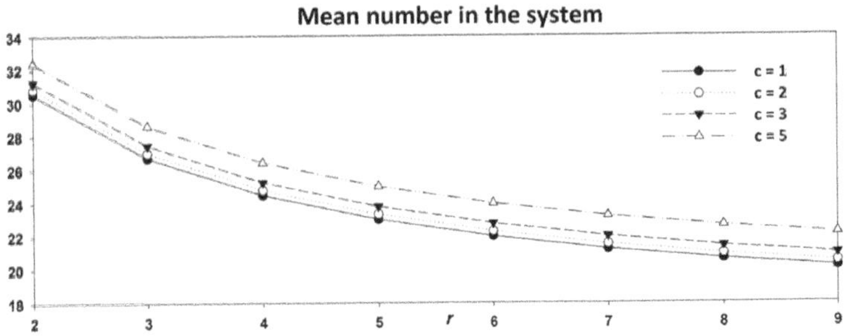

Figure 5.33. *Mean number in the system for example 5.15*

EXAMPLE 5.16.– This example is similar to example 5.13 except we model the inter-arrival times using a positively correlated arrival process, labeled as PC_r, that was considered in example 5.1. We vary r from 2 through 9. All other parameter values remain the same as in example 5.10. However, these BMAPs will be normalized so that the (group) arrival rate will be $\frac{1}{2.25}$. The standard deviations of these 8 distributions, in that order, after normalization (to have the rate be $\frac{1}{2.25}$) are, respectively, given by 2.3383, 2.2955, 2.2776, 2.2685, 2.2633, 2.26, 2.2578 and 2.2562. The 1-lag correlation coefficients are, respectively, given by 0.3267, 0.4804, 0.5786, 0.6454, 0.6935, 0.7296, 0.7577 and 0.7802. Recall $\mu = \dfrac{1}{0.95\,c}$.

The purpose is to see how the system performance measures are affected when the (positive) correlation coefficients are varied. We use the same parameter values of example 5.11 but here we vary $c = 1, 2, 3$, and $c = 5$. In Figure 5.34 and 5.35, we display the plots of a few measures as functions of the type, PC_r, of a positively correlated process. These graphs reveal some interesting observations and shed light on the impact of the variability as well as the 1-lag correlation in the inter-arrival times.

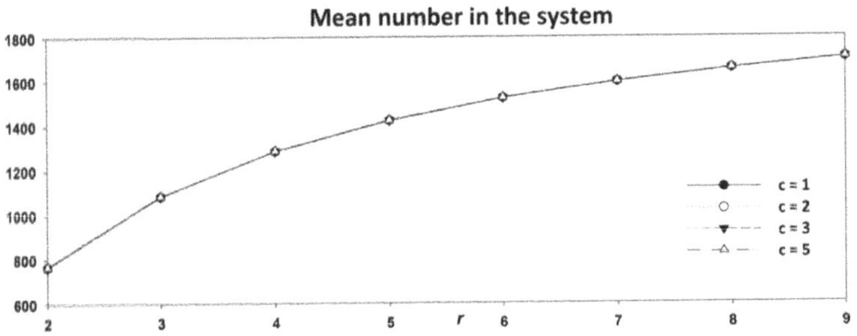

Figure 5.34. *Mean number in the system for example 5.16*

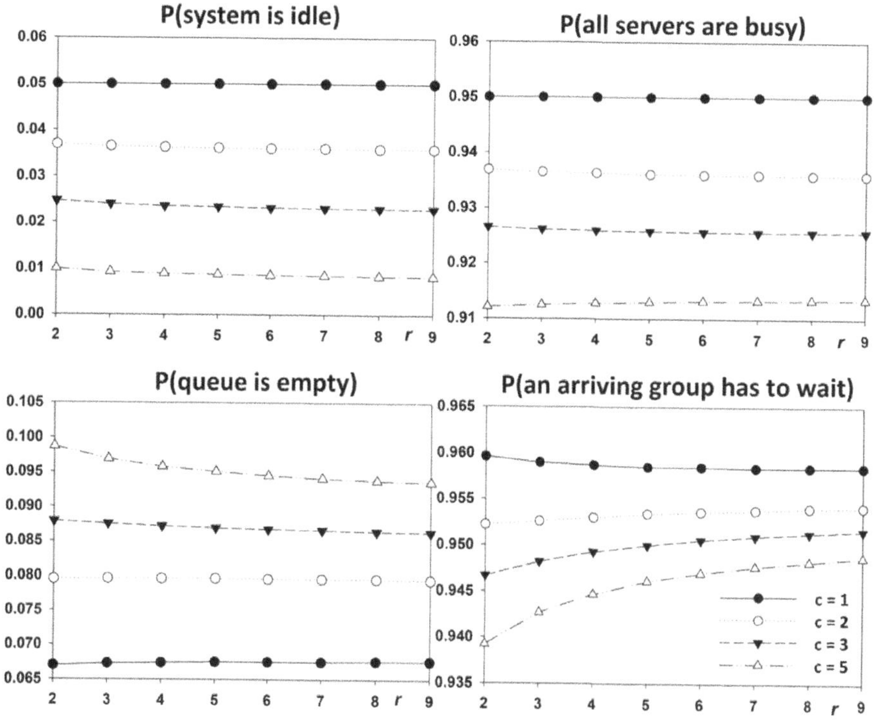

Figure 5.35. *Selected measures for example 5.16*

5.3.1. MAP/M/c *queue*

The queueing model $MAP/M/c$ referring to a multi-server system with MAP arrivals and exponential services can be studied as a QBD process in continuous-time. Here, we assume that the servers are identical. However, this assumption can be relaxed to include heterogeneous servers but the state space needs to be enlarged to accommodate for keeping track of which servers are busy as the rates depend on the server. Let the representation of the MAP be given by (D_0, D_1) of order m. We assume that the matrix $D = D_0 + D_1$ is an irreducible matrix. The customers are served on an FCFS basis and the service times are assumed to be exponential with parameter μ.

Letting $N(t)$ and $J(t)$ denote the number of customers and the phase of the arrival process at time t, respectively, the model can be studied as a QBD process on the state space:

$$\Omega = \{(i, j) : i \geq 0, 1 \leq j \leq m\}.$$

and with the infinitesimal generator given by:

$$Q = \begin{pmatrix} D_0 & D_1 & & & & & & \\ \mu I & D_0 - \mu I & D_1 & & & & & \\ & 2\mu I & D_0 - 2\mu I & D_1 & & & & \\ & & \ddots & \ddots & \ddots & & & \\ & & & (c-1)\mu I & D_0 - (c-1)\mu I & D_1 & & \\ & & & & c\mu I & D_0 - c\mu I & D_1 & \\ & & & & & c\mu I & D_0 - c\mu I & D_1 \\ & & & & & & \ddots & \ddots & \ddots \end{pmatrix}. \qquad [5.30]$$

Under the stability condition that $\lambda < c\mu$ (which can be established through Neuts's results on a matrix-geometric solution), the steady-state probability vector, $\boldsymbol{y}$, of Q is of matrix-geometric type. Note that $\boldsymbol{y}_i$ of dimension m gives the steady-state probability vector that there are i customers in the system and the arrival process in one of m phases. Using Neuts's results using a matrix-geometric method (Neuts 1981), it is known that the steady-state probability vector $\boldsymbol{y}$ is given by:

$$\boldsymbol{y}_{c-1+i} = \boldsymbol{y}_{c-1} R^i, \quad i \geq 0, \qquad [5.31]$$

where the rate matrix R is the minimal non-negative solution to the matrix-quadratic equation given by:

$$c\mu R^2 + R(D_0 - c\mu I) + D_1 = 0, \qquad [5.32]$$

and the vectors $\boldsymbol{y}_i$, $0 \leq i \leq c$, are obtained by solving:

$$\begin{aligned} &\boldsymbol{y}_0 D_0 + \mu \boldsymbol{y}_1 = \mathbf{0}, \\ &\boldsymbol{y}_{i-1} D_1 + \boldsymbol{y}_i (D_0 - i\mu I) + (i+1)\mu \boldsymbol{y}_{i+1} = \mathbf{0}, \ 1 \leq i \leq c-2, \qquad [5.33] \\ &\boldsymbol{y}_{c-2} D_1 + \boldsymbol{y}_{c-1} [D_0 - c\mu I + c\mu R] = \mathbf{0}, \end{aligned}$$

subject to the normalizing condition: $\sum_{i=0}^{c-2} \boldsymbol{y}_i \boldsymbol{e} + \boldsymbol{y}_{c-1}(I-R)^{-1}\boldsymbol{e} = 1.$

REMARK 5.5.– The model, $MAP/M/c$, is studied here as a QBD process and hence the results, including the ones on a busy period, seen in section 8.6 of Volume 1 and section 4.6 in this volume, can be applied.

The System Performance Measures: The qualitative analysis of any queueing model requires one or more system performance measures. While one can develop a number of such measures, here we give a few and leave the rest to interested readers.

1) *Probability that the system is idle.* The probability, P_{idle}, that the system is idle (there are no customers in the system) is given by $P_{idle} = \boldsymbol{y}_0\mathbf{e} = 1 - \frac{\lambda}{\mu}$.

2) *Mean queue length.* The mean, L_q, queue length is given by $L_q = \boldsymbol{y}_{c-1}R^2(I - R)^{-2}\mathbf{e}$.

3) *P(an arrival has to wait).* The probability, P_{wait}, that an arriving customer has to wait to receive service is given by $P_{wait} = \frac{1}{\lambda}\boldsymbol{y}_{c-1}R(I - R)^{-1}D_1\mathbf{e}$.

4) *Mean number in system.* The mean, L_s, number of customers in the system is given by $L_s = \boldsymbol{y}_{c-1}R^2(I - R)^{-2}\mathbf{e} + \frac{\lambda}{\mu} = L_q + \frac{\lambda}{\mu}$.

5) *Mean busy period and the mean number served during a busy period.* The mean busy period and the mean number served during a busy period are obtained for the current model using equation [4.58].

EXAMPLE 5.17.– Here, we revisit example 5.11 but with single arrivals. Note that this model is $M/M/c$ and explicit results are available (see section below). However, we use the code developed for a $BMAP/M/c$ case as a way to check the computational results and treat this as an another accuracy check. All parameters are the same as in example 5.11.

First, note that the G, of dimension c, will have only one non-zero column, which is a unit vector occurring in the last column. The rate matrix R here is an interesting pattern. This is described as follows. It is worth recalling the way the state space is grouped into levels and here, except for level $\mathbf{0}$, which is of dimension m, all other levels are of dimension cm. Furthermore, for this example, $m = 1$. Thus, R is of dimension c. Start with $c = 1$, for which case we observe that $R = \rho = 0.95$. For $c = 2$, the matrix R is given by:

$$R = \begin{pmatrix} 0 & 0 \\ 0.950 & 0.902 \end{pmatrix}.$$

A quick glance at R shows that $R(2,1) = \rho$ and $R(2,2) = \rho^2$. It can be proved that, in general, the matrix R, which has only one non-zero row that occurs in the cth row, is explicitly obtained as:

$$R = \begin{pmatrix} 0 & 0 & 0 & 0 & \cdots & 0 \\ 0 & 0 & 0 & 0 & \cdots & 0 \\ \vdots & \vdots & \vdots & \vdots & \cdots & \vdots \\ \rho & \rho^2 & \rho^3 & \rho^4 & \cdots & \rho^c \end{pmatrix} = \begin{pmatrix} 0 & 0 & 0 & 0 & \cdots & 0 \\ 0 & 0 & 0 & 0 & \cdots & 0 \\ \vdots & \vdots & \vdots & \vdots & \cdots & \vdots \\ 0.950 & 0.903 & 0.857 & 0.815 & \cdots & 0.599 \end{pmatrix}.$$

In Table 5.3, we display the selected measures to compare various c-server systems.

c	P_{idle}	L_s	μ_{BS}	P_{busy}	PQ_0	L_q	P_{wait}
1	0.0500	19.0000	0.95	0.9500	0.0975	18.0500	0.9500
2	0.0256	19.4872	1.90	0.9256	0.1206	17.5872	0.9256
3	0.0118	20.0832	2.85	0.9070	0.1383	17.2332	0.9070
5	0.0022	21.4282	4.75	0.8778	0.1661	16.6782	0.8778
10	0.0000	25.1861	9.50	0.8256	0.2157	15.6861	0.8256

Table 5.3. *Selected measures for example 5.17*

EXAMPLE 5.18.– Refer to example 5.12 but use single arrivals and discuss the behavior of the selected measures as c is varied. In Table 5.4, we display the selected measures to compare various c-server systems.

c	P_{idle}	L_s	μ_{BS}	P_{busy}	PQ_0	L_q	P_{wait}
1	0.0500	712.8609	0.95	0.9500	0.0879	711.9109	0.9601
2	0.0268	713.2774	1.90	0.9268	0.0946	711.3774	0.9531
3	0.0134	713.7525	2.85	0.9151	0.0982	710.9025	0.9493
5	0.0032	714.7665	4.75	0.9041	0.1019	710.0165	0.9454
10	0.0001	717.4287	9.50	0.8958	0.1060	707.9287	0.9410

Table 5.4. *Selected measures for example 5.18*

EXAMPLE 5.19.– In this example, we look at a c-server queueing system with MAP arrivals and exponential services. For arrivals, we look at three renewal processes whose inter-arrival times are modeled using: (1) Erlang of order 3 with parameter 3 in each state; (2) Exponential with parameter 1; (3) hyperexponential with parameters 2.1, 1.05 and 0.21, with mixing probabilities, respectively, given by 0.7, 0.2 and 0.1; (4) NC_r, for $2 \le r \le 9$ (see example 5.1); and (5) PC_r, for $2 \le r \le 9$ (see example 5.1). The service times are exponential with parameter $\mu = \dfrac{1}{c\rho}$ and vary $c = 1, 2, 3$ and 4 and $\rho = 0.8, 0.95$, and 0.99. In Figures 5.36 and 5.37, respectively, we plot a couple of selected measures: Ln(mean busy period) and P_{wait}, and the spectral radius.

First, note that the multi-server system is offered the same load as the single-server system since we have taken the service rate to be $\frac{1}{c\rho}$. This results in the spectral radius being the same for all c and hence we display only the common values; however, the spectral radius depends on other parameters like the arrival process and the traffic load. It is interesting to see that while the spectral radius for the positively correlated arrivals shows upsurges in the queue length even for a moderate value of ρ whereas that is not the case even for a high variability arrivals like hyperexponential. This illustrates the significance of the correlation in modeling queueing systems.

For the busy period, it is interesting that, for the renewal arrivals, the mean busy period (and hence the mean number served during a busy period) remains the same for all values of c. However, we cannot say the same thing about the correlated arrival processes. This fact, to the best of our knowledge, is not reported or noticed in the literature thus far. Among the positively correlated arrivals, we notice a significant difference in the mean busy period as c is increased and this is not surprising in view of the upsurges in the queue length pointed out earlier.

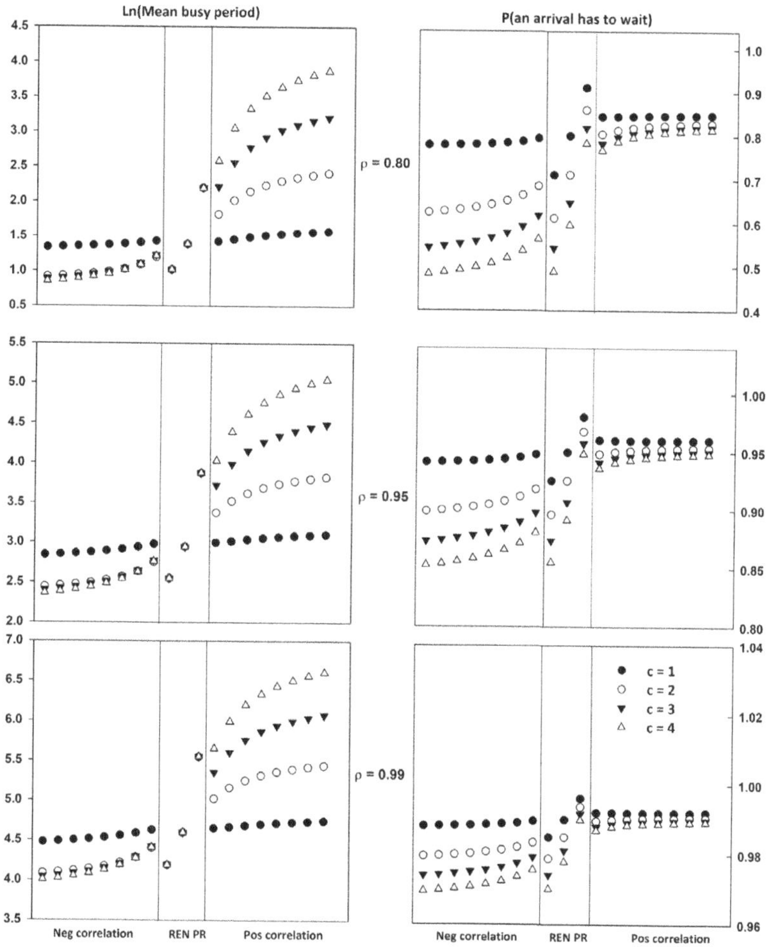

Figure 5.36. *Ln(mean busy period) and P_{wait} for example 5.19*

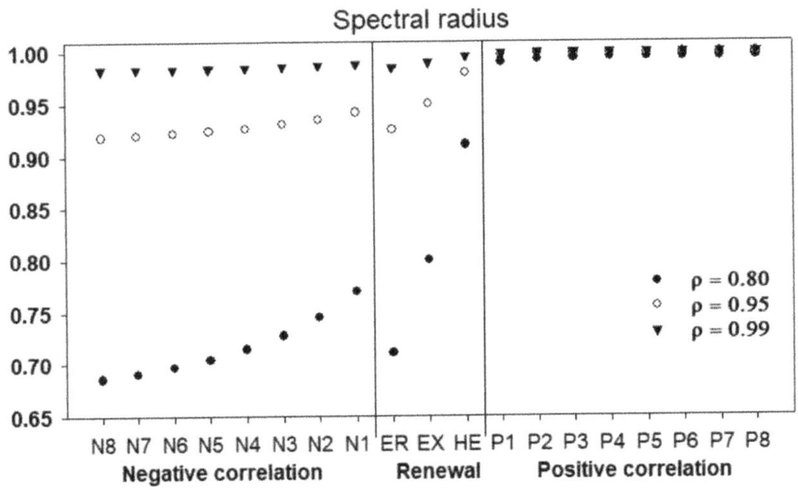

Figure 5.37. *Spectral radius under various scenarios for example 5.19*

5.3.2. M/M/c *queue*

In this model, there are c servers and arrivals occur according to a Poisson process with rate λ and the service times are exponentially distributed with rate μ. This model is another simple example of a QBD process whose generator is of the form:

$$Q = \begin{pmatrix} -\lambda & \lambda & & & & & & \\ \mu & -(\lambda+\mu) & \lambda & & & & & \\ & 2\mu & -(\lambda+2\mu) & \lambda & & & & \\ & & 3\mu & -(\lambda+3\mu) & \lambda & & & \\ & & & \ddots & \ddots & \ddots & & \\ & & & & c\mu & -(\lambda+c\mu) & \lambda & \\ & & & & & c\mu & -(\lambda+c\mu) & \lambda \\ & & & & & & \ddots & \ddots \;\ddots \end{pmatrix}. \quad [5.34]$$

Taking $\rho = \dfrac{\lambda}{c\mu}$ and noting that the stability condition is $\rho < 1$, the steady-state probability vector, $\boldsymbol{y} = (y_0, y_1, \cdots)$, can be obtained explicitly as follows:

$$y_k = \begin{cases} y_0 \dfrac{(c\rho)^k}{k!}, & 0 \le k \le c, \\ y_0 \dfrac{\rho^k c^c}{c!}, & k \ge c, \end{cases}$$

where:

$$y_0 = \left[\sum_{i=0}^{c-1} \frac{(c\rho)^k}{k!} + \left(\frac{(c\rho)^c}{c!} \right) \left(\frac{1}{1-\rho} \right) \right]^{-1}.$$

It is very easy to verify the following results.

The mean, μ_{BS}, number of busy servers is given by:

$$\mu_{BS} = \sum_{i=1}^{c-1} i y_i + c \sum_{i=c}^{\infty} y_i = c\rho = \frac{\lambda}{\mu}, \qquad [5.35]$$

which is to be expected.

The mean, L_s, number of customers in the system is given by:

$$L_s = \sum_{i=1}^{\infty} i y_i = c\rho + y_0 \frac{\rho\,(c\rho)^c}{c!(1-\rho)^2} = \mu_{BS} + L_q, \qquad [5.36]$$

where L_q is the mean number of customers in the queue.

REMARK 5.6.– The results given in equations [5.35] and [5.36], which are explicitly known in terms of the input parameters, can be used as accuracy checks in the numerical computations of the queues discussed in the previous sections.

EXAMPLE 5.20.– In this example, we look at an $M/M/c$ queue by fixing $\lambda = 1$ and consider the following scenarios by varying $\rho = 0.8, 0.9, 0.95, 0.99$ and $c = 1, 2, \cdots, 10$. We take the service rate to be $\mu = \frac{1}{c\rho}$. In Figure 5.38, we plot the steady-state probability vector of the number of customers in the system under the scenarios mentioned above.

5.4. *GI/PH/c* queue

In his classic book, Neuts (1981) mentions that $GI/PH/c$ queues with heterogeneous servers possess modified matrix-geometric solutions. Furthermore, he says, "The strictly positive matrix R arising in these results, however, is of such high dimension that the direct numerical implementation of the resulting algorithms is limited to very small values c and to particular PH distributions of low order." However, he also points out that "the *$GI/PH/c$ queue with identical servers* allows a major simplification in the description of the state space."

In the case of identical servers, as pointed before, Mayhugh and McCormick (1968) analyzed $M/E_k/r$ with r identical servers using the approach that reduced the dimension of the state space. By the same token, Ramaswami and Lucantoni (1985) adopted that approach to analyze $GI/PH/c$ queues with identical servers. Recall that this approach of keeping track of the number of servers serving in phase i is possible only in the case of a homogeneous service facility due to having a common set of phases; otherwise, there is no option but to keep track of the service phases of all servers. This curse of dimensionality is inherent in the model and is not due to the choice of the description of the state space.

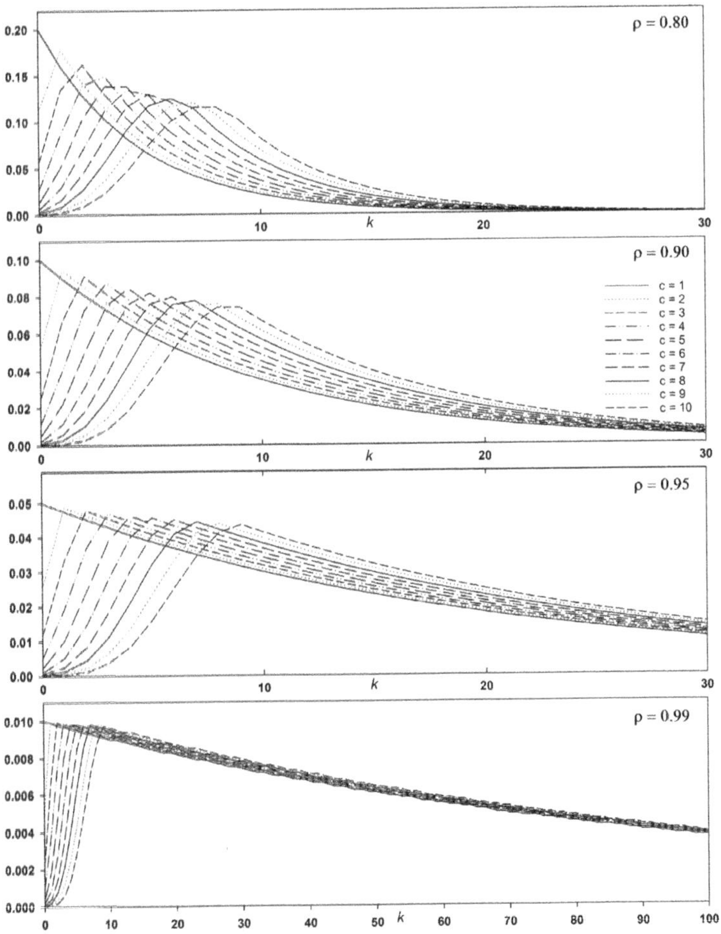

Figure 5.38. *Steady-state probability vector of the number in system for example 5.20*

We discuss examples dealing with $GI/PH/c$ models via simulation in the next chapter. However, readers interested in an analytical approach to this model in the case of identical servers are encouraged to refer to Ramaswami and Lucantoni (1985), and for a detailed description of $GI/PH/2$ queues using an embedded approach in the case of heterogeneous servers, refer to Neuts (1981).

5.5. Exercises

Exercise 5.1. Referring to the heterogeneous model $BMAP/PH/2$ queue, whose generator is given in equation [5.1], write down the steady-state equation by exploiting the special structure of the coefficient matrices. Develop a code and analyze the model after identifying a set of key system performance measures.

Exercise 5.2. Consider a heterogeneous model $BMAP/PH/3$ queue and analyze this model by first getting the generator along the lines discussed for the two-server case. Develop a code and analyze the model after identifying a set of key system performance measures.

Exercise 5.3. Independently set up the three-server heterogeneous $MAP/PH/3$ queueing system and verify the entries as given in section 5.2.

Exercise 5.4. Prove the expressions given in equations [5.20] and [5.21].

Exercise 5.5. Prove the expressions given in equations [5.22] and [5.23].

Exercise 5.6. Show that the number of ways to put c balls in n boxes without any restriction is given by $\binom{c+n-1}{n-1}$.

Exercise 5.7. Referring to example 5.3, point out key findings based on the graphs of selected measures.

Exercise 5.8. Referring to example 5.4, point out key findings based on the graphs of selected measures.

Exercise 5.9. Referring to example 5.5, point out key findings based on the graphs of selected measures.

Exercise 5.10. Referring to example 5.6, point out key findings based on the graphs of selected measures.

Exercise 5.11. Referring to example 5.7, point out key findings based on the graphs of selected measures.

Exercise 5.12. Referring to example 5.8, point out key findings based on the graphs of selected measures.

Exercise 5.13. Referring to example 5.9, point out key findings based on the graphs of selected measures.

Exercise 5.14. Suppose that in a $BMAP/M/c$ queueing system, let the maximum batch size, say, N, be at most c. Identify the (block) entries of the generator Q for this case.

Exercise 5.15. Referring to example 5.14, point out key findings based on the graphs of selected measures.

Exercise 5.16. Referring to example 5.15 point out key findings based on the graphs of selected measures.

Exercise 5.17. Referring to example 5.16, point out key findings based on the graphs of selected measures.

Exercise 5.18. For the $MAP/PH/3$ model, prove the expressions given in equations [5.20] and [5.21].

Exercise 5.19. For the $MAP/PH/3$ model, prove the expression given in equation [5.22].

Exercise 5.20. For the $MAP/PH/3$ model, prove the expression given in equation [5.23].

Exercise 5.21. Prove by appropriately identifying the needed matrices that a $BMAP/M/c$ queue in arbitrary time with a (finite) bound on the batch size can be studied as a QBD process in continuous-time. Derive the needed expressions by looking at the results from section 3.1.

Exercise 5.22. Set up the $BMAP/M/c$ queue in arbitrary time where the maximum batch size is assumed to be less than c. Identify the entries of the generator governing the queueing system.

Exercise 5.23. Referring to example 5.17, prove that the rate matrix R is of the form mentioned there.

Exercise 5.24. For the $BMAP/M/c$ queue, show that the performance measure listed in item 5 is equal to $\frac{\lambda}{\mu}$.

Exercise 5.25. Revisiting example 5.13 by considering single arrivals, implement an algorithm to analyze this model numerically. Study the effect of the order, k, of Erlang, on selected system performance measures.

Exercise 5.26. Revisiting example 5.14 by considering single arrivals, implement an algorithm to analyze this model numerically. Study the effect of the order, k, of Erlang, on selected system performance measures.

Exercise 5.27. Revisiting example 5.15 by considering single arrivals, implement an algorithm to analyze this model numerically. Study the effect of the order, k, of Erlang, on selected system performance measures.

Exercise 5.28. Revisiting example 5.16 by considering single arrivals, implement an algorithm to analyze this model numerically. Study the effect of the order, k, of Erlang, on selected system performance measures.

Exercise 5.29. Prove the expressions given in equations [5.35] and [5.36].

Exercise 5.30. Look at the $M/M/c$ queue by fixing $\lambda = 1$ and consider the following scenarios by varying $\rho = 0.8, 0.9, 0.95, 0.99$, $c = 1, 2, \cdots, 10$. We take the service rate to be $\mu = \frac{1}{c\rho}$. Compute a few selected measures and plot them as seen in the examples.

Exercise 5.31. Suppose that the customers arrive singly to a three-server homogeneous system in which the inter-arrival times are Erlang of order 5 with parameter 5 in each state, and the service times are Erlang of order 3 with parameter 1.05 in each state. The services are based on an FCFS basis and there is an infinite waiting room. Set up this model as a QBD process and answer the following questions:

a) Compute the mean and the standard deviation of the inter-arrival times.

b) Compute the mean and the standard deviation of the service times of a customer.

c) Compute the measures: $P_{idle}, P^{(r)}_{busy}, P_{busy}, P_{wait}, \mu_q, \sigma_q, \mu'_{BP}$ (Note: These denote the probability that the system is idle, the probability that only server $r, r = 1, 2, 3$ is busy, the probability that all three servers are busy, the probability that an arriving customer has to wait to receive a service, the mean queue length, the standard deviation of the queue length and the mean busy period, respectively).

Exercise 5.32. Referring to exercise 5.31, suppose that the service times are exponential with parameter 0.35. All other parameters are identical to exercise 5.31. Set up this model as a QBD process and answer the following questions:

a) Compute the mean and the standard deviation of the inter-arrival times.

b) Compute the mean and the standard deviation of the service times of a customer.

c) Compute the measures: $P_{idle}, P_{busy}^{(r)}, P_{busy}, P_{wait}, \mu_q, \sigma_q$ and μ'_{BP}.

Exercise 5.33. Referring to exercise 5.31, suppose that the service times are hyperexponential with parameters 2.555, 0.2555 and 0.02555, with mixing probabilities 0.8, 0.15 and 0.05, respectively. All other parameters are identical to exercise 5.31. Set up this model as a QBD process and answer the following questions:

a) Compute the mean and the standard deviation of the inter-arrival times.

b) Compute the mean and the standard deviation of the service times of a customer.

c) Compute the measures: $P_{idle}, P_{busy}^{(r)}, P_{busy}, P_{wait}, \mu_q, \sigma_q$ and μ'_{BP}.

Exercise 5.34. Suppose that the customers arrive singly according to a Poisson process to a three-server homogeneous system to receive services on an FCFS basis. Assuming that there is an infinite waiting room, set up this model as a QBD process and answer the following questions:

a) Compute the mean and the standard deviation of the inter-arrival times.

b) Compute the mean and the standard deviation of the service times of a customer.

c) Compute the measures: $P_{idle}, P_{busy}^{(r)}, P_{busy}, P_{wait}, \mu_q, \sigma_q$ and μ'_{BP}.

Exercise 5.35. Referring to exercise 5.34, suppose that the customers arrive singly according to a renewal process with parameters 2.18, 1.09, 0.545, 0.2725 and 0.13625 with mixing probabilities 0.5, 0.30, 0.15, 0.04 and 0.01, respectively. All other parameters and assumptions are identical as in exercise 5.35. Set up this model as a QBD process and answer the following questions:

a) Compute the mean and the standard deviation of the inter-arrival times.

b) For the three scenarios mentioned in exercise 5.34, compute the measures: $P_{idle}, P_{busy}^{(r)}, P_{busy}, P_{wait}, \mu_q, \sigma_q$ and μ'_{BP}.

Exercise 5.36. Referring to exercise 5.34, suppose that the customers arrive singly according to a MAP with representation given by:

$$D_0 = \begin{pmatrix} -1.75 & 1.75 & 0 & 0 \\ 0 & -1.75 & 1.75 & 0 \\ 0 & 0 & -1.75 & 0 \\ 0 & 0 & 0 & -3.5 \end{pmatrix}, \quad D_1 = \begin{pmatrix} 0 & 0 & 0 & 0 \\ 0 & 0 & 0 & 0 \\ 0.0175 & 0 & 0 & 1.7325 \\ 3.465 & 0 & 0 & 0.035 \end{pmatrix}.$$

All other parameters and assumptions are identical as in exercise 5.35. Set up this model as a QBD process and answer the following questions:

a) Compute the mean and the standard deviation of the inter-arrival times.

b) For the three scenarios mentioned in exercise 5.34, compute the measures: $P_{idle}, P_{busy}^{(r)}, P_{busy}, P_{wait}, \mu_q, \sigma_q$ and μ'_{BP}.

Exercise 5.37. Referring to exercise 5.34, suppose that the customers arrive singly according to a MAP with representation given by:

$$D_0 = \begin{pmatrix} -1.75 & 1.75 & 0 & 0 \\ 0 & -1.75 & 1.75 & 0 \\ 0 & 0 & -1.75 & 0 \\ 0 & 0 & 0 & -3.5 \end{pmatrix}, \quad D_1 = \begin{pmatrix} 0 & 0 & 0 & 0 \\ 0 & 0 & 0 & 0 \\ 1.7325 & 0 & 0 & 0.0175 \\ 0.035 & 0 & 0 & 3.465 \end{pmatrix}.$$

All other parameters and assumptions are identical as in exercise 5.35. Set up this model as a QBD process and answer the following questions:

a) Compute the mean and the standard deviation of the inter-arrival times.

b) For the three scenarios mentioned in exercise 5.34, compute the measures: $P_{idle}, P_{busy}^{(r)}, P_{busy}, P_{wait}, \mu_q, \sigma_q$ and μ'_{BP}.

Exercise 5.38. Suppose that we consider a three-server heterogeneous system in which the customers arrive according to a renewal process that has Erlang of order 5 with parameter 5 in each state for its inter-renewal times. Server 1 (S_1) offers services that follow an Erlang (E) of order 3 with parameter 1.5 in each state; server 2 (S_2) offers services that follow an exponential (X) distribution with parameter 0.35; and server 3 (S_3) offers services that follow an hyperexponential (H) distribution with parameters 1.46, 0.146 and 0.0146 with mixing probabilities 0.8, 0.15 and 0.05, respectively. The services are offered on an FCFS basis and there is an infinite waiting room. Analyze this model and answer the following questions:

a) Compute the mean and the standard deviation of the inter-arrival times.

b) Under the six scenarios (each corresponding to a heterogeneous system obtained through a permutation of the three service time distribution), compute the measures: $P_{idle}, P_{busy}^{(r)}, P_{busy}, P_{wait}, \mu_q, \sigma_q$ and μ'_{BP}.

Exercise 5.39. Referring to example 5.38, suppose that the arrivals occur according to a Poisson process with rate 1. All other parameters remain the same as in exercise 5.38. Analyze this model and compute $P_{idle}, P_{busy}^{(r)}, P_{busy}, P_{wait}, \mu_q, \sigma_q$ and μ'_{BP}.

Exercise 5.40. Referring to example 5.38, suppose that the inter-arrival times are modeled using a hyperexponential distribution with parameters 2.18, 1.09, 0.545, 0.2725 and 0.13625 with mixing probabilities 0.5, 0.30, 0.15, 0.04 and 0.01, respectively. All other parameters remain the same as in exercise 5.38. Analyze this model and compute $P_{idle}, P^{(r)}_{busy}, P_{busy}, P_{wait}, \mu_q, \sigma_q$ and μ'_{BP}.

Exercise 5.41. Referring to example 5.38, suppose that the customers arrive according to a MAP with representation (D_0, D_1) where:

$$D_0 = \begin{pmatrix} -1.75 & 1.75 & 0 & 0 \\ 0 & -1.75 & 1.75 & 0 \\ 0 & 0 & -1.75 & 0 \\ 0 & 0 & 0 & -3.5 \end{pmatrix}, \quad D_1 = \begin{pmatrix} 0 & 0 & 0 & 0 \\ 0 & 0 & 0 & 0 \\ 0.0175 & 0 & 0 & 1.7325 \\ 3.465 & 0 & 0 & 0.035 \end{pmatrix}.$$

All other parameters and assumptions are identical as in exercise 5.38. Analyze this model and compute $P_{idle}, P^{(r)}_{busy}, P_{busy}, P_{wait}, \mu_q, \sigma_q$ and μ'_{BP}.

Exercise 5.42. Referring to example 5.38, suppose that the customers arrive according to a MAP with representation (D_0, D_1) where:

$$D_0 = \begin{pmatrix} -1.75 & 1.75 & 0 & 0 \\ 0 & -1.75 & 1.75 & 0 \\ 0 & 0 & -1.75 & 0 \\ 0 & 0 & 0 & -3.5 \end{pmatrix}, \quad D_1 = \begin{pmatrix} 0 & 0 & 0 & 0 \\ 0 & 0 & 0 & 0 \\ 1.7325 & 0 & 0 & 0.0175 \\ 0.035 & 0 & 0 & 3.465 \end{pmatrix}.$$

All other parameters and assumptions are identical as in exercise 5.38. Analyze this model and compute $P_{idle}, P^{(r)}_{busy}, P_{busy}, P_{wait}, \mu_q, \sigma_q$ and μ'_{BP}.

Note: For exercises 5.43 through 5.47, we use the following notations: P_{idle} is for the system being idle; μ_{NS} denotes the mean number of customers in the system; μ_{busy} denotes the mean number of busy servers; P_{busy} is the probability that all servers are busy; P_{empty} denotes the probability that the queue is empty; and P_{wait} is for the probability that an arriving customer has to wait to get service. Furthermore, UFB, CTB, ABB, POB and GMB denote the batch sizes that are uniform on $\{1, 2, 3, 4\}$, constant batch size with a value of 4, having an arbitrary probability distribution with mass at 1, 2, 3 and 4 with probabilities, respectively, given by 0.5, 0.3, 0.15 and 0.05, Poisson distribution with parameter 1.25 and geometric distribution with parameter 0.8, respectively. The arrival rate of the groups will be denoted by λ_g which will depend on the type of batch distribution so that the arrival rate, λ, of the customers will be set at 1.0.

Exercise 5.43. Suppose that the customers arrive according to a BMAP with inter-arrival times of the groups following an Erlang of order 5 with parameter $5\lambda_g$ such that the rate of (group) arrivals is λ_g. Consider the above-mentioned five scenarios for the batch size distribution. There are c homogeneous servers offering services on an FCFS basis and the service rate of each server is $1/(0.95\,c)$ so that the traffic intensity is 0.95. The service times are exponential. Analyze this model and compute λ_g, P_{idle}, μ_{NS}, μ_{busy}, P_{busy}, P_{empty} and P_{wait}, under these scenarios by taking $c = 2, 5$, and 10.

Exercise 5.44. Referring to exercise 5.43, suppose that the customers arrive in groups and that the inter-arrival times of the groups are exponentially distributed with parameter λ_g. All other parameters are identical to exercise 5.43. Analyze the model and compute λ_g, P_{idle}, μ_{NS}, μ_{busy}, P_{busy}, P_{empty} and P_{wait}, under these scenarios by taking $c = 2, 5$, and 10.

Exercise 5.45. Referring to exercise 5.43, suppose that the customers arrive in groups and that the inter-arrival times of the groups are hyperexponential distributed with parameters $2.18\lambda_g$, $1.09\lambda_g$, $0.545\lambda_g$, $0.2725\lambda_g$ and $0.13625\lambda_g$ with mixing probabilities 0.5, 0.30, 0.15, 0.04 and 0.01, respectively. All other parameters are identical to exercise 5.43. Analyze the model and compute λ_g, P_{idle}, μ_{NS}, μ_{busy}, P_{busy}, P_{empty} and P_{wait}, under these scenarios by taking $c = 2, 5$, and 10.

Exercise 5.46. Referring to exercise 5.43, suppose that the customers arrive according to a BMAP with representation:

$$D_0 = \lambda_g \begin{pmatrix} -1.75 & 1.75 & 0 & 0 \\ 0 & -1.75 & 1.75 & 0 \\ 0 & 0 & -1.75 & 0 \\ 0 & 0 & 0 & -3.5 \end{pmatrix},$$

$$D_k = \lambda_g\, a_k \begin{pmatrix} 0 & 0\,0 & 0 \\ 0 & 0\,0 & 0 \\ 0.0175 & 0\,0 & 1.7325 \\ 3.465 & 0\,0 & 0.035 \end{pmatrix}, k \geq 1,$$

and $\{a_k\}$ has one of the five batch size distributions as indicated earlier. All other parameters are identical to exercise 5.43. Analyze the model and compute λ_g, P_{idle}, μ_{NS}, μ_{busy}, P_{busy}, P_{empty} and P_{wait}, under these scenarios by taking $c = 2, 5$, and 10.

Exercise 5.47. Referring to exercise 5.43, suppose that the customers arrive according to a BMAP with representation:

$$D_0 = \lambda_g \begin{pmatrix} -1.75 & 1.75 & 0 & 0 \\ 0 & -1.75 & 1.75 & 0 \\ 0 & 0 & -1.75 & 0 \\ 0 & 0 & 0 & -3.5 \end{pmatrix},$$

$$D_k = \lambda_g\, a_k \begin{pmatrix} 0 & 0\,0 & 0 \\ 0 & 0\,0 & 0 \\ 1.7325 & 0\,0 & 0.0175 \\ 0.035 & 0\,0 & 3.465 \end{pmatrix}, k \geq 1,$$

and $\{a_k\}$ has one of the five batch size distributions as indicated earlier. All other parameters are identical to exercise 5.43. Analyze the model and compute λ_g, P_{idle},

μ_{NS}, μ_{busy}, P_{busy}, P_{empty} and P_{wait}, under these scenarios by taking $c = 2, 5$, and 10.

Exercise 5.48. Suppose that the customers arrive singly to a c-server queueing system. The homogeneous servers offer services on an FCFS basis. The inter-arrival times are Erlang of order 5 with parameter 5 in each state. The service times are exponential with parameter $1/0.95\,c$ so that the traffic intensity is 0.95. Analyze the model and compute P_{idle}, μ_{NS}, μ_{busy}, P_{busy}, P_{empty} and P_{wait} under various scenarios by varying $c = 2, 5, 10, 20$, and 50.

Exercise 5.49. Suppose that the customers arrive singly to a c-server queueing system. The homogeneous servers offer services on an FCFS basis. The inter-arrival times are exponential parameter 1. The service times are exponential with parameter $1/0.95\,c$ so that the traffic intensity is 0.95. Analyze the model and compute P_{idle}, μ_{NS}, μ_{busy}, P_{busy}, P_{empty} and P_{wait} under various scenarios by varying $c = 2, 5, 10, 20$, and 50.

Exercise 5.50. Suppose that the customers arrive singly to a c-server queueing system. The homogeneous servers offer services on an FCFS basis. The inter-arrival times are hyperexponential with parameters 2.18, 1.09, 0.545, 0.2725 and 0.13625 mixing probabilities 0.5, 0.30, 0.15, 0.04 and 0.01, respectively. The service times are exponential with parameter $1/0.95\,c$ so that the traffic intensity is 0.95. Analyze the model and compute P_{idle}, μ_{NS}, μ_{busy}, P_{busy}, P_{empty} and P_{wait} under various scenarios by varying $c = 2, 5, 10, 20$, and 50.

Exercise 5.51. Suppose that the customers arrive singly to a c-server queueing system according to a MAP with representation:

$$D_0 = \begin{pmatrix} -1.75 & 1.75 & 0 & 0 \\ 0 & -1.75 & 1.75 & 0 \\ 0 & 0 & -1.75 & 0 \\ 0 & 0 & 0 & -3.5 \end{pmatrix}, \quad D_1 = \begin{pmatrix} 0 & 0 & 0 & 0 \\ 0 & 0 & 0 & 0 \\ 0.0175 & 0 & 0 & 1.7325 \\ 3.465 & 0 & 0 & 0.035 \end{pmatrix}.$$

The homogeneous servers offer services on an FCFS basis. The service times are exponential with parameter $1/0.95\,c$ so that the traffic intensity is 0.95. Analyze the model and compute P_{idle}, μ_{NS}, μ_{busy}, P_{busy}, P_{empty} and P_{wait} under various scenarios by varying $c = 2, 5, 10, 20$, and 50.

Exercise 5.52. Suppose that the customers arrive singly to a c-server queueing system according to a MAP with representation:

$$D_0 = \begin{pmatrix} -1.75 & 1.75 & 0 & 0 \\ 0 & -1.75 & 1.75 & 0 \\ 0 & 0 & -1.75 & 0 \\ 0 & 0 & 0 & -3.5 \end{pmatrix}, \quad D_1 = \begin{pmatrix} 0 & 0 & 0 & 0 \\ 0 & 0 & 0 & 0 \\ 1.7325 & 0 & 0 & 0.0175 \\ 0.035 & 0 & 0 & 3.465 \end{pmatrix}.$$

The homogeneous servers offer services on an FCFS basis. The service times are exponential with parameter $1/0.95\,c$ so that the traffic intensity is 0.95. Analyze the model and compute P_{idle}, μ_{NS}, μ_{busy}, P_{busy}, P_{empty} and P_{wait} under various scenarios by varying $c = 2, 5, 10, 20$ and 50.

Exercise 5.53. For the $MAP/M/c$ queueing model, derive an expression for the mean busy period and the mean number of customers served during a busy period. [Hint: Use equation [4.58] to get started on this.]

Exercise 5.54. Consider a three-server heterogeneous queueing system in which customers arrive according to a BMAP. Assume that the service times are of phase type. An arriving customer finding at least one idle server will pick one at random to enter into service. If all servers are busy, the arriving customer will enter into a buffer of infinite capacity. Set up this model by looking at it in arbitrary time. Develop a code to find the steady-state probability vector and other system performance measures.

Exercise 5.55. In section 5.2, a heterogeneous $MAP/PH/3$ queueing model is studied in which an arriving customer finding at least one free server will get service from the lowest numbered server. Suppose that a server is assigned randomly among the free servers to an arriving customer. How does this assumption change the generator and hence the system performance measures? Investigate this by implementing a code and use the same set of examples presented in that section.

6

Finite-Capacity Queues

Finite-capacity queues are very common in practice due to limited resources that all service providers face in real life. This results in the waiting space (whether it is real or virtual) being finite. In this chapter, the focus is on the finite-capacity queueing systems. Normally, the finite capacity systems are difficult to analyze mainly due to having two sets of boundary conditions (one near 0 and the other near the maximum size of the queue). Hence, one has to exploit the special structure of the matrices appearing in the generator. This is an integral part of matrix-analytic methods as envisioned and taught by Neuts.

This chapter is organized as follows. We split the study into two parts. The first one deals with the single-server case and is presented in section 6.1. The second part dealing with the multi-server case is considered in section 6.2. Within these sections different types of finite-capacity queueing models are studied.

We start with the most general setup for the queueing models and subsequently look at special cases. Some of these special cases might allow explicit solutions. The illustrative numerical examples will further assess the qualitative nature of the models.

6.1. Finite-capacity queues with single server

Here, we assume that the system has a finite capacity with only one server available for serving the customers and consider different scenarios for the arrivals and the services.

6.1.1. BMAP/PH/1/K *queue*

Earlier we studied an infinite capacity $BMAP/PH/1$ queue and in this section we assume that the system has a finite capacity leading to the possibility of the loss of customers at arrival epochs. We briefly describe the model and set up the generator of the Markov process governing the system. Using the same notation set earlier for $BMAP/PH/1$ queue, we analyze the model.

Model assumptions: The arrival process representation is given by $\{D_k\}$ of dimension m. Assume the underlying generator, $\tilde{Q} = \sum_{k=0}^{\infty} D_k$, is irreducible. The service times are of phase type with representation $(\boldsymbol{\beta}, S)$ of order n.

The system capacity is given by $K, 1 \leq K < \infty$. That is, at any given time the maximum number of customers that can be present in the system is K with one in service and $K - 1$ waiting in the queue.

An arriving batch will be admitted fully or partially depending on the size of the batch and the available space in the system. While we assume in this model that a batch can be admitted partially, it is easy to modify considering only a full batch admission depending on the application. For example, in telecommunications, generally, it is not a good idea to transmit only a partial message which arrives in packets of size one or more. Thus, in this application admitting only full batches should be considered.

The vector, $\boldsymbol{\eta}$, is the invariant vector of $\tilde{Q}$. That is:

$$\boldsymbol{\eta}\tilde{Q} = \mathbf{0}, \quad \boldsymbol{\eta}\, \boldsymbol{e} = 1.$$

The arrival rate, λ_g, of the batches, the arrival rate, λ, of the customers per unit of time and the service rate, μ, are given by:

$$\lambda_g = \boldsymbol{\eta}(-D_0)\boldsymbol{e}, \quad \lambda = \boldsymbol{\eta} \sum_{k=1}^{\infty} k D_k \boldsymbol{e}, \quad \mu = [-\boldsymbol{\beta}(-S)^{-1}\boldsymbol{e}]^{-1}.$$

For use in the following, we define:

$$\tilde{D}_k = \sum_{r=k}^{\infty} D_r, \quad k \geq 1.$$

The process $\{(N(t), J_1(t), J_2(t))\} : t \geq 0$, where $N(t), J_1(t)$, and $J_2(t)$ are the number of customers in the system, the phase of the service process (if the server is busy) and the phase of the arrival process at time t, respectively, is a Markov process on the state space $\Omega = \{(0,k) : 1 \le k \le m\} \bigcup \{(i,j,k) : 1 \le j \le n,\ 1 \le k \le m,\ 1 \le i \le K\}$. The generator Q is given by:

$$Q = \begin{pmatrix} D_0 & \boldsymbol{\beta} \otimes D_1 & \boldsymbol{\beta} \otimes D_2 & \boldsymbol{\beta} \otimes D_3 & \cdots & \boldsymbol{\beta} \otimes D_{K-1} & \boldsymbol{\beta} \otimes \tilde{D}_K \\ \mathbf{S^0} \otimes I & S \oplus D_0 & I \otimes D_1 & I \otimes D_2 & \cdots & I \otimes D_{K-2} & I \otimes \tilde{D}_{K-1} \\ 0 & \mathbf{S^0}\,\boldsymbol{\beta} \otimes I & S \oplus D_0 & I \otimes D_1 & \cdots & I \otimes D_{K-3} & I \otimes \tilde{D}_{K-2} \\ \vdots & \vdots & \vdots & \vdots & \cdots & \vdots & \vdots \\ 0 & 0 & 0 & 0 & \cdots & S \oplus D_0 & I \otimes \tilde{D}_1 \\ 0 & 0 & 0 & 0 & \cdots & \mathbf{S^0}\,\boldsymbol{\beta} \otimes I & S \oplus \tilde{Q} \end{pmatrix},$$

Let $\boldsymbol{y} = (\boldsymbol{y}_0, \boldsymbol{y}_1, \cdots, \boldsymbol{y}_K)$ denote the steady-state probability vector of Q. Note that the vector $\boldsymbol{y}_0$ of dimension m gives the steady-state probability vector that the system is idle with the arrival process in one of m states; the vector $\boldsymbol{y}_i$, $1 \le i \le K$, of dimension mn denotes the steady-state probability vector that the system has i customers in the system with the arrival and the service processes in various states.

As pointed out earlier, if the software and the computer can handle sufficiently large m and n, one can solve the system of linear equations directly; otherwise, one has to exploit the structure of the coefficient matrices. The steps are similar to the ones described in section 5.2. Without exploiting the structure of the coefficient matrices and noting that the matrices $S \oplus D_0$ and $S \oplus \tilde{Q}$ are non-singular, the steady-state probability vector $\boldsymbol{y}$ can be solved using, say, the (block) Gauss-Seidel method, the following equations:

$$\boldsymbol{y}_0 = \boldsymbol{y}_1(\mathbf{S^0} \otimes I)(-D_0))^{-1},$$

$$\boldsymbol{y}_i = \left[\boldsymbol{y}_0(\boldsymbol{\beta} \otimes D_i) + \sum_{j=1}^{i-1} \boldsymbol{y}_j(I \otimes D_{i-j}) + \boldsymbol{y}_{i+1}(\mathbf{S^0}\,\boldsymbol{\beta} \otimes I)\right]\Big(- (S \oplus D_0)^{-1}\Big),$$
$1 \le i \le K-1$,

$$\boldsymbol{y}_K = \left[\boldsymbol{y}_0(\boldsymbol{\beta} \otimes \tilde{D}_K) + \sum_{j=1}^{K-1} \boldsymbol{y}_j(I \otimes \tilde{D}_{K-j})\right]\Big(- (S \oplus \tilde{Q})^{-1}\Big),$$

and the normalizing condition $\boldsymbol{y}_0\boldsymbol{e} + \sum_{j=1}^{K} \boldsymbol{y}_j\boldsymbol{e} = 1$.

In order to study the queueing model under study, we need to have one or more system performance measures. We list a few along with their formulae below:

1) The probability that the server (and hence the system) is idle is obtained as:

$\mathrm{P(idle)} = \boldsymbol{y}_0\boldsymbol{e}$.

2) The probability that the server is busy is obtained as:

$$P(\text{busy}) = 1 - \boldsymbol{y}_0\boldsymbol{e}.$$

3) The probability mass function (PMF), $\{z_i\}$, of the number of customers seen immediately after an arrival (of a batch) is given by:

$$z_i = \begin{cases} \dfrac{1}{\lambda_g}\Big[\boldsymbol{y}_0 D_i \boldsymbol{e} + \sum\limits_{j=1}^{i-1} \boldsymbol{y}_j D_{i-j}\boldsymbol{e}\Big], & 1 \le i \le K, \\ \\ \dfrac{1}{\lambda_g}\Big[\boldsymbol{y}_0 D_i \boldsymbol{e} + \sum\limits_{j=1}^{K} \boldsymbol{y}_j D_{i-j}\boldsymbol{e}\Big], & i \ge K+1. \end{cases}$$

Note that when the batch size has a maximum limit, say, k^*, the second summation in the above equation will be written as $K+1 \le i \le K+k^*$. Furthermore, of those customers seen immediately after an arrival (of a batch) some will be lost if the buffer is filled up with some of the customers in the same batch.

4) The probability of a loss (if an arriving batch has one or more customers lost) is given by $P(loss) = \sum\limits_{i=1}^{\infty} z_{K+i}$. As indicated earlier, if the batch size is restricted to not exceed k^*, then the loss probability will be $\sum\limits_{i=1}^{k^*} z_{K+i}$.

5) With the PMF for the number of customers in the system at an arbitrary time and the number of customers lost at an arrival epoch, we can find the mean, the median, the mode, the standard deviation and other measures.

6) Based on the PMF of the number of customers lost after an arrival of a batch, one can compute the conditional mean number, say, μ_{CL}, of lost customers given that a loss occurs.

EXAMPLE 6.1.– In this example, we revisit example 1.9 by putting a restriction that the buffer space is of finite size given by $K = 20$. Thus, at most 20 customers may be present in the system at any time. In Figure 6.1, we display the plots of the conditional PMF of the number of lost customers at arrivals given that at least one customer is lost, under various scenarios.

Note that the standard deviations of these five arrival processes labeled ERA, EXA, HEA, NCA and PCA are, respectively, given by 1.29904, 2.25, 4.5, 2.33827 and 2.33827. We see an interesting pattern in the plot. When comparing the renewal arrivals, there appears to be a cut-off point, say, k^*, such that the loss probability appears to behave one way when the number of losses increases from 1 to k^* and then the trend reverses from $k^* + 1$ onward. The same pattern is also seen when comparing the correlated arrivals. This not only shows the impact of the variability but also the 1-lag correlation coefficient on this measure.

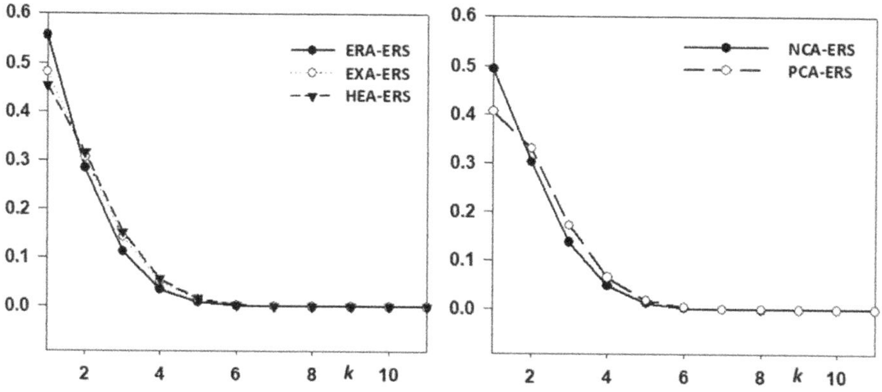

Figure 6.1. *Conditional PMF of the number of lost customers for example 6.1*

In Table 6.1, we display some selected measures under various scenarios. As is to be expected, all the measures appear to increase with increasing variability in the inter-arrival times when looking at the renewal arrivals. In the case of correlated arrivals, we see that the positive correlated arrivals incur a higher value for all but L_s. Note that these two correlated processes have the same standard deviation (for the inter-arrival times). This shows the role of the correlation present in the inter-arrival times.

ToA	P_{idle}	P_{loss}	μ_{CL}	L_s
ERA	0.10919	0.01378	1.66749	6.16834
EXA	0.13378	0.04632	1.82336	7.32912
HEA	0.20010	0.13310	1.88018	8.60951
NCA	0.12390	0.03327	1.79595	7.10990
PCA	0.32361	0.28229	1.98031	5.44991

Table 6.1. *Selected performance measures for example 6.1*

EXAMPLE 6.2.– In this example, we revisit example 1.10 by putting a restriction that the buffer space is of finite size given by $K = 20$. Thus, at most 20 customers may be present in the system at any time. In Figure 6.2, we display the plots of the conditional PMF of the number of lost customers at arrivals given that at least one customer is lost, under various scenarios.

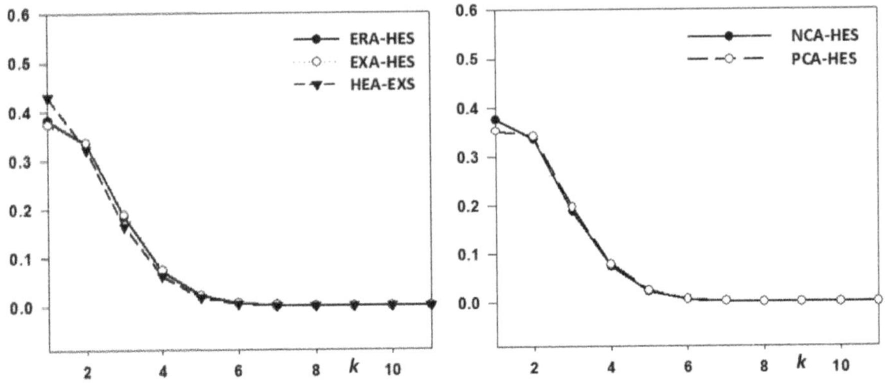

Figure 6.2. *Conditional PMF of the number of lost customers for example 6.2*

ToA	P_{idle}	P_{loss}	μ_{CL}	L_s
ERA	0.17271	0.08903	2.04171	7.76508
EXA	0.19422	0.11439	2.05929	7.91943
HEA	0.24820	0.17858	2.07479	8.25746
NCA	0.18648	0.10518	2.05562	7.90596
PCA	0.33493	0.27908	2.10447	6.23645

Table 6.2. *Selected performance measures for example 6.2*

Similar to the previous example, the figure and the table reveal interesting patterns and results.

6.1.2. MAP/PH/1/K *queue*

This is a special case of section 6.1.1 and corresponds to the case of single arrivals by taking $D_k = 0, k \geq 2$ in the batch Markovian arrival process (BMAP) representation. Thus, the generator is a (finite) quasi-birth-and-death (QBD) process and is given by:

$$Q = \begin{pmatrix} D_0 & \boldsymbol{\beta} \otimes D_1 & 0 & 0 & \cdots & 0 & 0 \\ \boldsymbol{S^0} \otimes I & S \oplus D_0 & I \otimes D_1 & 0 & \cdots & 0 & 0 \\ 0 & \boldsymbol{S^0}\, \boldsymbol{\beta} \otimes I & S \oplus D_0 & I \otimes D_1 & \cdots & 0 & 0 \\ \vdots & \vdots & \vdots & \vdots & \cdots & \vdots & \vdots \\ 0 & 0 & 0 & 0 & \cdots & S \oplus D_0 & I \otimes D_1 \\ 0 & 0 & 0 & 0 & \cdots & \boldsymbol{S^0}\, \boldsymbol{\beta} \otimes I & S \oplus \tilde{Q} \end{pmatrix}. \quad [6.1]$$

The steady-state vector of Q can be solved in a straightforward way. The waiting time (in the system) distribution of an admitted customer is of continuous phase type (CPH) with representation (γ, L) of order Kn, where:

$$\gamma = c\Big(\boldsymbol{y}_0 D_1 \boldsymbol{e}\boldsymbol{\beta}, \boldsymbol{y}_1(I \otimes D_1\boldsymbol{e}), \cdots, \boldsymbol{y}_{K-1}(I \otimes D_1\boldsymbol{e})\Big),$$

$$L = \begin{pmatrix} S & & & & & \\ \boldsymbol{S}^0\boldsymbol{\beta} & S & & & & \\ & \boldsymbol{S}^0\boldsymbol{\beta} & S & & & \\ & & \boldsymbol{S}^0\boldsymbol{\beta} & S & & \\ & & & \ddots & \ddots & \\ & & & & \boldsymbol{S}^0\boldsymbol{\beta} & S \end{pmatrix},$$

where c is the normalizing constant and is given by:

$$c = \lambda\Big[\lambda - \boldsymbol{y}_K D_1 \boldsymbol{e}\Big]^{-1}.$$

The mean waiting time in the system, μ'_{WTS}, is verified to be:

$$\mu'_{WTS} = \frac{\rho}{\lambda - \boldsymbol{y}_K D_1 \boldsymbol{e}}\left[\boldsymbol{y}_0 D_1 \boldsymbol{e} + \sum_{k=1}^{K-1} \boldsymbol{y}_k\big((-S)^{-1}\boldsymbol{e} \otimes \boldsymbol{e}\big)\right]. \qquad [6.2]$$

The system performance measures listed in section 6.1.1 can be similarly defined for this model and are left as an exercise to derive them.

EXAMPLE 6.3.– In this example, we revisit example 1.16 by putting in a restriction that the buffer space is of finite size given by $K = 20$. Thus, at most 20 customers may be present at any time. Furthermore, we look at two scenarios in which two service rates are considered. In the first one, we have $\mu = \frac{1}{0.90}$, and in the second one $\mu = \frac{1}{0.99}$. Note that the queueing model under consideration is always stable. However, we are looking at the cases where the service rates are larger than the arrival rate so as to compare with the corresponding infinite server queueing models.

In Table 6.3, we display some selected measures, and in Figures 6.3 and 6.4, respectively, we display the plots of the steady-state probability vectors of the number in system for the cases $\mu = \frac{1}{0.90}$ and $\mu = \frac{1}{0.99}$.

ToA	$\mu^{-1} = 0.90$				$\mu^{-1} = 0.99$			
	P_{idle}	P_{loss}	L_s	σ_s	P_{idle}	P_{loss}	L_s	σ_s
NC_1	0.1072	0.0080	5.9103	5.0505	0.0441	0.0345	9.6175	5.9362
NC_2	0.1051	0.0056	5.5915	4.8249	0.0400	0.0303	9.5977	5.8793
NC_3	0.1040	0.0044	5.3777	4.6662	0.0375	0.0278	9.5807	5.8430
NC_4	0.1033	0.0037	5.2251	4.5491	0.0359	0.0261	9.5666	5.8179
NC_5	0.1029	0.0032	5.1110	4.4594	0.0347	0.0250	9.5550	5.7996
NC_6	0.1026	0.0028	5.0225	4.3885	0.0339	0.0241	9.5454	5.7856
NC_7	0.1023	0.0026	4.9520	4.3312	0.0332	0.0234	9.5373	5.7746
NC_8	0.1022	0.0024	4.8945	4.2839	0.0327	0.0229	9.5304	5.7657
PC_1	0.2906	0.2118	5.5758	7.3133	0.2418	0.2341	6.1610	7.4137
PC_2	0.3427	0.2697	4.5766	6.8943	0.2933	0.2862	5.0412	7.0107
PC_3	0.3723	0.3025	3.9928	6.5213	0.3226	0.3157	4.4111	6.6554
PC_4	0.3913	0.3236	3.6137	6.2241	0.3413	0.3347	4.0078	6.3733
PC_5	0.4045	0.3383	3.3484	5.9875	0.3544	0.3479	3.7278	6.1493
PC_6	0.4142	0.3491	3.1525	5.7960	0.3641	0.3577	3.5220	5.9685
PC_7	0.4217	0.3574	3.0020	5.6383	0.3715	0.3651	3.3644	5.8198
PC_8	0.4275	0.3639	2.8828	5.5063	0.3773	0.3710	3.2398	5.6957

Table 6.3. *Selected performance measures for example 6.3*

It is clear from the table and from the figures the role played by the correlation, especially the positive ones.

EXAMPLE 6.4.– In this example, we look at various scenarios by considering Erlang of orders 2 through 9 as well as hyperexponentials labeled HE_2 through HE_9 (see example 5.14) for the inter-arrival times. We consider the same service time as in example 1.16, which is a CPH obtained as the minimum of three distributions: exponential with parameter 1, Erlang of order 2 with parameter 2 and an hyperexponential distribution with parameters 10 and 1 with mixing probabilities 0.9 and 0.1. In Figures 6.5 and 6.6, respectively, we display the plots of the steady-state probability vectors of the number in system for the cases $K = 20$, $\mu = \frac{1}{0.90}$ and $\mu = \frac{1}{0.99}$. Selected performance measures are displayed in Table 6.4.

EXAMPLE 6.5.– Referring to example 6.3, plot the (cumulative) distribution function (CDF) of the waiting time in the system of an admitted customer by taking $\mu = \frac{1}{0.95}$. Also, compile some key statistics related to this case. First, note that the waiting time is of CPH as pointed out earlier. Using the form of the CPH, and applying the technique for obtaining the CDF numerically seen in section 4.1 of Volume 1, the plots of the CDF of the waiting time are shown in Figure 6.7. In Table 6.5, selected performance measures are displayed.

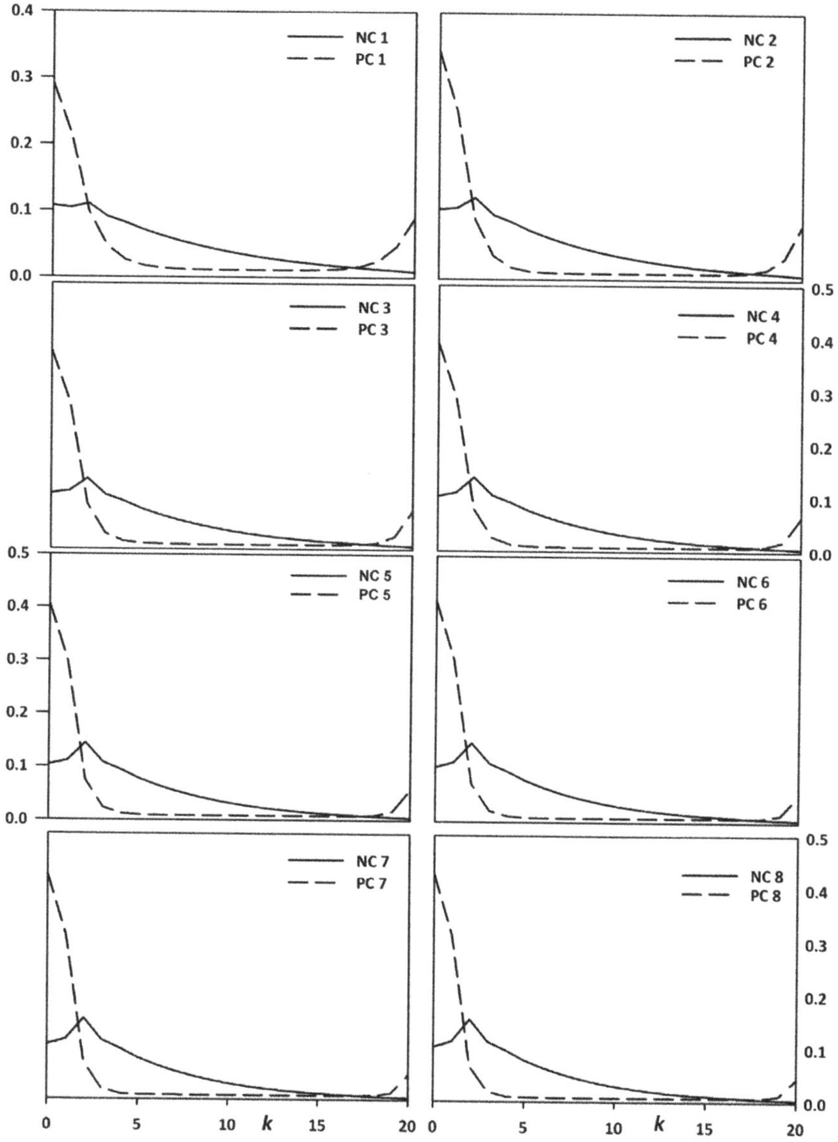

Figure 6.3. *PMF of the number in system when $\mu = \frac{1}{0.90}$ for example 6.3*

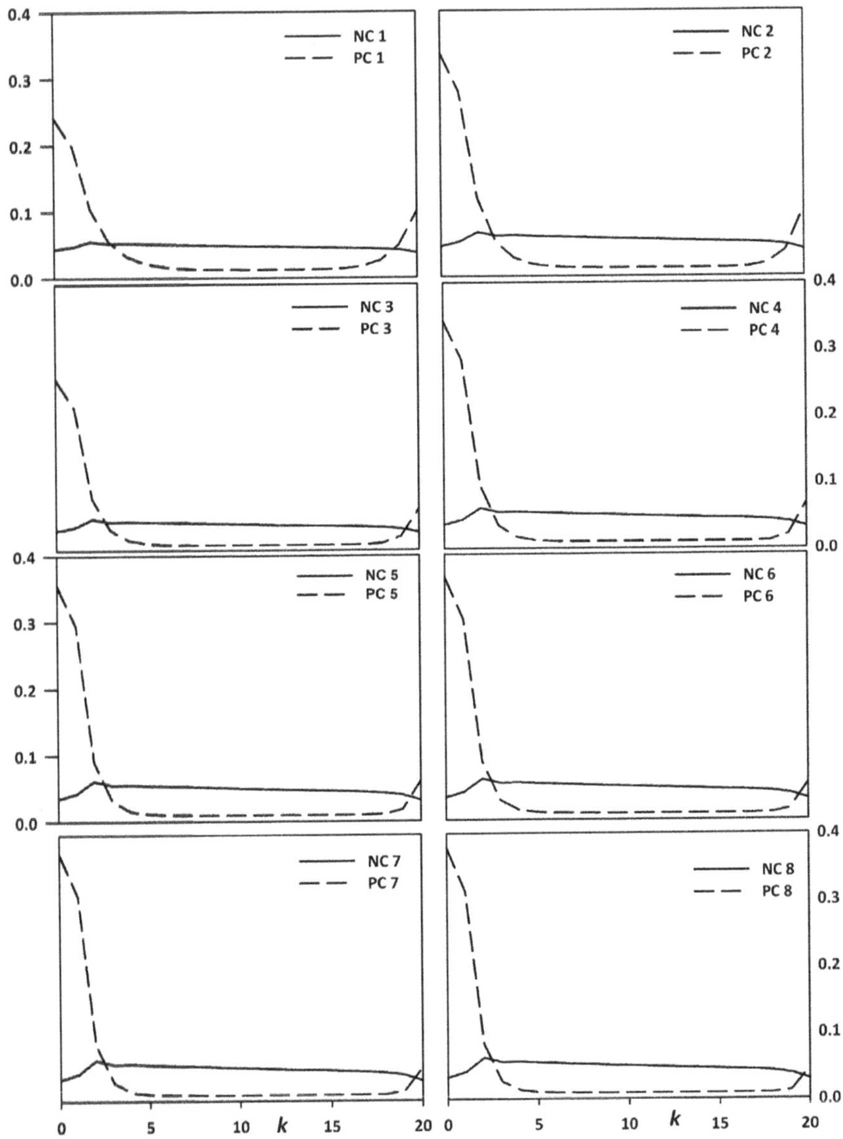

Figure 6.4. *PMF of the number in system when $\mu = \frac{1}{0.99}$ for example 6.3*

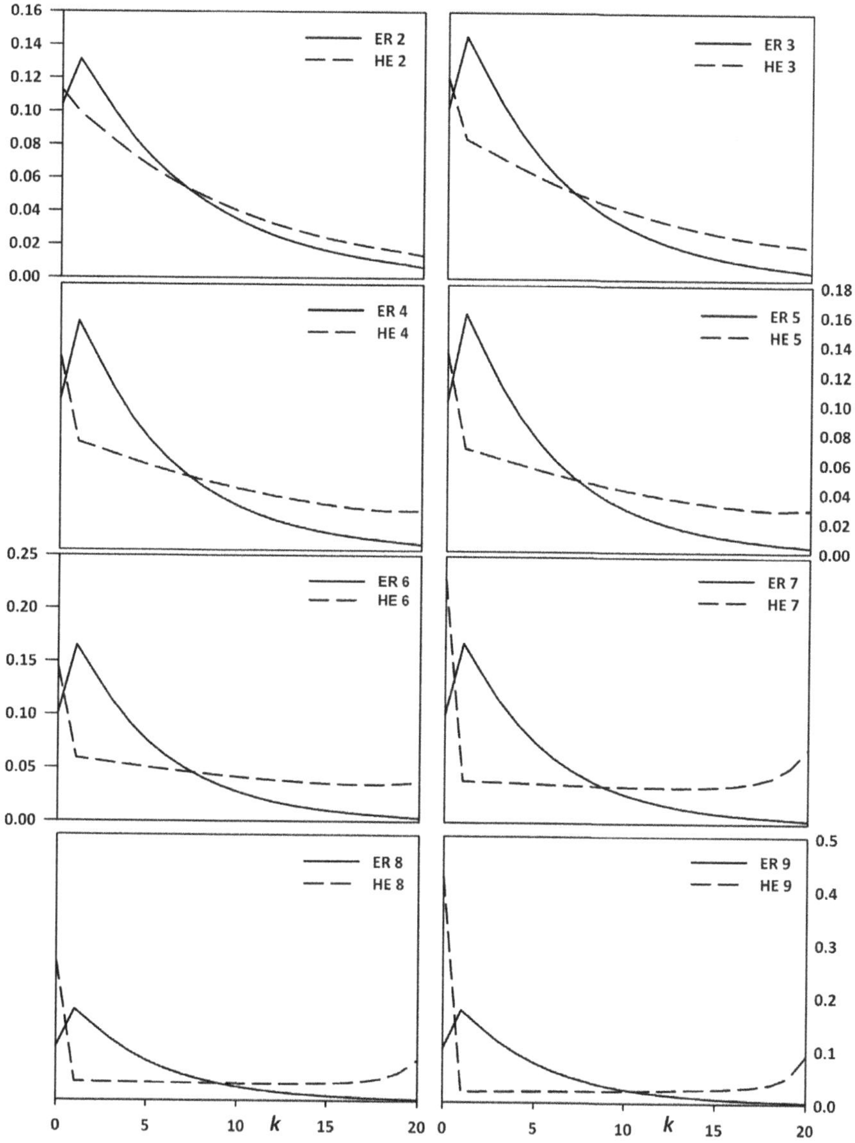

Figure 6.5. *PMF of the number in system when $\mu = \frac{1}{0.90}$ for example 6.4*

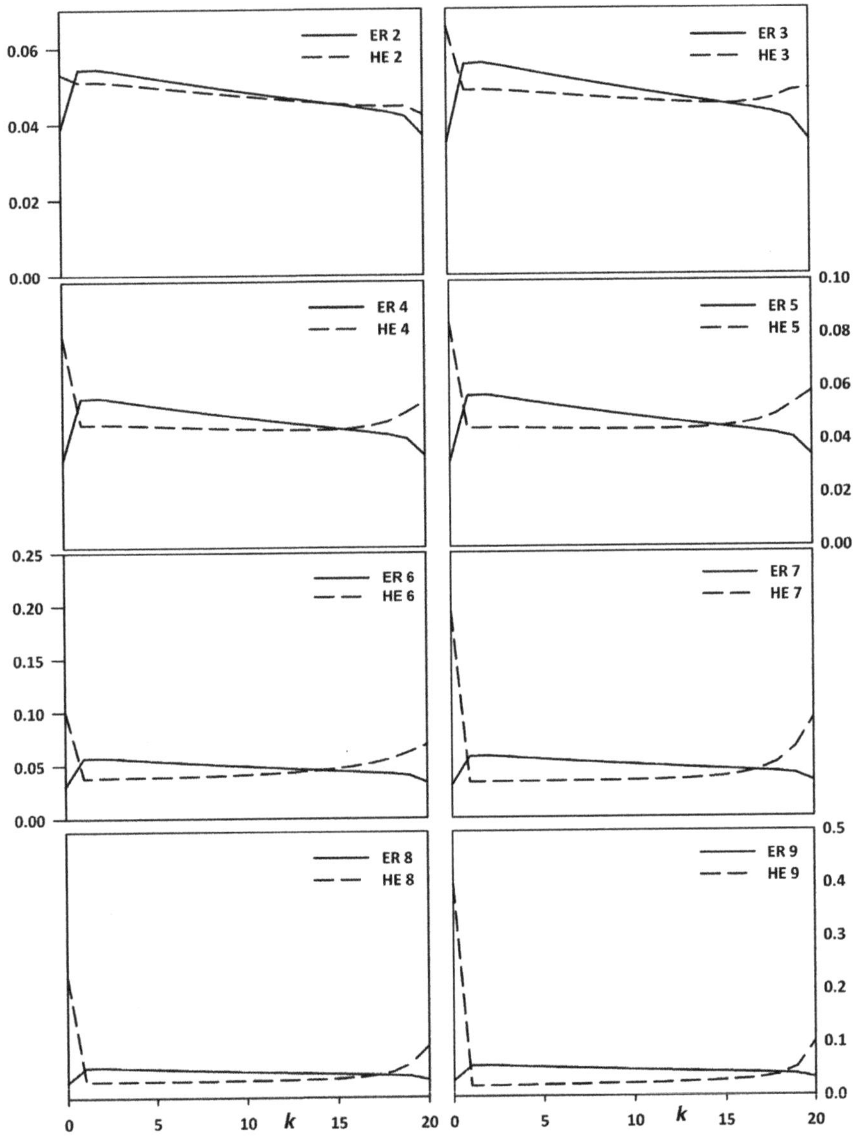

Figure 6.6. *PMF of the number in system when* $\mu = \frac{1}{0.99}$ *for example 6.4*

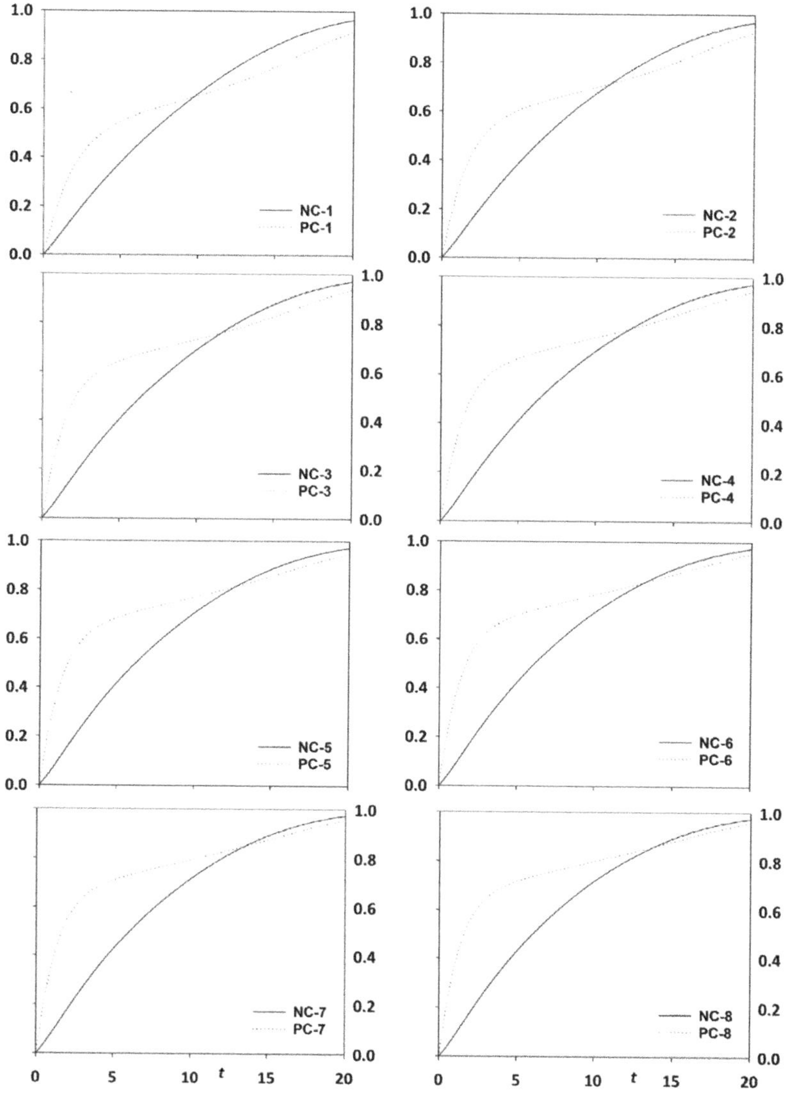

Figure 6.7. *Plots of the CDF of the waiting time in the system of an admitted customer when* $\mu = \frac{1}{0.95}$ *for example 6.5*

ToA	$\mu^{-1} = 0.90$				$\mu^{-1} = 0.99$			
	P_{idle}	P_{loss}	L_s	σ_s	P_{idle}	P_{loss}	L_s	σ_s
ER_2	0.1045	0.0050	5.4282	4.7910	0.1125	0.0139	6.4540	5.4132
ER_3	0.1029	0.0032	5.0375	4.5265	0.1205	0.0228	7.0149	5.7365
ER_4	0.1022	0.0025	4.8165	4.3681	0.1310	0.0344	7.4620	5.9988
ER_5	0.1019	0.0021	4.6750	4.2632	0.1339	0.0377	7.6218	6.0847
ER_6	0.1017	0.0018	4.5769	4.1887	0.1439	0.0488	8.1279	6.3344
ER_7	0.1015	0.0017	4.5049	4.1331	0.2284	0.1427	8.5402	7.1060
ER_8	0.1014	0.0015	4.4499	4.0901	0.2616	0.1795	8.3765	7.2735
ER_9	0.1013	0.0014	4.4064	4.0559	0.4241	0.3601	7.2101	7.7343
HE_2	0.0387	0.0290	9.5783	5.9236	0.0531	0.0436	9.6546	6.0534
HE_3	0.0349	0.0251	9.5421	5.8895	0.0654	0.0559	9.7352	6.1871
HE_4	0.0329	0.0232	9.5171	5.8716	0.0795	0.0702	9.7709	6.3196
HE_5	0.0318	0.0220	9.4992	5.8604	0.0842	0.0750	9.8835	6.3833
HE_6	0.0310	0.0212	9.4858	5.8528	0.1003	0.0912	10.2785	6.5718
HE_7	0.0304	0.0207	9.4756	5.8472	0.1930	0.1849	9.5778	7.2109
HE_8	0.0300	0.0202	9.4675	5.8430	0.2277	0.2199	9.2390	7.3664
HE_9	0.0297	0.0199	9.4609	5.8397	0.4000	0.3940	7.6894	7.8437

Table 6.4. *Selected performance measures for example 6.4*

ToA	P_{idle}	P_{loss}	L_s	σ_s	μ_{WSA}	σ_{WSA}
NC_1	0.0684	0.0194	7.8703	5.7039	8.0256	5.8152
NC_2	0.0649	0.0157	7.6746	5.5944	7.7967	5.7248
NC_3	0.0629	0.0135	7.5352	5.5177	7.6387	5.6596
NC_4	0.0616	0.0122	7.4314	5.4610	7.5230	5.6106
NC_5	0.0607	0.0112	7.3511	5.4172	7.4345	5.5724
NC_6	0.0600	0.0105	7.2872	5.3825	7.3647	5.5417
NC_7	0.0595	0.0100	7.2352	5.3541	7.3082	5.5167
NC_8	0.0591	0.0096	7.1921	5.3306	7.2616	5.4958
PC_1	0.2633	0.2245	5.8903	7.3738	7.5956	7.6004
PC_2	0.3151	0.2791	4.8257	6.9603	6.6939	7.4374
PC_3	0.3445	0.3100	4.2171	6.5960	6.1121	7.2210
PC_4	0.3634	0.3299	3.8251	6.3068	5.7086	7.0228
PC_5	0.3766	0.3438	3.5520	6.0769	5.4128	6.8521
PC_6	0.3863	0.3540	3.3508	5.8911	5.1868	6.7065
PC_7	0.3937	0.3618	3.1966	5.7383	5.0086	6.5821
PC_8	0.3996	0.3680	3.0746	5.6106	4.8644	6.4750

Table 6.5. *Selected performance measures for example 6.5*

EXAMPLE 6.6.– Referring to example 6.4, plot the CDF of the waiting time in the system of an admitted customer by taking $\mu = \frac{1}{0.95}$. Also, compile some key statistics related to this case. First, note that the waiting time is of CPH as pointed out earlier. Using the form of the CPH, and applying the technique for obtaining the CDF numerically seen in section 4.1 (see Volume 1), the plots of the CDF of the waiting time are shown in Figure 6.8. Selected performance measures are displayed in Table 6.6.

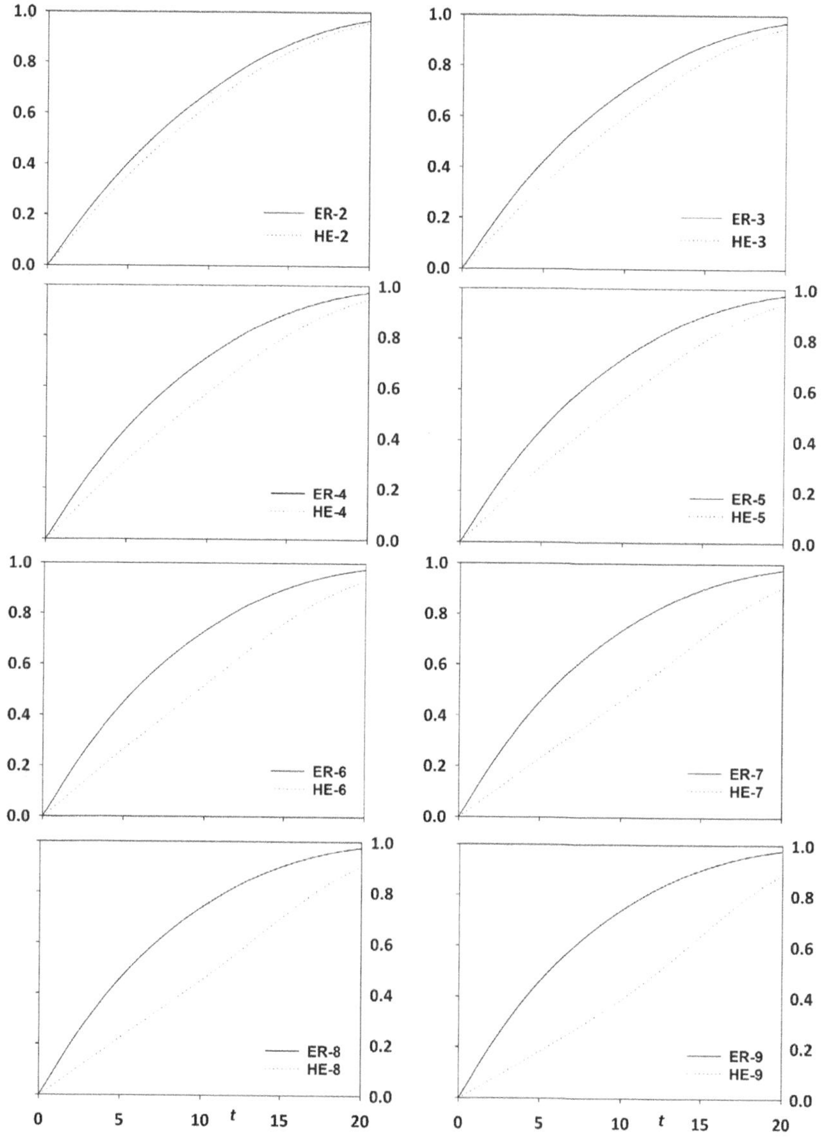

Figure 6.8. *Plots of the CDF of the waiting time in the system of an admitted customer when* $\mu = \frac{1}{0.95}$ *for example 6.6*

ToA	P_{idle}	P_{loss}	L_s	σ_s	μ_{WSA}	σ_{WSA}
ER_2	0.0639	0.0146	7.5721	5.6141	7.6844	5.7473
ER_3	0.0608	0.0114	7.3098	5.5016	7.3938	5.6511
ER_4	0.0593	0.0098	7.1509	5.4324	7.2216	5.5901
ER_5	0.0584	0.0089	7.0446	5.3854	7.1077	5.5481
ER_6	0.0579	0.0083	6.9687	5.3514	7.0269	5.5174
ER_7	0.0575	0.0078	6.9118	5.3256	6.9665	5.4940
ER_8	0.0572	0.0075	6.8675	5.3054	6.9197	5.4756
ER_9	0.0569	0.0073	6.8321	5.2891	6.8823	5.4608
HE_2	0.0762	0.0276	8.1869	5.8929	8.4195	5.9628
HE_3	0.0871	0.0390	8.5197	6.0789	8.8657	6.0904
HE_4	0.1000	0.0527	8.7554	6.2441	9.2422	6.1868
HE_5	0.1040	0.0569	8.8959	6.3149	9.4322	6.2286
HE_6	0.1175	0.0710	9.3606	6.5232	10.0765	6.3252
HE_7	0.2079	0.1662	9.1344	7.1797	10.9549	6.5529
HE_8	0.2421	0.2022	8.8690	7.3364	11.1164	6.5930
HE_9	0.4104	0.3793	7.4819	7.7995	12.0546	6.5998

Table 6.6. *Selected performance measures for example 6.6*

EXAMPLE 6.7.– The purpose of this example is to see the effect of changing the buffer size K. Toward this end, we look at four arrival processes: (1) Erlang of order 5 (ER); (2) hyperexponential with parameters $20, 10, 5, 2.5$ and 1.25 with mixing probabilities, respectively, given by 0.5, 0.3, 0.15, 0.04 and 0.01; (3) negatively correlated, as given in example 1.12, corresponding to $k = 5$ and labeled as NC_4; and (4) positively correlated, as given in example 1.12, corresponding to $k = 5$ and labeled as PC_4. The service times are of phase type obtained as the minimum of exponential, Erlang and hyperexponential as mentioned in example 1.16. We look at $K = 40$ and $K = 80$. The arrival rate is taken to be 1 (and hence the input parameters are normalized accordingly) and the mean service time is taken to be $\frac{1}{0.95}$. The plots of the CDF of the waiting time for ER, HE and NC, PC arrivals are shown in Figures 6.9 and 6.10, respectively, and selected performance measures are displayed in Table 6.7.

6.1.3. PH/M/1/K *queue*

In this special case of $MAP/PH/1/K$, we have the inter-arrival times of the customers to be of phase type with representation given by $(\boldsymbol{\alpha}, T)$ of order m. The arrival rate is given by $\lambda = [\boldsymbol{\alpha}(-T)^{-1}\mathbf{e}]^{-1}$. The service times are assumed to be exponentially distributed with rate μ. A number of simplifications occur that lead to getting explicit results for the steady-state probability vector.

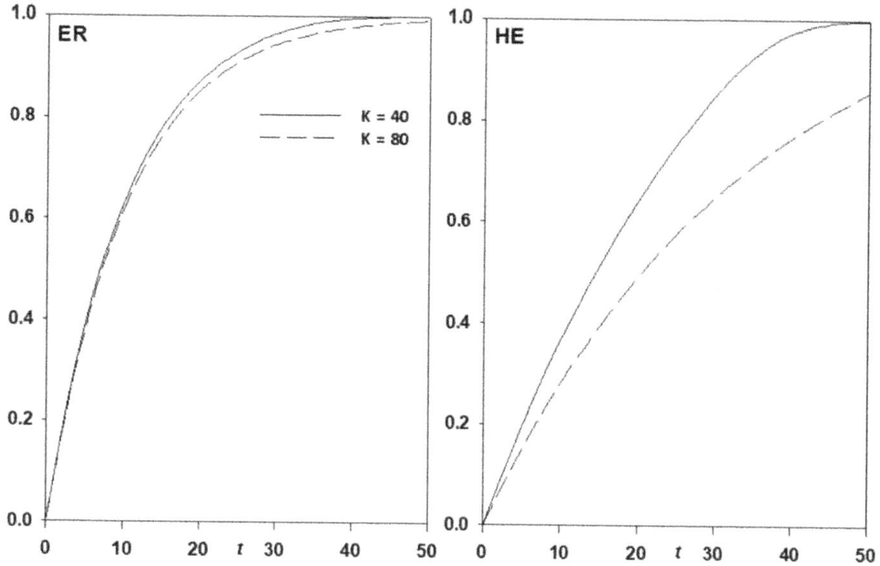

Figure 6.9. *Plots of the CDF of the waiting time in the system of an admitted customer for ER and HE when $\mu = \frac{1}{0.95}$ for example 6.7*

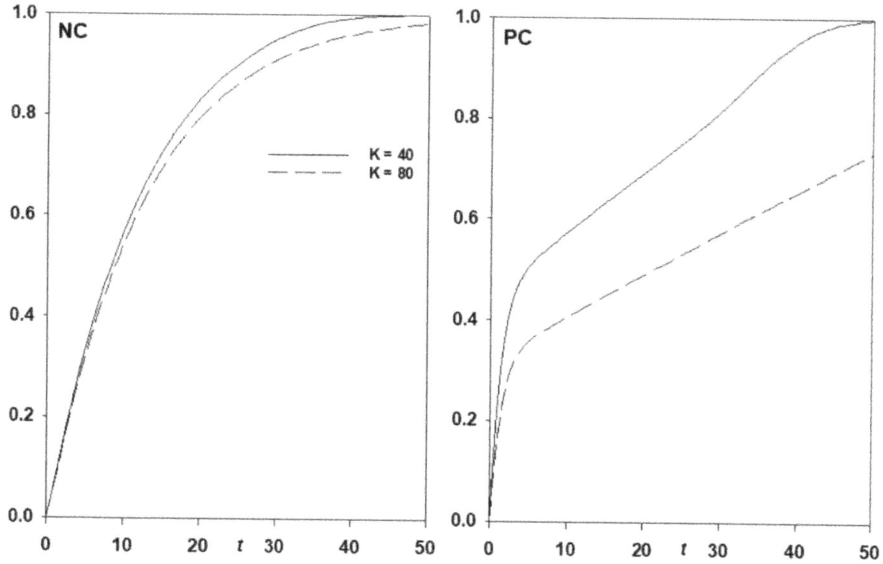

Figure 6.10. *Plots of the CDF of the waiting time in the system of an admitted customer for NC and PC when $\mu = \frac{1}{0.95}$ for example 6.7*

K	ToA	P_{idle}	P_{loss}	L_s	σ_s	μ_{WSA}	σ_{WSA}
40	ER	0.0511	0.0012	9.5875	8.6001	9.5986	8.5825
	HE	0.0709	0.0221	15.9741	11.8517	16.3343	11.4536
	NC	0.0522	0.0023	10.9454	9.3487	10.9710	9.2795
	PC	0.2917	0.2544	9.6697	13.7981	12.9687	14.2959
80	ER	0.0500	0	10.4351	10.3634	10.4354	10.2620
	HE	0.0553	0.0056	25.3576	21.0725	25.4994	20.2388
	NC	0.0501	0.0001	12.5809	12.2632	12.5821	12.0612
	PC	0.2256	0.1848	23.2449	27.9933	28.5152	27.4534

Table 6.7. *Selected performance measures for example 6.7*

First, verify that the generator for the current model is given by:

$$Q = \begin{bmatrix} T & \boldsymbol{T^0\alpha} & & & & & \\ \mu I & T-\mu I & \boldsymbol{T^0\alpha} & & & & \\ & \mu I & T-\mu I & \boldsymbol{T^0\alpha} & & & \\ & & \mu I & T-\mu I & \boldsymbol{T^0\alpha} & & \\ & & & \ddots & \ddots & & \\ & & & & \mu I & T-\mu I & \boldsymbol{T^0\alpha} \\ & & & & & \mu I & T+\boldsymbol{T^0\alpha}-\mu I \end{bmatrix}. \quad [6.3]$$

Defining:

$$\tilde{M} = \mu[\mu I - \mu \boldsymbol{e\alpha} - T]^{-1}, \quad [6.4]$$

the steady-state probability vector is obtained as (see Neuts (1981)):

$$\boldsymbol{y}_0 = c\mu\boldsymbol{\alpha}\tilde{M}^K(-T)^{-1},$$
$$\boldsymbol{y}_i = c\boldsymbol{\alpha}\tilde{M}^{K+1-i}, \; 1 \le i \le K, \quad [6.5]$$

where c is the normalizing constant.

The system performance measures listed in section 6.1.1 can be similarly defined for this model and are left as exercises.

EXAMPLE 6.8.– Suppose that the inter-arrival times of customers are modeled using a CPH. We assume in this example the CPH are modeled using various Erlang as well as hyperexponential distributions. Specifically we look at Erlang of order k, labeled as ER_k, with a parameter chosen such that the arrival rate is 1. For hyperexponential inter-arrival times, we look at the ones labeled as HE_k (see example 5.14). Again, the CPH representations will be normalized so the arrival rate will be 1. The service times are assumed to be exponential with parameter $\mu = \frac{1}{0.95}$ and $K = 20$.

Selected measures are displayed in Table 6.8 and the plots of the CDF of the waiting time are shown in Figure 6.11.

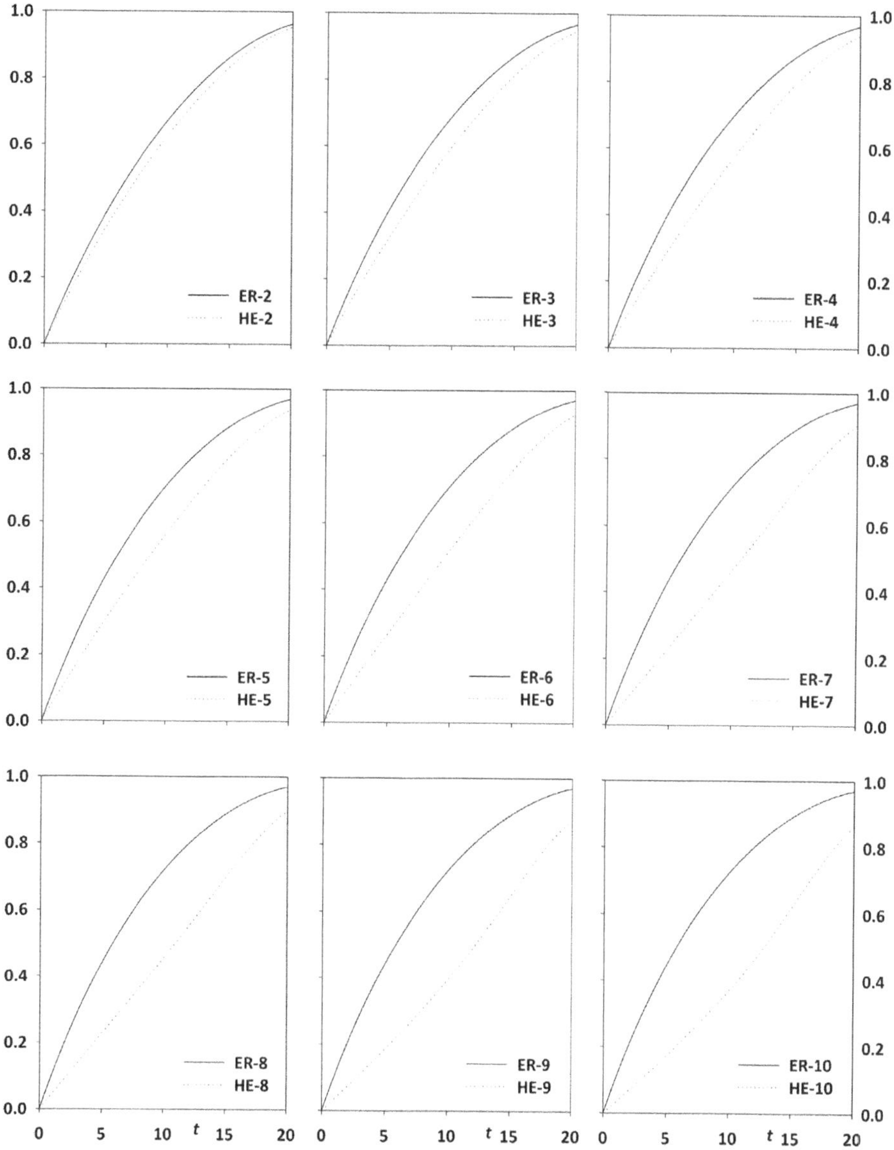

Figure 6.11. *Plots of the CDF of the waiting time in the system of an admitted customer when* $\mu = \frac{1}{0.95}$ *for example 6.8*

ToA	P_{idle}	P_{loss}	L_s	σ_s	μ_{WSA}	σ_{WSA}
ER_2	0.06599	0.01683	7.72098	5.68920	7.85317	5.88776
ER_3	0.06283	0.01351	7.49834	5.59431	7.60104	5.80770
ER_4	0.06130	0.01189	7.36477	5.53726	7.45341	5.75797
ER_5	0.06039	0.01094	7.27600	5.49905	7.35648	5.72413
ER_6	0.05980	0.01031	7.21280	5.47165	7.28797	5.69963
ER_7	0.05938	0.00987	7.16554	5.45102	7.23698	5.68108
ER_8	0.05906	0.00954	7.12888	5.43494	7.19755	5.66655
ER_9	0.05882	0.00929	7.09961	5.42204	7.16616	5.65487
ER_{10}	0.05863	0.00908	7.0757	5.41147	7.14057	5.64526
HE_2	0.07845	0.02994	8.25640	5.93749	8.51126	6.07683
HE_3	0.08927	0.04133	8.55690	6.11223	8.92582	6.19591
HE_4	0.10217	0.05492	8.77238	6.27073	9.28215	6.28822
HE_5	0.10614	0.05910	8.90553	6.33912	9.46487	6.32898
HE_6	0.11958	0.07324	9.34796	6.54178	10.08673	6.42527
HE_7	0.20926	0.16764	9.11944	7.18780	10.95618	6.65249
HE_8	0.24323	0.20340	8.85746	7.34338	11.11908	6.69337
HE_9	0.41086	0.37985	7.47920	7.80499	12.06038	6.70943
HE_{10}	0.42068	0.39019	7.59966	7.96288	12.46241	6.72925

Table 6.8. *Selected performance measures for example 6.8*

EXAMPLE 6.9.– The purpose of this example is to see the effect of varying K by referring to example 6.8 in the sense we consider the input parameters to be the same. However, we vary K from 5 to 80 in increments of 5. The plots of some key measures are shown in Figure 6.12.

6.1.4. *M/PH/1/K queue*

Here, we consider an $M/PH/1$ queue but with the buffer size assumed to be of finite size $K-1,\ K \geq 1$. Thus, at any time there can be a maximum of K customers present in the system. Any arrival finding the buffer full will be lost. Suppose we assume that the arrival rate is λ and the service rate is μ. Suppose that the arrivals occur according to a Poisson process with rate λ and the service times are of phase type with representation $(\boldsymbol{\beta}, S)$ of order n. Recall that the service rate is given by $\mu = [\boldsymbol{\beta}(-S)^{-1}\mathbf{e}]^{-1}$. Denote $N(t)$ and $J(t)$ to be, respectively, the number of customers in the system and the phase of the service at time t. The process $\{(N(t), J(t)) : t \geq 0\}$

completely describes the queueing model under study and this process is a continuous-time Markov chain (CTMC) on state space given by $\Delta = \{*\} \bigcup \{(i,j) : 1 \leq j \leq m, 1 \leq i \leq K\}$ and with the generator given by:

$$Q = \begin{bmatrix} -\lambda & \lambda\boldsymbol{\beta} & & & & & \\ \boldsymbol{S^0} & S-\lambda I & \lambda I & & & & \\ & \boldsymbol{S^0}\boldsymbol{\beta} & S-\lambda I & \lambda I & & & \\ & & \boldsymbol{S^0}\boldsymbol{\beta} & S-\lambda I & \lambda I & & \\ & & & \ddots & \ddots & & \\ & & & & \boldsymbol{S^0}\boldsymbol{\beta} & S-\lambda I & \lambda I \\ & & & & & \boldsymbol{S^0}\boldsymbol{\beta} & S \end{bmatrix}, \quad [6.6]$$

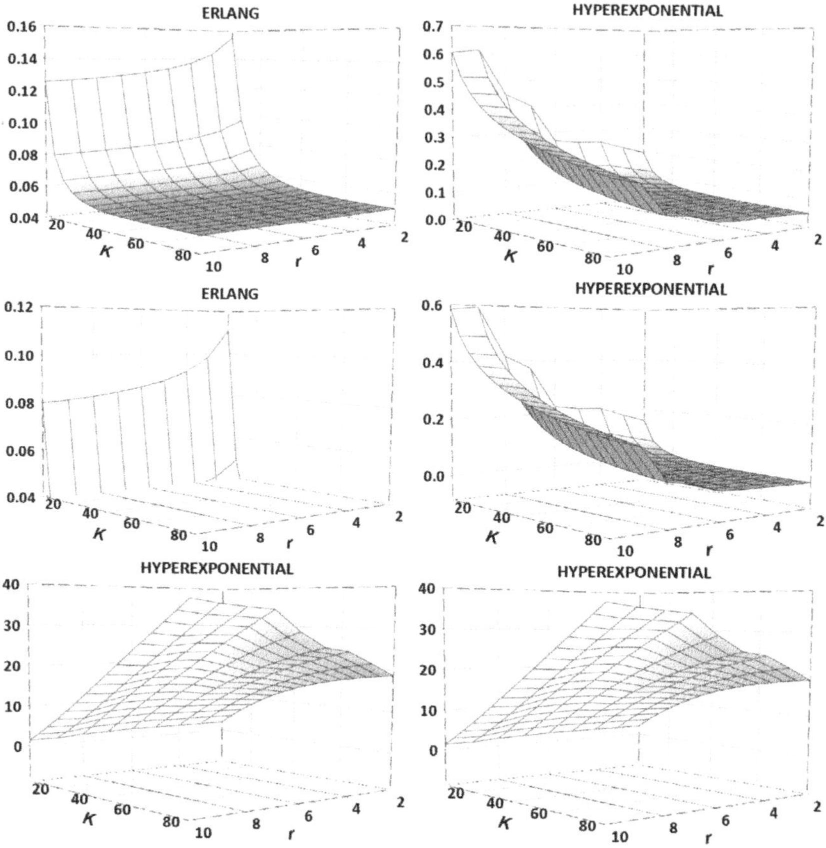

Figure 6.12. *Plots of P(idle), P(loss) and L_s for example 6.9*

The steady-state probability vector, $\boldsymbol{y}$, can be obtained as (Neuts (1981)):

$$\boldsymbol{y}_i = y_0 \boldsymbol{\beta} M^i,\ 1 \le i \le K-1, \quad \boldsymbol{y}_K = \lambda y_0 \boldsymbol{\beta} M^{K-1}(-S)^{-1},$$
$$y_0 = \left[\boldsymbol{\beta} \left(\sum_{k=0}^{K-1} M^k + \lambda M^{K-1}(-S)^{-1} \right) \boldsymbol{e} \right]^{-1}, \qquad [6.7]$$

where $M = \lambda(\lambda I - \lambda \boldsymbol{e}\,\boldsymbol{\beta} - S)^{-1}$.

The system performance measures listed in section 6.1.1 can be similarly defined for this model. The waiting time distribution obtained for a $MAP/PH/1/K$ model can be particularized for the current case.

EXAMPLE 6.10.– Suppose that the customers arrive according to a Poisson process with a rate of 1 per unit of time. For services we look at various Erlangs as well as hyperexponential distributions. Specifically we look at Erlang of order k, labeled as ER_k, with parameter chosen such that the mean service time is $\frac{1}{0.95}$. For hyperexponential services, we look at the ones labeled as HE_k (see example 5.14). Again, the CPH representations will be normalized to have the service rate be $\mu = \frac{1}{0.95}$ and $K = 20$.

The plots of the CDF of the waiting time are shown in Figure 6.13 and selected measures are displayed in Table 6.9.

ToA	P_{idle}	P_{loss}	L_s	σ_s	μ_{WSA}	σ_{WSA}
ER_2	0.06653	0.01740	7.73451	5.64610	7.87147	5.50754
ER_3	0.06348	0.01419	7.52758	5.54001	7.63593	5.30743
ER_4	0.06198	0.01261	7.40509	5.47867	7.49967	5.20024
ER_5	0.06109	0.01168	7.32424	5.43848	7.41077	5.13302
ER_6	0.06051	0.01106	7.26691	5.41006	7.34817	5.08680
ER_7	0.06009	0.01062	7.22415	5.38887	7.30171	5.05303
ER_8	0.05978	0.01030	7.19104	5.37247	7.26585	5.02726
ER_9	0.05954	0.01004	7.16465	5.35938	7.23734	5.00694
ER_{10}	0.05935	0.00984	7.14312	5.34870	7.21412	4.99050
HE_2	0.07854	0.03004	8.20397	5.94582	8.45804	6.19918
HE_3	0.08924	0.04130	8.35227	6.15389	8.71210	6.80556
HE_4	0.10140	0.05411	8.44843	6.34869	8.93170	7.47829
HE_5	0.10592	0.05887	8.36122	6.40575	8.88421	7.85223
HE_6	0.12103	0.07477	8.0226	6.5566	8.6709	9.41709
HE_7	0.18888	0.14618	8.3973	7.421	9.83503	13.30345
HE_8	0.2116	0.1701	8.48995	7.65673	10.23011	14.8117
HE_9	0.3084	0.27199	8.63323	8.50805	11.85876	27.55083
HE_{10}	0.31405	0.27794	8.50221	8.57133	11.77497	32.91065

Table 6.9. *Selected performance measures for example 6.10*

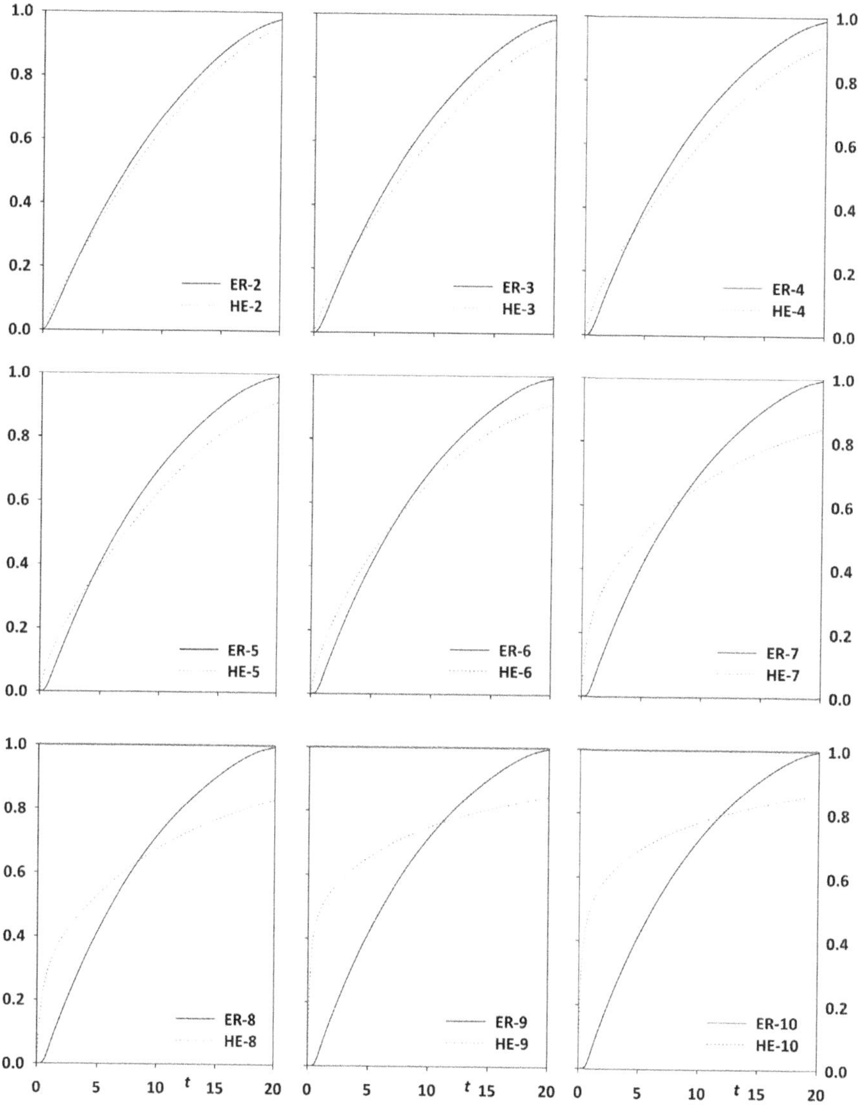

Figure 6.13. *Plots of the CDF of the waiting time in the system of an admitted customer when* $\mu = \frac{1}{0.95}$ *for example 6.10*

EXAMPLE 6.11.– The purpose of this example is to see the effect of varying K by referring to example 6.10 in the sense we consider the input parameters to be the same. However, we vary K from 5 to 80 in increments of 5. The plots of some key measures are shown in Figure 6.14.

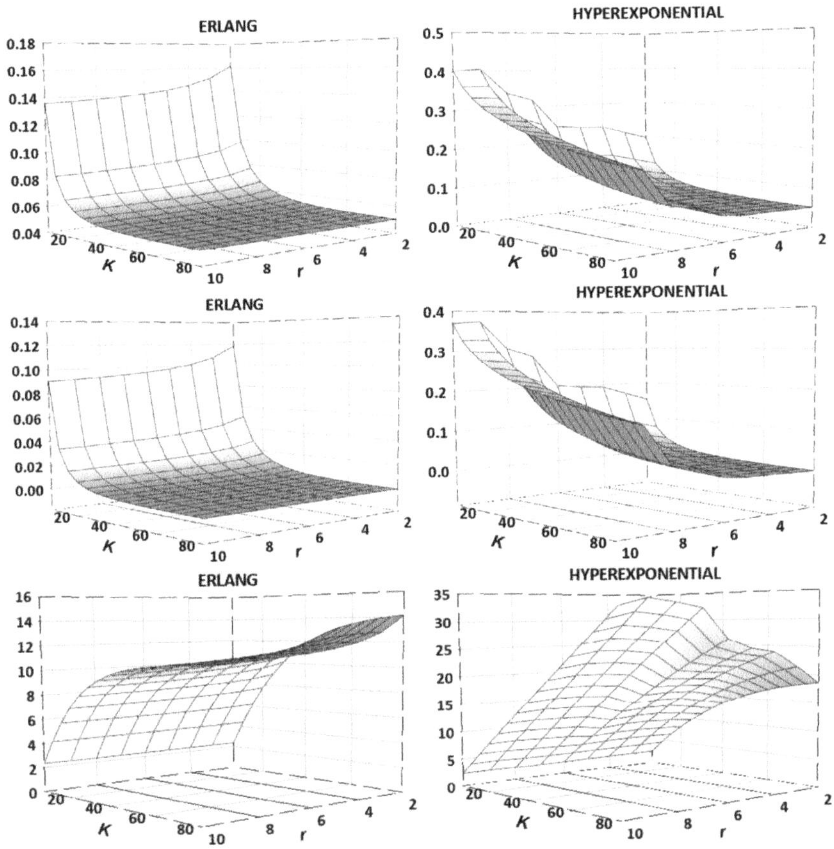

Figure 6.14. *Plots of P(idle), P(loss) and L_s for example 6.11*

6.1.5. *M/M/1/K queue*

Here, we are dealing with a finite capacity $M/M/1$ queue such that at any given time the maximum number of customers in the system cannot exceed K.

The generator $\tilde{Q}$ of this CTMC is given by:

$$\tilde{Q} = \begin{pmatrix} -\lambda & \lambda & & & & \\ \mu & -(\lambda+\mu) & \lambda & & & \\ & \mu & -(\lambda+\mu) & \lambda & & \\ & & \ddots & \ddots & \ddots & \\ & & & \mu & -(\lambda+\mu) & \lambda \\ & & & & \mu & -\mu \end{pmatrix}. \qquad [6.8]$$

It is easy to verify that the queue is always stable and the steady-state probability vector can be explicitly obtained as follows. Taking $\rho = \frac{\lambda}{\mu}$, verify that $x_0 = (1 - \rho^{K+1})^{-1}(1-\rho)$ and $(1-\rho^{K+1})^{-1}(1-\rho)\rho^k$, $1 \leq k \leq K$.

The system performance measures listed in section 6.1.1 can be similarly defined for this model and are left as exercises. The waiting time distribution obtained for a $MAP/PH/1/K$ model can be particularized for the current case.

EXAMPLE 6.12.– In this example, we look at an $M/M/1/K$ queueing model in which we consider five values for the buffer capacity, namely, $K = 10, 20, 30, 40$ and 50. We fix the arrival rate to be 1 and the service rate is taken to be $\mu = \frac{1}{0.95}$ even though the queue is always stable. The CDF of the waiting time in the system of an admitted customer is shown in Figure 6.15 and selected measures are displayed in Table 6.10.

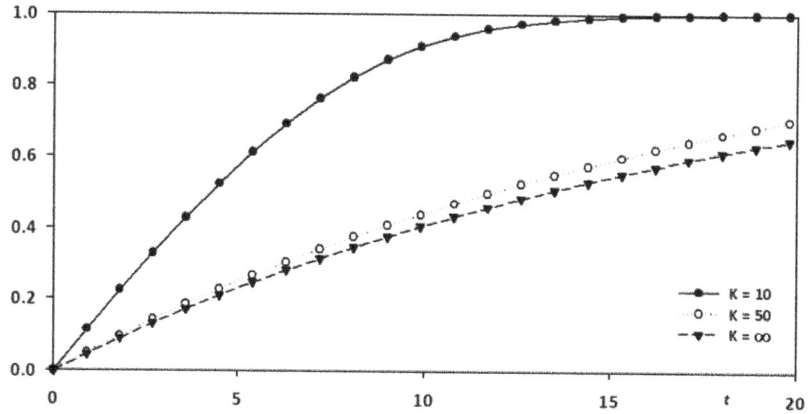

Figure 6.15. *Plots of the CDF of the waiting time in the system of an admitted customer when $\mu = \frac{1}{0.95}$ for example 6.12*

$Buffer\ Size$	P_{idle}	P_{loss}	L_s	σ_s	μ_{WSA}	σ_{WSA}
$K = 10$	0.11596	0.06943	4.48979	3.13712	4.82476	3.45416
$K = 20$	0.07582	0.02718	8.15472	5.88464	8.38257	6.03758
$K = 30$	0.06281	0.01348	11.05983	8.41479	11.21097	8.42282
$K = 40$	0.05695	0.00732	13.29836	10.66414	13.39641	10.55177
$K = 50$	0.05394	0.00415	14.97802	12.59772	15.04045	12.38877

Table 6.10. *Selected performance measures for example 6.12*

6.2. Finite-capacity queues with multiple servers

Here, we assume that the system has a finite capacity with multiple (identical) servers available for serving the customers. We consider different scenarios for the arrival processes. The service times are assumed to be exponentially distributed.

6.2.1. BMAP/M/c/K *queue*

In section 5.3, we looked at the infinite capacity $BMAP/M/c$ queues. Here, we restrict the number of customers in the system at any given time to not exceed $K + c$, where $0 \leq K < \infty$. Thus, the buffer capacity is restricted to be finite of size K. Let the arrival process parameter matrices be $\{D_k\}$ with the steady-state vector of the generator, $Q^* = \sum_k D_k$, of the underlying BMAP be denoted by $\boldsymbol{\eta}$ so that the arrival rate of batches is given by $\lambda_g(-D_0)\boldsymbol{e}$ and the arrival rate of the customers is given by $\lambda = \boldsymbol{\eta} \sum_{i=1}^{\infty} kD_k\boldsymbol{e}$. Let the service times be exponential with mean μ. There are c homogeneous servers in the system. Define $\tilde{D}_k = \sum_{r=k}^{\infty} D_r, \ k \geq 1$.

Letting $N(t)$ and $J(t)$ denote, respectively, the number of customers and the phase of the arrival process at time t, the model can be studied as a QBD process on the state space $\Omega = \{(i, j) : 1 \leq j \leq m, \ 0 \leq i \leq K + c\}$. It is easy to see that the infinitesimal generator, Q, of the Markov process governing the queueing system under study is given by:

$$Q = \begin{pmatrix} D_0 & D_1 & D_2 & \cdots & D_c & D_{c+1} & \cdots & D_{c+K-1} & \tilde{D}_{c+K} \\ \mu I & D_0 - \mu I & D_1 & \cdots & D_{c-1} & D_c & \cdots & D_{c+K-2} & \tilde{D}_{c+K-1} \\ 0 & 2\mu I & D_0 - 2\mu I & \cdots & D_{c-2} & D_{c-1} & \cdots & D_{c+K-3} & \tilde{D}_{c+K-2} \\ \vdots & \vdots & \vdots & \cdots & \vdots & \vdots & \cdots & \vdots & \vdots \\ 0 & 0 & 0 & \cdots & D_0 - c\mu I & D_1 & \cdots & D_{K-1} & \tilde{D}_K \\ 0 & 0 & 0 & \cdots & c\mu I & D_0 - c\mu I & \cdots & D_{K-2} & \tilde{D}_{K-1} \\ 0 & 0 & 0 & \cdots & 0 & c\mu I & \cdots & D_{K-3} & \tilde{D}_{K-2} \\ \vdots & \vdots & \vdots & \cdots & \vdots & \vdots & \cdots & \vdots & \vdots \\ 0 & 0 & 0 & \cdots & 0 & 0 & \cdots & D_0 - c\mu I & \tilde{D}_1 \\ 0 & 0 & 0 & \cdots & 0 & 0 & \cdots & c\mu I & Q^* - c\mu I \end{pmatrix}. \quad [6.9]$$

Noting that the queue is always stable, the steady-state equations for obtaining the steady-state probability vector, $\boldsymbol{y} = (\boldsymbol{y}_0, \boldsymbol{y}_1, \cdots, \boldsymbol{y}_{K+c})$, of Q are as follows:

$$\boldsymbol{y}_0 = \mu \boldsymbol{y}_1 (-D_0)^{-1},$$

$$\boldsymbol{y}_i = \Big[\boldsymbol{y}_0 D_i + \sum_{j=1}^{i-1} \boldsymbol{y}_j D_{i-j} + \min\{i+1, c\}\mu \boldsymbol{y}_{i+1}\Big]\big(\min\{i, c\}\mu I - D_0\big)^{-1},$$

for $1 \leq i \leq K + c - 1$, and:

$$\boldsymbol{y}_{K+c} = \Big[\sum_{j=0}^{K+c-1} \boldsymbol{y}_j \tilde{D}_{K+c-j}\Big]\big(c\mu I - Q^*\big)^{-1},$$

along with the normalizing equation: $\sum_{i=0}^{K+c} \boldsymbol{y}_i \boldsymbol{e} = 1$. In addition to the system performance measures listed in section 6.1.1, which for the current model can be similarly defined and are left as exercises, we can include the PMF of the number of busy servers.

PMF of the number of busy servers: Suppose we define $\{a_j\}_{j=0}^{c}$ to be the PMF of the number of busy servers for the model under study. It can be verified that this function is computed as follows:

$$a_j = \begin{cases} \boldsymbol{y}_0 \boldsymbol{e}, & i = 0, \\ \boldsymbol{y}_j \boldsymbol{e}, & 1 \leq j \leq c-1, \\ \sum_{r=c}^{K+c} \boldsymbol{y}_r \boldsymbol{e}, & i = c. \end{cases}$$

Once the PMF is obtained we can compute various associated measures such as the mean, the median, the mode and the standard deviation.

EXAMPLE 6.13.– Suppose we look at five different arrival processes: Erlang of order 5; exponential; hyperexponential with parameters 20, 10, 5, 2.5 and 1.25 with mixing probabilities 0.5, 0.30, 0.15, 0.04 and 0.01, respectively; negatively correlated process, labeled as NC_3; and positively correlated process, labeled as PC_3, as given in example 1.12. We look at different scenarios by varying $c = 2, 3, 4, 5$ and $K = 20, 40, 60, 80$. Arrivals occur in batches that are modeled using a Poisson distribution with parameter 1.25. The arrival rate of the batches will be taken as $\frac{1}{2.25}$ so that the rate of arrivals of customers will be 1. The service times are exponential with parameter μ. In order to properly compare various scenarios, we take $\mu = \frac{1}{0.95\,c}$.

In Figure 6.16, we display the conditional PMF of the number of customers lost at an arrival epoch given that a loss occurs.

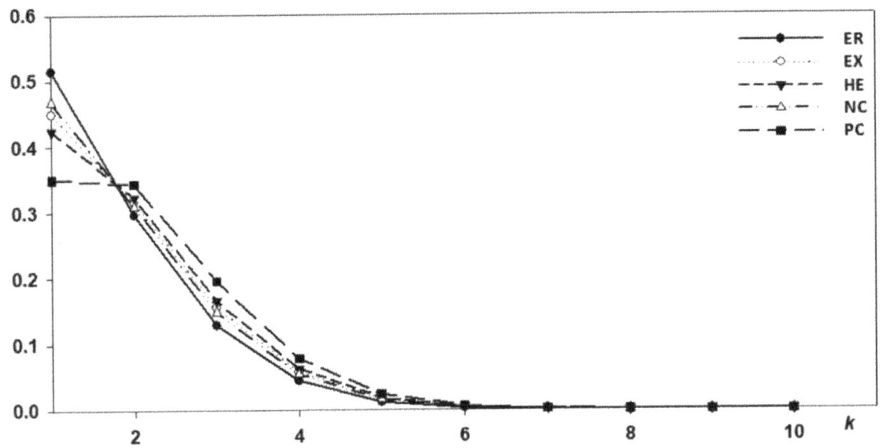

Figure 6.16. *Conditional PMF of the number lost for example 6.13*

It is interesting to point out (for the choice of the rate of service chosen, which makes the total service rate for the c-server system constant) that the conditional PMF of the number lost does not depend on the buffer size but does depend on the type of the arrival process. The emphasis here is on the "conditional" and the loss. However, if one looks at the unconditional probabilities of the number lost, the story is different.

In Figures 6.17 through 6.21, we plot the PMF of the number of customers in the system in steady-state under various scenarios.

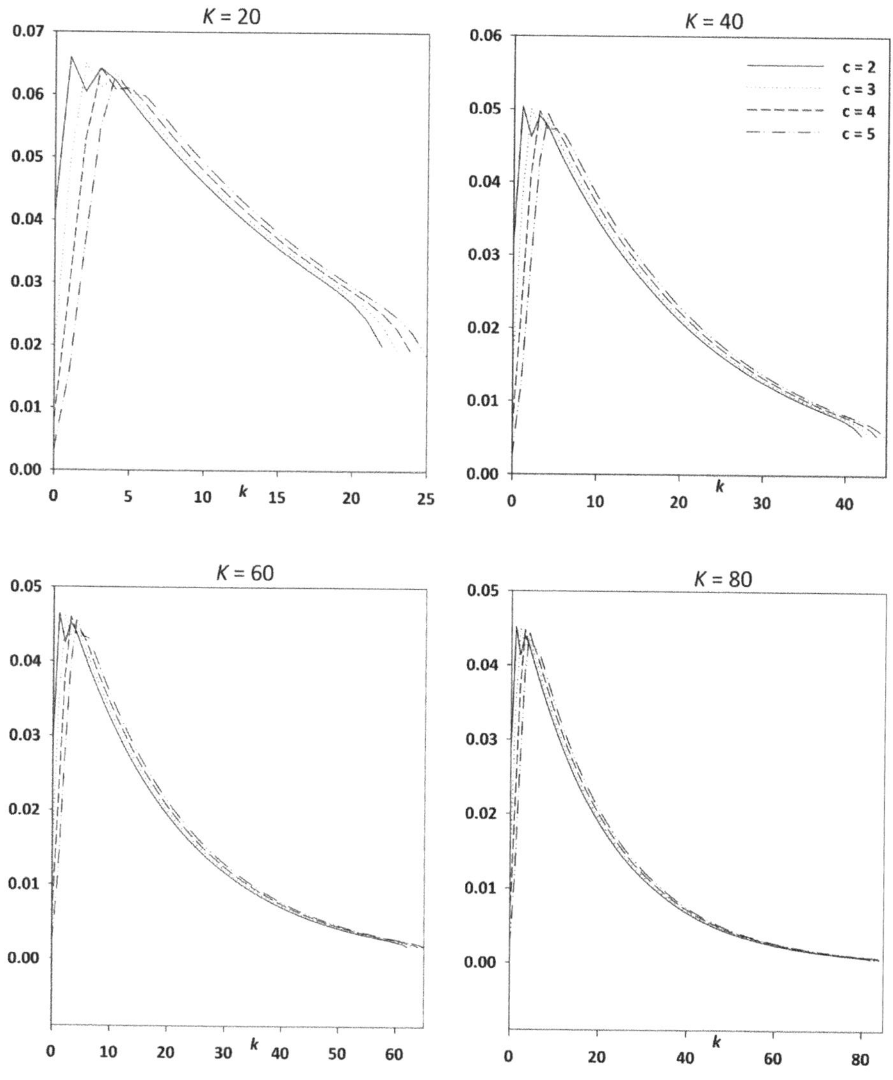

Figure 6.17. *Plot of the PMF of the number in system for Erlang arrivals for example 6.13*

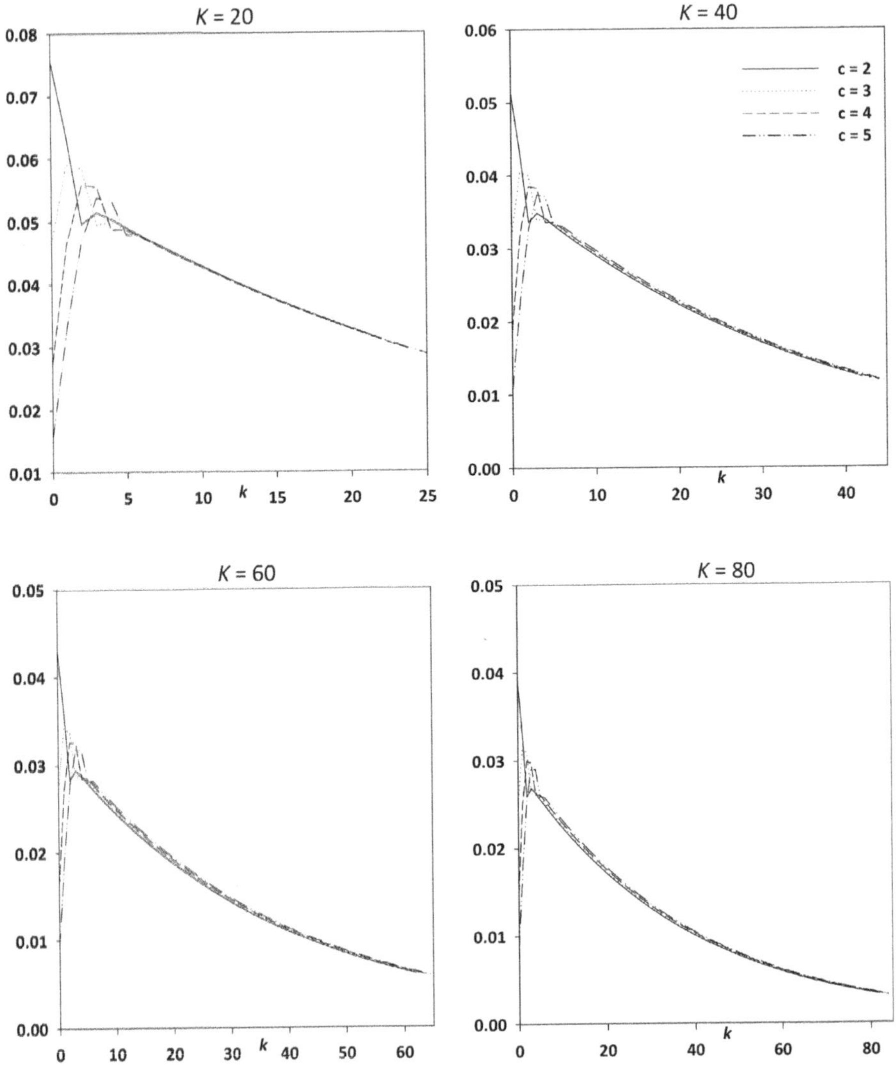

Figure 6.18. *Plot of the PMF of the number in system for Poisson arrivals for example 6.13*

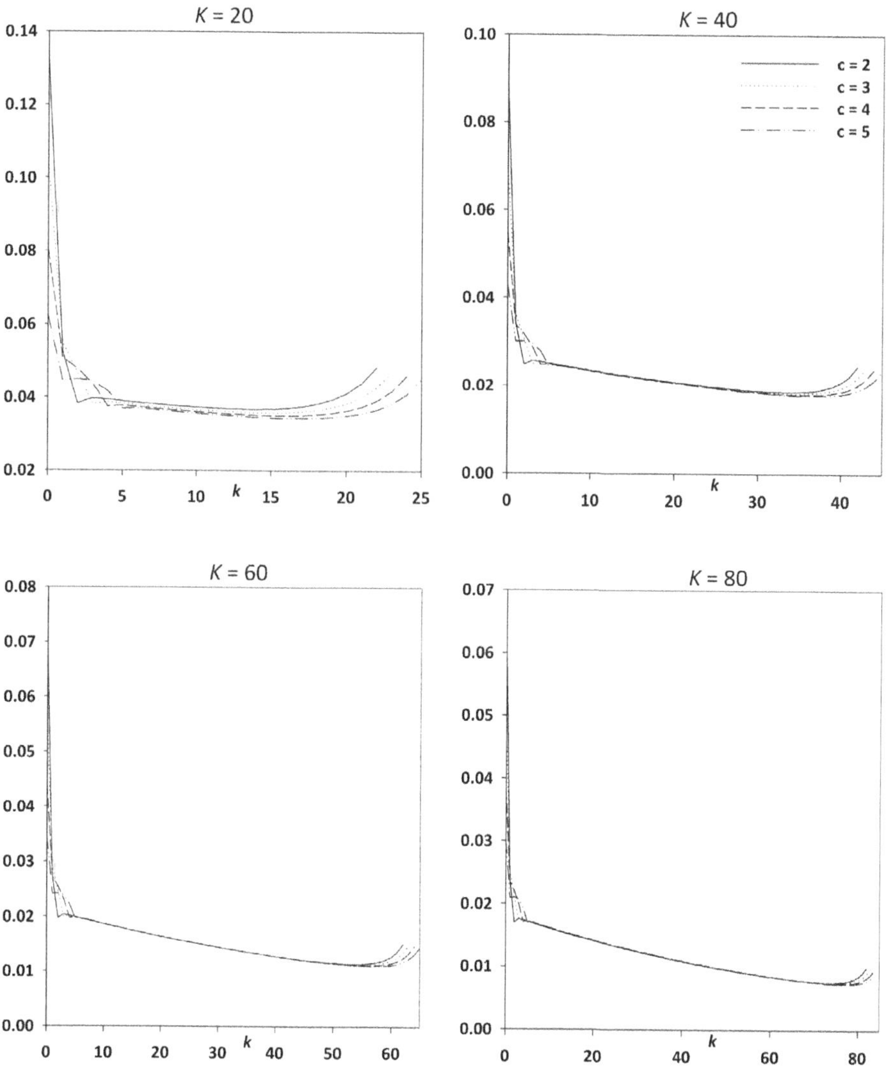

Figure 6.19. *Plot of the PMF of the number in system for hyperexponential arrivals for example 6.13*

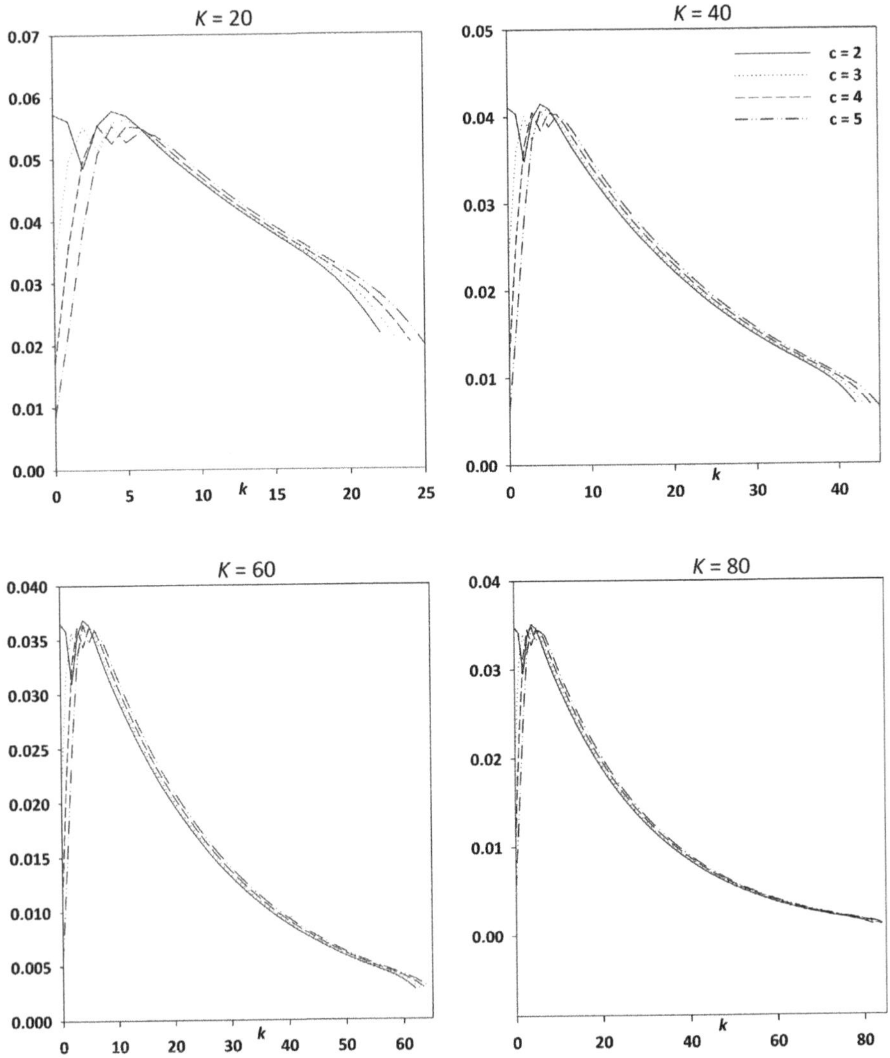

Figure 6.20. *Plot of the PMF of the number in system for negatively correlated arrivals for example 6.13*

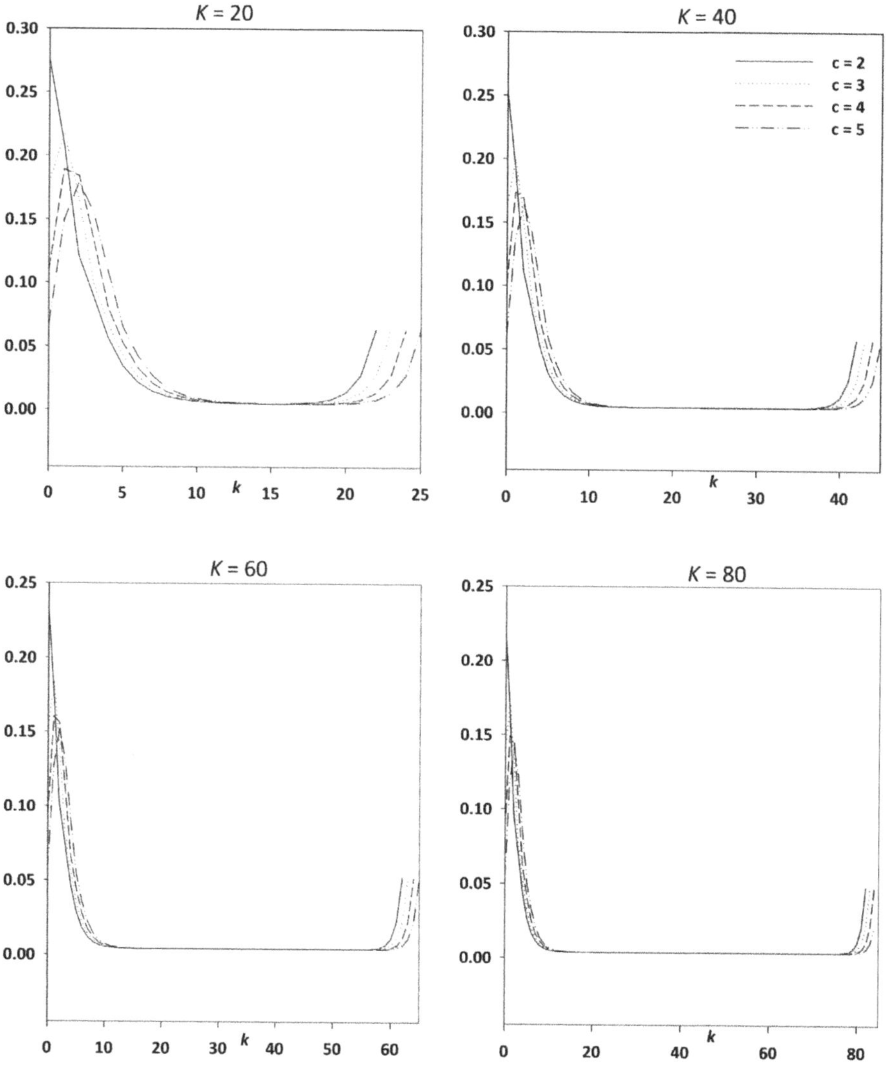

Figure 6.21. *Plot of the PMF of the number in system for positively correlated arrivals for example 6.13*

6.2.2. MAP/M/c/K *queue*

Here, the arrivals occur one at a time and the representation of the MAP governing the arrival process is given by (D_0, D_1) of order m. Let the service times be exponential with parameter μ. Assume that the c servers in the system are homogeneous. Keeping track of the number of customers in the system and the phase of the arrival process, the generator governing the $MAP/M/c/K$ is then given by:

$$Q = \begin{pmatrix} D_0 & D_1 & 0 & \cdots & 0 & 0 & \cdots & 0 & 0 \\ \mu I & D_0 - \mu I & D_1 & \cdots & 0 & 0 & \cdots & 0 & 0 \\ 0 & 2\mu I & D_0 - 2\mu I & \cdots & 0 & 0 & \cdots & 0 & 0 \\ \vdots & \vdots & \vdots & \cdots & \vdots & \vdots & \cdots & \vdots & \vdots \\ 0 & 0 & 0 & \cdots & c\mu I & D_0 - c\mu I & \cdots & 0 & 0 \\ \vdots & \vdots & \vdots & \cdots & \vdots & \vdots & \cdots & \vdots & \vdots \\ 0 & 0 & 0 & \cdots & 0 & 0 & \cdots & D_0 - c\mu I & D_1 \\ 0 & 0 & 0 & \cdots & 0 & 0 & \cdots & c\mu I & Q^* - c\mu I \end{pmatrix},$$

where $Q^* = D_0 + D_1$. The steady-state vector of Q can be solved in a straightforward way. Like in the case of an $MAP/PH/1/K$ model, we can show that the waiting time in the system of an admitted customer to be of phase type with representation type (γ, L) of order $K + 1$, where:

$$\gamma = (z_0, z_1, \cdots, z_K), \; L = \begin{pmatrix} -\mu & & & & & \\ c\mu & -c\mu & & & & \\ & c\mu & -c\mu & & & \\ & & c\mu & -c\mu & & \\ & & & \ddots & \ddots & \\ & & & & c\mu & -c\mu \end{pmatrix},$$

where $z_0 = d \sum_{k=0}^{c-1} \boldsymbol{y}_k D_1 \boldsymbol{e}$ and $z_{j-c+1} = d\,\boldsymbol{y}_j D_1 \boldsymbol{e},\; c \le j \le K + c - 1$, and d is the normalizing constant and is given by:

$$d = \lambda \Big[\lambda - \boldsymbol{y}_{K+c} D_1 \boldsymbol{e}\Big]^{-1}.$$

The mean waiting time in the system, μ'_{WTS}, is verified to be:

$$\mu'_{WTS} = \frac{1}{\mu}\left[1 + \frac{1}{c}\sum_{k=1}^{K} k z_k\right]. \qquad [6.10]$$

REMARK 6.1.– The expression in equation [6.10] has a nice interpretation. The mean waiting time in the system of an arriving customer is the sum of the mean service time and the mean time needed to serve all those customers seen at the time of the arrival of the customer.

EXAMPLE 6.14.– In this example, we revisit example 6.13 but with the restriction that the arrivals occur singly. Also, we consider only the correlated arrivals for this example. All other parameters are kept the same. The plots of the CDF of the waiting time in the system of an admitted customer are displayed in Figures 6.22 and 6.23, and in Figure 6.24 we display three key measures under varying scenarios.

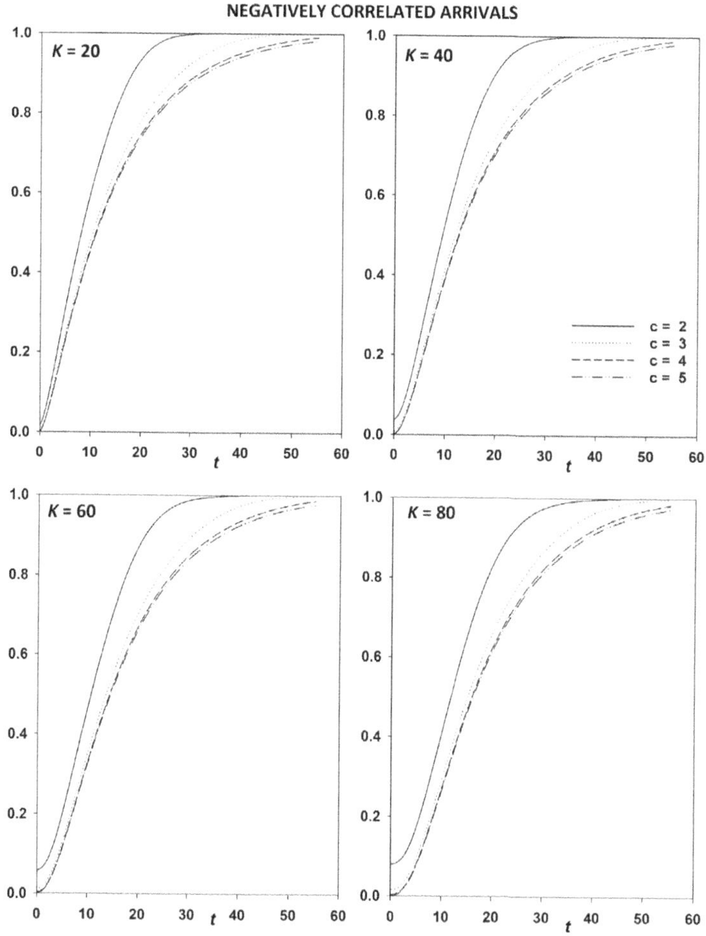

Figure 6.22. *Plot of the CDF of the waiting time in the system of an admitted customer for negatively correlated arrivals for example 6.14*

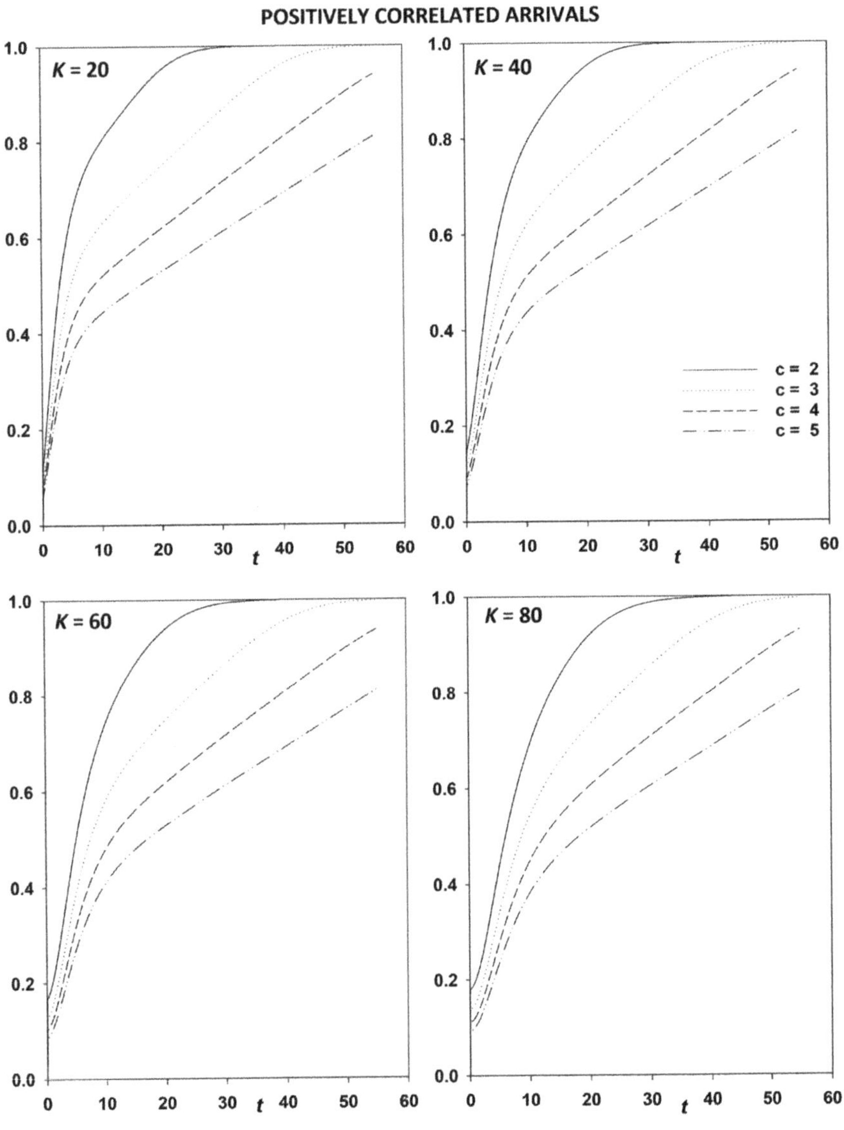

Figure 6.23. *Plot of the CDF of the waiting time in the system of an admitted customer for positively correlated arrivals for example 6.14*

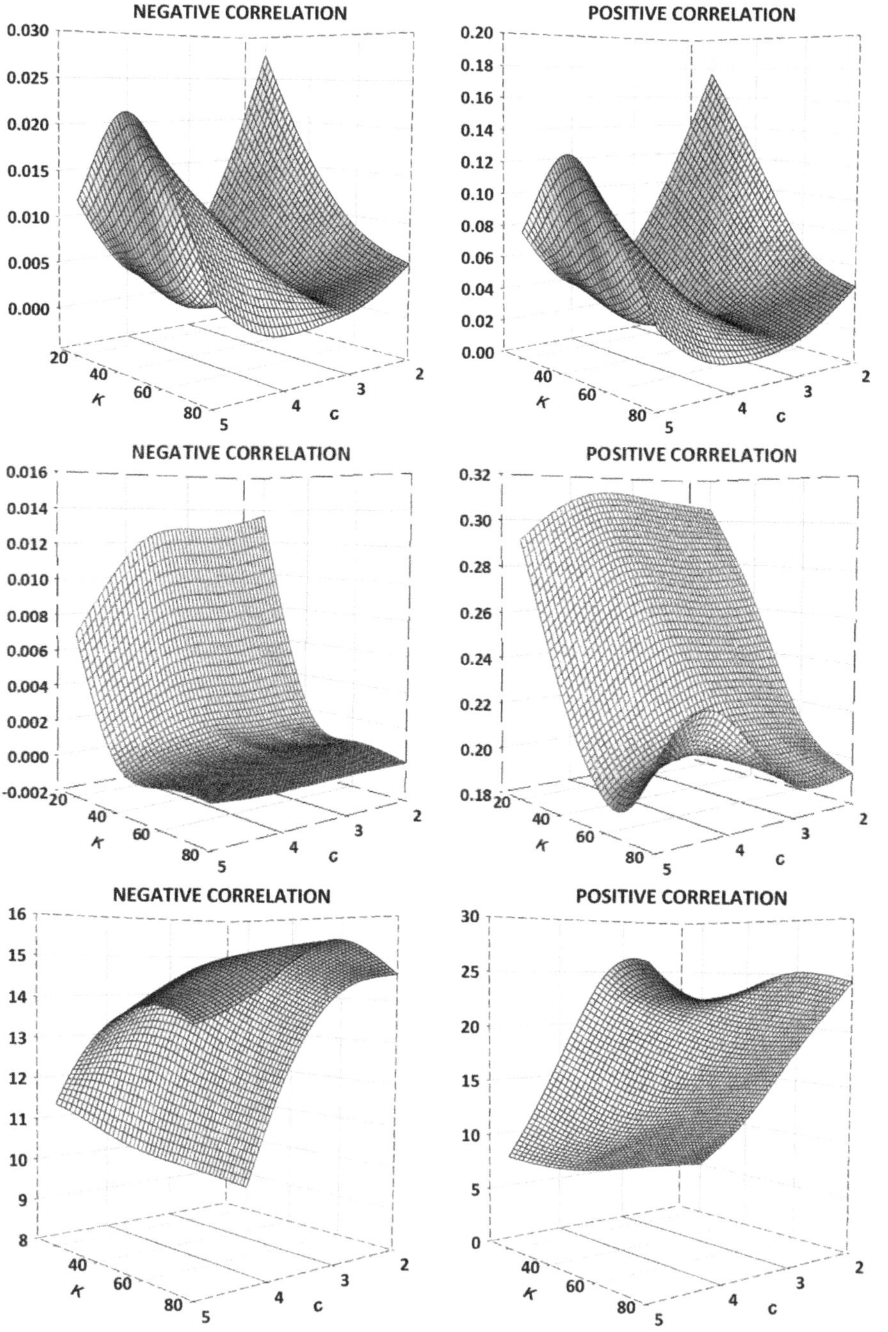

Figure 6.24. *Plots of P(idle) (top plot), P(loss) (middle plot) and L_s for example 6.14*

6.2.3. PH/M/c/K *queue*

This model, with single arrivals, has the inter-arrival times modeled using a CPH, and using the setup of section 6.2.2 we can obtain the steady-state probability vector explicitly in terms of the input parameters (see Neuts (1981)). Suppose that the CPH representation of the arrival process is of dimension m and is given by $(\boldsymbol{\alpha}, T)$. Taking $D_0 = T$ and $D_1 = \boldsymbol{T^0}\,\boldsymbol{\alpha}$ of order m, the generator, Q, for the current model is of the form (note that $Q^* = D_0 + D_1$):

$$Q = \begin{pmatrix} T & \boldsymbol{T^0}\,\boldsymbol{\alpha} & 0 & \cdots & 0 & 0 & \cdots & 0 & 0 \\ \mu I & T-\mu I & \boldsymbol{T^0}\,\boldsymbol{\alpha} & \cdots & 0 & 0 & \cdots & 0 & 0 \\ 0 & 2\mu I & T-2\mu I & \cdots & 0 & 0 & \cdots & 0 & 0 \\ \vdots & \vdots & \vdots & \cdots & \vdots & \vdots & \cdots & \vdots & \vdots \\ 0 & 0 & 0 & \cdots & c\mu I & T-c\mu I & \cdots & 0 & 0 \\ \vdots & \vdots & \vdots & \cdots & \vdots & \vdots & \cdots & \vdots & \vdots \\ 0 & 0 & 0 & \cdots & 0 & 0 & \cdots & T-c\mu I & \boldsymbol{T^0}\,\boldsymbol{\alpha} \\ 0 & 0 & 0 & \cdots & 0 & 0 & \cdots & c\mu I & Q^*-c\mu I \end{pmatrix}, \quad [6.11]$$

where $Q^* = T + \boldsymbol{T^0}\,\boldsymbol{\alpha}$. Defining $M_i = (i+1)\mu\,(i\,\mu I - i\,\mu\boldsymbol{e}\,\boldsymbol{\alpha} - T)^{-1},\ 1 \le i \le c-1$, and $M_c = c\mu(c\mu I - c\mu\boldsymbol{e}\,\boldsymbol{\alpha} - T)^{-1}$, we see that the steady-state vector, $\boldsymbol{y} = (\boldsymbol{y}_0, \cdots, \boldsymbol{y}_{K+c})$, of Q is:

$$\begin{aligned} \boldsymbol{y}_0 &= \mu\boldsymbol{y}_1(-T)^{-1}, \\ \boldsymbol{y}_i &= \boldsymbol{y}_{i+1}M_i,\ 1 \le i \le c-1, \\ \boldsymbol{y}_i &= \boldsymbol{y}_{i+1}M_c,\ c \le i \le K+c-1, \\ \boldsymbol{y}_{K+c} &= d\,\boldsymbol{\alpha}M_c, \end{aligned} \quad [6.12]$$

where d is the normalizing constant.

EXAMPLE 6.15.– In this example, we consider the two renewal arrivals, namely, the ones dealing with Erlang and hyperexponential, as mentioned in example 6.13, but with the restriction that arrivals occur singly. All other parameters remain the same. The plots of the CDF of the waiting time in the system of an admitted customer are displayed in Figures 6.25 and 6.26, and in Figure 6.27 we display the three key measures under varying scenarios.

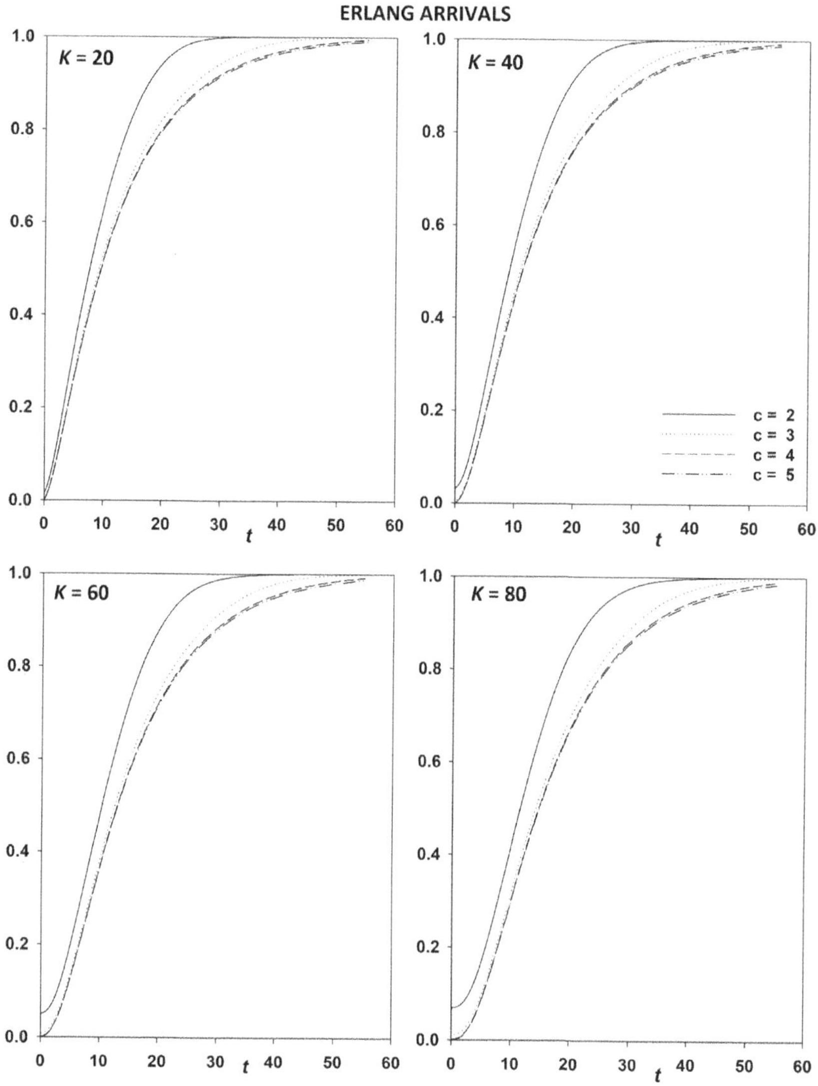

Figure 6.25. *Plot of the CDF of the waiting time in the system of an admitted customer for Erlang arrivals for example 6.15*

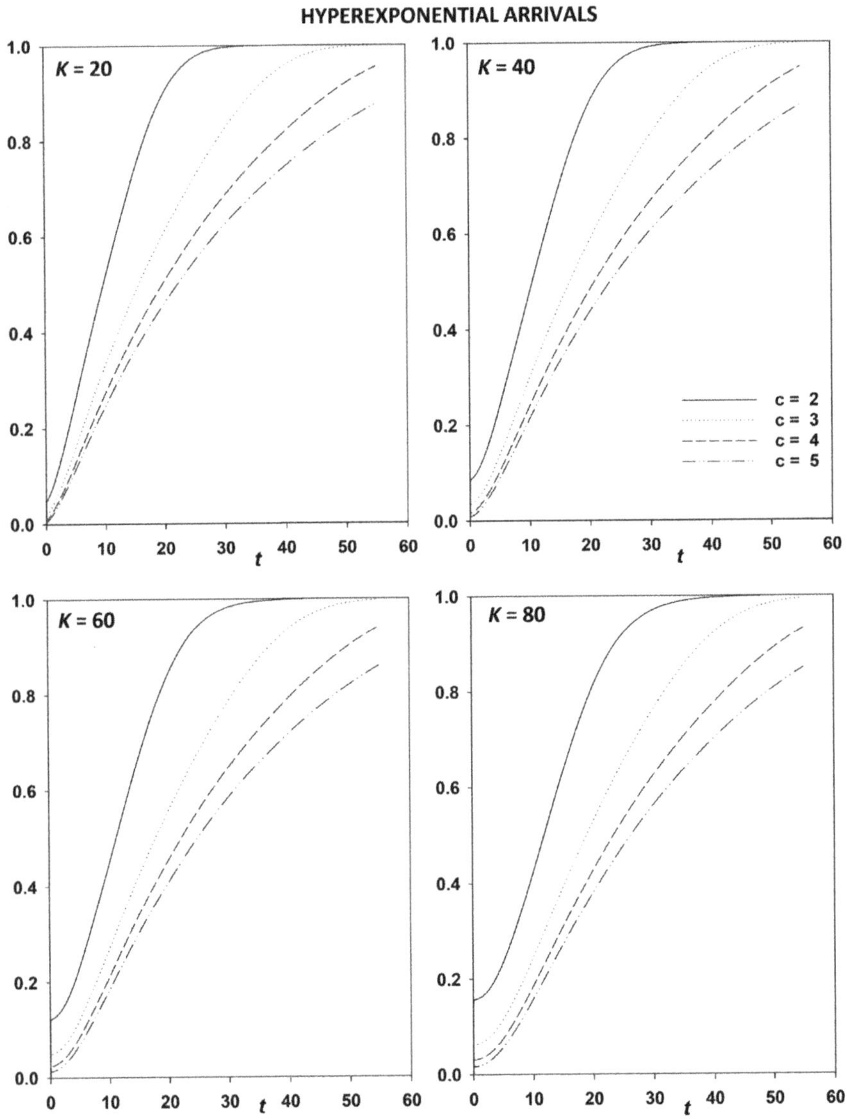

Figure 6.26. *Plot of the CDF of the waiting time in the system of an admitted customer for hyperexponential arrivals for example 6.15*

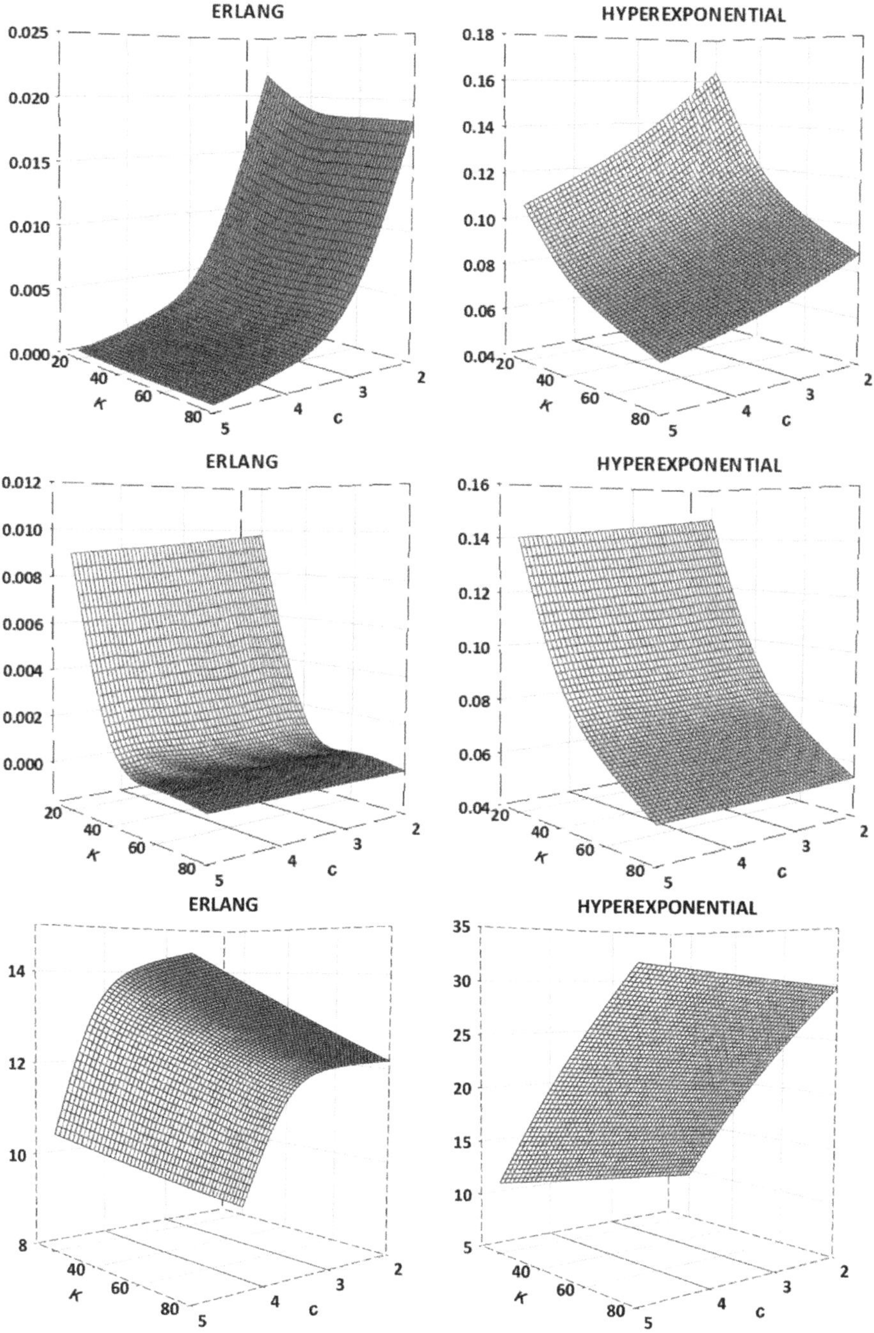

Figure 6.27. *Plots of P(idle) (top plot), P(loss) (middle plot) and L_s for example 6.15*

6.2.4. *M/M/c/K queue*

The model $M/M/c/K$ is the simplest among all the finite-capacity queueing models (in continuous-time). Here, the generator of the system (using the same convention as in previous sections with regard to the state space description) is given by:

$$Q = \begin{pmatrix} -\lambda & \lambda & 0 & \cdots & 0 & 0 & \cdots & 0 & 0 \\ \mu & -\lambda-\mu & \lambda & \cdots & 0 & 0 & \cdots & 0 & 0 \\ 0 & 2\mu & -\lambda-2\mu & \cdots & 0 & 0 & \cdots & 0 & 0 \\ \vdots & \vdots & \vdots & \cdots & \vdots & \vdots & \cdots & \vdots & \vdots \\ 0 & 0 & 0 & \cdots & c\mu & -\lambda-c\mu & \cdots & 0 & 0 \\ \vdots & \vdots & \vdots & \cdots & \vdots & \vdots & \cdots & \vdots & \vdots \\ 0 & 0 & 0 & \cdots & 0 & 0 & \cdots & -\lambda-c\mu & \lambda \\ 0 & 0 & 0 & \cdots & 0 & 0 & \cdots & c\mu & -c\mu \end{pmatrix}. \quad [6.13]$$

We can obtain the steady-state vector, say, $\boldsymbol{y} = (y_0, \cdots, y_{K+c})$, of Q either directly by solving the corresponding system of linear equations or we can particularize the expressions for this model using the $PH/M/c/K$ model. By particularizing, we get:

$$y_0 = \frac{\mu}{\lambda} y_1, \; y_i = \frac{(i+1)\mu}{\lambda} y_{i+1}, \; 1 \le i \le c-1,$$
$$y_i = \frac{c\mu}{\lambda} y_{i+1}, \; c \le i \le K+c-1, \;\; y_{K+c} = d\,\frac{c\mu}{\lambda}, \quad [6.14]$$

where d is the normalizing constant.

EXAMPLE 6.16.– In this example, we consider the customers to arrive singly according to a Poisson process with a rate of 1. All other parameters remain the same as in example 6.13. In Figure 6.28, we display the three key measures under varying scenarios, and the plots of the CDF of the waiting time in the system of an admitted customer are displayed in Figure 6.29.

These figures reveal some interesting observations and are left as exercises.

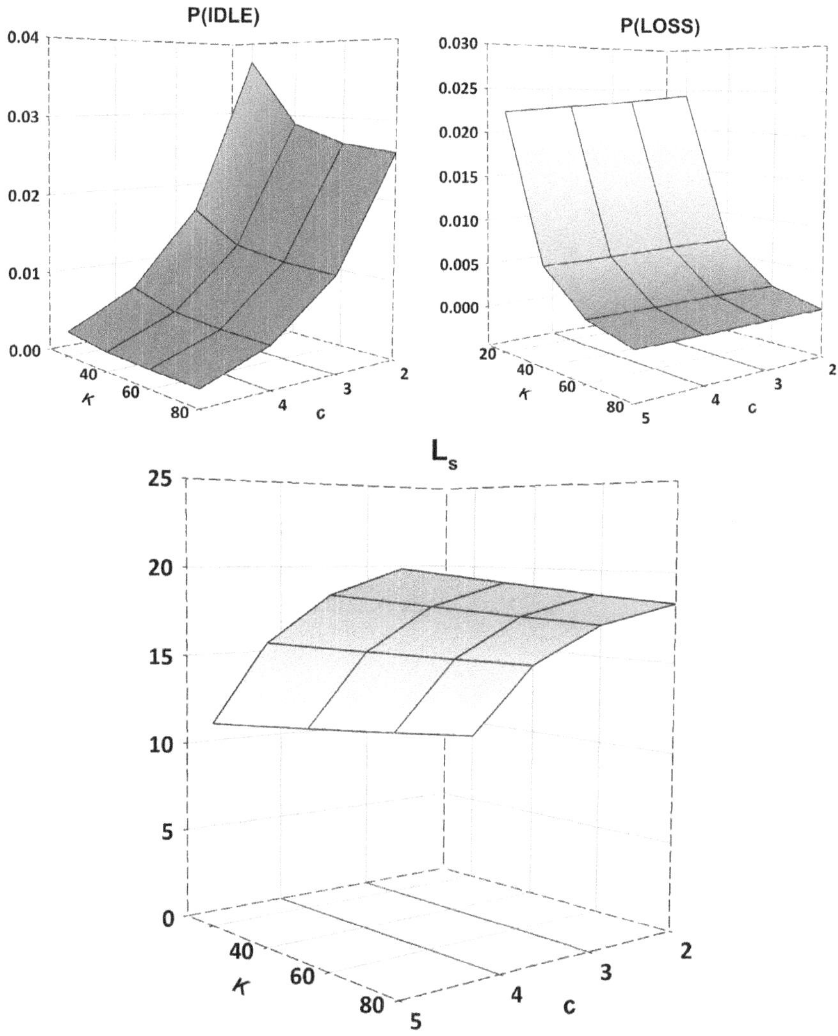

Figure 6.28. *Plots of three key measures for the Poisson arrivals for example 6.16*

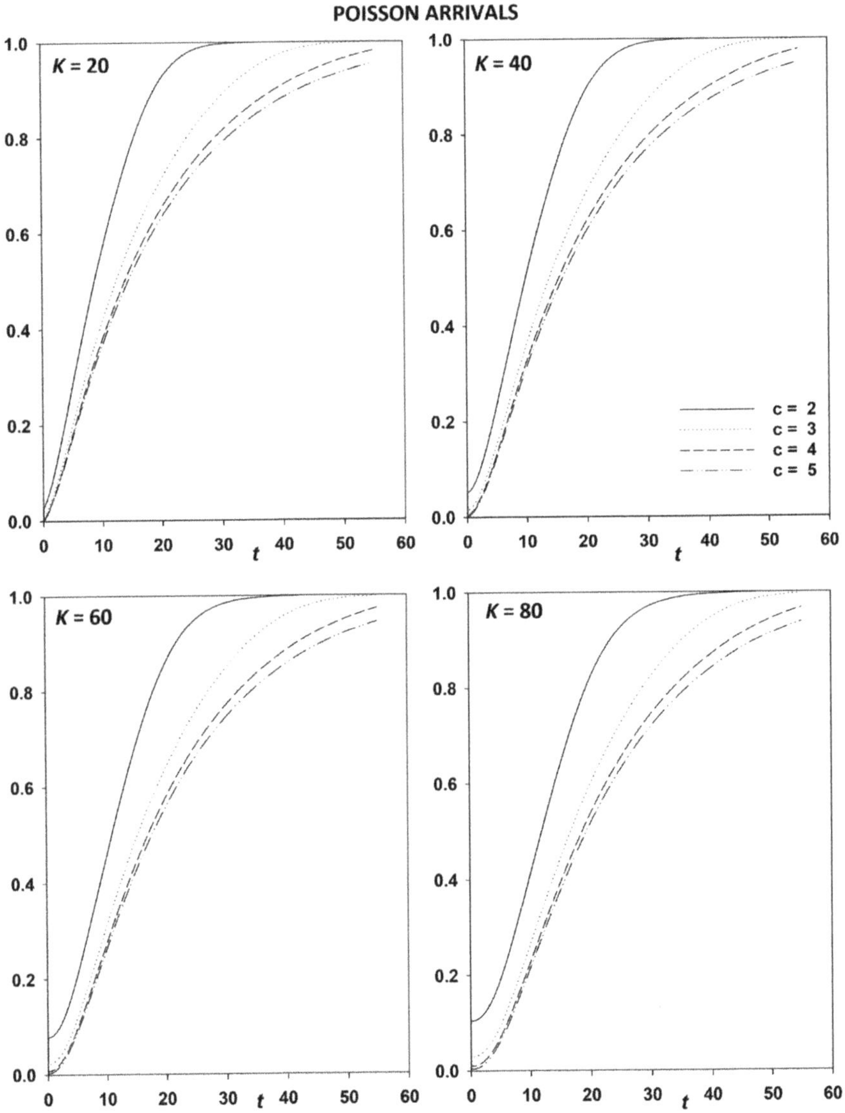

Figure 6.29. *Plots of the CDF of the waiting time in the system of an admitted customer for Poisson arrivals for example 6.16*

6.3. Exercises

Exercise 6.1. Referring to example 6.1, look at the counterpart infinite-capacity queue and compare the results. This requires developing codes for both finite and infinite capacity systems.

Exercise 6.2. Referring to example 6.2, highlight key points inferred from the plot and the table given there.

Exercise 6.3. Referring to example 6.2, look at the counterpart infinite-capacity queue and compare the results. This requires developing codes for both finite and infinite capacity systems.

Exercise 6.4. Referring to the $MAP/PH/1/K$ model with generator given in equation [6.1], develop a code to find the steady-state probability vector as well as key system performance measures. Verify your code and the results with the results of examples 6.3 through 6.7.

Exercise 6.5. Referring to example 6.3, look at the counterpart infinite-capacity queue and compare the results. This requires developing codes for both finite and infinite capacity systems. Summarize key observations from these two systems.

Exercise 6.6. Referring to example 6.4, look at the counterpart infinite-capacity queue and compare the results. This requires developing codes for both finite and infinite capacity systems. Summarize key observations from these two systems.

Exercise 6.7. Referring to example 6.5, look at the counterpart infinite-capacity queue and compare the results. This requires developing codes for both finite and infinite capacity systems. Summarize key observations from these two systems.

Exercise 6.8. Referring to example 6.6, look at the counterpart infinite-capacity queue and compare the results. This requires developing codes for both finite and infinite capacity systems. Summarize key observations from these two systems.

Exercise 6.9. Referring to example 6.7, look at the counterpart infinite-capacity queue and compare the results. This requires developing codes for both finite and infinite capacity systems. Summarize key observations from these two systems.

Exercise 6.10. For the $BMAP/M/c/K$ queueing model, show that the conditional PMF of the number lost is independent of K.

Exercise 6.11. For the $MAP/M/c/K$ queueing model, prove that the mean waiting time in the system is as given in equation [6.10].

Exercise 6.12. Referring to example 6.14, summarize key observations based on the graphs displayed in that example.

Exercise 6.13. Referring to example 6.15, summarize key observations based on the graphs displayed in that example.

Exercise 6.14. Rewrite the steady-state equations involving the vector $\boldsymbol{y}$ for the $BMAP/PH/1/K$ queue by exploiting the coefficient matrices. Develop a code implementing these equations and verify the results of examples 6.1 and 6.2.

Exercise 6.15. For the $MAP/PH/1/K$ queue, write down the steady-state equations by exploiting the coefficient matrices. Develop a code implementing these equations and verify the results of examples 6.3 and 6.4.

Exercise 6.16. Prove the expression given in equation [6.2] for the mean waiting time in the system for the $MAP/PH/1/K$ queue.

Exercise 6.17. Prove the expression given in equation [6.10] for the mean waiting time in the system for the $MAP/M/c/K$ queue.

Exercise 6.18. Prove the expression given in equation [6.12] for the $PH/M/c/K$ queue.

Exercise 6.19. Prove the expression given in equation [6.14] for the mean waiting time in the system for the $M/M/c/K$ queue.

Exercise 6.20. Show that for a $BMAP/PH/1/K$ queueing model with geometric batch size, the conditional probability mass function of the number of lost customers is also geometric, and the parameters are identical.

Exercise 6.21. Show that for a $BMAP/M/c/K$ queueing model with geometric batch size, the conditional probability mass function of the number of lost customers is also geometric, and the parameters are identical.

Exercise 6.22. Suppose that the customers arrive according to a BMAP in which the inter-arrival times of the groups are Erlang of order 5 with parameter $5\,\lambda_g$ in each state. The group size is assumed to follow a Poisson distribution (POB) with parameter 1.25. Taking the group arrival rate, $\lambda_g = 1/2.25$ (in this case normalize the parameter matrices of the BMAP accordingly), verify that the arrival rate λ of the customers is 1. The system has a finite buffer such that the system can have at most K customers in the system. There is a single server offering services on an FCFS basis. The service rate is denoted by μ and the service times are modeled using one of the three CPH distributions: (i) ERS: Erlang of order 4 with parameter $4\,\mu$; (ii) EXS: Exponential with rate μ; and (iii) HES: Hyperexponential with parameters

$121\mu/60, 121\mu/120, 121\mu/30$ and $121\mu/500$. Analyze this finite-capacity queueing system assuming $\mu = 1/0.9$, vary $K = 20, 30, 50$, and compute $P(idle)$, $P(loss)$, μ_{CL}, L_s, σ_s and l_i, $1 \leq i \leq 5$, where l_i stands for the conditional probability that i customers are lost given that a loss occurs at an arrival epoch.

Exercise 6.23. Referring to exercise 6.22, suppose that the batch size is now assumed to follow a geometric distribution (GEO) with parameter 0.8 such that the mean batch size is 1.25. Take $\lambda_g = 0.8$ and analyze the model. Compute $P(idle)$, $P(loss)$, μ_{CL}, L_s, σ_s and l_i, $1 \leq i \leq 5$.

Exercise 6.24. Referring to exercise 6.22, suppose that the batch size is now assumed to follow a uniform distribution (UFB) on $\{1, 2, 3, 4\}$. Noting that the mean batch size is 2.5, taking $\lambda_g = 0.4$, verify that $\lambda = 1$. Analyze the model and compute $P(idle)$, $P(loss)$, μ_{CL}, L_s, σ_s and l_i, $1 \leq i \leq 5$.

Exercise 6.25. Referring to exercise 6.22, suppose that the batch size is now assumed to follow a DPH distribution (DPB) with representation:

$$\alpha = (0.73970, 0.26030), \quad T = \begin{pmatrix} 0.1 & 0.2 \\ 0.1521 & 0.2341 \end{pmatrix}.$$

Verify that the mean batch size is 1.5. Hence, taking $\lambda_g = 2/3$, verify that $\lambda = 1$. Analyze the model and compute $P(idle)$, $P(loss)$, μ_{CL}, L_s, σ_s and l_i, $1 \leq i \leq 5$.

Exercise 6.26. Referring to exercises 6.22 through 6.25, suppose that the inter-arrival times of the groups are exponential with parameter λ_g. Compute first the values of λ_g which depend on the type of batch size distribution. Second, compute $P(idle)$, $P(loss)$, μ_{CL}, L_s, σ_s and l_i, $1 \leq i \leq 5$ under various batch sizes.

Exercise 6.27. Referring to exercises 6.22 through 6.25, suppose that the inter-arrival times of the groups are hyperexponential with parameters 10, 5, 2.5, 1.25, and 0.625 with mixing probabilities, respectively, given by 0.5 0.3, 0.15, 0.04, and 0.01. This hyperexponential will be normalized so that a desired value for λ_g is obtained. That is, λ_g is such that given the type of batch size distribution and its parameter(s), the average arrival rate of customers, $\lambda = 1$. Compute first the values of λ_g. Second, compute $P(idle)$, $P(loss)$, μ_{CL}, L_s, σ_s and l_i, $1 \leq i \leq 5$ under various batch sizes.

Exercise 6.28. Referring to exercises 6.22 through 6.25, suppose that the customers arrive according to a BMAP with representation:

$$D_0 = \lambda_g \begin{pmatrix} -1.75 & 1.75 & 0 & 0 \\ 0 & -1.75 & 1.75 & 0 \\ 0 & 0 & -1.75 & 0 \\ 0 & 0 & 0 & -3.5 \end{pmatrix},$$

$$D_k = \lambda_g\, a_k \begin{pmatrix} 0 & 0\;0 & 0 \\ 0 & 0\;0 & 0 \\ 0.0175 & 0\;0 & 1.7325 \\ 3.465 & 0\;0 & 0.035 \end{pmatrix}, \; k \geq 1,$$

where $\{a_k\}$ is the batch size distribution. Compute first the values of λ_g which depend on the type of batch size distribution. Second, compute $P(idle)$, $P(loss)$, μ_{CL}, L_s, σ_s and l_i, $1 \leq i \leq 5$ under various batch sizes.

Exercise 6.29. Referring to exercises 6.22 through 6.25, suppose that the customers arrive according to a BMAP with representation:

$$D_0 = \lambda_g \begin{pmatrix} -1.75 & 1.75 & 0 & 0 \\ 0 & -1.75 & 1.75 & 0 \\ 0 & 0 & -1.75 & 0 \\ 0 & 0 & 0 & -3.5 \end{pmatrix},$$

$$D_k = \lambda_g\, a_k \begin{pmatrix} 0 & 0\;0 & 0 \\ 0 & 0\;0 & 0 \\ 1.7325 & 0\;0 & 0.0175 \\ 0.035 & 0\;0 & 3.465 \end{pmatrix}, \; k \geq 1,$$

where $\{a_k\}$ is the batch size distribution. Verify first the values of λ_g which depend on the type of batch size distribution. Second, compute $P(idle)$, $P(loss)$, μ_{CL}, L_s, σ_s and l_i, $1 \leq i \leq 5$ under various batch sizes.

Exercise 6.30. Referring to exercises 6.22 through 6.25, suppose that the customers arrive singly (i.e. $\lambda_g = \lambda = 1$). Consider the same three scenarios for the services. Denoting by W the waiting time in the system of an admitted customer, L_s, σ_s, μ_W and σ_W to be, respectively, the mean and the standard deviation of the number in the system and the mean and the standard deviation of the waiting time in the system of an admitted customer, compute these measures as well as the CDF, $F_W(t) = P(W \leq t)$ of W, at 0.1, 0.5, 1, 10, 20, 50 and 100.

Exercise 6.31. Referring to exercise 6.26, suppose that the customers arrive singly (i.e. $\lambda_g = \lambda = 1$). Consider the same three scenarios for the services. Denoting by W the waiting time in the system of an admitted customer, L_s, σ_s, μ_W and σ_W to be, respectively, the mean and the standard deviation of the number in the system and the mean and the standard deviation of the waiting time in the system of an admitted customer, compute these measures as well as the CDF, $F_W(t) = P(W \leq t)$ of W, at 0.1, 0.5, 1, 10, 20, 50 and 100.

Exercise 6.32. Referring to exercise 6.27, suppose that the customers arrive singly (i.e. $\lambda_g = \lambda = 1$). Consider the same three scenarios for the services. Denoting by W the waiting time in the system of an admitted customer, L_s, σ_s, μ_W and σ_W to be, respectively, the mean and the standard deviation of the number in the system and the mean and the standard deviation of the waiting time in the system of an admitted customer, compute these measures as well as the CDF, $F_W(t) = P(W \leq t)$ of W, at $0.1, 0.5, 1, 10, 20, 50$ and 100.

Exercise 6.33. Referring to exercise 6.28, suppose that the customers arrive singly (i.e. $\lambda_g = \lambda = 1$). Consider the same three scenarios for the services. Denoting by W the waiting time in the system of an admitted customer, L_s, σ_s, μ_W and σ_W to be, respectively, the mean and the standard deviation of the number in the system and the mean and the standard deviation of the waiting time in the system of an admitted customer, compute these measures as well as the CDF, $F_W(t) = P(W \leq t)$ of W, at $0.1, 0.5, 1, 10, 20, 50$ and 100.

Exercise 6.34. Referring to exercise 6.29, suppose that the customers arrive singly (i.e. $\lambda_g = \lambda = 1$). Consider the same three scenarios for the services. Denoting by W the waiting time in the system of an admitted customer, L_s, σ_s, μ_W and σ_W to be, respectively, the mean and the standard deviation of the number in the system and the mean and the standard deviation of the waiting time in the system of an admitted customer, compute these measures as well as the CDF, $F_W(t) = P(W \leq t)$ of W, at $0.1, 0.5, 1, 10, 20, 50$ and 100.

Exercise 6.35. Suppose that the customers arrive in Poisson batches with parameter 1.25 and that the inter-arrival times are Erlang of order 5 with parameter $20/9$ in each state. There are c servers and the service times are exponential with parameter $1/(0.99\,c)$. There is a limited waiting capacity of size K and hence at most $K + c$ customers may be present in the system. An arriving batch may partially be admitted subject to the available space. Study this model and compute $P(idle)$, $P(loss)$, μ_{CL}, L_s and σ_s when $K = 10, 15$, and $c = 2, 3$ and 4.

Exercise 6.36. Referring to exercise 6.35 suppose that the inter-arrival times are Erlang of order 3 with parameter $12/9$. There are c servers and the service times are exponential with parameter $1/(0.99\,c)$. There is a limited waiting capacity of size K and hence at most $K + c$ customers may be present in the system. An arriving batch may partially be admitted subject to the available space. Study this model and compute $P(idle)$, $P(loss)$, μ_{CL}, L_s and σ_s when $K = 10, 15$ and $c = 2, 3$, and 4.

Exercise 6.37. Referring to exercise 6.35, suppose that the inter-arrival times are exponential with parameter $4/9$. All other parameters remain the same as in that exercise. Study this model and compute $P(idle)$, $P(loss)$, μ_{CL}, L_s and σ_s when $K = 10, 15$, and $c = 2, 3$ and 4.

Exercise 6.38. Referring to exercise 6.35, suppose that the inter-arrival times are hyperexponential with parameters $6.4/9, 3.2/9$ and $1.28/9$, respectively, with mixing probabilities 0.7, 0.2 and 0.1. All other parameters remain the same as in that exercise. Study this model and compute $P(idle)$, $P(loss)$, μ_{CL}, L_s and σ_s when $K = 10, 15$, and $c = 2, 3$ and 4.

Exercise 6.39. Referring to exercise 6.35, suppose that the inter-arrival times are hyperexponential with parameters $8.72/9, 4.36/9, 2.18/9, 1.09/9$ and $0.545/9$, respectively, with mixing probabilities 0.5, 0.3, 0.15, 0.04 and 0.01. All other parameters remain the same as in that exercise. Study this model and compute $P(idle)$, $P(loss)$, μ_{CL}, L_s and σ_s when $K = 10, 15$, and $c = 2, 3$ and 4.

Exercise 6.40. Referring to exercise 6.35, suppose that the customers arrive according to a BMAP with representation:

$$D_0 = \frac{4}{9}\begin{pmatrix} -1.25 & 1.25 & 0 \\ 0 & -1.25 & 0 \\ 0 & 0 & -2.5 \end{pmatrix}, \quad D_k = \frac{4\,a_k}{9}\begin{pmatrix} 0 & 0 & 0 \\ 0.0125 & 0 & 1.2375 \\ 2.475 & 0 & 0.025 \end{pmatrix}, \quad k \geq 1,$$

where $\{a_k\}$ follows a Poisson distribution with parameter 1.25. All other parameters remain the same as in that exercise. Study this model and compute $P(idle)$, $P(loss)$, μ_{CL}, L_s and σ_s when $K = 10, 15$ and $c = 2, 3$ and 4.

Exercise 6.41. Referring to exercise 6.35, suppose that the customers arrive according to a BMAP with representation:

$$D_0 = \begin{pmatrix} -1 & 1 & 0 & 0 & 0 \\ 0 & -1 & 1 & 0 & 0 \\ 0 & 0 & -1 & 1 & 0 \\ 0 & 0 & 0 & -1 & 0 \\ 0 & 0 & 0 & 0 & -2 \end{pmatrix}, \quad D_k = a_k\begin{pmatrix} 0 & 0 & 0 & 0 & 0 \\ 0 & 0 & 0 & 0 & 0 \\ 0 & 0 & 0 & 0 & 0 \\ 0.01 & 0 & 0 & 0 & 0.99 \\ 1.98 & 0 & 0 & 0 & 0.02 \end{pmatrix}, \quad k \geq 1,$$

where $\{a_k\}$ follows a Poisson distribution with parameter 1.25. All other parameters remain the same as in that exercise. Study this model and compute $P(idle)$, $P(loss)$, μ_{CL}, L_s and σ_s when $K = 10, 15$ and $c = 2, 3$ and 4.

Exercise 6.42. Referring to exercise 6.35, suppose that the customers arrive according to a BMAP with representation:

$$D_0 = \frac{4}{9}\begin{pmatrix} -1.25 & 1.25 & 0 \\ 0 & -1.25 & 0 \\ 0 & 0 & -2.5 \end{pmatrix}, \quad D_k = \frac{4\,a_k}{9}\begin{pmatrix} 0 & 0 & 0 \\ 1.2375 & 0 & 0.0125 \\ 0.025 & 0 & 2.475 \end{pmatrix}, \quad k \geq 1,$$

where $\{a_k\}$ follows a Poisson distribution with parameter 1.25. All other parameters remain the same as in that exercise. Study this model and compute $P(idle)$, $P(loss)$, μ_{CL}, L_s and σ_s when $K = 10, 15$ and $c = 2, 3$ and 4.

Exercise 6.43. Referring to exercise 6.35, suppose that the customers arrive according to a BMAP with representation:

$$D_0 = \begin{pmatrix} -1 & 1 & 0 & 0 & 0 \\ 0 & -1 & 1 & 0 & 0 \\ 0 & 0 & -1 & 1 & 0 \\ 0 & 0 & 0 & -1 & 0 \\ 0 & 0 & 0 & 0 & -2 \end{pmatrix}, \quad D_k = a_k \begin{pmatrix} 0 & 0\,0\,0 & 0 \\ 0 & 0\,0\,0 & 0 \\ 0 & 0\,0\,0 & 0 \\ 0.99 & 0\,0\,0 & 0.01 \\ 0.02 & 0\,0\,0 & 1.98 \end{pmatrix}, \quad k \geq 1,$$

where $\{a_k\}$ follows a Poisson distribution with parameter 1.25. All other parameters remain the same as in that exercise. Study this model and compute $P(idle)$, $P(loss)$, μ_{CL}, L_s and σ_s when $K = 10, 15$, and $c = 2, 3$ and 4.

Exercise 6.44. Suppose that the customers arrive singly to a finite capacity c-server queueing system. That is, at most there can be $(K + c)$ customers in the system. The homogeneous servers offer services on an FCFS basis. The inter-arrival times are Erlang of order 5 with parameter 5 in each state. The service times are exponential with parameter $1/0.95\,c$ so that the traffic intensity is 0.95. Analyze the model and compute $P(idle)$, $P(loss)$, L_s, σ_s, μ_W, σ_W and the CDF of W at time points: 0.1, 0.5, 1, 10, 20, 50 and 100, by varying $K = 20, 40, 60$ and 80, and $c = 2, 3, 4$ and 5.

Exercise 6.45. Suppose that the customers arrive singly to a finite capacity c-server queueing system. That is, at most there can be $(K + c)$ customers in the system. The homogeneous servers offer services on an FCFS basis. The inter-arrival times are exponential parameter 1. The service times are exponential with parameter $1/0.95\,c$ so that the traffic intensity is 0.95. Analyze the model and compute $P(idle)$, $P(loss)$, L_s, σ_s, μ_W, σ_W, and the CDF of W at time points: 0.1, 0.5, 1, 10, 20, 50 and 100, by varying $K = 20, 40, 60$ and 80, and $c = 2, 3, 4$ and 5.

Exercise 6.46. Suppose that the customers arrive singly to a finite capacity c-server queueing system. That is, at most there can be $(K + c)$ customers in the system. The homogeneous servers offer services on an FCFS basis. The inter-arrival times are hyperexponential with parameters 2.18, 1.09, 0.545, 0.2725 and 0.13625 mixing probabilities 0.5, 0.30, 0.15, 0.04, and 0.01, respectively. The service times are exponential with parameter $1/0.95\,c$ so that the traffic intensity is 0.95. Analyze the model and compute $P(idle)$, $P(loss)$, L_s, σ_s, μ_W, σ_W, and the CDF of W at time points: 0.1, 0.5, 1, 10, 20, 50, and 100, by varying $K = 20, 40, 60$ and 80, and $c = 2, 3, 4$ and 5.

Exercise 6.47. Suppose that the customers arrive singly to a finite capacity c-server queueing system according to a MAP with representation:

$$D_0 = \begin{pmatrix} -1.75 & 1.75 & 0 & 0 \\ 0 & -1.75 & 1.75 & 0 \\ 0 & 0 & -1.75 & 0 \\ 0 & 0 & 0 & -3.5 \end{pmatrix}, \quad D_1 = \begin{pmatrix} 0 & 0\,0 & 0 \\ 0 & 0\,0 & 0 \\ 0.0175 & 0\,0 & 1.7325 \\ 3.465 & 0\,0 & 0.035 \end{pmatrix}.$$

That is, at most there can be $(K+c)$ customers in the system. The homogeneous servers offer services on an FCFS basis. The service times are exponential with parameter $1/0.95\,c$ so that the traffic intensity is 0.95. Analyze the model and compute $P(idle)$, $P(loss)$, L_s, σ_s, μ_W, σ_W, and the CDF of W at time points: 0.1, 0.5, 1, 10, 20, 50 and 100, by varying $K = 20, 40, 60$, and 80, and $c = 2, 3, 4$ and 5.

Exercise 6.48. Suppose that the customers arrive singly to a finite capacity c-server queueing system according to a MAP with representation:

$$D_0 = \begin{pmatrix} -1.75 & 1.75 & 0 & 0 \\ 0 & -1.75 & 1.75 & 0 \\ 0 & 0 & -1.75 & 0 \\ 0 & 0 & 0 & -3.5 \end{pmatrix}, \quad D_1 = \begin{pmatrix} 0 & 0 & 0 & 0 \\ 0 & 0 & 0 & 0 \\ 1.7325 & 0 & 0 & 0.0175 \\ 0.035 & 0 & 0 & 3.465 \end{pmatrix}.$$

That is, at most there can be $(K+c)$ customers in the system. The homogeneous servers offer services on an FCFS basis. The service times are exponential with parameter $1/0.95\,c$ so that the traffic intensity is 0.95. Analyze the model and compute $P(idle)$, $P(loss)$, L_s, σ_s, μ_W, σ_W and the CDF of W at time points: 0.1, 0.5, 1, 10, 20, 50 and 100, by varying $K = 20, 40, 60$ and 80, and $c = 2, 3, 4$ and 5.

Exercise 6.49. For the $MAP/PH/1/K$ queueing model, develop a code to compute the steady-state probability vector.

Exercise 6.50. For the $PH/M/1/K$ queueing model, prove the equations given in [6.5]. Develop a code to compute the steady-state probability vector.

Exercise 6.51. For the $M/M/1/K$ queueing model, find the steady-state probability vector of the generator given in [6.8]. Develop a code to compute the steady-state probability vector.

Exercise 6.52. For the $M/PH/M/1/K$ queueing model, prove the equations given in [6.7]. Develop a code to compute the steady-state probability vector.

Exercise 6.53. For the $MAP/M/c/K$ queueing model, develop a code to compute the steady-state probability vector.

7

Simulation

Traditionally, simulation has been the last resort for the study of a (complex) stochastic model. However, the simulation tools have evolved over the years to a point that it is easy (in the sense of not having to write very complicated codes in most simulation software) to build models by dragging and dropping, modifying the parameters, validating the models and generating examples. For such complex models, the theoretical results are hard to establish, if they are at all feasible. Now simulation can be thought of as one of the earlier steps to analyze complex models. Thus, in this chapter, using one of the popular simulation software, namely, ARENA, we discuss a variety of models. This author has published a number of papers using simulation (with ARENA) and arrived at some interesting results. In this chapter, we present some of those results as well as give a brief but self-contained introduction to ARENA in the process. This will enable the readers to get started on simulating basic to advanced stochastic models. The references are given for those readers interested in acquiring deeper understanding of this powerful software.

It should be pointed out that even though we focus on ARENA, the basic ideas remain the same should a reader wish to use a different software. This is keeping in mind the syntax may differ from one software to another.

7.1. Introduction to ARENA

In this section, we display the "look" of ARENA based on Version 16. We refer the reader to the instructions provided with the software for installation. The opening screen of ARENA is as shown in Figure 7.1.

There are various tabs available in the project bar and these are displayed in Figures 7.2–7.6. Each has its own use. A brief description of the purpose of these project bars is given in these figures.

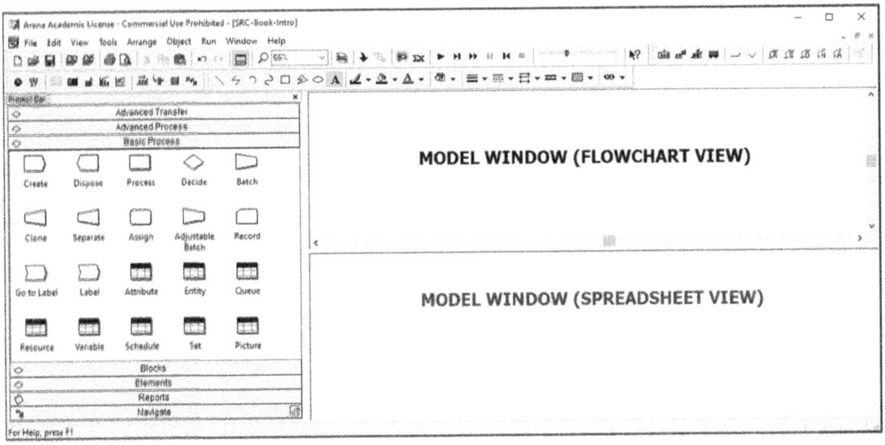

Figure 7.1. *ARENA model windows*

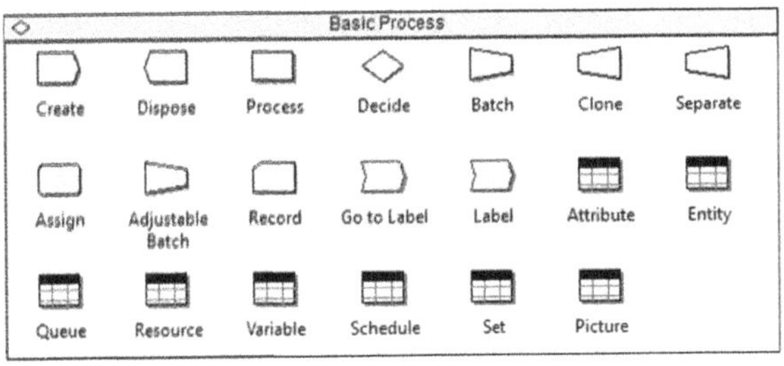

Basic Process

1. Highest level of modeling.
2. Ease of creating the model.
3. Greater flexibility.
4. For novice users.

Figure 7.2. *ARENA basic process*

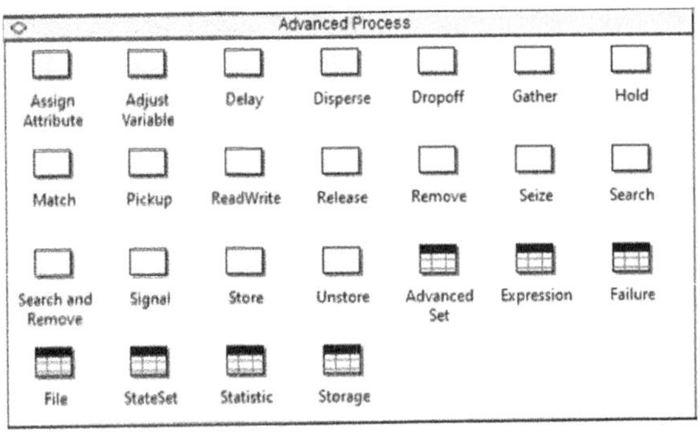

Advanced Basic Process

1. Supplements the basic one by providing additional and more detailed modules.
2. Suitable for experienced modelers.

Figure 7.3. *ARENA advanced process*

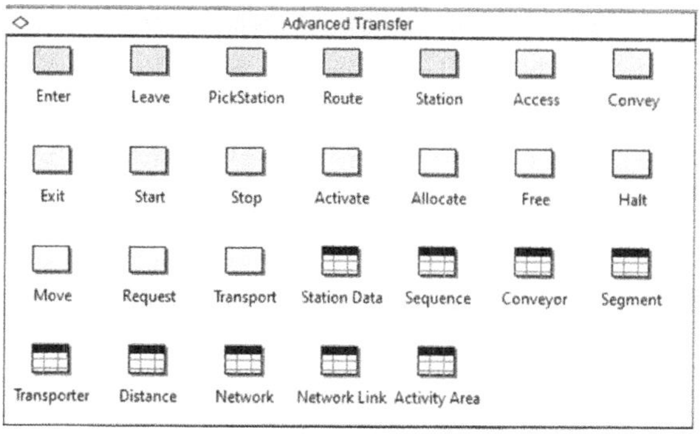

Advanced Transfer

1. Provides modules needed to transfer the entities from one place to another.
2. Has the modules for materials handling activities such as conveyors, AGVs, etc.
3. Adds flexibility to modeling.

Figure 7.4. *ARENA advanced transfer*

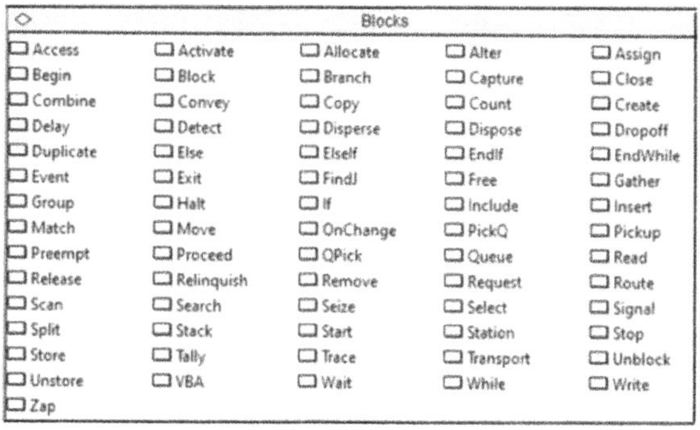

Blocks

1. Has basic functions used in creating modules in Basic Process, Advanced Process, and Advanced Transfer.
2. Has lower level of modeling capability.
3. Has special-purpose constructs not available in the highest-level modules.

Figure 7.5. *ARENA blocks*

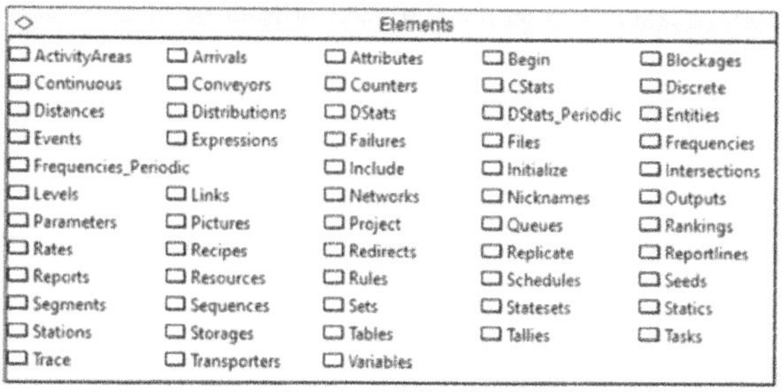

Elements

1. Has experiment frame modules.
2. Rarely need this as higher-level panels take care of these features.
3. Provides for modeling continuous systems.

Figure 7.6. *ARENA elements*

7.2. Model building using ARENA

Discussion of all the modules in ARENA will need its own book. However, in this section, we present the details of those modules that are used to generate the examples in this chapter. For additional details on these and other modules, we refer the reader to Kelton et al. (2010). Furthermore, one can visit the website of the software developer Rockwell Automation.

Before we discuss the selected modules, some key points related to ARENA are listed below.

1) The module name, which is optional but highly recommended to keep track of the modules, must be unique, not only among the same type, like CREATE, but also among different types. Thus, one cannot use the same name for both CREATE and PROCESS modules.

2) There are certain built-in names or identifiers within ARENA and these cannot be used as user-defined names for attributes or variables. We list a few of these built-in names that are used in this chapter. It should be pointed out that one does not need to remember these as they can be pulled from the pull-down menu when using the "Build Expression" option.

a) NQ – This stands for number in queue and returns the number of entities in the specified queue.

b) NR – This returns the number of busy resources at that instant of calling.

c) MR – This returns the capacity (integer value) of the specified resource.

d) TAVG – Gives the average value of the tally variable used.

e) TFIN - This stands for the time that moment simulation ends.

f) TNOW – This stands for the time at that moment an entity gets into the module that calls this. That is, this gives the current simulation time.

g) TSTD – Gives the standard deviation of the tally variable used.

h) TVALUE – This is to be used when one is interested to get the last value recorded in a "Record" module defined prior to calling this.

3) The readers are highly encouraged to proceed further by opening up their ARENA software to have a full understanding. This is especially needed for novice users of ARENA.

Description of selected modules

In Figure 7.7, we display the CREATE module and the needed input parameters for this module. This is an important module to start building a simulation model. Note that the entities, which correspond to customers or jobs or patients or machines or whatever the situation for which a simulation model is being built around, are created in this module. Additionally, when simulating a closed queueing system, such as a reliability system consisting of a finite number of machines that are subject to failures and repairs, we need this module to start the simulation.

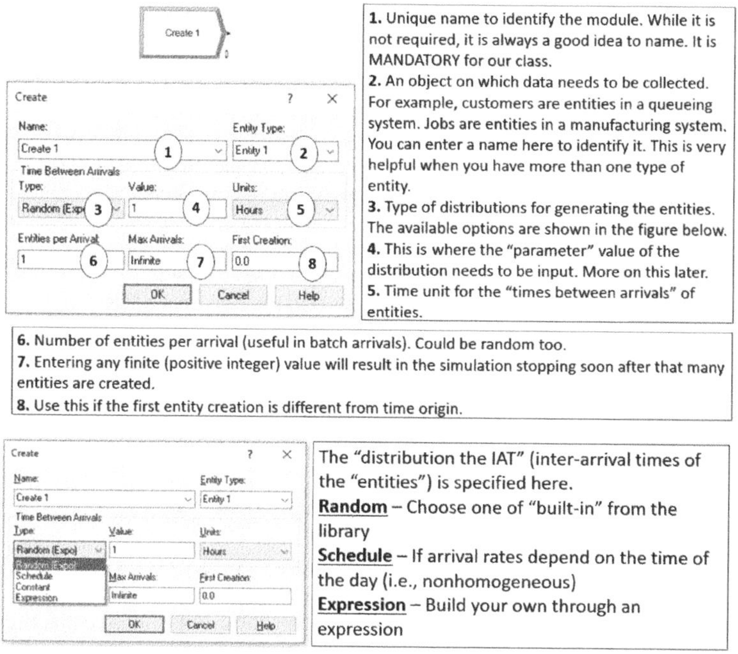

Figure 7.7. *CREATE module description*

Since batch arrivals as well as non-homogeneous arrivals are very common in practice, we briefly point out how to incorporate this in "CREATE" module. Suppose that the arrivals occur in batches of size 1, 2 and 4 with probabilities, respectively, given by 0.6, 0.25 and 0.15. To simulate this all one needs to do is insert the command DISC(0.6,1,0.85,2,1,4) in the window labeled 6 in Figure 7.7. What does DISC stands for? In ARENA's syntax, this represents a discrete probability function. The syntax for this DISC function is as follows: $DISC(p_1, v_1, p_2, v_2, \cdots)$, where p_i is the cumulative probability corresponding to the value v_i. Thus, if N denotes the random variable representing the batch size for the arrivals, then $P(N \leq v_i) = p_i$. If the batch size is to be modeled using, say, a Poisson random variable, entering POIS(3) +1 in that window, the probability mass function of N is then given by $P(N = n) = \dfrac{e^{-3}3^{n-1}}{(n-1)!}$, $n \geq 1$. It is worth pointing out that the arrival batch has to have at least one entity, and hence we need to add 1 to POIS(3); otherwise, a zero value (with probability e^{-3}) may be chosen.

In Figure 7.8, we display the instructions for creating schedule based arrivals to model non-homogeneous Poisson arrivals.

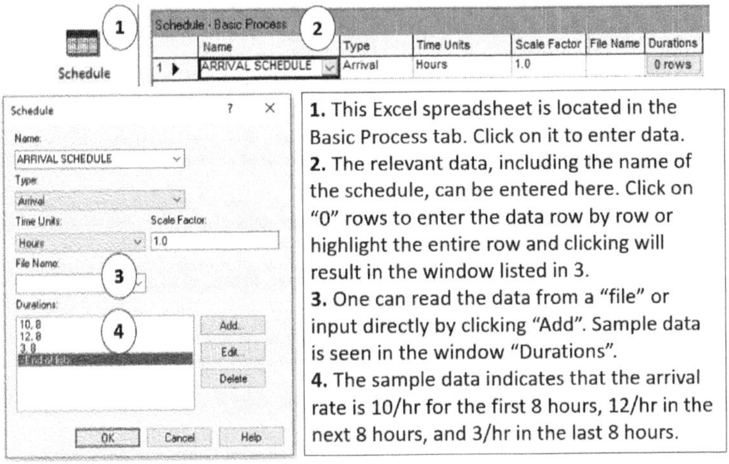

Figure 7.8. *SCHEDULE based CREATE module description*

However, if one needs to model non-homogenous non-Poisson arrivals, then using schedule option under "other" category, as shown in Figure 7.9, will be needed.

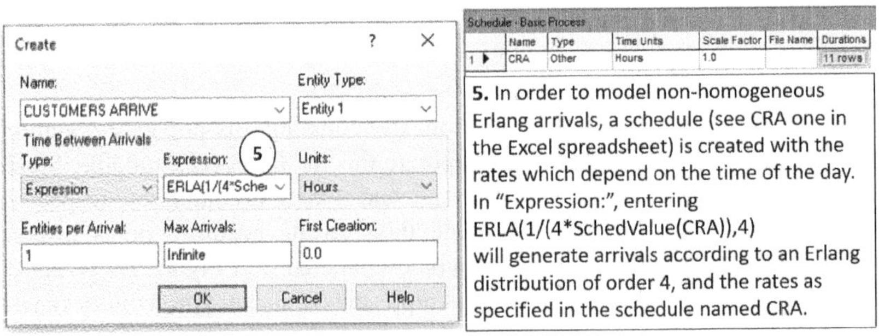

Figure 7.9. *SCHEDULE based CREATE module description for non-homogeneous arrivals*

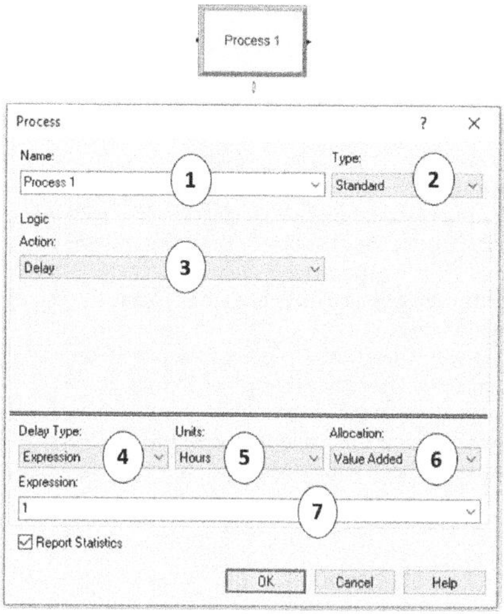

1. Unique name to identify the module.
2. Standard (signifies that all logic will be stored within this module) Submodel (logic will be hierarchically defined in a Submodel which can include any number of logic modules). For now, use only "Standard".
3. Available options are: "Delay", "Seize Delay", "Seize Delay Release", and "Delay Release". Each one has its own applications. Except for "Delay", one needs to specify the type of "Resources" to be seized and released.
4. Type of distribution for the delay in this module. In this window other options are: Constant, Normal, Triangular, and Uniform.
5. Time units for the delay time.
6. Ways to determine the processing time/costs to be allocated to the entity.
7. Here one can choose a variety of distributions that ARENA provides or one can build a new one.

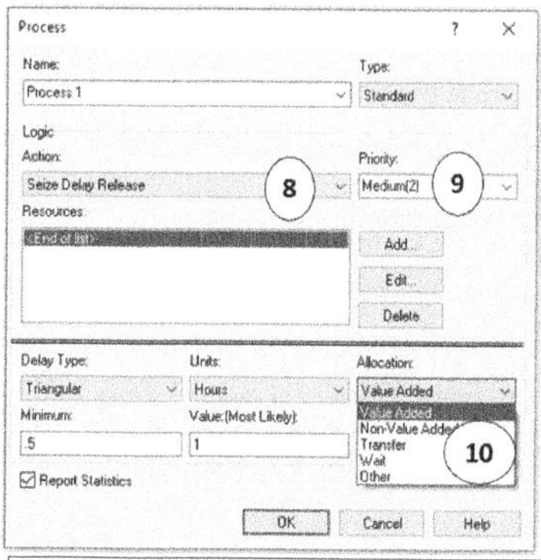

8. For this specific "Action", the newer window "Resources:" will appear.
9. Priority for the entity in getting service. Seen only for Seize Delay or Seize Delay Release option.
10. Generally, this is "Value Added" as getting service is the main reason the entities arrive to this module.

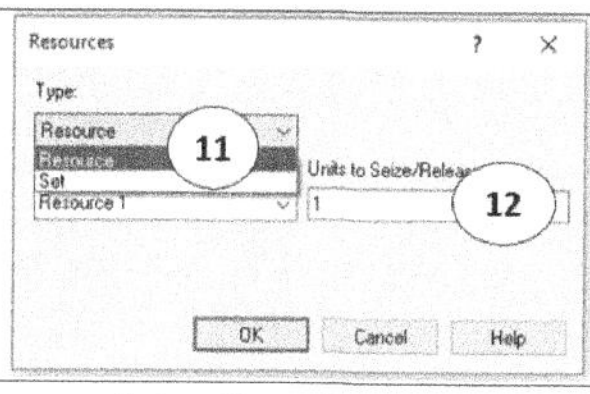

11. In this "Resources" window, select either Resource or Set as Type. Resource option requires a resource be mentioned (like server or operator) or use the default one, Resource 1. "Set" Type requires defining this Set with names of resources in a different place.
12. In this window specify how many "units" (or "resources") an entity needs to get processed. The capacity of the resources in the system needs to be specified in an Excel spreadsheet and not here.

13. In the "Type", when fixed capacity is chosen, the "Capacity" is where one inputs the number of resources available in the system. If "Based on Schedule" is chosen, then a schedule for the resources needs to be defined.

Figure 7.10. *PROCESS module description*

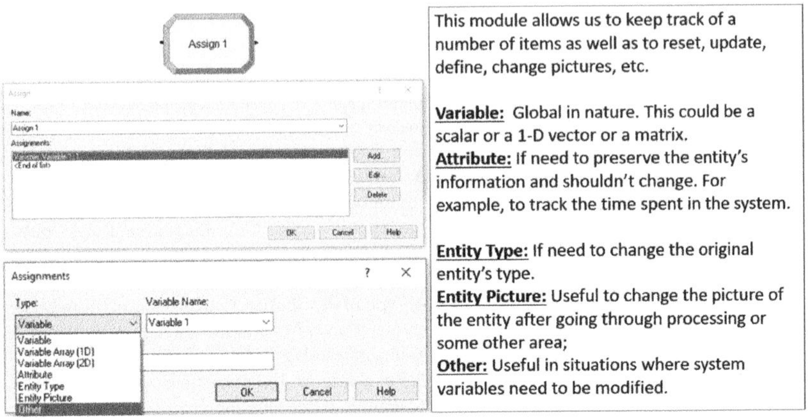

Figure 7.11. *ASSIGN module description*

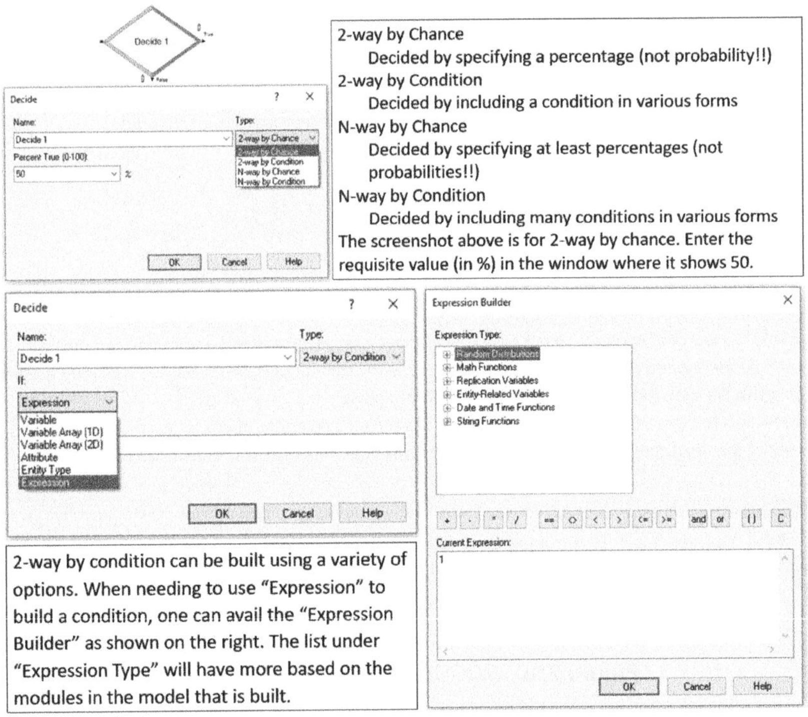

Figure 7.12. *DECIDE module description*

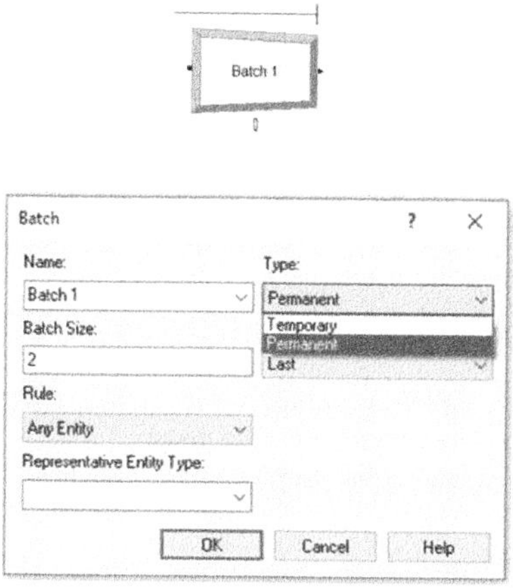

"Batch" module, as the name suggests, is used to batch entities.

There are two types: Temporary and Permanent. Batches of entities can be permanently or temporarily grouped.

Temporary batches must later be split using the Separate module.

Batches may be made with any specified number of entering entities or may be matched together based on an attribute.

Entities arriving at the Batch module are placed in a queue until the required number of entities has accumulated. Once accumulated, a new representative entity is created.

The entity type of the outgoing entity may be changed by specifying a representative entity type.

Rule: "Any Entity" will batch with no regard to entities; "By Attribute" will make batches based on indicated criteria that will pop up in a new window.

Figure 7.13. *BATCH module description*

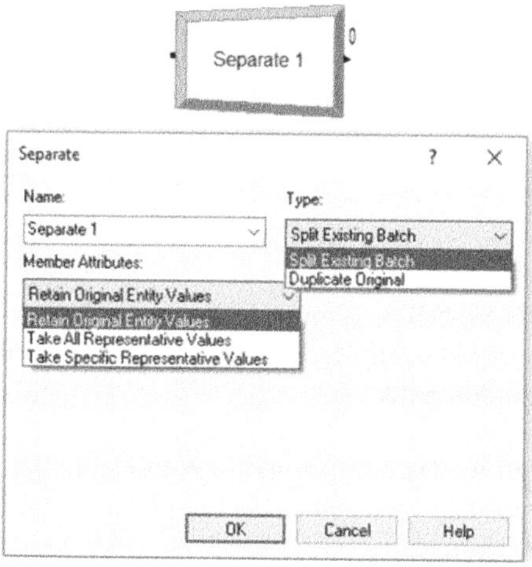

On occasions we need to separate the entities either at points of arrivals or at a later point in time when these entities are temporarily grouped for some activity.

Each of the two has its own function.

"Split Existing Batch" will separate the arriving entity (or batch) that was temporarily batched.

"Duplicate Original" is needed to duplicate (as many copies as required) original entities.

Figure 7.14. *SEPARATE module description*

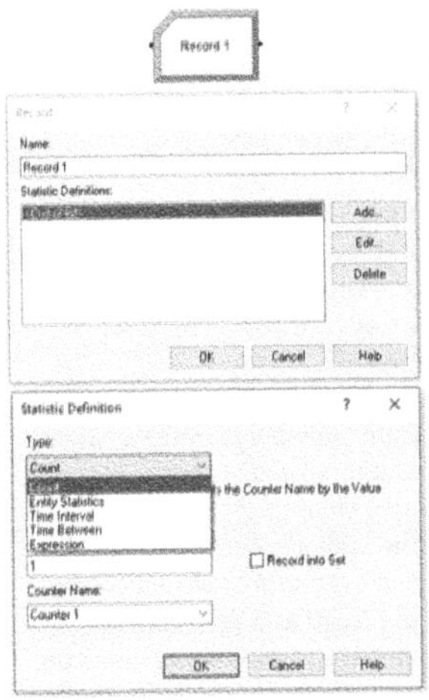

This module is important to capture key statistics. The diagram on the right shows the various options available. Each "Type" has its purpose and is described briefly below.

COUNT

Every time an entity passes through this module, the value will be increased by the amount given in "Value".

ENTITY STATISTICS

Generate statistics related to the entities

TIME INTERVAL

When attribute is used to keep track of time in system

TIME BETWEEN

This is used when needing to find the time between entities entering this

EXPRESSION

Need this when an expression is to be captured

Recording into Set is very helpful when dealing with more than one type of entities and statistics for each type that needs to be recorded.

Figure 7.15. *RECORD module description*

Dispose Module is needed to dispose entities unless there is a feedback for these. Not having the dispose module will accumulate the entities and may result in the simulation ending/crashing without running fully. So, it is important to keep this in mind for every model that is simulated.

Figure 7.16. *DISPOSE module description*

In Figure 7.17, we display the "RUN SETUP" commands. Some of the key tabs are explained below. The reader is encouraged to explore each and every one of these using their own software running on their computer.

Before running the simulation model, we need to set up a few things in the Run Setup. Choose this option from Run menu. The ones that are self-explanatory (such as Number of Replications) will NOT be explained.

Initialize between replications: When running more than one replication – (1) to erase all statistics up until now check the box in Statistics (default mode); and (2) to clean the system by removing all entities – the system will be brought back to the state when the model was originally simulated. Not to get the same simulated results you need to uncheck system.

Warm-up period: By specifying positive value here, ARENA will run the simulation for this duration before the replication starts. This is useful when trying to run the model in steady-state or when the machine needs to settle (like burn-in period).

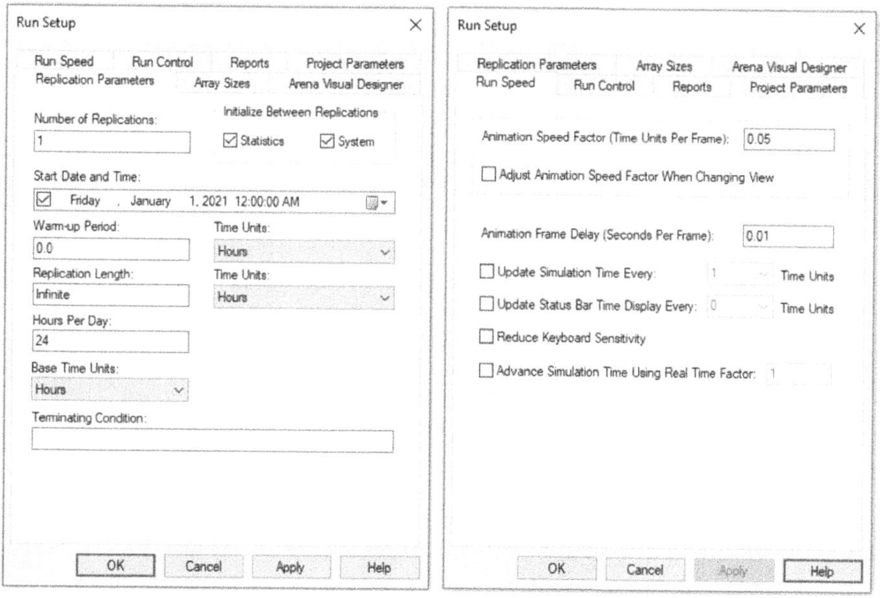

Figure 7.17. *RUN SETUP module description*

Hours per day: When modeling shifts or stores are not open for 24 hours, we change this to the required one.

Time units: Make sure the units are consistent with the model units. If minutes are used, then the base time units have to be changed to minutes. Replication length units need not be changed to the model units.

Terminating condition: Specify how you want to terminate the simulation. Even though the length of the replication is specified sometimes it is convenient to terminate when, say, there are 40 jobs served or when the queue size exceeds 50, etc.

Run speed: Here you can specify the "Animation Speed Factor" and the "Animation Frame Delay". The former is for the animation of the entities in the model developed (the smaller the number the slower the speed will be to see how they move); the latter is useful when animation is shifted to a new place. This is needed when the model is complex and you want to hide it from the animation. The larger the number the slower the entity moves.

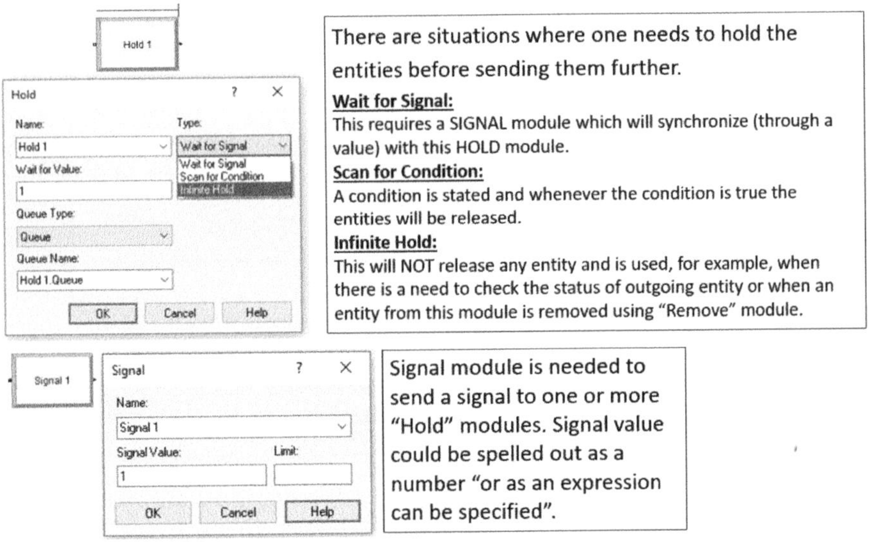

Figure 7.18. *HOLD & SIGNAL modules description*

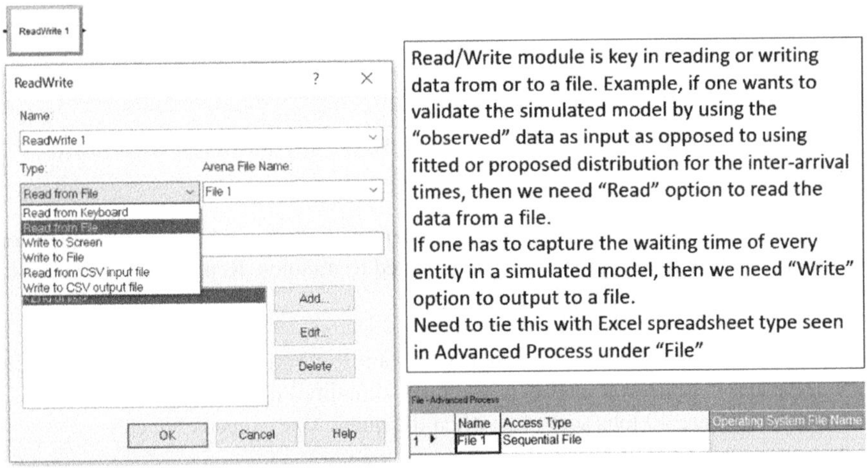

Figure 7.19. *READWRITE module description*

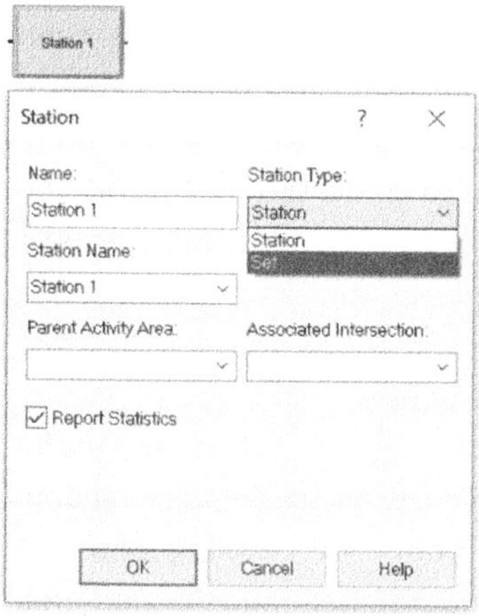

The "Station" module available through "Advanced Transfer" is very useful in a number of situations including production and manufacturing systems where entities are to be moved from one area (station) to another area (station).

The "Set" option for "Station Type" allows this station to be listed as multiple stations, like Station 1, Station 2, etc. All entities transferred to any one of these stations will be sent to this module.

Also very handy if one has to simulate the movements of the entities from one processing to another.

Figure 7.20. *STATION module description*

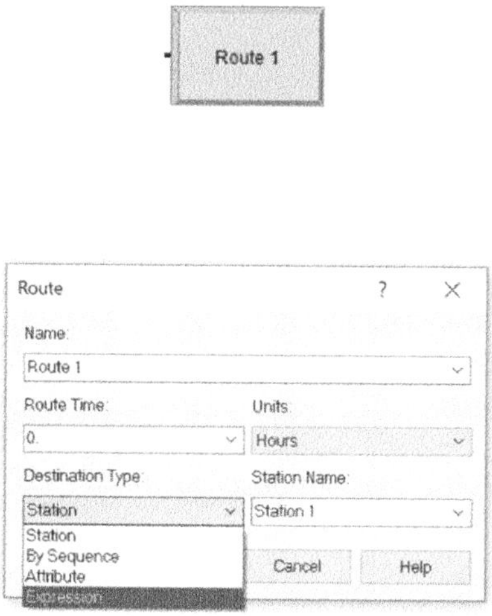

The "Route" module, in "Advanced Transfer", is needed to route the entity from its current area (station) to another area (station). The routing can be done either to a station (so all entities go to the same station as indicated) or "By Sequence" (so that entities will go through the sequence as specified for each entity).

Station: The entities arriving at this "Route" module will be sent to the station defined in "Station Name"

By Sequence: The entities arriving at this "Route" module will be sent to the next station in the sequence (of stations) identified for each of those entities.

Attribute: The entities arriving at this "Route" module will be sent to the station identified in the "Attribute".

Expression: The entities arriving at this "Route" module will be sent to the station number which is obtained by calculating the expression (truncated to an integer).

Figure 7.21. *ROUTE module description*

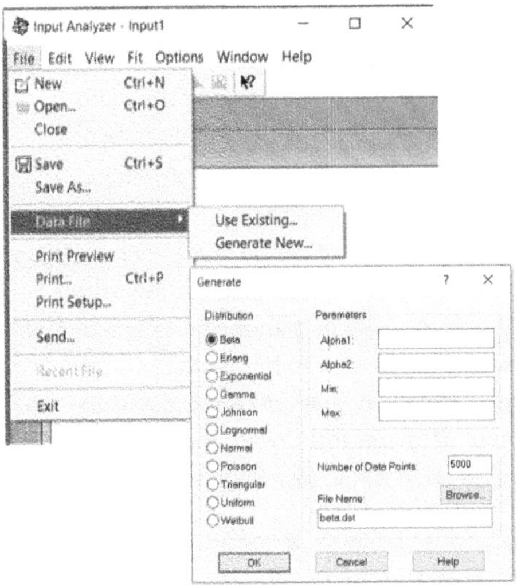

"Use Existing" – This is to be used when you want to open an already created file or to create a new fitted plot based on the text file that has the data. So, if you have not fitted a curve yet, choose this and point to the directory that has the text file that has the data.

"Generate New" – This is very useful when you want to generate random data from an existing list of distributions (see to the left the available ones).

Figure 7.22. *INPUT ANALYZER module description*

There are situations where one has to use Seize Delay and Release separately. In this situation, the modules, shown on the left available under Advanced Process tab, should be used.

Figure 7.23. *SEIZE, DELAY and RELEASE modules description*

REMARK 7.1.– It should be pointed out that one can use "Input Analyzer" to identify the type of probability distribution for any variable such as the waiting time in the system and the busy period (BP), by first writing the output data to a file and then to use that file to fit a curve.

7.3. *GI/G/c* queues

This section is devoted to the study of general independent and identically distributed inter-arrival times and generally distributed services via simulation. First, we discuss the building of the simulation model in various stages, especially for novice readers, starting with the basic model and then add additional modules to capture some of the system performance measures discussed in previous chapters. Second, we validate the models by using the corresponding (but available) analytical models discussed earlier. Finally, we use these simulation models to analyze complicated models for which the analytical results are either unknown or difficult to implement numerically. It is worth pointing out that the models developed are with the (novice) readers in mind. Readers who are familiar with ARENA can simplify a lot of modules and their commands.

Thus, we assume in this section that the arrivals occur according to a general point process with the inter-arrival times forming an $i.i.d$ process. The service times have a general distribution and there are c homogeneous servers to process the arriving customers. Let $F(.)$ and $H(.)$, respectively, denote the (cumulative) distribution function (CDF) of the inter-arrival times and the service times. Furthermore, assume that the means of $F(.)$ and $H(.)$ are finite and are, respectively, denoted by $\frac{1}{\lambda}$ and $\frac{1}{\mu}$. Recall that the traffic intensity is given by $\rho = \frac{\lambda}{c\mu}$ and that $\rho < 1$ is assumed to have a stable queue.

7.3.1. *ARENA modules for* GI/G/c *queues*

In Figure 7.24, the ARENA simulation model for $GI/G/c$-type queues is displayed. It should be pointed out that these modules are sample ones and are developed so as to present them in a more transparent way to help readers who are new to ARENA (or as a refresher after a break from learning the software). However, when presenting validated results as well as for illustrative examples in later sections, these modules and their input parameters will be modified to include the situation for which the models are used.

A brief explanation of the modules is given below for readers to create the simulation model on their own. It should be pointed out that for a novice user it is advisable to use the same notation and nomenclature to avoid debugging errors.

First, we display in Figure 7.25 the variables used in this simulation example. The purpose of these variables is explained briefly in "comments" column. To enter a specific value for a variable, one has to click the corresponding row under "Initial Values". Prior to entering a new value, one will see "0 rows" under "Initial Values". As an example, for item 12, the seven values entered are: $TV(1) = 1, TV(2) = 2, TV(3) = 4, TV(4) = 8, TV(5) = 16, TV(6) = 40$ and $TV(7) = 80$.

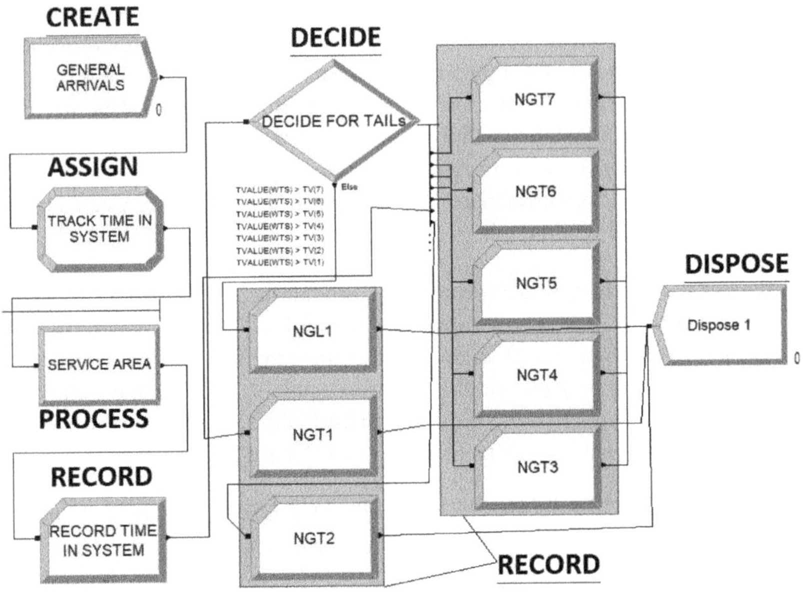

Figure 7.24. *$GI/G/c$ queue*

Variable - Basic Process

	Name	Comment	Rows	Columns	Data Type	Clear Option	File Name	Initial Values
1	OrA	Order for Erlang Arrivals			Real	System		1 rows
2	XLA	Arrival Rate			Real	System		1 rows
3	OrS	Order for Erlang Services			Real	System		1 rows
4	XMU	Service Rate			Real	System		1 rows
5	AA	Arrival Uniform Lower End			Real	System		1 rows
6	AB	Arrival Uniform Upper End			Real	System		1 rows
7	SA	Service Uniform Lower End			Real	System		1 rows
8	SB	Service Uniform Upper End			Real	System		1 rows
9	ToA	Identifier for Arrival Type			Real	System		1 rows
10	ToS	Identifier for Service Type			Real	System		1 rows
11	RHOQ	Traffic Intensity			Real	System		1 rows
12	TV	Vector of Time Points	7		Real	System		7 rows

Figure 7.25. *List of variables used in $GI/G/c$ queue*

The "Create" module used here is displayed in Figure 7.26. In the "Expression:" window enter the following:

(ToA==1)*(1/XLA)+(ToA==2)*UNIF(AA/XLA,AB/XLA)
+ (ToA==3)*EXPO(1/XLA)+(ToA==4)*ERLA(1/(OrA*XLA),OrA)
+(ToA==5)*DISC(0.9,EXPO(1/(1.9*XLA)),1,EXPO(1/(0.19*XLA)))
+(ToA==6)*WEIB(0.5/XLA, 0.5),

and appropriate distribution will be chosen based on the value of ToA specified in the "Variable" sheet in "Basic Process" tab (see Figure 7.25).

It is worth pointing out that the six distributions considered for the arrival process are: (1) Constant; (2) Uniform on (AA, BB); (3) Exponential with parameter λ; (4) Erlang of order OrA with rate λ in each stage; (5) Hyperexponential with mixing probabilities 0.9 and 0.1, respectively, with rates 1.9λ and 0.19λ; and (6) Weibull with shape parameter of 0.5 and scale parameter of $\frac{0.5}{\lambda}$. We refer the reader to section 1.1.4 in Volume 1 for details on these probability distributions.

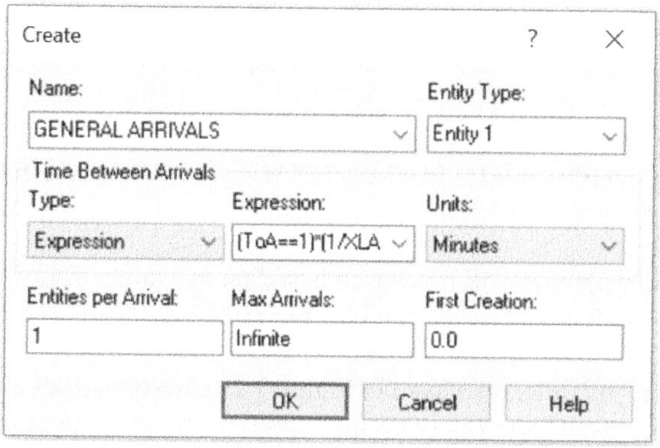

Figure 7.26. *CREATE module in $GI/G/c$ queue*

The "Assign" module is displayed in Figure 7.27 and as can be seen from this figure, two items are assigned. The first one is an "Attribute", TiS, needed to track the time in system of every entity entering and leaving the system. The second is a "Variable", XMU, which is set as $XLA/(RHOQ * MR(SERVER))$. Note that the value assigned to XMU is the ratio of the arrival rate to the product of the traffic intensity and the number of servers (c) in the system. This way, for a given arrival rate, traffic intensity and the number of servers, the service rate can be set to a desired value of the traffic intensity.

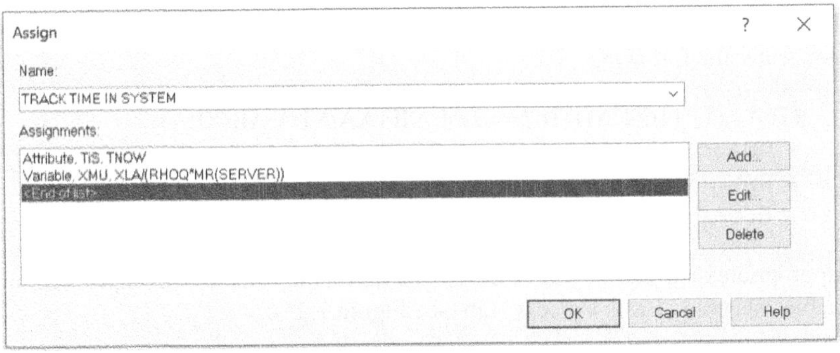

Figure 7.27. *ASSIGN module in $GI/G/c$ queue*

This "Process" module is for processing (in queueing terminology it is the service area). The resource is called "SERVER" and the type of Action is "Seize Delay Release". This action choice is due to the fact that in this type of queueing models, soon after a service completion the server is available for the next in line, if any. In the "Expression:" window enter the following:

```
(ToS==1)*(1/XMU)+(ToS==2)*UNIF(SA/XMU,SB/XMU)
+ (ToS==3)*EXPO(1/XMU)+(ToA==4)*ERLA(1/(OrS*XMU),OrS)
+(ToS==5)*DISC(0.9,EXPO(1/(1.9*XMU)),1,EXPO(1/(0.19*XMU)))
+(ToS==6)*WEIB(0.5/XMU, 0.5),
```

and appropriate distribution will be chosen based on the value of ToS specified in the "Variable" sheet in "Basic Process" tab (see Figure 7.25).

In the "Record" module, displayed in Figure 7.29, two quantities are captured. The first one is of the type "Time Interval" for capturing the time spent in the system by an entity. The attribute, TiS, defined in the Assign module earlier, is recalled here. This Record module computes the time since the last time stamp, TNOW. The second type is of the type "Time Between" used to register the time between departures of two successive entities.

The "Decide" module, based on "N-way by condition" displayed in Figure 7.30, helps to send the entities to the right Record module to capture the tail probabilities (at 7 time points in TV(.)) of the sojourn time in the system. Recall that the Attribute, WTS, is the name given in the previous Record module for the "Time Interval" type storing the entities' sojourn time in the system. TVALUE is ARENA's terminology for the last recorded value.

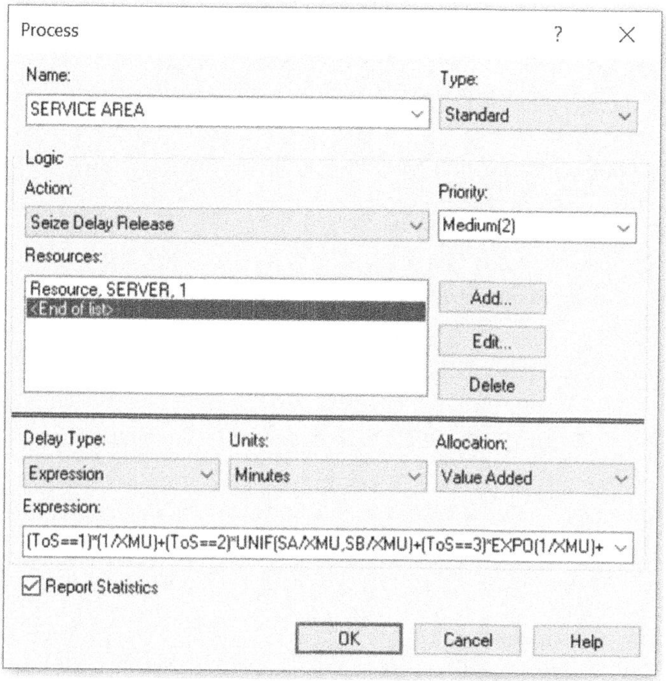

Figure 7.28. *PROCESS module in $GI/G/c$ queue*

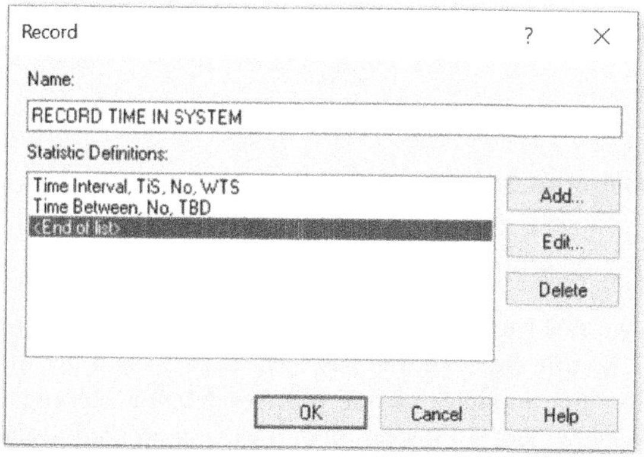

Figure 7.29. *RECORD module in $GI/G/c$ queue for tracking times*

In the second "Record" module displayed in Figure 7.31, the entities arriving are counted so as to calculate the tail probabilities. Thus, the type used is "Count" and the value is 1. There are seven such modules; however, this one is a representative.

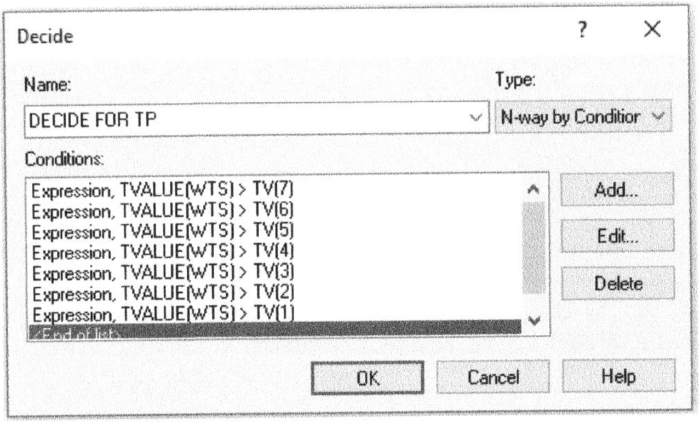

Figure 7.30. *DECIDE module in $GI/G/c$ queue*

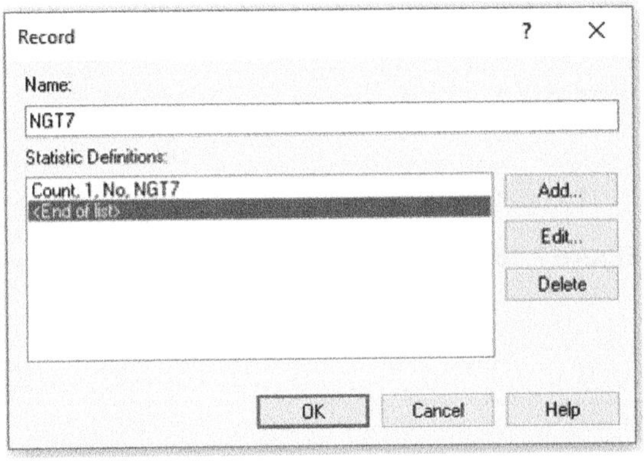

Figure 7.31. *RECORD module in $GI/G/c$ queue for tracking tail probabilities*

Suppose that we need to compute the recurrence time to the system becoming empty as well as the idle time. In this case, we need to add the modules as shown in Figure 7.32. The five modules on the top row are for keeping track of the idle durations of the system. The additional modules (as compared to the ones displayed for $GI/G/c$ queue in Figure 7.24) are needed to compute the recurrence time for the system becoming empty.

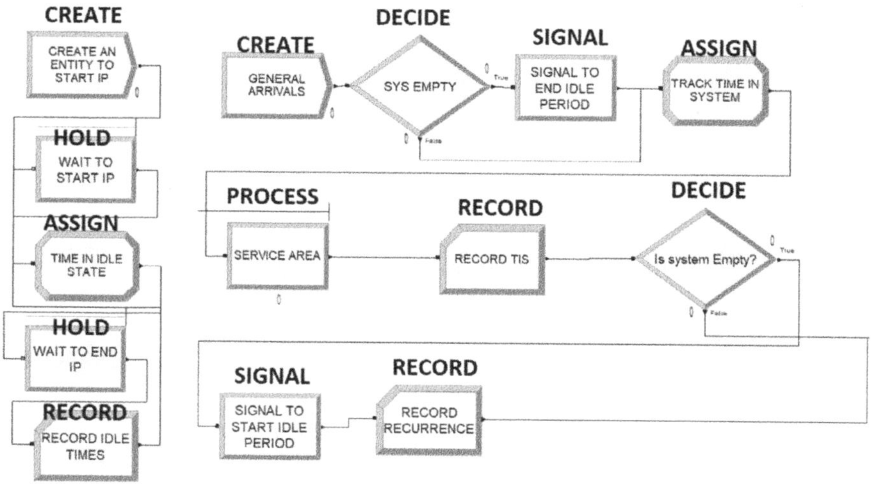

Figure 7.32. *Modules needed for capturing the recurrence time in $GI/G/c$ queue*

Suppose that we need to compute the BP statistics. Recall from section 4.6 that the BP in the context of a multiple-server system is defined in two different ways: partial and full BPs. While a *full BP* is defined as the duration in which the system starts with all servers becoming busy until at least one server becomes free, a *partial BP* is one in which the BP starts with an arrival of a customer to an empty system and ends with the system becoming empty again at the departure of a customer. These two will be the same in the single-server systems. In Chakravarthy (2019a), the BP in the context of $GI/G/c$ and $MAP/G/c$ queues is discussed in detail with a number of illustrative numerical examples based on simulation. In this section, we follow along the same lines and include some additional examples.

The additional modules needed in compiling statistics on the BP are displayed in Figures 7.33 and 7.34. These figures contain the left and right set of modules in that one needs to connect the left side (namely, the connector coming out of "Separate 1" module seen in Figure 7.33 to the connector seen in "Decide" module in Figure 7.34).

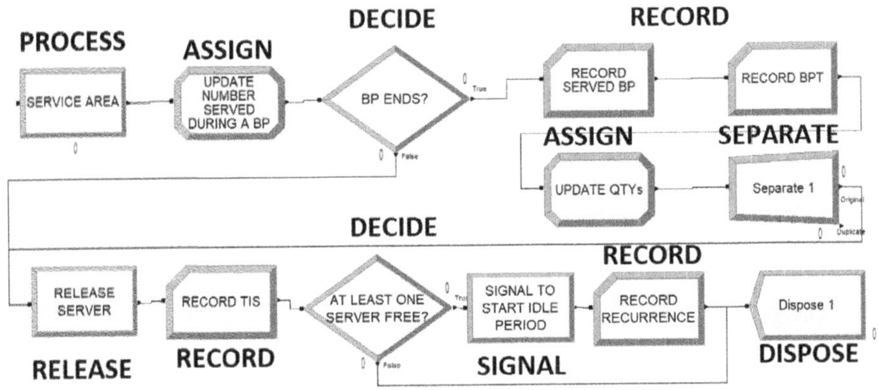

Figure 7.33. *Modules needed for capturing busy period statistics in $GI/G/c$ queue (left side)*

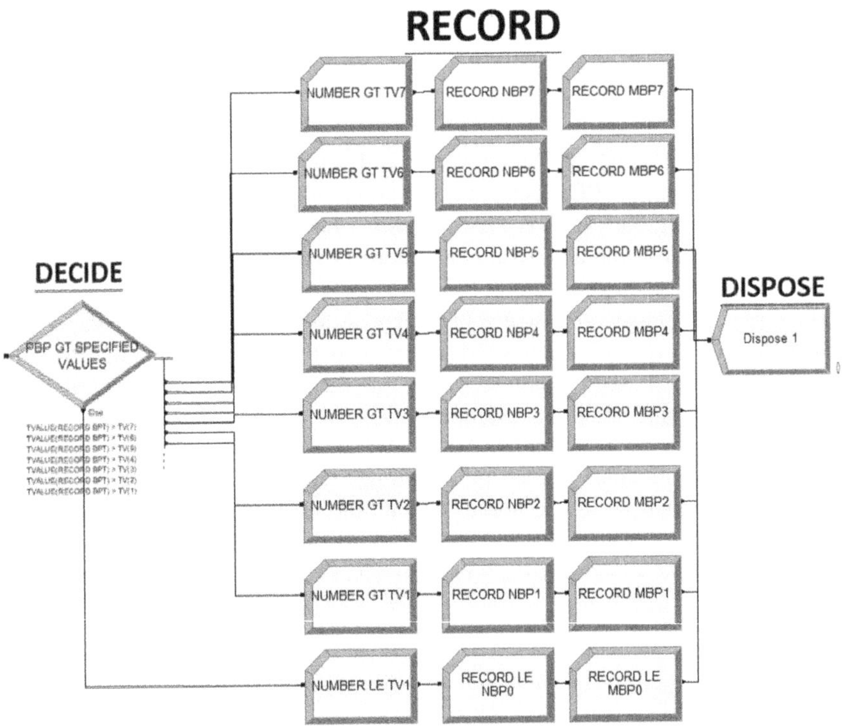

Figure 7.34. *Modules needed for capturing busy period statistics in $GI/G/c$ queue (right side)*

REMARK 7.2.– For some models the simulation needs different run lengths and possibly more replications so as to have the error percentages (when validating against known models) to a desired level.

After developing the simulation model, it is important to construct key system performance measures and to capture them. Validated results in the context of classical queueing models under various scenarios are presented in Chakravarthy (2019a). Since this author developed those models, we refer the reader to those results. Here, for the sake of completeness, we present simulated results for the models that are not considered in Chakravarthy (2019a).

7.3.2. *Validation of GI/G/1 queues*

In this section, we validate the simulation models in the context of $GI/G/1$ queues against the analytical models discussed in earlier chapters. Toward this end, one can (1) look at the error percentages of using the simulation model against the analytical model for a particular measure; (2) use scatter plots of simulated results against the corresponding theoretical results (whenever possible) or the corresponding approximations (whenever possible, but here one has to be careful as approximations are subject to errors and hence the term "approximation"); and (3) fit a regression model for the analytical result based on the simulated model. This is like calibrating a piece of equipment or a measuring device that is commonly used in engineering sciences. When using error percentages, one can adopt the following definition:

$$Error\ \% \ = 100\left|\frac{\text{analytical} - \text{simulated}}{\text{analytical}}\right|\%.$$

In this book, we illustrate the above three methods as we go through validating the simulated models.

EXAMPLE 7.1.– In this example, we validate the model under various scenarios. We look at (1) constant arrivals; (2) arbitrary arrivals; and (3) uniform arrivals. For services, we look at three phase type (PH) distributions: (1) exponential; (2) Erlang of order 5; and (3) hyperexponential. Specifically, we have:

vCTA: Constant arrivals with parameter 1.

vABA: Inter-arrival times are discrete with masses at points 0.5, 1.0 and 2.5 with probabilities, respectively, given by 0.3, 0.6 and 0.1.

vUFA: Arrivals have a uniform distribution on $(0.1, 1.9)$.

All these arrival distributions have a mean of 1 but they are qualitatively different. For services, we consider the following three PH distributions.

vEXS: Exponential with parameter μ.

vERS: Service distribution is Erlang of order 5 with parameter μ.

vHES: Services are assumed to be hyperexponential distribution with parameters $2.4\mu, 1.2\mu, 0.6\mu, 0.24\mu$ and 0.048μ with mixing probabilities, respectively, given by 0.6, 0.25, 0.10, 0.04 and 0.01.

The parameter μ will be chosen so as to get a specific ρ. We look at $\rho = 0.5, 0.8, 0.95$ and 0.99.

For all the scenarios, the number of replications is at least 10 and the length of each replicate is at least 1,000,000 units of time. There are some scenarios for which the number of replicates has to be increased to more than 10 so as to achieve more consistent results (in agreement with the theoretical ones). These scenarios include the distributions (for arrival and/or services) that have a large variability and/or when the traffic intensity is large. Of course, there are situations where as few as five replicates with 100,000 units of time per replicate suffice to generate the results.

The scatter plots of the simulated results (in y-axis) against the corresponding theoretical results are displayed in Figure 7.35.

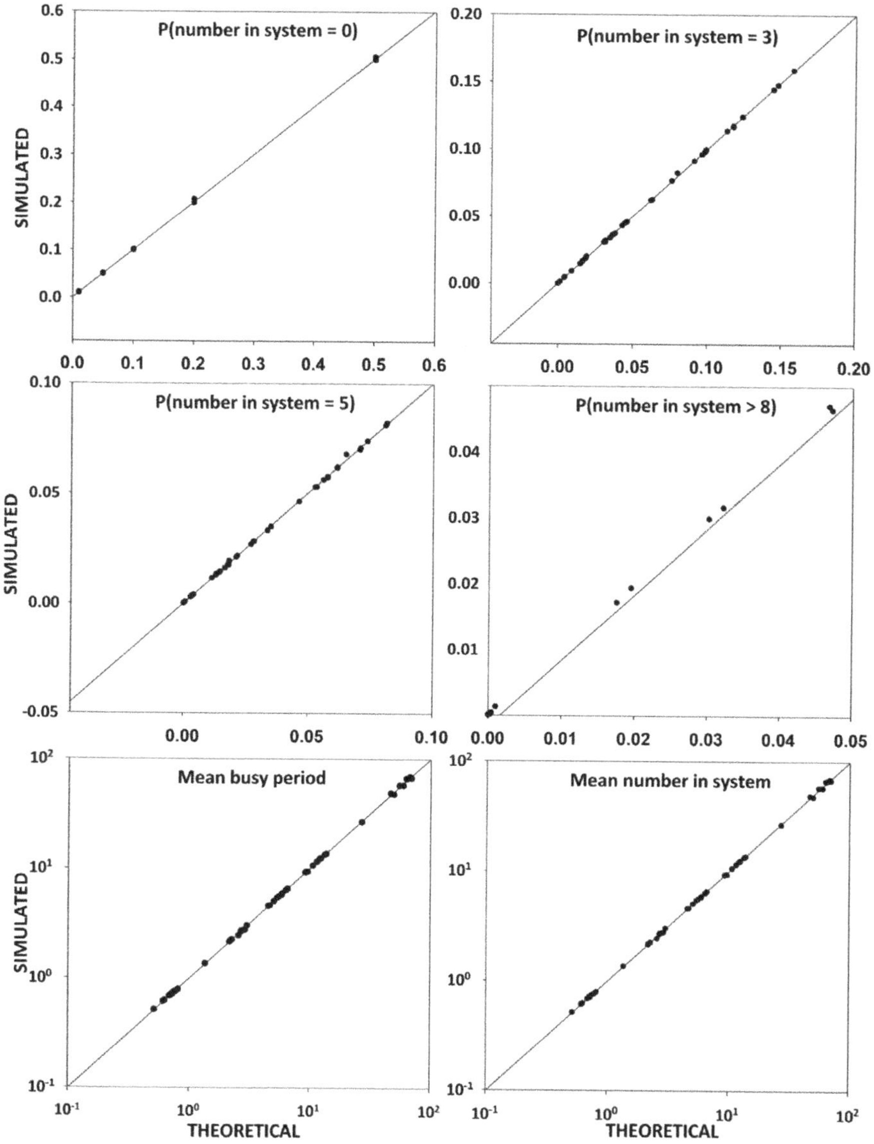

Figure 7.35. *Plots of the simulated results against theoretical values for example 7.1*

We chose the scatter plots (these are similar to probability plots) due to a number of scenarios that need to be displayed. Thus, any significant (at 5% level) deviation from the $45°$ line indicates the simulated results to be different from the corresponding theoretical results. However, the plots seen in Figure 7.35 appear to show the simulated results to be very close or equal to the theoretical results. We also did a pair-wise student's T-test for the simulated results against their counterpart theoretical results and in every single scenario the P-values were greater than 0.05 indicating that statistically there is no evidence (at 5% level) to say that the simulated results differ from the theoretical results. Of course, any simulation is subject to random error including sampling and thus one has to look at individual scenarios and modify the number of replicates as well as the length of each replicate. Our experience indicates that most scenarios having 5 replicates with 1,000,000 units of time, and for others having 20 replicates with 1,500,000 or more units of time per replicate, appear to be sufficient. As is known, having more replicates and a large replicate length result in longer simulation time.

EXAMPLE 7.2.– In this example, we validate the model under various scenarios. We look at (1) Poisson arrivals; (2) Erlang of order 5 arrivals; and (3) hyperexponential arrivals. For services, we look at three PH distributions: (1) exponential; (2) Erlang of order 5; and (3) hyperexponential. Specifically, we have:

vEXA: Poisson arrivals with parameter λ.

vERA: Inter-arrival times are Erlang of order 5 with parameter λ.

vHEA: Inter-arrival times are assumed to be hyperexponential distribution with parameters 2.18λ, 1.09λ, 0.545λ, 0.2725λ and 0.13625λ with mixing probabilities, respectively, given by 0.5, 0.3, 0.15, 0.04 and 0.01.

All these arrival distributions have a mean of 1 but they are qualitatively different. For services, we have:

vEXS: Exponential with parameter μ.

vERS: Service distribution is Erlang of order 5 with parameter μ.

vHES: Services are assumed to be hyperexponential distribution with parameters 2.4μ, 1.2μ, 0.6μ, 0.24μ and 0.048μ with mixing probabilities, respectively, given by 0.6, 0.25, 0.10, 0.04 and 0.01.

All these arrival distributions have a mean of 1 but they are qualitatively different. The parameter μ will be chosen so as to get a specific ρ. We look at $\rho = 0.5, 0.8, 0.95$, and 0.99. In Figure 7.36, we display the simulated results against the theoretical results for selected measures under various scenarios.

Like in example 7.1, here also we notice the simulated results to be close enough to the known analytical results.

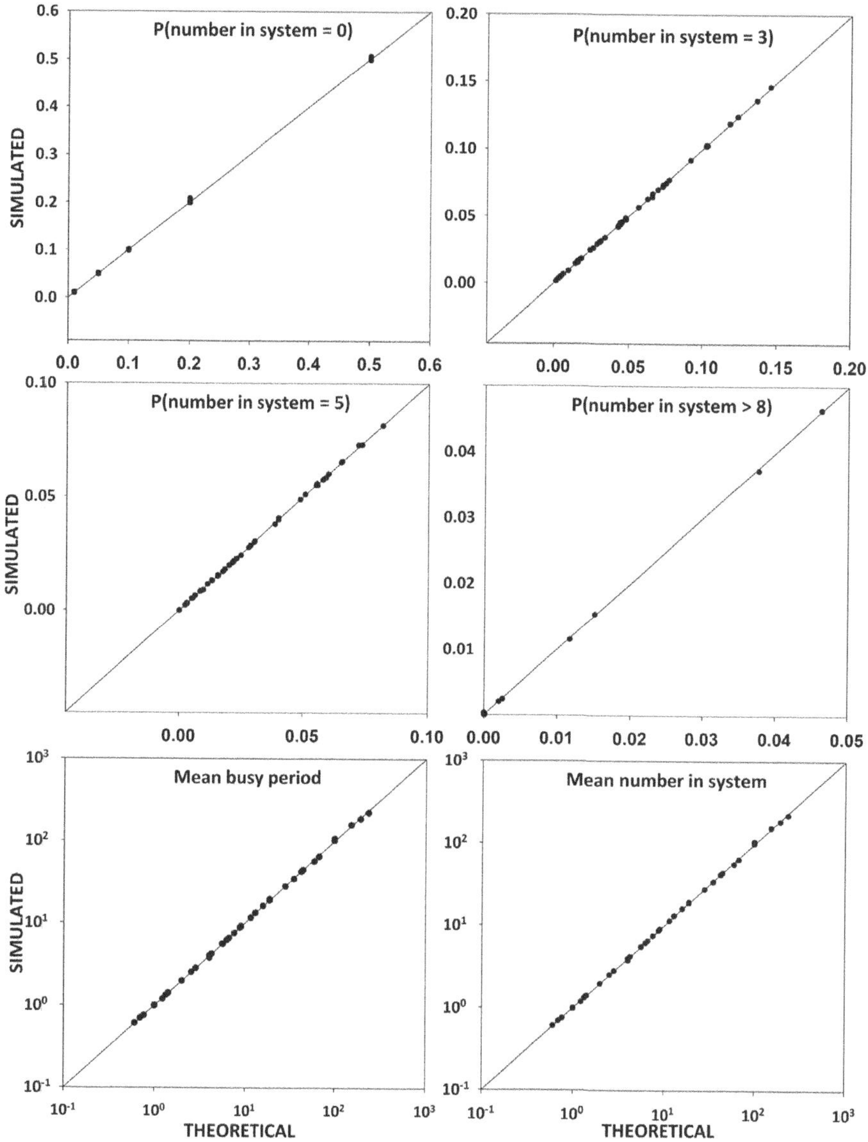

Figure 7.36. *Plots of the simulated results against theoretical values for example 7.2*

7.3.3. *Simulated examples for GI/G/1 queues*

Having validated the simulated $GI/G/1$ model under various combinations of arrival and service processes, we will discuss a few examples based on the simulation model for the $GI/G/1$ queues. For all the examples considered in this section, we look at the following arrival processes all of which give a mean time between two successive arrivals to be 1. However, these are qualitatively different and cover a number of scenarios in practice. Simulation was carried out with eight replicates with 1,000,000 units of times for each replicate.

CTA: Constant arrivals with parameter 1.

ABA: Inter-arrival times are discrete with masses at points 0.5, 1.0 and 2.5 with probabilities, respectively, given by 0.3, 0.6 and 0.1.

UFA: Arrivals have a uniform distribution on $(0.1, 1.9)$.

EXA: Poisson arrivals with parameter 1.

ERA: Erlang of order 5 with parameter 5.

HEA: This is hyperexponential distribution with parameters 2.18, 1.09, 0.545, 0.2725 and 0.13625 with mixing probabilities, respectively, given by 0.5, 0.3, 0.15, 0.04 and 0.01.

WBA: This corresponds to the inter-arrival times having a Weibull distribution with parameters $\alpha = \beta = 0.5$ (see section 1.1.4 of Volume 1). Verify that the standard deviations of the above arrival processes are, respectively, given by 0, 0.54772, 0.51962, 1, 0.44721, 1.69710 and 2.23607.

For services, we look at the following ones. The service rate will be chosen so that it will be equal to a given traffic intensity value. That is, $\mu = \frac{1}{\rho}$ since we fixed $\lambda = 1$.

EXS: Service times follow an exponential distribution with parameter μ.

ERS: Service distribution is Erlang of order 5 with parameter μ.

HES: Services are assumed to be hyperexponential distribution with parameters 2.4μ, 1.2μ, 0.6μ, 0.24μ and 0.048μ with mixing probabilities, respectively, given by 0.6, 0.25, 0.10, 0.04 and 0.01.

Verify that the ratio of the standard deviations of the above service times with respect to that of the Erlang (ERS) are, respectively, given by 2.23607 and 7.13462.

EXAMPLE 7.3.– Here, we look at a number of scenarios using the seven arrival processes, three service times and two values for ρ as listed above and for a single-server system. In Table 7.1, we display the mean values along with their half-widths (needed to compute the 95% confidence level). Note that we have included the scenarios for which we have the analytical results so that one can see the differences in various measures when we compare them with WBA arrivals.

ToA	$Measure$	EXS	ERS	HES
CTA	μ'_Y	9.667 ± 0.1489	5.4201 ± 0.0283	11.167 ± 0.7166
	μ_{N_Y}	10.175 ± 0.1489	5.7064 ± 0.0282	11.774 ± 0.7164
	μ'_{RT}	10.175 ± 0.1489	5.7064 ± 0.0282	11.774 ± 0.7164
	μ'_I	0.5086 ± 0.0009	0.2863 ± 0.0003	0.607 ± 0.0005
	L_s	9.6686 ± 0.2869	2.5626 ± 0.0174	88.972 ± 7.5972
ABA	μ'_Y	12.488 ± 0.3158	10.619 ± 0.1202	13.427 ± 0.807
	μ_{N_Y}	13.145 ± 0.3180	11.178 ± 0.1245	14.133 ± 0.8111
	μ'_{RT}	13.143 ± 0.3154	11.18 ± 0.1214	14.131 ± 0.8069
	μ'_I	0.655 ± 0.0014	0.561 ± 0.0023	0.704 ± 0.001
	L_s	12.654 ± 0.2834	5.2477 ± 0.0701	92.66 ± 5.0626
UFA	μ'_Y	12.224 ± 0.2047	9.1887 ± 0.0592	13.432 ± 0.6078
	μ_{N_Y}	12.865 ± 0.2054	9.6723 ± 0.0588	14.155 ± 0.607
	μ'_{RT}	12.867 ± 0.2044	9.6751 ± 0.059	14.158 ± 0.6076
	μ'_I	0.6431 ± 0.0014	0.4865 ± 0.0004	0.7258 ± 0.0015
	L_s	12.324 ± 0.3403	5.0823 ± 0.0629	92.474 ± 7.7385
EXA	μ'_Y	18.912 ± 0.5488	18.893 ± 0.3689	19.804 ± 1.3986
	μ_{N_Y}	19.914 ± 0.5664	19.891 ± 0.3835	20.805 ± 1.4061
	μ'_{RT}	19.91 ± 0.5471	19.893 ± 0.3708	20.803 ± 1.3971
	μ'_I	0.997 ± 0.0047	0.9992 ± 0.0033	0.9982 ± 0.0026
	L_s	19.038 ± 0.7681	11.702 ± 0.3302	104.17 ± 9.7676
ERA	μ'_Y	11.749 ± 0.1464	8.6997 ± 0.0844	13.368 ± 0.6972
	μ_{N_Y}	12.361 ± 0.145	9.1575 ± 0.087	14.049 ± 0.6977
	μ'_{RT}	12.356 ± 0.1465	9.1578 ± 0.0844	14.05 ± 0.6984
	μ'_I	0.6074 ± 0.0012	0.4582 ± 0.0008	0.6816 ± 0.0019
	L_s	11.787 ± 0.2713	4.4103 ± 0.0608	92.255 ± 7.054
HEA	μ'_Y	35.299 ± 0.5807	42.217 ± 0.9126	27.522 ± 1.8328
	μ_{N_Y}	37.16 ± 0.6109	44.448 ± 0.959	28.993 ± 1.8471
	μ'_{RT}	37.168 ± 0.5821	44.457 ± 0.9156	29.014 ± 1.8306
	μ'_I	1.8677 ± 0.0197	2.239 ± 0.0181	1.4912 ± 0.0087
	L_s	35.469 ± 0.8544	27.006 ± 0.6351	116.33 ± 12.038
WBA	μ'_Y	52.936 ± 1.7692	64.785 ± 2.7577	45.926 ± 3.3793
	μ_{N_Y}	55.764 ± 1.8099	68.177 ± 2.8881	48.389 ± 3.46
	μ'_{RT}	55.843 ± 1.7889	68.128 ± 2.7591	48.345 ± 3.3855
	μ'_I	2.9059 ± 0.0334	3.3383 ± 0.0226	2.4168 ± 0.0303
	L_s	51.309 ± 2.1054	49.624 ± 3.1215	144.26 ± 13.025

Table 7.1. *Selected measures for* $c = 1, \rho = 0.95$ *for example 7.3*

Obviously, the simulated results indicate that an increasing variability in the services (i.e. going from Erlang to exponential to hyperexponential) increases the values of the performance measures. This is the case for all but WBA arrivals. For WBA arrivals, we see a decreasing trend in all but L_s measures as the variability of the service times increases. For WBA arrivals, we notice that L_s increases with increasing variability in the services.

7.3.4. *Validation of* GI/G/c *queues*

Here, we validate our simulated multi-server queues against a few reported analytical ones using 20 replicates of length 100,000 units per replicate.

EXAMPLE 7.4.– In this example, we validate our simulation model with the analytical result as reported in Ramaswami and Lucantoni (1985). In this paper, the authors look at four arrival processes in which the inter-arrival times are assumed to be (1) Erlang of order 5; (2) Erlang of order 2; (3) exponential; and (4) hyperexponential with parameters 24.52270384 and 2.477296157 with mixing probabilities, respectively, given by 0.90824829 and 0.09175171.

Note that the mean of the hyperexponential arrivals is 0.074074, resulting in a rate of 13.5 arrivals per unit of time. The parameters of the first three arrival processes are normalized to have the same rate of 13.5 per unit of time.

For services, the authors look at hyperexponential distribution with parameters 2.535183758 and 0.232408121 with mixing probabilities 0.845061253 and 0.154938747, respectively. Note that this distribution has a mean of 1. The number of servers is taken to be $c = 15$ so that the traffic intensity is 0.9.

The measures used in Ramaswami and Lucantoni (1985) are: (1) Probability an arrival sees zero customers in the system (PASZ); (2) Mean number of customers in the system at an arrival $L_s^{(a)}$; (3) Probability that an arrival will get into service immediately (PSGS); (4) Mean waiting time in the queue (μ'_{W_q}); and (5) Variance of the waiting time in the queue ($\sigma^2_{W_q}$).

In Table 7.2, we display the analytical results of Ramaswami and Lucantoni (1985) (within parentheses) and our simulated results. Looking at these results, we see that overall our simulated values are agree very much with the analytical results. The one measure, namely, PSGS, which is very close to zero for all practical purposes, appears to differ significantly under some scenarios. It is worthy of further investigation by simulating longer periods of time.

	ER_5	ER_2	EX	HE_2
$PASZ$	0.0000011852 (0.00000016195)	0.0000015553 (0.00000037987)	0.0000016308 (0.00000098739)	0.000017703 (0.00001771)
$L_s^{(a)}$	25.105 (25.19336)	26.657 (26.45329)	28.461 (28.56676)	45.104 (44.90133)
$PSGS$	0.43753 (0.434622)	0.40729 (0.409876)	0.37815 (0.375084)	0.24292 (0.242164)
μ'_{W_q}	0.88934 (0.8954)	0.9922 (0.97773)	1.1091 (1.11606)	2.2069 (2.19318)
$\sigma^2_{W_q}$	2.3791 (2.39621)	2.76 (2.71572)	3.2798 (3.29048)	9.994 (9.42434)

Table 7.2. *Selected performance measures for example 7.2*

7.3.5. *Simulated examples of* GI/G/c *queues*

In this section, we present some illustrative examples based on simulating $GI/G/c$ queues under a variety of scenarios. Simulation was carried out with 10 replicates with each of length of 1,500,000 units of time. It is worth pointing out that the simulated model was validated in the previous section for the single-server case. In Chakravarthy (2019a) and Chakravarthy (2021c), several multi-server queues, such as $GI/G/c$ and $BMAP/G/c$, were simulated and interesting results were noted. Since this author was involved in developing those simulated models and since some of those models are used to generate the results, we refer the reader to those papers for additional examples. Some interesting results related to the mean BP and the mean waiting time in the system were reported in those papers, and some of these are briefly stated below. These shed additional light for further research both analytically and via simulation.

1) For models of the type $GI/ER/c$ and $GI/SXP/c$ (where ER stands for Erlang and SXP stands for shifted exponentials), the mean (partial) BP is greater than the mean waiting time in the system.

2) For $M/M/c$ type models, when $c > 1$, the mean (partial) BP is greater than the mean waiting time in the system. [Note: For $M/M/1$, these two means are equal.]

3) For models of the type $GI/HE/c$ and $GI/WB/c$ (where HE stands for hyperexponential and WB stands for Weibull), some interesting patterns are noted. For $c = 1$ and $c = 2$, the mean (partial) BP is less than the mean waiting in the system; however, for $c = 5$ it is the other way.

4) Intuitively, one can explain why the mean BP may be larger than the mean waiting time under some scenarios for a single-server case involving non-exponential distributions. Consider a situation where the customers arrive regularly such that their waiting time in the queue is very marginal but the server is kept busy resulting in a large value for the mean BP.

ToA	$Measure$	EXS	ERS	HES
CTA	μ'_Y	26.968 ± 0.50986	51.977 ± 0.50192	23.255 ± 1.0704
	μ_{N_Y}	27.397 ± 0.50928	52.156 ± 0.50198	23.762 ± 1.0708
	μ'_{RT}	27.398 ± 0.50937	52.158 ± 0.50197	23.763 ± 1.0709
	μ'_I	0.42922 ± 0.0011	0.17924 ± 0.00035152	0.50742 ± 0.00086206
	L_s	10.299 ± 0.22447	3.4868 ± 0.02217	88.57 ± 4.2161
ABA	μ'_Y	30.701 ± 0.394	39.49 ± 0.62162	26.46 ± 1.2202
	μ_{N_Y}	31.323 ± 0.39314	40.045 ± 0.6411	27.121 ± 1.2204
	μ'_{RT}	31.319 ± 0.39302	40.041 ± 0.6211	27.112 ± 1.2194
	μ'_I	0.61721 ± 0.00181	0.5519 ± 0.0029	0.6523 ± 0.0027
	L_s	13.133 ± 0.2101	6.0995 ± 0.1102	93.078 ± 7.2501
UFA	μ'_Y	26.46 ± 1.2202	30.108 ± 0.42924	37.096 ± 0.41653
	μ_{N_Y}	27.121 ± 1.2204	30.687 ± 0.43789	37.451 ± 0.43198
	μ'_{RT}	27.112 ± 1.2194	30.68 ± 0.42877	37.45 ± 0.41614
	μ'_I	0.65227 ± 0.0027	0.57102 ± 0.00268	0.3543 ± 0.00102
	L_s	93.078 ± 7.2501	13.133 ± 0.25827	5.895 ± 0.06078
EXA	μ'_Y	37.805 ± 0.48149	40.89 ± 0.60743	33.35 ± 2.4391
	μ_{N_Y}	38.803 ± 0.51324	41.89 ± 0.62466	34.356 ± 2.4261
	μ'_{RT}	38.805 ± 0.48098	41.886 ± 0.6176	34.345 ± 2.4314
	μ'_I	0.99891 ± 0.00756	1.0019 ± 0.00417	0.9981 ± 0.00351
	L_s	19.559 ± 0.78068	12.533 ± 0.23906	96.469 ± 8.1412
ERA	μ'_Y	29.37 ± 0.50593	39.379 ± 0.52365	25.169 ± 0.81462
	μ_{N_Y}	29.921 ± 0.5145	39.768 ± 0.53731	25.777 ± 0.81697
	μ'_{RT}	29.917 ± 0.50642	39.76 ± 0.52449	25.777 ± 0.81526
	μ'_I	0.54587 ± 0.00122	0.38066 ± 0.00066125	0.6078 ± 0.00141
	L_s	12.159 ± 0.38215	5.214 ± 0.05134	86.603 ± 4.2023
HEA	μ'_Y	59.308 ± 1.5789	72.984 ± 1.686	47.468 ± 2.6624
	μ_{N_Y}	61.424 ± 1.6502	75.807 ± 1.779	49.171 ± 2.7345
	μ'_{RT}	61.522 ± 1.5715	75.719 ± 1.7013	49.181 ± 2.6637
	μ'_I	2.2101 ± 0.02239	2.736 ± 0.02198	1.71 ± 0.0142
	L_s	35.31 ± 1.9137	28.97 ± 0.70623	116.09 ± 11.425
WBA	μ'_Y	81.115 ± 1.1146	89.382 ± 4.3705	71.302 ± 4.2296
	μ_{N_Y}	84.389 ± 1.1977	93.206 ± 4.5868	74.011 ± 4.2468
	μ'_{RT}	84.406 ± 1.0979	93.14 ± 4.3571	74.036 ± 4.248
	μ'_I	3.2838 ± 0.02817	3.7687 ± 0.05972	2.731 ± 0.02715
	L_s	53.772 ± 1.6474	48.092 ± 2.6097	133.34 ± 8.7871

Table 7.3. *Selected measures for* $c = 2, \rho = 0.95$ *for example 7.5*

EXAMPLE 7.5.– In this example, we look at a homogeneous two-server queueing system in which the above-mentioned seven different arrival processes and three service times (see example 7.3) are used. In order to compare the systems properly, we fix the service rate to be $\frac{\lambda}{2\rho}$, where $\rho = 0.95$. In Table 7.3, we display various measures along with their half-widths at 95% level. It should be pointed out that the BP statistics are related to the partial BP.

EXAMPLE 7.6.– This example is similar to example 7.5 but with a homogeneous three-server system. We fix the service rate to be $\frac{\lambda}{3\rho}$, where $\rho = 0.95$. Table 7.4 displays various measures along with their half-widths at 95% level. The BP statistics are related to the partial BP.

ToA	$Measure$	EXS	ERS	HES
CTA	μ'_Y	92.229 ± 1.807	851.42 ± 34.944	57.051 ± 2.3306
	μ_{N_Y}	92.614 ± 1.8065	851.54 ± 34.945	57.504 ± 2.3307
	μ'_{RT}	92.622 ± 1.8067	851.41 ± 34.788	57.507 ± 2.3308
	μ'_I	0.38492 ± 0.00206	0.12279 ± 0.00257	0.4534 ± 0.00161
	L_s	10.981 ± 0.20172	4.4011 ± 0.01969	84.645 ± 5.981
ABA	μ'_Y	85.726 ± 2.0543	145.2 ± 2.0254	60.567 ± 3.6128
	μ_{N_Y}	86.306 ± 2.0603	145.69 ± 2.054	61.203 ± 3.6123
	μ'_{RT}	86.32 ± 2.059	145.68 ± 2.0227	61.198 ± 3.6118
	μ'_I	0.59366 ± 0.00441	0.47752 ± 0.00339	0.6284 ± 0.00395
	L_s	13.574 ± 0.29308	6.9131 ± 0.08384	86.172 ± 8.8278
UFA	μ'_Y	86.108 ± 1.981	210.01 ± 3.3876	58.025 ± 2.4885
	μ_{N_Y}	86.618 ± 1.9916	210.25 ± 3.4331	58.602 ± 2.4891
	μ'_{RT}	86.642 ± 1.9805	210.31 ± 3.4047	58.626 ± 2.4891
	μ'_I	0.52687 ± 0.00225	0.27902 ± 0.00239	0.59753 ± 0.00231
	L_s	13.48 ± 0.36879	6.7417 ± 0.07544	87.022 ± 6.2963
EXA	μ'_Y	84.774 ± 1.1663	93.645 ± 1.3685	69.196 ± 3.5862
	μ_{N_Y}	85.813 ± 1.1833	94.678 ± 1.4313	70.249 ± 3.6072
	μ'_{RT}	85.769 ± 1.1642	94.635 ± 1.3987	70.189 ± 3.5946
	μ'_I	1.0003 ± 0.00584	1.0047 ± 0.01166	1.001 ± 0.00664
	L_s	20.232 ± 0.39334	13.372 ± 0.28212	93.124 ± 8.6313
ERA	μ'_Y	86.349 ± 1.5525	220.7 ± 4.7162	58.242 ± 2.6847
	μ_{N_Y}	86.877 ± 1.5746	221.08 ± 4.7573	58.799 ± 2.6926
	μ'_{RT}	86.864 ± 1.5529	221.08 ± 4.7227	58.809 ± 2.6838
	μ'_I	0.50898 ± 0.00288	0.34311 ± 0.00406	0.56502 ± 0.00378
	L_s	12.584 ± 0.18645	6.0954 ± 0.07378	89.772 ± 6.7057
HEA	μ'_Y	106.62 ± 3.6408	109.4 ± 2.2155	93.955 ± 3.6017
	μ_{N_Y}	109.17 ± 3.8036	112.43 ± 2.3158	95.926 ± 3.5937
	μ'_{RT}	109.22 ± 3.6218	112.64 ± 2.2136	95.921 ± 3.6098
	μ'_I	2.5937 ± 0.03669	3.2374 ± 0.04362	1.9582 ± 0.02077
	L_s	36.114 ± 1.5299	28.085 ± 0.68028	113.53 ± 5.0727
WBA	μ'_Y	124.7 ± 5.6783	117.55 ± 3.4929	118.82 ± 10.14
	μ_{N_Y}	128.38 ± 5.9414	121.5 ± 3.6735	121.87 ± 10.39
	μ'_{RT}	128.39 ± 5.7034	121.67 ± 3.5039	121.84 ± 10.142
	μ'_I	3.6768 ± 0.06082	4.144 ± 0.06847	3.0119 ± 0.04183
	L_s	54.267 ± 2.08	47.009 ± 1.6185	138.44 ± 27.333

Table 7.4. *Selected measures for $c = 3, \rho = 0.95$ for example 7.6*

EXAMPLE 7.7.– In this example, we look at $GI/PH/c$ models in which there are seven homogeneous servers and the inter-arrival times are modeled using the following processes. The arrival rate is taken to be 6.65 so that the traffic intensity is $\rho = 0.95$.

1) Constant times (CTA) with inter-arrival times given by $\frac{1}{\lambda}$.

2) Arbitrary distribution ABA with masses at $\frac{0.5}{\lambda}, \frac{1}{\lambda}$ and $\frac{2.5}{\lambda}$ with probabilities, respectively, given by 0.3, 0.6 and 0.1.

3) Uniform times (UFA) on the interval $(\frac{0.5}{\lambda}, \frac{1.5}{\lambda})$.

4) Weibull (WBA) inter-arrival times with shape parameter to be 0.5 and the scale parameter to be $\frac{0.5}{\lambda}$.

For services, we look at different PH distributions (with all of them having the same mean of 1) as follows:

1) EX: Exponential with parameter 1.

2) ER_2: Erlang of order 2 with parameter 2 (in each phase).

3) ER_6: Erlang of order 6 with parameter 6 (in each phase).

4) ER_{10}: Erlang of order 10 with parameter 10 (in each phase)

5) HE_2: Hyperexponential with parameters 1.1 and 0.55 with mixing probabilities, respectively, given by 0.9 and 0.1.

6) HE_6: Hyperexponential with parameters 2.29, 1.145, $\frac{2.29}{3}$, $\frac{2.29}{4}$, $\frac{2.29}{6}$ and $\frac{2.29}{30}$ with mixing probabilities, respectively, given by 0.5, 0.2, 0.15, 0.10, 0.04 and 0.01.

7) HE_{10}: Hyperexponential with parameters 13.115, 6.5575, $\frac{13.115}{2.5}$, $\frac{13.115}{4}$, $\frac{39.345}{20}$, 1.3115, 0.65575, 0.13115, 0.065575 and 0.013115 with mixing probabilities, respectively, given by 0.3, 0.2, 0.15, 0.11, 0.09, 0.07, 0.04, 0.025, 0.010 and 0.005.

We ran the simulation using eight replicates and each replicate is of length 250,000 units of time. We look at 28 scenarios (four arrival processes and seven service times – note that the seven servers are identical and so we need to identify just one service time distribution) and the simulated results for selected measures are displayed in Table 7.5. It should be pointed out that BP statistics are for partial BP.

PH	GI	μ'_Y	μ'_{N_Y}	L_s	μ'_{WS}
EX	CTA	3367.8 ± 380.85	22396 ± 2532.6	14.05 ± 0.16	2.11 ± 0.02
	ABA	1193.8 ± 106.21	7941.7 ± 709.06	16.76 ± 0.12	2.52 ± 0.02
	UFA	2586.4 ± 308.74	17198 ± 2051.2	14.84 ± 0.2	2.23 ± 0.03
	WBA	94.64 ± 4.27	634.35 ± 29.43	56.83 ± 1.77	8.55 ± 0.26
ER_2	CTA	47389 ± 18663	315140 ± 124110	10.29 ± 0.06	1.55 ± 0.01
	ABA	2802.1 ± 166.36	18631 ± 1105.9	12.67 ± 0.06	1.91 ± 0.01
	UFA	19610 ± 6934.4	130410 ± 46116	11 ± 0.07	1.65 ± 0.01
	WBA	73.88 ± 3.56	496.78 ± 24.88	55.71 ± 3.59	8.37 ± 0.52
ER_6	CTA	*	*	7.85 ± 0.01	1.18 ± 0
	ABA	32988 ± 10002	219360 ± 66493	10.15 ± 0.04	1.53 ± 0.01
	UFA	*	*	8.48 ± 0.02	1.27 ± 0
	WBA	57.1 ± 2.08	385.53 ± 14.48	52.7 ± 2.39	7.92 ± 0.35
ER_{10}	CTA	*	*	7.41 ± 0.01	1.11 ± 0
	ABA	45440 ± 65579	302200 ± 436150	9.64 ± 0.06	1.45 ± 0.01
	UFA	*	*	8 ± 0.01	1.2 ± 0
	WBA	52.21 ± 1.75	352.77 ± 12.16	50.72 ± 1.31	7.62 ± 0.19
HE_2	CTA	2755.1 ± 286.08	18321 ± 1902.4	15.22 ± 0.26	2.29 ± 0.04
	ABA	1116.8 ± 58.33	7427.6 ± 387.57	17.96 ± 0.44	2.7 ± 0.07
	UFA	2249.7 ± 233.14	14962 ± 1550.9	15.96 ± 0.25	2.4 ± 0.04
	WBA	100.41 ± 4.12	672.58 ± 28.43	61.68 ± 2.95	9.28 ± 0.43
HE_6	CTA	1051.4 ± 66.71	6992.7 ± 443.63	40.61 ± 1.31	6.11 ± 0.2
	ABA	632.01 ± 44.71	4204.6 ± 297.72	43.4 ± 1.91	6.52 ± 0.29
	UFA	834.84 ± 53.19	5551.9 ± 353.61	41.72 ± 1.53	6.27 ± 0.23
	WBA	141.72 ± 8.96	948.63 ± 60.64	90.24 ± 7.57	13.54 ± 1.12
HE_{10}	CTA	260.01 ± 36.29	1729.6 ± 241.34	504.63 ± 53.81	75.87 ± 8.11
	ABA	239.01 ± 33.64	1589.7 ± 223.6	504.3 ± 174.23	75.85 ± 26.2
	UFA	277.74 ± 77.15	1847.4 ± 513.06	517.12 ± 64.82	77.76 ± 9.75
	WBA	411.8 ± 91.33	2741.9 ± 606.38	621.21 ± 117.95	93.33 ± 17.62

Note: The cells that contain "*" indicate that the partial busy period did not end.

Table 7.5. *Selected measures and their half-widths for example 7.7*

Note the possibility of a partial BP not being over during a simulation run due to having at least one server becoming idle with at least one server remaining busy during that time. This is indicated with an * in the above table. This fact will be clearer when looking at the full BP (see below).

EXAMPLE 7.8.– Referring to example 7.7, we now compute the full BP statistics. The measures are displayed in Table 7.6. This is not only to explain the * seen in Table 7.5 but also to show the significant differences seen in full and partial BPs as it should be since partial BP includes the full BP by definition.

Note that when comparing partial BPs to full BPs (of course, in the case of $c > 1$), the mean full BP will always be less than the partial BP (for the same model). This is due to the fact that the partial BP contains the full BP and additional periods during which at least one server is busy and at least one server is idle. Thus, as c is increased one will see an increase in the mean (partial) BP.

PH	GI	μ'_Y	μ'_{N_Y}
EX	CTA	1.4554 ± 0.0169	10.187 ± 0.112
	ABA	1.8695 ± 0.036	13.084 ± 0.2406
	UFA	1.5636 ± 0.0164	10.946 ± 0.1107
	WBA	8.2542 ± 0.2488	57.801 ± 1.7438
ER_2	CTA	0.9882 ± 0.0092	7.0582 ± 0.0608
	ABA	1.4866 ± 0.017	10.529 ± 0.1138
	UFA	1.123 ± 0.0121	7.9981 ± 0.0821
	WBA	9.1609 ± 0.2005	64.025 ± 1.393
ER_6	CTA	0.5312 ± 0.0033	4.0038 ± 0.0221
	ABA	1.1213 ± 0.0089	8.0942 ± 0.0615
	UFA	0.6877 ± 0.0038	5.0909 ± 0.0255
	WBA	10.068 ± 0.4112	70.242 ± 2.8667
ER_{10}	CTA	0.4136 ± 0.003	3.2153 ± 0.0196
	ABA	1.0321 ± 0.0088	7.5 ± 0.0599
	UFA	0.5734 ± 0.0038	4.326 ± 0.0255
	WBA	10.053 ± 0.2107	70.119 ± 1.4757
HE_2	CTA	1.5148 ± 0.0339	10.584 ± 0.2254
	ABA	1.9481 ± 0.0354	13.611 ± 0.2359
	UFA	1.638 ± 0.026	11.443 ± 0.1724
	WBA	8.1589 ± 0.291	57.146 ± 1.9877
HE_6	CTA	2.2702 ± 0.0696	15.625 ± 0.4627
	ABA	2.6005 ± 0.0675	17.96 ± 0.4509
	UFA	2.3209 ± 0.0746	16.005 ± 0.4954
	WBA	7.704 ± 0.2621	53.963 ± 1.8328
HE_{10}	CTA	5.1732 ± 0.4065	35.002 ± 2.7032
	ABA	5.2213 ± 0.6793	35.435 ± 4.5191
	UFA	4.9999 ± 0.3833	33.887 ± 2.5479
	WBA	6.4371 ± 0.8038	44.965 ± 5.375

Table 7.6. *Full busy period statistics along with their half-widths for example 7.8*

EXAMPLE 7.9.– In this example, we consider $GI/D/c$ queues to further explain the differences in the partial and BP statistics. Toward this end, we look at the four arrival processes of example 7.7 and assume the services to be constant with a value of 1. We fix the arrival rate to be $\lambda = 6.65$, the number of (identical) servers to be $c = 7$ and hence the traffic intensity is 0.95. In Table 7.7, we display a few selected measures for both partial and full BP scenarios. Note that the measures, L_s and μ'_{WS}, do not change when going from full to partial but we chose to display them anyway for both the cases.

$Measure$	GI	PBP	FBP
μ'_Y	CTA	0 ± 0	0.0977 ± 0.00000012405
	ABA	0 ± 0	0.8417 ± 0.00726
	UFA	0 ± 0	0.2747 ± 0.00096235
	WBA	43.565 ± 0.6683	10.473 ± 0.1851
μ'_{N_Y}	CTA	0 ± 0	1 ± 0
	ABA	0 ± 0	6.2705 ± 0.05074
	UFA	0 ± 0	2.3338 ± 0.00652
	WBA	295.17 ± 4.6863	72.976 ± 1.2971
L_s	CTA	6.6499 ± 0.0000079393	6.6499 ± 0.0000079393
	ABA	8.7474 ± 0.02836	8.7474 ± 0.02836
	UFA	7.1201 ± 0.00446	7.1201 ± 0.00446
	WBA	50.78 ± 1.5429	50.78 ± 1.5429
μ'_{WS}	CTA	1 ± 0	1 ± 0
	ABA	1.3152 ± 0.00383	1.3152 ± 0.00383
	UFA	1.0708 ± 0.00054127	1.0708 ± 0.00054127
	WBA	7.6327 ± 0.23019	7.6327 ± 0.23019

Table 7.7. *Selected measures along with their half-widths for both partial and full busy periods for example 7.9*

EXAMPLE 7.10.– In this example, we revisit example 7.7 and make the seven servers heterogeneous with each of them offering services according to the seven service time distributions mentioned in that example. An arriving customer will pick one of the servers at random when more than one server is available to serve at that time. All other parameters such as arrival process, arrival and service rates are the same. In Table 7.8, we display a few selected measures for the seven-server heterogeneous systems under four types of arrivals. The BP statistics correspond to the partial BP.

$Measure$	CTA	ABA	UFA	WBA
μ'_Y	446.77 ± 18.388	289.33 ± 13.658	403.91 ± 25.774	82.861 ± 3.7627
μ_{N_Y}	2916.3 ± 122.84	1823.1 ± 81.724	2602.4 ± 171.15	492.89 ± 24.5
μ'_I	0.05559 ± 0.00086472	0.08591 ± 0.00163	0.06229 ± 0.00131	0.64497 ± 0.01216
μ'_{RT}	437.84 ± 17.979	274.06 ± 12.012	391.3 ± 25.858	74.156 ± 3.624
L_s	64.673 ± 2.0774	74.38 ± 4.5273	71.166 ± 7.1088	120.74 ± 9.95
μ'_{WS}	10.151 ± 0.31274	11.611 ± 0.68275	11.127 ± 1.0692	18.589 ± 1.4798

Table 7.8. *Selected measures along with their half-widths for example 7.10*

7.4. *BMAP/G/c* queues

This section deals with the study of batch Markovian arrival process (BMAP) arrivals and generally distributed services in the context of multi-server systems. Similar to the section on simulating $GI/G/c$ queues (see section 7.3.5) we first validate a few models for which analytical results are known. This is followed by illustrative examples based on simulation.

We assume that the arrivals occur according to a BMAP with representation matrices given by $\{D_k\}$ of dimension m. The service times have a general distribution and there are c homogeneous servers to process the customers. We assume that the arrival rate of customers to the system is given by λ and the service rate is given by μ with the assumption that $\lambda < c\mu$ so as to make the queueing model under study to be stable. Let the traffic intensity be denoted by $\rho = \frac{\lambda}{c\mu}$.

The simulation studies involving Markovian arrival process (MAP) and BMAP processes are relatively new in the literature (see Buchholz and Felko (2014)). In fact, in Buchholz and Felko (2014), the authors point out that "MAPs, like PHDs, can also be integrated in simulation models but again this approach is not really supported by available simulation tools because first approaches describing the integration of MAPs in simulation models have been published only recently" and refer to the papers by Bause and Kriege (2009); Brickner et al. (2011); Heckmüller and Wolfinger (2009). Since then quite a few authors including this one have incorporated MAP and BMAP in simulating complicated models for which analytical results are not known or are very complex to be implemented numerically. Some of those papers include Chakravarthy (2012a,b, 2013, 2015); Chakravarthy and Maurina (2015); Chakravarthy (2016); Chakravarthy and Ozkar (2017); Chakravarthy and Rumyantsev (2018); Chakravarthy (2019b); Chakravarthy and Rumyantsev (2019); Chakravarthy (2020, 2021a,b, 2022b).

7.4.1. *ARENA modules for* BMAP/G/c *queues*

The ARENA modules that need to be discussed here are those generating the BMAP arrival process as other modules, such as process, assign and record, and are similar to the ones for $GI/G/c$ discussed in section 7.3.1. We leave to the readers the needed integration of all the modules to get the full simulation model as part of the exercises.

Simulating a BMAP depends on how the parameter matrices $\{D_k\}$ need to be set up. In this section, we discuss two methods of simulating a BMAP and both of these assume that $D_k = a_k D_1, \ k \geq 1$. Otherwise, one has to enter the data for all the parameter matrices assuming that the infinite sequence is suitably truncated. Thus, it is not the limitation of the simulation but rather a tedious job to input more than two matrices of dimension m. Except for this laborious coding, the rest of the simulation is similar to what is presented below.

In the first one, we assume that D_0 and D_1 of dimension m are given along with the batch size probability function, $\{a_k\}$, and hence these are input as parameters in ARENA. In the second one, we use the idea of generating MAP as spelled out in section 6.1 of Volume 1.

Typically, the probability function $\{a_k\}$ is modeled using Poisson, geometric, arbitrary (on a finite support) and (discrete) uniform (on a finite support). We focus mainly on Poisson and leave the others as exercises.

In Figure 7.37, we display the ARENA modules for generating MAP using the first approach in which we are given D_0 and D_1 matrices along with the batch size distribution (here we assume that ARENA has built-in functions for such a distribution; otherwise, one has to use another set of modules to simulate such a distribution).

The labels 1 through 5 shown in Figure 7.37 are explained below. This is with the assumption that $m = 3$.

1) The modules here are the starting point for simulating BMAP using the approach in which we are given D_0 and D_1 matrices. These matrices are entered through "variable" seen in "Basic Process" tab (see label 5). The CREATE module here has "Max Arrivals" set at 1 so as to start the simulation. According to the initial probability vector (referred to as "IPV" here) the process will start in one of the three states. We use "STATION" and "ROUTE" modules to model here mainly to get a clear picture as opposed to having too many connecting lines distracting the model development. Once a state is chosen, the ROUTE module directs the entity to that STATION. It is worth to point out that the ROUTE module has a place to put the time to send the item to the indicated STATION; however, here it is set at 0 since STATION/ROUTE is used to display the modules without using the connecting lines.

2) Here, the MAP process is simulated taking into account the duration spent in each of the three states and then routing from the current state to the other states (without an arrival) and to the other states including the state from where the absorption occurs (with an arrival of a batch). The DECIDE module determines whether the transition is with or without an arrival (of a batch) and where to route back. Due to the similarity of the construction of the modules for Station 2 and Station 3 areas, and for clarity we display only those modules coming out of Station 1 area in the box labeled 3. It should be pointed out that we use "PROCESS" module as opposed to "DELAY" for the duration of the time spent whenever that state is visited. If one needs to obtain the steady-state probability vector of the underlying continuous-time Markov chain (CTMC) governing the BMAP process, using process modules will help to capture the statistics using built-in expression as opposed to building one when using delay modules.

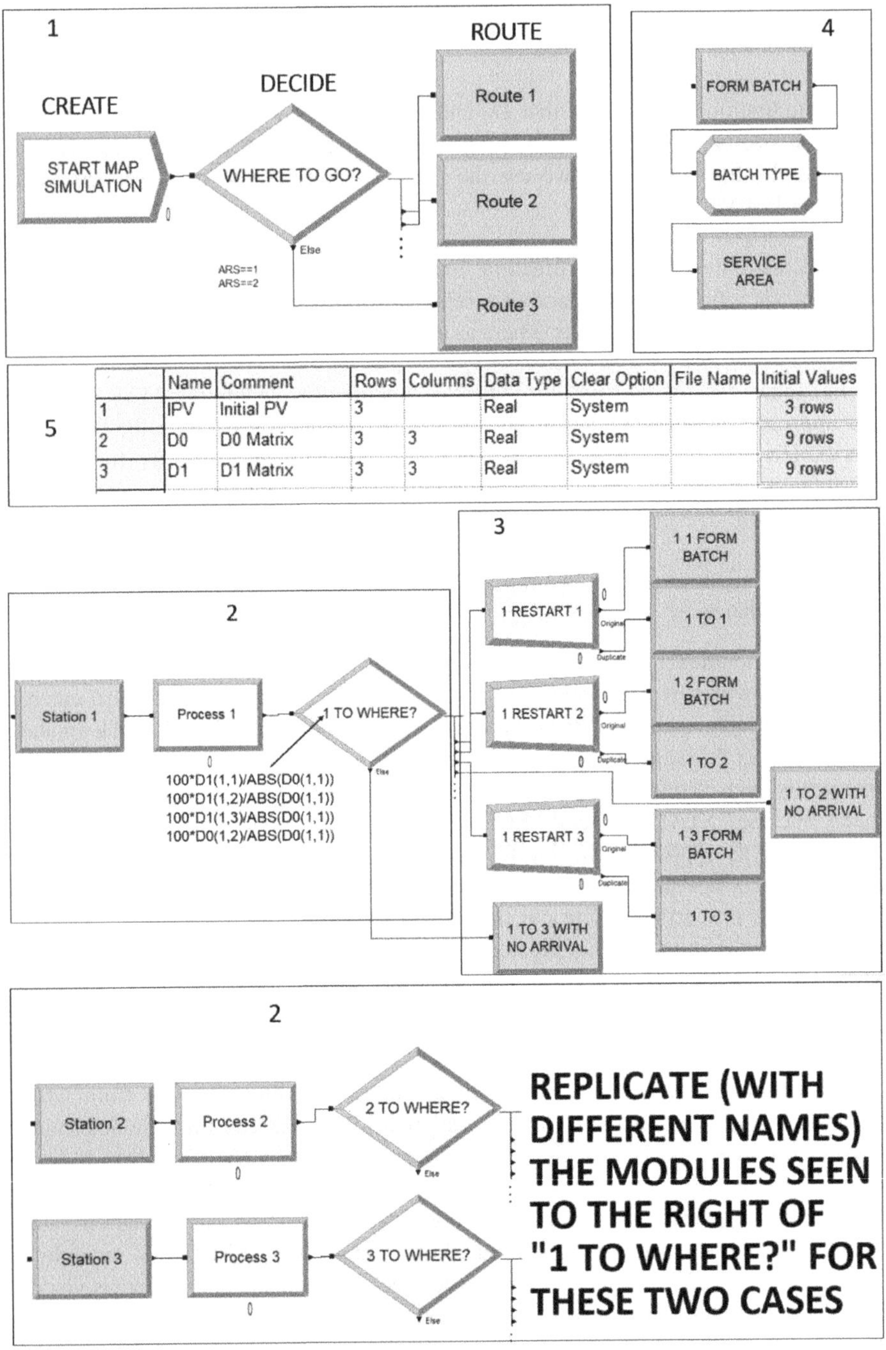

Figure 7.37. *Modules for simulating BMAP – Method 1*

3) At the end of a sojourn in State 1, either an arrival occurs (through an absorption in the underlying CTMC governing the MAP process) or a transition to another state occurs. This results in having a SEPARATE module to duplicate the entity and route accordingly.

4) If the transition results in an arrival (of a batch), here the batch is formed based on the probability distribution specified and the ASSIGN module takes care of it. Here, our ASSIGN module has the following quantities defined:

- An attribute named TIS is used to track the time an entity spends in the system (i.e. this is for waiting time in the system).

- Variables, VAR1 and VAR2, are defined for Poisson batch size and geometric batch size, respectively.

- An attribute named ToB is used to identify whether the batches follow Poisson or Geometric. The user has to specify this before the simulation begins. This is done through "variable" seen in "Basic Process" tab.

5) A partial picture of the "variable" seen in "Basic Process" tab is seen here. The variables are IPV (needed for starting the simulation of MAP process – see section 6.1 for details on this), and D0 and D1 are for D_0 and D_1 matrices. Note that IPV is a vector, and D_0 and D_1 are matrices. For our example, here we have taken $m = 3$ and hence we need three rows for IPV and three rows and three columns for both D_0 and D_1.

Suppose that one chooses to generate a BMAP using the technique described in "Generation of CMAP" in section 6.1 in Volume 1. In this case, the number of modules needed are much less compared to the previous method. Furthermore, the input parameters are also kept to a minimum level as we use Erlang and exponential to generate BMAP here. The parameters needed (see equation [6.41] in Volume 1) are the order and the parameter of the Erlang, the parameter for the exponential and the two probabilities that determine the choice of going back to their respective distributions.

In Figure 7.38, we display the modules needed to generate the BMAP using this approach. The modules are self-explanatory and we leave the details to the astute readers.

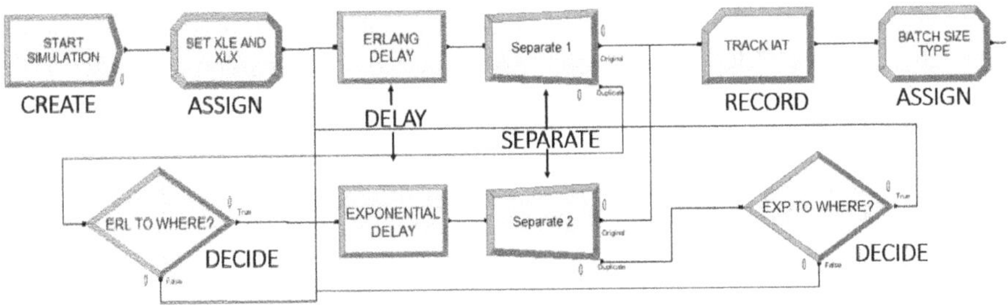

Figure 7.38. *Modules for simulating BMAP – Method 2*

7.4.2. *Validation of BMAP/G/1 queues*

The purpose of this section is to validate the simulated models against the known theoretical results for $BMAP/G/1$-type queues. In Chakravarthy (2019a), a number of models of $MAP/M/c$-type and $MAP/PH/1$-type queues have been validated against the known theoretical/approximation-based models in the literature. To that set of models, we add a few more (in the context of batch arrivals). It should be pointed out that this author developed the simulated models in all the papers cited with his name. We recommend that the readers refer to those papers for additional details along with examples and key observations. Here, we validate a few additional models. Some of these may overlap with the ones reference in those papers, especially with the ones discussed in Chakravarthy (2019a).

Here, we look at a number of scenarios comprising five different arrival processes and six different service distributions. For all of these scenarios, we assume Poisson batch size with parameter 1.25. That is, we assume that $\{a_k\}$ follows a Poisson distribution with an average value of 1.25. Since an arrival should have at least one customer and since Poisson has masses at non-negative integer values, we note that the average batch size is given by 2.25 by taking the arrival size to be 1 with probability $e^{-1.25}$. One can also verify this by noticing $D_k = a_{k-1}D,\ k \geq 1$, where (D_0, D) is the representation for the MAP process, and this combined with the (Poisson) batches means we get the BMAP process to model the arrivals. We look at two values for the traffic intensity, $\rho = 0.8$ and $\rho = 0.95$. Since we are looking at the batch arrivals, let λ_g denote the rate of batch arrivals, and let λ be the rate of the customers arriving per unit of time.

vERA: Inter-arrival times of batches follow an Erlang of order 5 with parameter $\frac{5}{\lambda_g}$.

vEXA: Inter-arrival times of batches are exponential with parameter λ_g.

vHEA: This is for hyperexponential inter-arrival times for the batches with parameters $\frac{2.18}{\lambda_g}$, $\frac{1.09}{\lambda_g}$, $\frac{0.545}{\lambda_g}$, $\frac{0.2725}{\lambda_g}$ and $\frac{0.13625}{\lambda_g}$ and their respective mixing probabilities given by 0.5, 0.3, 0.15, 0.04 and 0.01.

vNCA: This corresponds to the inter-arrival times whose 1-lag correlation between two successive arrivals of batches is -0.32667 and the matrices D_0 and D_1 are given by:

$$D_0 = \begin{pmatrix} -1.25 & 1.25 & 0 \\ 0 & -1.25 & 1.25 \\ 0 & 0 & -2.5 \end{pmatrix}, \quad D_1 = \begin{pmatrix} 0 & 0 & 0 \\ 0.0125 & 0 & 1.2375 \\ 2.475 & 0 & 0.0025 \end{pmatrix}.$$

vPCA: This corresponds to the inter-arrival times whose 1-lag correlation between two successive arrivals of batches is 0.32667 and the matrices D_0 and D_1 are given by:

$$D_0 = \begin{pmatrix} -1.25 & 1.25 & 0 \\ 0 & -1.25 & 1.25 \\ 0 & 0 & -2.5 \end{pmatrix}, \quad D_1 = \begin{pmatrix} 0 & 0 & 0 \\ 1.2375 & 0 & 0.0125 \\ 0.0025 & 0 & 2.475 \end{pmatrix}.$$

All these arrival distributions, which are qualitatively different and cover a wide variety of scenarios in practice, will be normalized so as to have $\lambda = 1$. That is, we take $\lambda_g = \frac{1}{2.25}$.

For services, we consider the following distributions. Since we fixed the arrival rate to be 1 and in order to achieve a given traffic intensity, we take the service rate, $\mu = \frac{1}{\rho}$.

vCTS: Constant services with the value ρ.

vABS: Service times have an arbitrary distribution. When $\rho = 0.8$, the distribution has masses at 0.6, 1.3 and 2.5 with probabilities, respectively, given by 0.80, 0.15 and 0.05; when $\rho = 0.95$, the distribution has masses at 0.7, 1.5 and 3.3 with probabilities, respectively, given by 0.80, 0.15 and 0.05.

vUFS: Services are assumed to be uniform on the interval $(\rho - 0.5, \rho + 0.5)$, for $\rho = 0.8, 0.95$.

vERS: Service distribution is Erlang of order 5 with parameter $\frac{5}{\rho}$.

vEXA: Inter-arrival times of batches are exponential with parameter $\frac{1}{\rho}$.

vHES: Services are assumed to be hyperexponential distribution with parameters $\frac{2.4}{\rho}$, $\frac{1.2}{\rho}$ and $\frac{0.6}{\rho}$ with mixing probabilities, respectively, given by 0.6, 0.3 and 0.1.

For all the scenarios, the number of replications is at least 15 and the length of each replicate is at least 1,000,000 units of time. There are some scenarios for which the number of replicates has to be increased to more than 15 so as to achieve more consistent results (in agreement with the theoretical ones). These scenarios include the distributions (for arrival and/or services) that have a large variability and/or when the traffic intensity is large.

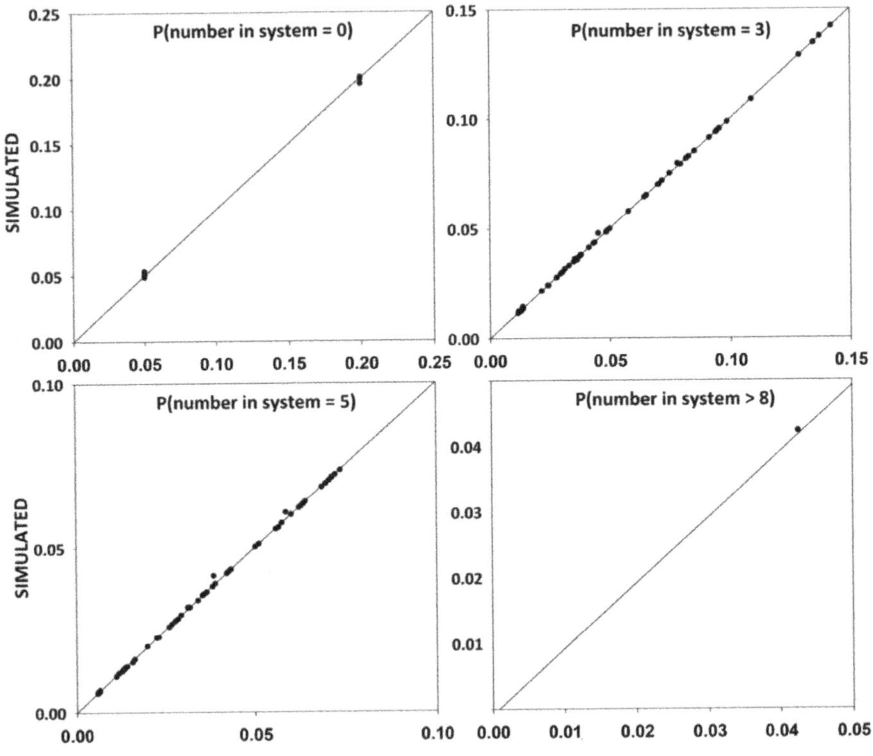

Figure 7.39. *Plots of the simulated results against theoretical values for example 7.11*

EXAMPLE 7.11.– In this validation example, we look at $BMAP/G/1$ queues with the above-mentioned five arrival processes and six service times under the assumption of Poisson batch arrivals for two traffic intensity values. In Figure 7.40, we display the scatter plots of the simulated results against the theoretical results for selected measures under various scenarios.

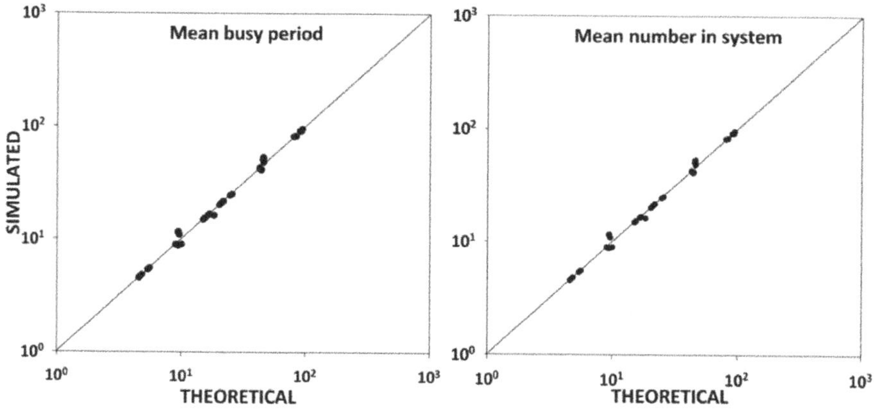

Figure 7.40. *Plots of the simulated results against theoretical values for example 7.11*

7.4.3. *Simulated examples for BMAP/G/1 queues*

Having validated the simulated $BMAP/G/1$ under various combinations of arrival and service processes, we present a few simulated examples of $BMAP/G/1$ queues.

EXAMPLE 7.12.– In this example, we look at the five arrival processes, labeled as $vERA$, $vEXA$, $vHEX$, $vNCA$ and $vPCA$, from section 7.4.2, all of which have the same (batch) arrival rate of λ_g. We assume that the batch size follows a Poisson distribution with parameter 1.25. Hence, we need to take $\lambda_g = \frac{1}{2.25}$ so that $\lambda = 1$. Suppose that the services follow a Weibull distribution with parameter $\alpha = 0.50$ and $\beta = 0.5\rho$. We look at three values for $\rho = 0.8, 0.9$ and $\rho = 0.95$.

In Table 7.9, we display selected measures along with their half-widths (needed to compute the 95% confidence level).

ToA	$Measure$	$\rho = 0.80$	$\rho = 0.90$	$\rho = 0.95$
$vERA$	μ'_Y	6.4144 ± 0.03672	14.214 ± 0.20524	29.477 ± 1.1306
	μ_{N_Y}	8.012 ± 0.03896	15.784 ± 0.20379	31.046 ± 1.1362
	μ'_{RT}	8.0119 ± 0.03692	15.784 ± 0.20434	31.026 ± 1.1331
	μ'_I	1.5975 ± 0.00202	1.5702 ± 0.00145	1.5581 ± 0.00309
	L_s	10.663 ± 0.07375	25.682 ± 0.61875	55.4 ± 2.8554
$vEXA$	μ'_Y	9.0206 ± 0.05884	20.519 ± 0.34751	43.383 ± 1.0402
	μ_{N_Y}	11.273 ± 0.0635	22.783 ± 0.36082	45.654 ± 1.0693
	μ'_{RT}	11.271 ± 0.0585	22.769 ± 0.3475	45.638 ± 1.0416
	μ'_I	2.251 ± 0.00318	2.2497 ± 0.00598	2.2553 ± 0.00808
	L_s	14.052 ± 0.16037	33.435 ± 0.79054	71.635 ± 3.0101
$vHEA$	μ'_Y	13.343 ± 0.12538	31.698 ± 0.45786	68.148 ± 1.9605
	μ_{N_Y}	16.675 ± 0.14507	35.226 ± 0.48067	71.782 ± 2.0609
	μ'_{RT}	16.662 ± 0.12466	35.211 ± 0.45616	71.766 ± 1.9623
	μ'_I	3.318 ± 0.01373	3.514 ± 0.01801	3.6199 ± 0.02711
	L_s	20.357 ± 0.25689	49.554 ± 1.0131	103.64 ± 4.2869
$vNCA$	μ'_Y	8.7688 ± 0.07646	19.61 ± 0.29605	41.701 ± 1.2608
	μ_{N_Y}	10.959 ± 0.08147	21.798 ± 0.30569	43.881 ± 1.2663
	μ'_{RT}	10.962 ± 0.0754	21.794 ± 0.29641	43.883 ± 1.2608
	μ'_I	2.1932 ± 0.00271	2.1838 ± 0.00486	2.1805 ± 0.00885
	L_s	12.94 ± 0.16243	30.402 ± 0.8139	65.464 ± 2.8359
$vPCA$	μ'_Y	12.092 ± 0.25767	27.524 ± 1.5414	57.578 ± 5.3583
	μ_{N_Y}	15.117 ± 0.3136	30.588 ± 1.6971	60.598 ± 5.6077
	μ'_{RT}	15.087 ± 0.25698	30.507 ± 1.5392	60.55 ± 5.3572
	μ'_I	2.9945 ± 0.00578	2.9826 ± 0.00711	2.9689 ± 0.01013
	L_s	150.93 ± 8.5745	375.97 ± 27.852	790.61 ± 110.08

Table 7.9. *Selected measures for example 7.12 for Weibull services*

7.4.4. *Validation of* BMAP/M/c *queues*

In this section, we validate the simulation model in two parts. First, we look at a single arrival case. That is, we look at a $MAP/M/c$ case and then look at $BMAP/M/c$ queues.

A. Validation using $MAP/M/c$ queues: Here, we look at the five arrival processes listed in section 7.4.2 and consider the following set of values for c and ρ. We fix the arrival rate to be 1 and take the service rate in such a way a given traffic intensity will be achieved. That is, $\mu = \dfrac{1}{c\rho}$. We vary $c = 1, 2, 3, 4$ and $\rho = 0.8, 0.95$ and validate the simulated model against the theoretical one discussed in section 5.30 using these measures: (1) mean number in system (L_s); (2) mean number of busy servers

($\mu_{BY} = \frac{\lambda}{\mu}$; (3) P_{wait} = P(an arrival has to wait to get into service); (4) PQ_0 = P(the queue is zero) = P(at most c customers are in the system); (5) mean (full) BP (μ'_Y); and (6) mean number served during a full BP (μ_{N_Y}). It is worth pointing out that we use full BP here to validate since analytical results for this multi-server case pertain to the full BP.

EXAMPLE 7.13.– In this example, we look at the five arrival processes mentioned in section 7.4.2 and consider the following set of values for $\rho = 0.8, 0.95$, and $c = 1, 2, 3, 4$. Under these various combinations, the simulated values are used as independent values and the theoretical values are dependent values. The fitted line along with the values of the adjusted coefficient of variation, commonly referred to as $adj - R^2$, are displayed below. In the following, we use the prefix "s" to indicate that the measure is based on simulation:

$$ln(L_s) = -0.01305 + 1.006\, ln(sLs), \quad adj(R^2) = 99.98\%,$$

$$\mu_{BY} = -0.000285 + 1.000\ {}_s\mu_{BY}, \quad adj(R^2) = 100.00\%,$$

$$P_{wait} = -0.001611 + 1.002\ {}_sP_{wait}, \quad adj(R^2) = 99.99\%,$$

$$PQ_0 = -0.000692 + 1.003\ {}_sPQ_0, \quad adj(R^2) = 99.99\%,$$

$$\mu'_Y = -0.3353 + 1.004\ {}_s\mu'_Y, \quad adj(R^2) = 99.16\%,$$

$$\mu_{N_Y} = -0.3858 + 1.005\ {}_s\mu_{N_Y}, \quad adj(R^2) = 99.16\%.$$

Obviously looking at the above statistics, it is clear that the simulated values are good predictors for the corresponding measures for the queueing model under consideration. Note that, except for L_s, the fitted coefficient for the slope is almost 1 with a near perfect adjusted R^2; for L_s we had to use a natural logarithm due to large values for HEA and PCA arrival processes.

B. Validation using $BMAP/M/c$ queues: Here, we validate the simulated models for the multi-server systems with BMAP arrivals and exponential services.

EXAMPLE 7.14.– In this example, we look at the same five arrival processes seen in example 7.13 but with group arrivals here. We assume that the batch size follows a Poisson distribution with parameter 1.25. We normalize the arrival process to have a unit arrival rate. We consider three values for the traffic intensity. The service times are exponential and the service rate is chosen so that the traffic intensity will be equal to the set value. The error percentages of the simulated and the theoretical values are displayed in Tables 7.10 and 7.11.

c	ρ	ToA	L_s	μ_{busy}	$P(empty\ queue)$	$P(all\ busy)$
$c=1$	0.8	ERA	0.19%	0.03%	0.08%	0.03%
		EXA	0.32%	0.04%	0.15%	0.04%
		HEA	0.23%	0.08%	0.31%	0.08%
		NCA	0.01%	0.02%	0.06%	0.02%
		PCA	0.41%	0.12%	0.44%	0.12%
	0.9	ERA	0.38%	0.03%	0.18%	0.03%
		EXA	0.64%	0.04%	0.39%	0.04%
		HEA	0.72%	0.05%	0.49%	0.05%
		NCA	0.28%	0.02%	0.20%	0.02%
		PCA	6.16%	0.36%	3.27%	0.36%
	0.95	ERA	0.19%	0.01%	0.28%	0.01%
		EXA	0.54%	0.03%	0.55%	0.03%
		HEA	0.75%	0.01%	0.15%	0.01%
		NCA	0.22%	0.05%	0.86%	0.05%
		PCA	3.00%	0.01%	0.18%	0.01%
$c=2$	0.8	ERA	0.02%	0.02%	0.10%	0.04%
		EXA	0.02%	0.03%	0.06%	0.03%
		HEA	0.12%	0.01%	0.04%	0.00%
		NCA	0.12%	0.01%	0.00%	0.01%
		PCA	1.32%	0.24%	0.96%	0.32%
	0.9	ERA	0.04%	0.02%	0.09%	0.01%
		EXA	0.27%	0.01%	0.02%	0.00%
		HEA	0.55%	0.02%	0.15%	0.02%
		NCA	0.04%	0.02%	0.26%	0.03%
		PCA	0.32%	0.04%	0.38%	0.05%
	0.95	ERA	0.09%	0.01%	0.22%	0.02%
		EXA	0.86%	0.03%	0.48%	0.04%
		HEA	2.52%	0.12%	2.27%	0.14%
		NCA	0.24%	0.00%	0.04%	0.00%
		PCA	4.51%	0.15%	2.77%	0.18%

Table 7.10. *Error percentages for example 7.14 when* $c = 1, 2$

c	ρ	ToA	L_s	μ_{busy}	$P(empty\ queue)$	$P(all\ busy)$
$c=3$	0.8	ERA	0.13%	0.03%	0.10%	0.05%
		EXA	0.08%	0.02%	0.04%	0.04%
		HEA	0.22%	0.01%	0.02%	0.01%
		NCA	0.32%	0.02%	0.11%	0.05%
		PCA	2.16%	0.24%	0.98%	0.37%
	0.9	ERA	0.57%	0.04%	0.32%	0.07%
		EXA	0.11%	0.03%	0.24%	0.04%
		HEA	0.13%	0.00%	0.01%	0.00%
		NCA	0.02%	0.02%	0.21%	0.03%
		PCA	0.81%	0.05%	0.54%	0.09%
	0.95	ERA	1.01%	0.02%	0.56%	0.05%
		EXA	0.20%	0.02%	0.42%	0.03%
		HEA	1.59%	0.07%	1.12%	0.09%
		NCA	0.80%	0.01%	0.23%	0.02%
		PCA	3.11%	0.02%	0.43%	0.03%

Table 7.11. *Error percentages for example 7.14 when* $c = 3$

7.4.5. *Simulated examples for* BMAP/G/c *queues*

Here, we see a few illustrative simulated examples, all of which were run with 10 replicates and each replicate is for 1,500,000 units of time.

EXAMPLE 7.15.– In this example, we look at several scenarios by varying the number of servers, the type of arrival process, the service times and the traffic intensity as follows. We consider $c = 2, 3, 5$ and $\rho = 0.8, 0.95$, and for arrival process we look at ERA, HEA, NCA, PCA and WBA, and for services we look at a uniform distribution on the interval $(0.3, 1.3)$ when $\rho = 0.8$ and on the interval $(0.45, 1.45)$ when $\rho = 0.95$. The first four processes are the ones discussed in example 7.14 and WBA is the Weibull distribution mentioned in section 7.3.3. The batch size is modeled using Poisson distribution with parameter 1.25. Note that the arrival process will be normalized so as to have the arrival rate, λ, of customers set at 1. Some selected measures along with their half-widths are displayed in Tables 7.12 and 7.13. It should be pointed out that the mean BP obtained is that of the partial BP.

c	ρ	ToA	μ'_I	μ'_Y	μ'_{RT}
2	0.8	ERA	1.3582 ± 0.0007	1.3625 ± 0.0007	2.7207 ± 0.0011
		HEA	3.0738 ± 0.0058	2.5201 ± 0.0025	5.5939 ± 0.006
		NCA	2.3079 ± 0.0029	2.0066 ± 0.0029	4.3145 ± 0.0046
		PCA	2.8014 ± 0.0054	2.3197 ± 0.0115	5.1211 ± 0.0113
		WBA	5.0911 ± 0.0146	3.8445 ± 0.004	8.9356 ± 0.0165
	0.95	ERA	1.2651 ± 0.0006	1.7595 ± 0.0007	3.0247 ± 0.0009
		HEA	3.2927 ± 0.0102	3.6002 ± 0.0088	6.8929 ± 0.0167
		NCA	2.3036 ± 0.0026	2.7071 ± 0.0048	5.0107 ± 0.0047
		PCA	2.8154 ± 0.0032	3.142 ± 0.0258	5.9574 ± 0.0253
		WBA	5.4456 ± 0.008	5.5329 ± 0.0236	10.978 ± 0.0234
3	0.8	ERA	1.4086 ± 0.0007	1.0972 ± 0.0004	2.5059 ± 0.0008
		HEA	2.8819 ± 0.0052	1.6693 ± 0.0013	4.5513 ± 0.0054
		NCA	2.3455 ± 0.0022	1.4597 ± 0.0008	3.8053 ± 0.0024
		PCA	2.5616 ± 0.0049	1.5334 ± 0.0055	4.0951 ± 0.0034
		WBA	4.6887 ± 0.0105	2.286 ± 0.0038	6.9748 ± 0.0123
	0.95	ERA	1.317 ± 0.0005	1.3416 ± 0.0005	2.6587 ± 0.0009
		HEA	3.0232 ± 0.0051	2.1711 ± 0.003	5.1944 ± 0.0056
		NCA	2.3478 ± 0.0018	1.8405 ± 0.0014	4.1883 ± 0.0026
		PCA	2.6335 ± 0.0067	1.9694 ± 0.0144	4.603 ± 0.01
		WBA	4.9563 ± 0.0153	3.0097 ± 0.0061	7.9661 ± 0.0186
5	0.8	ERA	1.4256 ± 0.0008	1.0236 ± 0.0002	2.4493 ± 0.0008
		HEA	2.7968 ± 0.0051	1.3576 ± 0.0007	4.1545 ± 0.0052
		NCA	2.3725 ± 0.0031	1.2489 ± 0.0006	3.6214 ± 0.0032
		PCA	2.4403 ± 0.0084	1.268 ± 0.0037	3.7083 ± 0.005
		WBA	4.4359 ± 0.0079	1.6155 ± 0.0013	6.0515 ± 0.0081
	0.95	ERA	1.3357 ± 0.0006	1.2259 ± 0.0004	2.5617 ± 0.0006
		HEA	2.9 ± 0.0048	1.6743 ± 0.0013	4.5743 ± 0.0053
		NCA	2.3834 ± 0.0024	1.5137 ± 0.0009	3.8971 ± 0.0028
		PCA	2.4696 ± 0.0075	1.5407 ± 0.0051	4.0103 ± 0.0032
		WBA	4.6376 ± 0.0104	1.9957 ± 0.0013	6.6334 ± 0.0109

Table 7.12. *Simulated values for $\mu'_I, \mu'_Y, \mu'_{RT}$ along with their half-widths for example 7.15*

c	ρ	ToA	L_s	PQ_0	P_{wait}
2	0.8	ERA	1.0429 ± 0.0007	0.8509 ± 0.0002	0.3262 ± 0.0003
		HEA	1.6767 ± 0.0039	0.7426 ± 0.0005	0.6363 ± 0.0005
		NCA	1.3187 ± 0.0026	0.7796 ± 0.0004	0.5345 ± 0.0006
		PCA	4.4716 ± 0.0825	0.7349 ± 0.0016	0.6401 ± 0.0018
		WBA	2.761 ± 0.0058	0.6861 ± 0.0005	0.7722 ± 0.0002
	0.95	ERA	1.2866 ± 0.0008	0.8057 ± 0.0002	0.3743 ± 0.0002
		HEA	2.31 ± 0.0072	0.6672 ± 0.0007	0.6989 ± 0.0008
		NCA	1.6938 ± 0.0044	0.7154 ± 0.0007	0.5893 ± 0.0007
		PCA	12.48 ± 0.3209	0.6628 ± 0.0029	0.6897 ± 0.0028
		WBA	3.9397 ± 0.0319	0.6083 ± 0.0014	0.8134 ± 0.0009
3	0.8	ERA	0.8617 ± 0.0004	0.9573 ± 0.0001	0.1137 ± 0.0003
		HEA	1.0787 ± 0.002	0.8914 ± 0.0002	0.3836 ± 0.0004
		NCA	0.9695 ± 0.0011	0.9134 ± 0.0002	0.2871 ± 0.0004
		PCA	1.2246 ± 0.008	0.8904 ± 0.001	0.3887 ± 0.002
		WBA	1.5346 ± 0.0056	0.8368 ± 0.0004	0.5798 ± 0.0009
	0.95	ERA	1.0343 ± 0.0004	0.9433 ± 0.0001	0.1348 ± 0.0003
		HEA	1.3631 ± 0.0028	0.8543 ± 0.0003	0.4403 ± 0.0005
		NCA	1.1885 ± 0.0011	0.8848 ± 0.0002	0.3279 ± 0.0005
		PCA	1.8061 ± 0.0253	0.8474 ± 0.0024	0.4592 ± 0.0041
		WBA	2.0091 ± 0.0055	0.792 ± 0.0005	0.6264 ± 0.0007
5	0.8	ERA	0.804 ± 0.0003	0.9973 ± 0	0.0099 ± 0.0001
		HEA	0.849 ± 0.0013	0.977 ± 0.0001	0.1234 ± 0.0002
		NCA	0.8236 ± 0.0008	0.9858 ± 0.0001	0.0653 ± 0.0003
		PCA	0.8622 ± 0.0061	0.9762 ± 0.0004	0.1368 ± 0.0014
		WBA	0.9968 ± 0.0016	0.9441 ± 0.0001	0.3043 ± 0.0005
	0.95	ERA	0.9559 ± 0.0003	0.996 ± 0	0.0137 ± 0.0001
		HEA	1.0261 ± 0.0017	0.9663 ± 0.0001	0.1547 ± 0.0005
		NCA	0.9859 ± 0.001	0.9793 ± 0.0001	0.0811 ± 0.0003
		PCA	1.0533 ± 0.0078	0.9645 ± 0.0005	0.1757 ± 0.0016
		WBA	1.2319 ± 0.0028	0.9254 ± 0.0002	0.3448 ± 0.0004

Table 7.13. *Simulated values for L_s, PQ_0, P_{wait} along with their half-widths for example 7.15*

EXAMPLE 7.16.– This example is similar to example 7.15 but here the service times are assumed to be constant (CTS). That is, the service times are constant and given by

ρ as the mean of the services. Some selected measures along with their half-widths are displayed in Tables 7.14 and 7.15. It should be pointed out that the mean BP obtained is that of the partial BP.

c	ρ	ToA	μ'_I	μ'_Y	μ'_{RT}
2	0.8	ERA	0.9908 ± 0.0008	6.7912 ± 0.0187	7.782 ± 0.0185
		HEA	4.5516 ± 0.0195	20.823 ± 0.1393	25.375 ± 0.1437
		NCA	2.2457 ± 0.0067	11.666 ± 0.0357	13.912 ± 0.0336
		PCA	2.6334 ± 0.0045	12.838 ± 0.2938	15.47 ± 0.2921
		WBA	6.983 ± 0.0324	30.372 ± 0.2149	37.355 ± 0.2019
	0.95	ERA	0.9024 ± 0.0027	32.5 ± 0.3653	33.403 ± 0.3649
		HEA	5.4126 ± 0.0692	115.54 ± 3.2602	120.96 ± 3.2661
		NCA	2.2126 ± 0.0117	55.035 ± 0.8898	57.249 ± 0.8842
		PCA	2.5554 ± 0.0113	57.768 ± 5.2646	60.322 ± 5.2567
		WBA	7.7596 ± 0.074	159.34 ± 5.668	167.11 ± 5.6558
3	0.8	ERA	0.8321 ± 0.0014	12.275 ± 0.0319	13.107 ± 0.0316
		HEA	4.9625 ± 0.0189	26.068 ± 0.1698	31.031 ± 0.1715
		NCA	2.2628 ± 0.0072	15.584 ± 0.0435	17.847 ± 0.0377
		PCA	2.5263 ± 0.0059	15.814 ± 0.3538	18.34 ± 0.3529
		WBA	7.3287 ± 0.0317	34.725 ± 0.2672	42.054 ± 0.254
	0.95	ERA	0.7599 ± 0.0022	72.812 ± 0.7632	73.574 ± 0.7624
		HEA	5.9313 ± 0.0833	142.57 ± 3.8483	148.51 ± 3.8454
		NCA	2.2256 ± 0.014	74.869 ± 1.2504	77.097 ± 1.246
		PCA	2.44 ± 0.0162	72.629 ± 6.5437	75.067 ± 6.5356
		WBA	8.1595 ± 0.0905	181.19 ± 6.3613	189.37 ± 6.3522
5	0.8	ERA	0.6784 ± 0.0039	76.584 ± 0.3642	77.263 ± 0.3654
		HEA	5.812 ± 0.0292	39.795 ± 0.2625	45.607 ± 0.2674
		NCA	2.2532 ± 0.0087	26.733 ± 0.0979	28.986 ± 0.0931
		PCA	2.3351 ± 0.0062	27.433 ± 0.6059	29.768 ± 0.6057
		WBA	7.9967 ± 0.0448	45.062 ± 0.3719	53.06 ± 0.3631
	0.95	ERA	0.6468 ± 0.0077	772.88 ± 11.352	773.75 ± 11.317
		HEA	6.8952 ± 0.1268	209.05 ± 6.3905	215.78 ± 6.4636
		NCA	2.1917 ± 0.0157	135.84 ± 2.464	138.03 ± 2.4588
		PCA	2.2587 ± 0.0142	136.84 ± 12.613	139.1 ± 12.606
		WBA	8.8702 ± 0.1337	231.29 ± 8.5318	240.16 ± 8.5589

Table 7.14. *Simulated values for $\mu'_I, \mu'_Y, \mu'_{RT}$ along with their half-widths for example 7.16*

c	ρ	ToA	L_s	PQ_0	P_{wait}
2	0.8	ERA	3.3504 ± 0.0135	0.4546 ± 0.001	0.7062 ± 0.0008
		HEA	11.85 ± 0.1032	0.2704 ± 0.0015	0.9142 ± 0.0006
		NCA	5.5512 ± 0.0268	0.3366 ± 0.001	0.8362 ± 0.0005
		PCA	136.53 ± 7.6053	0.2989 ± 0.0058	0.8733 ± 0.003
		WBA	20.778 ± 0.2289	0.2421 ± 0.0018	0.9439 ± 0.0004
	0.95	ERA	11.172 ± 0.1893	0.1401 ± 0.0016	0.9213 ± 0.0009
		HEA	64.209 ± 3.1374	0.068 ± 0.0019	0.9817 ± 0.0005
		NCA	22.134 ± 0.5215	0.0935 ± 0.0015	0.9579 ± 0.0006
		PCA	696.47 ± 67.337	0.0848 ± 0.0075	0.9659 ± 0.0031
		WBA	106.8 ± 4.324	0.0612 ± 0.0023	0.9873 ± 0.0005
3	0.8	ERA	3.9636 ± 0.0137	0.5203 ± 0.0011	0.6475 ± 0.0009
		HEA	12.396 ± 0.1033	0.2917 ± 0.0016	0.8847 ± 0.0007
		NCA	6.1066 ± 0.0268	0.3776 ± 0.0012	0.776 ± 0.0007
		PCA	137.08 ± 7.6093	0.3251 ± 0.0063	0.8436 ± 0.0038
		WBA	21.281 ± 0.2308	0.2558 ± 0.0019	0.9224 ± 0.0006
	0.95	ERA	11.909 ± 0.1912	0.1626 ± 0.0018	0.9035 ± 0.0011
		HEA	64.851 ± 3.1357	0.0736 ± 0.002	0.9751 ± 0.0007
		NCA	22.789 ± 0.5203	0.1069 ± 0.0017	0.9416 ± 0.0009
		PCA	697.12 ± 67.335	0.0935 ± 0.0082	0.9575 ± 0.0039
		WBA	107.4 ± 4.322	0.065 ± 0.0024	0.9823 ± 0.0007
5	0.8	ERA	5.3432 ± 0.0143	0.5914 ± 0.0013	0.5459 ± 0.0012
		HEA	13.59 ± 0.1038	0.3257 ± 0.0017	0.8375 ± 0.0011
		NCA	7.3236 ± 0.0278	0.45 ± 0.0011	0.6891 ± 0.0009
		PCA	138.31 ± 7.6149	0.3532 ± 0.0068	0.8092 ± 0.0045
		WBA	22.369 ± 0.2342	0.2795 ± 0.0021	0.8868 ± 0.0009
	0.95	ERA	13.509 ± 0.1928	0.1927 ± 0.0022	0.87 ± 0.0016
		HEA	66.246 ± 3.1394	0.0829 ± 0.0022	0.9641 ± 0.001
		NCA	24.217 ± 0.5209	0.1321 ± 0.0021	0.9174 ± 0.0014
		PCA	698.55 ± 67.342	0.1033 ± 0.0091	0.9469 ± 0.0048
		WBA	108.66 ± 4.3221	0.0716 ± 0.0027	0.9739 ± 0.001

Table 7.15. *Simulated values for L_s, PQ_0, P_{wait} along with their half-widths for example 7.16*

EXAMPLE 7.17.– This example is similar to example 7.15 but here the service times are assumed to be hyperexponential with parameters $2.4\,\mu, 1.2\,\mu, 0.6\,\mu, 0.24\,\mu$ and $0.048\,\mu$ with mixing probabilities, respectively, given by 0.6, 0.25, 0.1, 0.04 and 0.01.

The value of μ will be chosen so as to achieve a given traffic intensity. Some selected measures along with their half-widths are displayed in Tables 7.16 and 7.17. It should be pointed out that the mean BP obtained is that of the partial BP.

c	ρ	ToA	μ'_I	μ'_Y	μ'_{RT}
2	0.8	ERA	1.3921 ± 0.002	9.0787 ± 0.0489	10.47 ± 0.0475
		HEA	3.5969 ± 0.0152	19.04 ± 0.2105	22.637 ± 0.2178
		NCA	2.2258 ± 0.0032	12.967 ± 0.0821	15.192 ± 0.0823
		PCA	2.8514 ± 0.0077	15.631 ± 0.4083	18.482 ± 0.4103
		WBA	5.8165 ± 0.0237	28.726 ± 0.4036	34.542 ± 0.3969
	0.95	ERA	1.3429 ± 0.0051	41.252 ± 1.7623	42.596 ± 1.7628
		HEA	3.9216 ± 0.0313	97.117 ± 6.7528	101.04 ± 6.7457
		NCA	2.2184 ± 0.0053	59.842 ± 2.7729	62.063 ± 2.7695
		PCA	2.8021 ± 0.0109	78.998 ± 5.6365	81.805 ± 5.6361
		WBA	6.2858 ± 0.0842	145.26 ± 8.0618	151.55 ± 8.0779
3	0.8	ERA	1.2921 ± 0.0019	16.408 ± 0.164	17.7 ± 0.1636
		HEA	3.9519 ± 0.0174	30.506 ± 0.2701	34.458 ± 0.2666
		NCA	2.2299 ± 0.0061	21.875 ± 0.2917	24.104 ± 0.2864
		PCA	2.7655 ± 0.0081	23.752 ± 0.4747	26.509 ± 0.4752
		WBA	6.2802 ± 0.0433	42.233 ± 0.6975	48.512 ± 0.7035
	0.95	ERA	1.2468 ± 0.0044	76.993 ± 3.3273	78.226 ± 3.3242
		HEA	4.3954 ± 0.0579	150.68 ± 4.6844	155.09 ± 4.6741
		NCA	2.2162 ± 0.0168	103.67 ± 3.5208	105.89 ± 3.5138
		PCA	2.727 ± 0.012	111.23 ± 6.9298	113.89 ± 6.886
		WBA	6.798 ± 0.0868	213.46 ± 12.867	220.23 ± 12.955
5	0.8	ERA	1.1649 ± 0.003	65.302 ± 0.5652	66.465 ± 0.5633
		HEA	4.7677 ± 0.0654	83.811 ± 1.0919	88.573 ± 1.0739
		NCA	2.2102 ± 0.0077	69.659 ± 1.3556	71.863 ± 1.3498
		PCA	2.6474 ± 0.0109	58.969 ± 1.5732	61.617 ± 1.5763
		WBA	7.1593 ± 0.0608	96.469 ± 1.3593	103.62 ± 1.3366
	0.95	ERA	1.1241 ± 0.0097	340.73 ± 16.659	341.91 ± 16.666
		HEA	5.4666 ± 0.1124	403.28 ± 16.889	408.47 ± 16.669
		NCA	2.1845 ± 0.0202	335.8 ± 15.81	337.96 ± 15.851
		PCA	2.6037 ± 0.0264	280.73 ± 15.722	283.36 ± 15.733
		WBA	7.8759 ± 0.1704	500.26 ± 30.524	508.17 ± 30.481

Table 7.16. *Simulated values for $\mu'_I, \mu'_Y, \mu'_{RT}$ along with their half-widths for example 7.17*

c	ρ	ToA	L_s	PQ_0	P_{wait}
2	0.8	ERA	16.216 ± 0.2945	0.3489 ± 0.0016	0.7695 ± 0.0011
		HEA	26.592 ± 0.5326	0.2772 ± 0.0021	0.8915 ± 0.001
		NCA	18.791 ± 0.454	0.3111 ± 0.0018	0.8362 ± 0.0009
		PCA	156.15 ± 4.5296	0.2898 ± 0.006	0.88 ± 0.0032
		WBA	35.021 ± 1.0895	0.2558 ± 0.003	0.9308 ± 0.0009
	0.95	ERA	101.85 ± 10.522	0.0939 ± 0.0037	0.9401 ± 0.0024
		HEA	152.34 ± 13.425	0.0707 ± 0.0046	0.9749 ± 0.0016
		NCA	107.8 ± 8.1171	0.0829 ± 0.0037	0.9582 ± 0.0019
		PCA	930.13 ± 136.09	0.0699 ± 0.0055	0.9721 ± 0.0023
		WBA	197.23 ± 14.543	0.065 ± 0.0036	0.9839 ± 0.0009
3	0.8	ERA	14.859 ± 0.3952	0.4049 ± 0.0032	0.6994 ± 0.0023
		HEA	24.684 ± 0.317	0.3119 ± 0.0022	0.8522 ± 0.0011
		NCA	17.351 ± 0.3637	0.3586 ± 0.0037	0.7778 ± 0.0023
		PCA	148.8 ± 8.3139	0.3177 ± 0.0056	0.8467 ± 0.0033
		WBA	33.504 ± 0.629	0.2818 ± 0.0039	0.9034 ± 0.0014
	0.95	ERA	92.588 ± 5.2028	0.1095 ± 0.0046	0.922 ± 0.0033
		HEA	147.34 ± 6.6486	0.0807 ± 0.0027	0.9654 ± 0.0011
		NCA	109.48 ± 9.0553	0.0937 ± 0.0029	0.9446 ± 0.0017
		PCA	793.59 ± 99.66	0.0835 ± 0.0056	0.9613 ± 0.0028
		WBA	200.9 ± 19.442	0.0709 ± 0.0038	0.9779 ± 0.0013
5	0.8	ERA	13.657 ± 0.2652	0.486 ± 0.0031	0.6018 ± 0.0027
		HEA	23.327 ± 0.4329	0.3649 ± 0.0034	0.7907 ± 0.0023
		NCA	15.933 ± 0.4407	0.4336 ± 0.0033	0.6872 ± 0.0028
		PCA	154.57 ± 11.432	0.3497 ± 0.0083	0.8081 ± 0.0056
		WBA	31.474 ± 1.1368	0.3234 ± 0.0038	0.8574 ± 0.002
	0.95	ERA	93.669 ± 6.5292	0.1348 ± 0.005	0.8945 ± 0.0041
		HEA	139.56 ± 9.7431	0.0978 ± 0.0035	0.9495 ± 0.0019
		NCA	102.32 ± 7.5777	0.1202 ± 0.005	0.9176 ± 0.0035
		PCA	769.07 ± 63.065	0.1007 ± 0.0052	0.9468 ± 0.0028
		WBA	201.83 ± 12.834	0.08 ± 0.0042	0.9682 ± 0.0017

Table 7.17. *Simulated values for L_s, PQ_0, P_{wait} along with their half-widths for example 7.17*

EXAMPLE 7.18.– This example is similar to example 7.15 but here the service times are assumed to be Weibull with shape parameter of 0.5 and scale parameter of $\frac{0.5}{\mu}$. The value of μ will be chosen so as to achieve a given traffic intensity. Some selected measures along with their half-widths are displayed in Tables 7.18 and 7.19. It should be pointed out that the mean BP obtained is that of the partial BP.

c	ρ	ToA	μ'_I	μ'_Y	μ'_{RT}
2	0.8	ERA	1.5042 ± 0.0015	10.747 ± 0.0683	12.251 ± 0.0689
		HEA	3.6095 ± 0.0187	19.863 ± 0.1492	23.472 ± 0.1403
		NCA	2.1878 ± 0.0042	13.84 ± 0.1256	16.028 ± 0.1283
		PCA	2.9131 ± 0.0068	16.815 ± 0.4791	19.728 ± 0.4802
		WBA	5.6671 ± 0.0304	28.866 ± 0.3321	34.533 ± 0.344
	0.95	ERA	1.4641 ± 0.0031	49.994 ± 1.1329	51.459 ± 1.1344
		HEA	3.9697 ± 0.0468	101.91 ± 3.5818	105.89 ± 3.5785
		NCA	2.1762 ± 0.0076	63.84 ± 1.8718	66.019 ± 1.8718
		PCA	2.8851 ± 0.0151	80.507 ± 15.356	83.397 ± 15.352
		WBA	6.1499 ± 0.0774	148.25 ± 8.3737	154.41 ± 8.3634
3	0.8	ERA	1.43 ± 0.0027	20.699 ± 0.0768	22.129 ± 0.0758
		HEA	3.9949 ± 0.0273	32.919 ± 0.3057	36.913 ± 0.3075
		NCA	2.1854 ± 0.0073	24.903 ± 0.186	27.088 ± 0.1839
		PCA	2.8428 ± 0.0071	25.99 ± 0.2054	28.831 ± 0.208
		WBA	6.2143 ± 0.0386	43.631 ± 0.3558	49.846 ± 0.3803
	0.95	ERA	1.3936 ± 0.005	96.703 ± 1.9309	98.099 ± 1.9301
		HEA	4.4203 ± 0.0774	167.13 ± 9.8155	171.33 ± 9.9142
		NCA	2.1589 ± 0.0154	118.52 ± 4.7389	120.64 ± 4.6784
		PCA	2.8159 ± 0.0121	130.07 ± 16.307	132.88 ± 16.303
		WBA	6.7517 ± 0.1048	222.42 ± 11.274	229.19 ± 11.301
5	0.8	ERA	1.3297 ± 0.0067	86.128 ± 1.13	87.458 ± 1.1311
		HEA	4.8286 ± 0.0257	91.513 ± 0.7634	96.338 ± 0.7663
		NCA	2.163 ± 0.0137	87.135 ± 0.6284	89.298 ± 0.622
		PCA	2.7531 ± 0.01	66.954 ± 1.7991	69.706 ± 1.8063
		WBA	7.2094 ± 0.0325	99.508 ± 1.1513	106.72 ± 1.1501
	0.95	ERA	1.3127 ± 0.0141	472.58 ± 14.428	473.9 ± 14.487
		HEA	5.402 ± 0.1095	469.02 ± 19.007	474.48 ± 19.036
		NCA	2.1456 ± 0.0206	459.16 ± 15.293	461.3 ± 15.217
		PCA	2.7109 ± 0.0298	356.46 ± 29.821	359.21 ± 29.85
		WBA	7.9769 ± 0.1044	511.31 ± 22.743	519.43 ± 22.768

Table 7.18. *Simulated values for $\mu'_I, \mu'_Y, \mu'_{RT}$ along with their half-widths for example 7.18*

c	ρ	ToA	L_s	PQ_0	P_{wait}
2	0.8	ERA	10.046 ± 0.091	0.3485 ± 0.0016	0.7823 ± 0.001
		HEA	19.54 ± 0.2687	0.2825 ± 0.002	0.8883 ± 0.0008
		NCA	12.165 ± 0.2133	0.3204 ± 0.0021	0.8325 ± 0.0013
		PCA	145.23 ± 6.2202	0.2921 ± 0.0071	0.8821 ± 0.0035
		WBA	28.089 ± 0.4896	0.2597 ± 0.0023	0.9266 ± 0.0008
	0.95	ERA	55.633 ± 2.5152	0.0918 ± 0.002	0.9446 ± 0.0012
		HEA	110.87 ± 4.3468	0.0714 ± 0.0026	0.9745 ± 0.0009
		NCA	63.082 ± 3.2227	0.0859 ± 0.0024	0.957 ± 0.0013
		PCA	902.19 ± 277.54	0.0775 ± 0.0123	0.9698 ± 0.005
		WBA	155.07 ± 11.899	0.0656 ± 0.0039	0.9831 ± 0.001
3	0.8	ERA	9.9609 ± 0.1092	0.4058 ± 0.0018	0.7127 ± 0.0013
		HEA	19.032 ± 0.3274	0.3193 ± 0.0028	0.8462 ± 0.0014
		NCA	11.888 ± 0.1014	0.3716 ± 0.0016	0.7722 ± 0.0011
		PCA	143.73 ± 5.9834	0.323 ± 0.0024	0.8466 ± 0.0013
		WBA	27.693 ± 0.4077	0.2882 ± 0.0016	0.8967 ± 0.0009
	0.95	ERA	52.473 ± 1.9999	0.1121 ± 0.0026	0.924 ± 0.0018
		HEA	109.28 ± 9.4494	0.0824 ± 0.0045	0.9641 ± 0.0021
		NCA	64.906 ± 3.0796	0.0999 ± 0.0036	0.9418 ± 0.0022
		PCA	842.16 ± 130.51	0.0839 ± 0.0092	0.9619 ± 0.0043
		WBA	156.97 ± 7.9935	0.073 ± 0.0034	0.9762 ± 0.0012
5	0.8	ERA	10.299 ± 0.0865	0.4889 ± 0.0017	0.6114 ± 0.0015
		HEA	19.083 ± 0.306	0.3758 ± 0.0025	0.7792 ± 0.0017
		NCA	12.071 ± 0.1588	0.4484 ± 0.0019	0.6794 ± 0.0016
		PCA	147.02 ± 8.7863	0.3522 ± 0.008	0.8075 ± 0.0056
		WBA	27.101 ± 0.5135	0.3358 ± 0.0029	0.8447 ± 0.0017
	0.95	ERA	53.789 ± 2.1013	0.1377 ± 0.0031	0.8958 ± 0.0023
		HEA	104.55 ± 4.4051	0.0992 ± 0.0033	0.9474 ± 0.0018
		NCA	65.739 ± 3.4974	0.1233 ± 0.0028	0.917 ± 0.0019
		PCA	772.71 ± 94.108	0.0966 ± 0.0074	0.9495 ± 0.0041
		WBA	146.91 ± 8.9888	0.0877 ± 0.0041	0.9635 ± 0.0016

Table 7.19. *Simulated values for L_s, PQ_0, P_{wait} along with their half-widths for example 7.17*

7.5. Exercises

Exercise 7.1. Develop a simulation model for $GI/G/1$ by incorporating waiting time in the queue, waiting time in the system, busy period and other measures. Run your model for 10 replications with 1,000,000 units of time for each replicate. Compare your simulated results with the reported ones in this chapter (see example 7.3). Write a short report on your findings.

Exercise 7.2. Develop a simulation model for $GI/G/c$ by incorporating waiting time in the queue, waiting time in the system, both partial and full busy periods and other measures. Run your model for 10 replications with 1,000,000 units of time for each replicate. Compare your simulated results with the reported ones in this chapter (see examples 7.5 through 7.10). Write a short report on your findings.

Exercise 7.3. Develop a simulation model for $BMAP/G/1$ by incorporating waiting time in the queue, waiting time in the system, busy period, recurrence time and other measures. Run your model for 10 replications with 1,000,000 units of time for each replicate. Compare your simulated results with the ones in example 7.12. Write a short report on your findings.

Exercise 7.4. Develop a simulation model for $BMAP/G/c$ by incorporating waiting time in the queue, waiting time in the system, busy period, recurrence time and other measures. Run your model for 10 replications with 1,000,000 units of time for each replicate. Compare your simulated results with the reported ones in this chapter (see examples 7.15 through 7.18). Write a short report on your findings.

Exercise 7.5. Develop a simulation model for $BMAP/PH/1/K$ as presented in examples 6.1 and 6.2. Run 10 replicates with a length of 1,000,000 units for each replicate and compare your simulated results for the measures given in those examples. Write a short report on your findings.

Exercise 7.6. Develop a simulation model for $MAP/PH/1/K$. Run 10 replicates with a length of 1,000,000 units for each replicate and compare your simulated results for the measures given in examples 6.3 through 6.7. Write a short report on your findings.

Exercise 7.7. Develop a simulation model for $PH/M/1/K$. Run 10 replicates with a length of 1,000,000 units for each replicate and compare your simulated results for the measures given in examples 6.8 and 6.9. Write a short report on your findings.

Exercise 7.8. Develop a simulation model for $M/PH/1/K$. Run 10 replicates with a length of 1,000,000 units for each replicate and compare your simulated results for the measures given in examples 6.10 and 6.11. Write a short report on your findings.

Exercise 7.9. Develop a simulation model for $M/M/1/K$. Run 10 replicates with a length of 1,000,000 units for each replicate and compare your simulated results for the measures given in example 6.12. Write a short report on your findings.

Exercise 7.10. Develop a simulation model for $BMAP/G/c/K + c$ by incorporating all key distributions as well as various performance measures. Run your model for 10 replications with 1,000,000 units of time for each replicate. Generate simulated results of this model under a variety of scenarios by varying the arrival process, the service times, c, and K. Summarize the key observations and if needed generate graphs and tables. Write a short report on your findings.

References

Alfa, A.S. (2010). *Queueing Theory for Telecommunications*. Springer, New York, USA.

Alfa, A.S. (2016). *Applied Discrete-Time Queues*, 2nd edition. Springer, New York, USA.

Artalejo, J.R. and Lopez-Herrero, M.J. (2001). Analysis of the busy period for the $M/M/c$ queue: An algorithmic approach. *Journal of Applied Probability*, 38, 209–222.

Asmussen, S. (2003). *Applied Probability and Queues*, 2nd edition. Springer-Verlag, New York, USA.

Bailey, N.T.J. (1954). A continuous time treatment of a simple queue using generating functions. *Journal of the Royal Statistical Society: Series B*, 16, 288–291.

Bause, F., Buchholz, P., Kriege, J. (2009). A comparison of Markovian arrival processes and AMRA/ARTA processes for the modelling of correlated input processes. In *Proceedings of the Winter Simulation Conference*, Rossetti, M.D., Hill, R.R., Johansson, B., Dunkin, A., Ingalls, R.G. (eds). 634–645.

Bertsimas, D. (1990). An analytic approach to a general class of $GI/G/s$ queueing systems. *Journal of Operations Research*, 38(1), 139–155.

Bhat, U.N. (2015). *An Introduction to Queueing Theory*, 2nd edition. Springer, New York, USA.

Bini, D.A., Latouche, G., Meini, B. (2005). *Numerical Methods for Structured Markov Chains (Numerical Mathematics and Scientific Computation)*. Oxford University Press, New York, USA.

Bladt, M. and Nielsen, B.F. (2017). *Matrix-Exponential Distributions in Applied Probability*. Springer, Boston, USA.

Breuer, L. and Baum, D. (2005). *An Introduction to Queueing Theory and Matrix-Analytic Methods*. Springer, Dordrecht, Netherlands.

Brickner, C., Indrawan, D., Williams, D., Chakravarthy, S.R. (2011). Simulation of a stochastic model for a service system. In *Proceedings of the 2010 Winter Simulation Conference*, Johansson, B., Jain, S., Montoya-Torres, J., Hugan, J., Yucesan, E. (eds), 1636–1647.

Brill, P.H. (2017). *Level Crossing Methods in Stochastic Models*, 2nd edition. Springer, Cham, Switzerland.

Buchholz, P.B., Krieje.J., Felko, I. (2014). *Input Modeling with Phase-Type Distributions and Markov Models – Theory and Applications*. Springer, Cham, Switzerland.

Chakravarthy, S.R. (2012a). Reliability, health care, and simulation. *Simulation Modelling Practice and Theory*, 29, 44–51.

Chakravarthy, S.R. (2012b). Stochastic modelling and simulation in healthcare system. In *Computational and Mathematical Modelling*, Nadarjan, R., Lekshmi, R., Krishnan, G.S.S. (eds). Narosa Publishing House, New Delhi, India.

Chakravarthy, S.R. (2013). Analysis of $MAP/PH/c$ retrial queue with phase type retrials – Simulation approach. In *Modern Probabilistic Methods for Analysis of Telecommunication Networks*, Dudin, A., Klimenok, V., Tsarenkov, G., Dudin, S. (eds). Springer, Berlin, Germany.

Chakravarthy, S.R. (2015). A multi-server queueing model with Markovian arrivals and phase type cooperative services – Simulation approach. In *Information Technologies and Mathematical Modelling*, Dudin, A., Nazarov, A., Yakupov, R., Gortsev, A. (eds). Springer, Cham, Switzerland.

Chakravarthy, S.R. (2016). A multi-server parallel system with pure space sharing and Markovian arrivals – A simulation approach. In *Proceedings of the 4th International Conference on Business Analytics and Intelligence*. Bangalore, India.

Chakravarthy, S.R. (2019a). Busy period analysis of $GI/G/c$ and $MAP/G/c$ queues. In *Performance Prediction and Analytics of Fuzzy, Reliability and Queuing Models. Asset Analytics (Performance and Safety Management)*, Deep, K., Jain, M., Salhi, S. (eds). Springer, Singapore.

Chakravarthy, S.R. (2019b). Service-storage system with random volume requirements and positive service times (in Russian). *Mathematical Game Theory and its Applications*, 11, 95–120.

Chakravarthy, S.R. (2020). Busy period analysis of multi-server retrial queueing systems. In *Applied Probability and Stochastic Processes*, Joshua, V.C., Varadhan, S., Vishnevsky, V. (eds). Infosys Science Foundation Series, Springer.

Chakravarthy, S.R. (2021a). A comparative study of vacation models under various vacation policies – A simulation approach. In *Mathematical Modeling and Computation of Real-Time Problems*, Kulshrestha, R., Shekhar, C., Jain, M., Chakravarthy, R. (eds). CRC Press, Boca Raton, USA.

Chakravarthy, S.R. (2021b). Queueing models in services – Analytical and simulation approach. In *Queuing Theory 2: Advanced Trends*, Anisimov, P.V. and Limnios, P.N. (eds). ISTE Ltd, London, UK, and John Wiley and Sons, New York, USA.

Chakravarthy, S.R. (2021c). Queueing models with MAP arrivals useful in service sectors. *Queueing Models and Service Management*, 4(1), 55–84.

Chakravarthy, S. (2022a). *Introduction to Matrix-Analytic Methods in Queues 1: Analytical and Simulation Approach – Basics*. ISTE Ltd, London, UK, and John Wiley and Sons, New York, USA.

Chakravarthy, S.R. (2022b). Applications of the batch Markovian arrival process in the Internet of Things. In *IoT Applications, Security, Threats, and Countermeasures*, Nayak, P., Ray, N., Ravichandran, P. (eds). CRC Press, Boca Raton, USA.

Chakravarthy, S.R. and Maurina, J. (2015). A stochastic model for a military entrance processing station. In *Proceedings of the 2015 Industrial and Systems Engineering Research Conference*, Cetinkaya, S. and Ryan, J.K. (eds). IEEE IESM 2015.

Chakravarthy, S.R. and Ozkar, S. (2017). Simulation of a queueing model useful in crowdsourcing. In *Proceedings of 31st European Conference on Modelling and Simulation*, Paprika, Z.Z. (ed.). European Council for Modelling and Simulation.

Chakravarthy, S.R. and Rumyantsev, A. (2018). Efficient redundancy techniques in cloud and desktop grid systems using $MAP/G/c$-type queues. *Open Engineering*, 8, 17–31.

Chakravarthy, S.R. and Rumyantsev, A. (2019). $BMAP/G/c$ queueing model with group clearance – A simulation approach. *International Journal of Innovative Technology and Exploring Engineering*, 8, 621–627.

Conolly, B.W. (1959). The busy period in relation to the queueing process $GI/M/1$. *Biometrika*, 46, 246–251.

Conolly, B.W. (1960). The busy period in relation to the single-server queueing system with general independent arrivals and Erlangian service-time. *Journal of the Royal Statistical Society. Series B (Methodological)*, 22(1), 89–96.

Daley, D.J. and Servi, L.D. (1998). Idle and busy periods in stable $M/M/k$ queues. *Journal of Applied Probability*, 35(4), 950–962.

Dudin, A.N., Klimenok, V.I., Vishnevsky, V.M. (2020). *The Theory of Queueing Systems with Correlated Flows*. Springer Nature, Switzerland.

Furmans, K. (2022). *Material Handling and Production Systems Modelling-based on Queuing Models*. Springer Berlin, Heidelberg, Germany.

Garimella, R.M. and Rumyantsev, A. (2018). On an exact solution of the rate matrix of G/M/1-type Markov process with small number of phases. *Journal of Parallel and Distributed Computing*, 119, 172–178 [Online]. Available at: http://linkinghub.elsevier.com/retrieve/pii/S074373151830282X [accessed March 26, 2022].

Gross, D. and Harris, C.M. (1998). *Fundamentals of Queueing Theory*, 3rd edition. Wiley, New York, USA.

He, Q.M. (2014). *Fundamentals of Matrix-Analytic Methods*. Springer-Verlag, New York, USA.

Heckmüller, S. and Wolfinger, B. (2009). Using load transformations for the specification of arrival processes in simulation and analysis. *Simulation*, 85(8), 485–496.

Kelton, W.D., Sadowski, R.P., Swets, N.B. (2010). *Simulation with ARENA*, 5th edition. McGraw-Hill, New York, USA.

Kherani, A. (2006). A direct approach to sojourn times in a busy period of an $M/M/1$ queue. *Queueing Systems*, 53, 159–169.

Kimura, T. (1991). Approximations for the waiting time in the queue GI/G/s queue. *Journal of Operations Research Society of Japan*, 34(2), 173–186.

Kinateder, K.K.J. and Lee, E.Y. (2000). A new approach to the busy period of the $M/M/1$ queue. *Queueing Systems*, 35, 105–115.

Kleinrock, L. (1975). *Queueing Systems, Volume I: Theory*. Wiley, New York, USA.

Kleinrock, L. (1976). *Queueing Systems. Volume 2: Computer Applications*. Wiley, New York, USA.

Latouche, G. and Ramaswami, V. (1999). *Introduction to Matrix Analytic Methods in Stochastic Modeling*. SIAM Press, Philadelphia, USA.

Leguesdron, P., Pellaumail, J., Rubio, G., Sericola, B. (1993). Transient analysis of the $M/M/1$ queue. *Advances in Applied Probability*, 25(3), 702–713.

Li, Quan-Lin (2010). *Constructive Computation in Stochastic Models with Applications – The RG-Factorization*. Springer-Verlag, Berlin, Germany.

Lindley, D.V. (1952). The theory of queues with a single server. *Mathematical Proceedings of the Cambridge Philosophical Society*, 48, 277–289.

Lucantoni, D.M. (1991). New results on the single server queue with a batch Markovian arrival process. *Communications in Statistics. Stochastic Models*, 7, 1–46.

Lucantoni, D.M. (1993). The $BMAP/G/1$ queue: A tutorial. In *Models and Techniques for Performance Evaluation of Computer and Communications Systems*, Donatiello, L. and Nelson, R. (eds). Springer-Verlag, Berlin, Germany.

Lucantoni, D.M. and Neuts, M.F. (1994). Simpler proofs of some properties of the fundamental period of the $MAP/G/1$ queue. *Journal of Applied Probability*, 31(1), 235–243.

Lucantoni, D.M., Meier-Hellstern, K.S., Neuts, M.F. (1990). A single-server queue with server vacations and a class of nonrenewal arrival processes. *Advances in Applied Probability*, 22(3), 676–705.

Mayhugh, J. and McCormick, R. (1968). Steady-state solution of the queue $M/E_k/r$. *Management Science*, 14, 693–712.

Narayana, S. and Neuts, M.F. (1992). The first two moment matrices of the counts for the Markovian arrival process. *Stochastic Models*, 8, 459–477.

Neuts, M.F. (1981). *Matrix-Geometric Solutions in Stochastics Models: An Algorithmic Approach*. Johns Hopkins University Press, Baltimore, USA.

Neuts, M.F. (1986). The caudal characteristic curve of queues. *Advances in Applied Probability*, 18(1), 221–254.

Neuts, M.F. (1989). *Structured Stochastic Matrices of $M/G/1$ Type and their Applications*. Marcel Dekker, New York, USA.

Neuts, M.F. (1995). *Algorithmic Probability: A Collection of Problems*. Chapman and Hall, New York, USA.

Neuts, M.F. and Li, J.M. (1996). An algorithm for the $P(n,t)$ matrices of a continuous $BMAP$. In *Matrix-Analytic Methods in Stochastics Models*, Chakravarthy, S.R. and Alfa, A.S. (eds). Marcel Dekker, New York, USA.

Nielsen. B.F., Nilsson, L.A.F., Thygesen, U.H., Beyer, J.E. (2007). Higher order moments and conditional asymptotics of the batch Markovian arrival process. *Stochastic Models*, 23, 1–26.

Ozawa, T. (2006). Sojourn time distributions in the queue defined by a general QBD process. *Queueing Systems*, 53, 203–211.

Prabhu, N.U. (1960). Some results for the queue with Poisson arrivals. *Journal of the Royal Statistical Society: Series B*, 22, 104–107.

Ramaswami, V. (1982). The busy period of queues which have a matrix-geometric steady state probability vector. *Opsearch*, 19, 238–261.

Ramaswami, V. (1985). Independent Markov processes in parallel. *Stochastic Models*, 1(3), 419–432.

Ramaswami, V. and Lucantoni, D. (1985). Algorithms for the multi-server queue with phase type service. *Stochastic Models*, 1(3), 393–417.

Rojas-Cessa, R. (2017). *Interconnections for Computer Communications and Packet Networks*. CRC Press, Boca Raton, USA.

Rossetti, M. (2016). *Simulation Modeling and Arena*, 2nd edition. Wiley, Hoboken, USA.

Sengupta, B. (1989). Markov processes whose steady state distribution is matrix-exponential with an application to the GI/PH/1 queue. *Advances in Applied Probability*, 21(1), 159–180.

Shore, H. (1988). Simple approximations for the GI/G/c queue II: The moments, the inverse distribution function and the loss function of the number in the system and of the queue delay. *Journal of Operational Research Society*, 39(4), 381–391.

Takacs, L. (1962). *Introduction to the Theory of Queues*. Oxford University Press, New York, USA.

Tijms, H.C. (1994). *Stochastic Models: An Algorithmic Approach*. Wiley, New York, USA.

Whitt, W. (1993). Approximations for the GI/G/m queue. *Production and Operations Management*, 2(2), 114–161.

Index

B

C, F, G, K

L, M

P, Q

R, S

U, V, W

Summary of Volume 1

Other titles from

iSTE

in

Mathematics and Statistics

2022

DE SAPORTA Benoîte, ZILI Mounir
Martingales and Financial Mathematics in Discrete Time

LESFARI Ahmed
Integrable Systems

2021

KOROLIOUK Dmitri, SAMOILENKO Igor
Random Evolutionary Systems: Asymptotic Properties and Large Deviations

MOKLYACHUK Mikhail
Convex Optimization: Introductory Course

POGORUI Anatoliy, SWISHCHUK Anatoliy, RODRÍGUEZ-DAGNINO Ramón M.
Random Motions in Markov and Semi-Markov Random Environments 1: Homogeneous Random Motions and their Applications
Random Motions in Markov and Semi-Markov Random Environments 2: High-dimensional Random Motions and Financial Applications

PROVENZI Edoardo
From Euclidean to Hilbert Spaces: Introduction to Functional Analysis and its Applications

2020

BARBU Vlad Stefan, VERGNE Nicolas
Statistical Topics and Stochastic Models for Dependent Data with Applications

CHABANYUK Yaroslav, NIKITIN Anatolii, KHIMKA Uliana
Asymptotic Analyses for Complex Evolutionary Systems with Markov and Semi-Markov Switching Using Approximation Schemes

KOROLIOUK Dmitri
Dynamics of Statistical Experiments

MANOU-ABI Solym Mawaki, DABO-NIANG Sophie, SALONE Jean-Jacques
Mathematical Modeling of Random and Deterministic Phenomena

2019

BANNA Oksana, MISHURA Yuliya, RALCHENKO Kostiantyn, SHKLYAR Sergiy
Fractional Brownian Motion: Approximations and Projections

GANA Kamel, BROC Guillaume
Structural Equation Modeling with lavaan

KUKUSH Alexander
Gaussian Measures in Hilbert Space: Construction and Properties

LUZ Maksym, MOKLYACHUK Mikhail
Estimation of Stochastic Processes with Stationary Increments and Cointegrated Sequences

MICHELITSCH Thomas, PÉREZ RIASCOS Alejandro, COLLET Bernard, NOWAKOWSKI Andrzej, NICOLLEAU Franck
Fractional Dynamics on Networks and Lattices

VOTSI Irene, LIMNIOS Nikolaos, PAPADIMITRIOU Eleftheria, TSAKLIDIS George
Earthquake Statistical Analysis through Multi-state Modeling
(Statistical Methods for Earthquakes Set – Volume 2)

2018

AZAÏS Romain, BOUGUET Florian
Statistical Inference for Piecewise-deterministic Markov Processes

IBRAHIMI Mohammed
Mergers & Acquisitions: Theory, Strategy, Finance

PARROCHIA Daniel
Mathematics and Philosophy

2017

CARONI Chysseis
First Hitting Time Regression Models: Lifetime Data Analysis Based on Underlying Stochastic Processes
(Mathematical Models and Methods in Reliability Set – Volume 4)

CELANT Giorgio, BRONIATOWSKI Michel
Interpolation and Extrapolation Optimal Designs 2: Finite Dimensional General Models

CONSOLE Rodolfo, MURRU Maura, FALCONE Giuseppe
Earthquake Occurrence: Short- and Long-term Models and their Validation
(Statistical Methods for Earthquakes Set – Volume 1)

D'AMICO Guglielmo, DI BIASE Giuseppe, JANSSEN Jacques, MANCA Raimondo
Semi-Markov Migration Models for Credit Risk
(Stochastic Models for Insurance Set – Volume 1)

GONZÁLEZ VELASCO Miguel, del PUERTO GARCÍA Inés, YANEV George P.
Controlled Branching Processes
(Branching Processes, Branching Random Walks and Branching Particle Fields Set – Volume 2)

HARLAMOV Boris
Stochastic Analysis of Risk and Management
(Stochastic Models in Survival Analysis and Reliability Set – Volume 2)

KERSTING Götz, VATUTIN Vladimir
Discrete Time Branching Processes in Random Environment
(Branching Processes, Branching Random Walks and Branching Particle Fields Set – Volume 1)

MISHURA YULIYA, SHEVCHENKO Georgiy
Theory and Statistical Applications of Stochastic Processes

NIKULIN Mikhail, CHIMITOVA Ekaterina
Chi-squared Goodness-of-fit Tests for Censored Data
(Stochastic Models in Survival Analysis and Reliability Set – Volume 3)

SIMON Jacques
Banach, Fréchet, Hilbert and Neumann Spaces
(Analysis for PDEs Set – Volume 1)

2016

CELANT Giorgio, BRONIATOWSKI Michel
Interpolation and Extrapolation Optimal Designs 1: Polynomial Regression and Approximation Theory

CHIASSERINI Carla Fabiana, GRIBAUDO Marco, MANINI Daniele
Analytical Modeling of Wireless Communication Systems
(Stochastic Models in Computer Science and Telecommunication Networks Set – Volume 1)

GOUDON Thierry
Mathematics for Modeling and Scientific Computing

KAHLE Waltraud, MERCIER Sophie, PAROISSIN Christian
Degradation Processes in Reliability
(Mathematial Models and Methods in Reliability Set – Volume 3)

KERN Michel
Numerical Methods for Inverse Problems

RYKOV Vladimir
Reliability of Engineering Systems and Technological Risks
(Stochastic Models in Survival Analysis and Reliability Set – Volume 1)

2015

DE SAPORTA Benoîte, DUFOUR François, ZHANG Huilong
Numerical Methods for Simulation and Optimization of Piecewise Deterministic Markov Processes

DEVOLDER Pierre, JANSSEN Jacques, MANCA Raimondo
Basic Stochastic Processes

LE GAT Yves
Recurrent Event Modeling Based on the Yule Process
(Mathematical Models and Methods in Reliability Set – Volume 2)

2014

COOKE Roger M., NIEBOER Daan, MISIEWICZ Jolanta
Fat-tailed Distributions: Data, Diagnostics and Dependence
(Mathematical Models and Methods in Reliability Set – Volume 1)

MACKEVIČIUS Vigirdas
Integral and Measure: From Rather Simple to Rather Complex

PASCHOS Vangelis Th
Combinatorial Optimization – 3-volume series – 2nd edition
Concepts of Combinatorial Optimization / Concepts and Fundamentals – volume 1
Paradigms of Combinatorial Optimization – volume 2
Applications of Combinatorial Optimization – volume 3

2013

COUALLIER Vincent, GERVILLE-RÉACHE Léo, HUBER Catherine, LIMNIOS Nikolaos, MESBAH Mounir
Statistical Models and Methods for Reliability and Survival Analysis

JANSSEN Jacques, MANCA Oronzio, MANCA Raimondo
Applied Diffusion Processes from Engineering to Finance

SERICOLA Bruno
Markov Chains: Theory, Algorithms and Applications

2012

BOSQ Denis
Mathematical Statistics and Stochastic Processes

CHRISTENSEN Karl Bang, KREINER Svend, MESBAH Mounir
Rasch Models in Health

DEVOLDER Pierre, JANSSEN Jacques, MANCA Raimondo
Stochastic Methods for Pension Funds

2011

MACKEVIČIUS Vigirdas
Introduction to Stochastic Analysis: Integrals and Differential Equations

MAHJOUB Ridha

Recent Progress in Combinatorial Optimization – ISCO2010

RAYNAUD Hervé, ARROW Kenneth
Managerial Logic

2010

BAGDONAVIČIUS Vilijandas, KRUOPIS Julius, NIKULIN Mikhail
Nonparametric Tests for Censored Data

BAGDONAVIČIUS Vilijandas, KRUOPIS Julius, NIKULIN Mikhail
Nonparametric Tests for Complete Data

IOSIFESCU Marius *et al.*
Introduction to Stochastic Models

VASSILIOU PCG
Discrete-time Asset Pricing Models in Applied Stochastic Finance

2008

ANISIMOV Vladimir
Switching Processes in Queuing Models

FICHE Georges, HÉBUTERNE Gérard
Mathematics for Engineers

HUBER Catherine, LIMNIOS Nikolaos *et al.*
Mathematical Methods in Survival Analysis, Reliability and Quality of Life

JANSSEN Jacques, MANCA Raimondo, VOLPE Ernesto
Mathematical Finance

2007

HARLAMOV Boris
Continuous Semi-Markov Processes

2006

CLERC Maurice
Particle Swarm Optimization

Printed and bound by CPI Group (UK) Ltd, Croydon, CR0 4YY

13/09/2026

14961041-0007